Reflecting on the Future of the Built Environment

Reflecting on the Future of the Built Environment

Editors

Elizelle Juanee Cilliers
Sarel Cilliers

MDPI

Basel • Beijing • Wuhan • Barcelona • Belgrade • Novi Sad • Cluj • Manchester

Editors
Elizelle Juanee Cilliers
University of Technology Sydney
Ultimo Sydney
Australia

Sarel Cilliers
North-West University
Potchefstroom
South Africa

Editorial Office
MDPI
St. Alban-Anlage 66
4052 Basel, Switzerland

This is a reprint of articles from the Special Issue published online in the open access journal *Land* (ISSN 2073-445X) (available at: https://www.mdpi.com/journal/land/special_issues/future_of_built_environment).

For citation purposes, cite each article independently as indicated on the article page online and as indicated below:

Lastname, A.A.; Lastname, B.B. Article Title. *Journal Name* **Year**, *Volume Number*, Page Range.

ISBN 978-3-7258-0915-8 (Hbk)
ISBN 978-3-7258-0916-5 (PDF)
doi.org/10.3390/books978-3-7258-0916-5

Contents

About the Editors

Elizelle Juanee Cilliers

Prof. Dr. Elizelle Juanee Cilliers is the Head of the School of Built Environment and Professor of Urban Planning at the University of Technology Sydney, Australia. She is also the current President of the Commonwealth Association of Planners and adjunct Professor at the North-West University in South Africa. With over two decades of professional experience, she holds professional registrations with both the South African Council for Planners and the Planning Institute of Australia. She is a winner of the South African Teaching Excellence Award and has been a finalist for the National Science and Technology Awards in South Africa, in addition to being a Woman in Science Awards nominee. She leads workshops, symposia, and industry training events that focus on sustainable city planning, green infrastructure, and community resilience, and her leadship positions in the realms of research and public discourse have made her a pivotal figure in shaping future sustainable urban environments globally.

Sarel Cilliers

Prof. Dr. Sarel Cilliers is a Professor in Plant Ecology at the North-West University, Potchefstroom Campus, South Africa. His research examines biodiversity patterns and processes in urban and rural settlements, keenly focusing on the ecological and socio-economic impacts on both fragmented natural areas and home and community gardens. Professor Cilliers' innovative work has integrated resilience thinking to help explore community gardens as complex social-ecological systems, and he has advanced the scientific community's understanding of urban biodiversity while establishing practical applications by collaborating with town planners and landscape architects, an interdisciplinary approach that has infused urban ecological insights into the creation of sustainable and resilient urban environments. Professor Cilliers is also renowned for his expertise in the dynamics of small and medium-sized urban areas, biodiversity and ecosystem services, as well as the multifaceted drivers of urban biodiversity, including social, cultural, and legacy effects. Applying his ecological principles to enhance urban living environments has positioned Professor Cilliers as a leader in the field of urban ecology.

Preface

In this special edition of "Reflecting on the Future of the Built Environment", we delve into shaping the future of urban communities, with a focus on adaptive planning, nature-based solutions, resilient social-ecological systems, and urban climate migration in particular, spotlighting the need for a transformative approach in the development of cityscapes. The COVID-19 global pandemic has catalyzed a re-evaluation of the term "business as usual", urging authorities and communities alike to reconsider our collective role in developing sustainable, equitable, and livable environments for our future. This shift underscores the importance of accessibility, flexibility, and safety, intertwining these with the ongoing technological evolution and a growing emphasis on quality of life. This collection will not only showcase the latest scientific advancements and theoretical reflections in the field, but it will also help chart a path that embraces efficiency, well-being, and social justice on multiple scales—from individual sites to entire cities.

This Reprint is directed at practitioners, researchers, policymakers, and anyone interested in the future of the Built Environment, exploring themes such as city planning, property development, green infrastructure, climate action, and the role of adaptive design in fostering sustainable communities and healthy spaces. We extend our gratitude to all the distinguished authors who have offered their contributions within these pages, and we also acknowledge the invaluable support and insights received from peers and mentors in the field.

Elizelle Juanee Cilliers and Sarel Cilliers
Editors

Hence, NBS are seen as deliberate interventions seeking to use the properties of *nature* to address *societal* challenges.

NBS have illuminated a substantial potential to respond to climate change [21,22], to adapt to climate change impacts [23–25], and to create co-benefits in urban areas. In this way, NBS are well placed to accelerate urban transitions and create additional and multiple health [26,27] and social [28] benefits, such as equity, fairness, and justice. The uptake in governance and policy arenas [29,30] are seen as conditional for their success. NBS are seen as a more efficient and cost-effective approach to design resilient landscapes and cities that turn out to be economically viable and reach beneficial social and environmental outcomes [31]. The implementation of NBS in cities, however, still requires significant attention to be aesthetically appealing to citizens, create new green urban commons, create trust in local government and the process of experimentation, secure diversity and social innovation, be collaborative, develop an inclusive narrative, and develop design that can be replicated in the long-term [32]. Therefore, "urban planners need to have an open approach to collaborative governance of nature-based solutions that allows learning with and about new appealing designs, perceptions and images of nature from different urban actors, allows forming of new institutions for operating and maintaining nature-based solutions to ensure inclusivity, livability and resilience" [32]. Furthermore, the importance of collaborative research is acknowledged as a successful way researchers and practitioners iteratively learn from coproducing knowledge using holistic, integrative approaches and systematizing how multiple types of data and knowledge can be smartly used to plan for adaptation and mitigation [33].

The all-encompassing meaning of NBS asks for an overarching concept, bringing the diversity of directions together in the form of nature-based thinking, to use "*inspiration by nature* as an outset for the development of more sustainable and inclusive cities, balancing anthropocentric and eco-centric values and acknowledging the importance of the social and governance dimensions in a more balanced socio-ecological perspective" [34]. The design of NBS is seen as one of the knowledge gaps requiring further investigation in how design knowledge and skills can be used to create multifunctional, environmental, and social urban environments [35]. In this context, the impact of design approaches to nature-based and nature-driven urbanism [36,37] could play an important role in discovering the benefits of integrative and future-oriented creative ways of thinking to overcome well-known barriers preventing implementation and upscaling of NBS, such as the fear of the unknowns, the disconnect between short-term actions and long-term goals, the discontinuity between short-term actions and long-term plans, sectoral silos, and the paradigm of growth [35].

This article begins by describing the context of the case study area, current approaches, and related problems to urban planning and design, and identifies three useful NBS frameworks for a design-led approach. In Section 3, the research problem and objective are formulated, followed by the methodology and research design. Section 5 presents the research findings, and the article is concluded with the discussion and conclusions.

2. Background

2.1. Case Study Area

The metropolitan area of Sydney is expected to grow from approximately five to over seven million people in the next 20 years [38]. Where new housing will be located and what supporting infrastructure are required is a major question that needs to be answered at the larger scale. The Greater Sydney Commission has therefore developed a long-term spatial strategy, dividing the total area as three complementary cities [39]. The majority of new housing is, in this strategy, planned in the so-called third city of the Western Sydney Parklands (Figure 1).

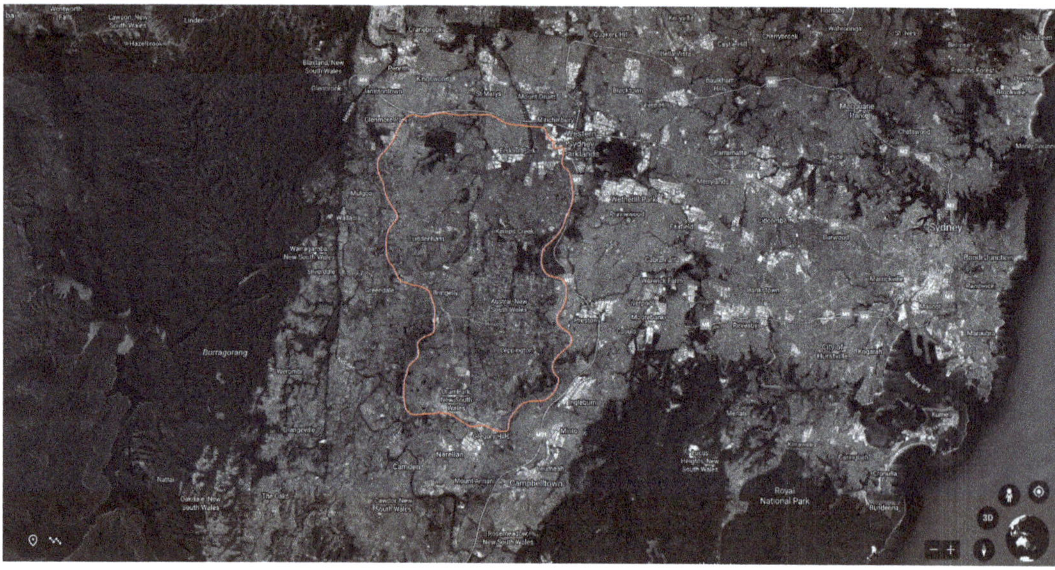

Figure 1. Western Sydney Parklands study site.

The Western Sydney City Deal formulates, as its main ambitions to deliver, around the anticipated Western Sydney airport, "a 30-minute city by delivering the North-South rail link, create 200,000 jobs capitalizing on the new Aerotropolis and the new agribusiness district, skilling residents in the region, respecting and building local character through a livability program, and coordinate and innovate through a planning partnership, continuing to consist of all three levels of government" [40]. This area will be home to over one million new residents and therefore requires a large-scale urban plan in which future resilience is the condition for newly built homes.

The Cumberland Plain landscape [41] is vulnerable to development due to its high ecological and aboriginal values, susceptibility to heat (with highs of 45 °C in summer), droughts (due to prolonged heat and low precipitation in summer), and flooding (up to nine meters of flooding during heavy rain and the rapid discharge of rainwater through the confined creek systems), and the risk of bushfires in times of strong western-northwestern winds combined with high temperatures [42]. The landscape is hilly, and the ground is fertile with semi-open vegetation and shrubs, and it is used as sparse or dense cattle fields and for chicken and mushroom farming. In this area a new airport is planned, surrounded by a so-called aerotropolis, consisting of logistics and agribusiness, with one million new residents, predicted to be living in approximately 350,000 new homes. The infrastructure consists of new trainlines connecting the new precinct with the existing urban area, bus connections, light-rail trams, and a new automobile grid with freeways and tangential roads.

If the development process is driven by the path-dependency connected to former urban typologies of urban sprawl, a vast sea of houses for one million people would result in overthrowing the current qualities and values, but also introduce a higher risk of bushfires, water scarcity, and water burden when it rains [43]. This context of rapid urbanization, in combination with a vulnerable landscape and climatic changes, makes this area an interesting research subject. Moreover, the way resources are generated, used, and related to each other is an essential feature of future urban design planning [44]. Finally, the local stakeholders and decision-makers have shown great interest in developing such a large-size urban area in the most resilient and sustainable way, taking its ecosystems as the base for their thinking.

A city that is in balance with nature and ecology, inclusive of green infrastructure [45,46], and urban ecology [47–49] is often constrained by neoliberal urbanism [50,51], in which market forces dominate the outcomes [52]. Despite the fact that consumers have often declared their desire to reside in green environments [53,54], developers and builders seem to follow their own market analyses which indicate that consumers demand the largest homes for the lowest prices [55–57]. While this is a generalization, the dominance of urban growth has long been characterized by fulfilling short-term needs and fast responses to housing demand [38]. When an ecological urban development is prioritized, a long-term urban plan is the only way forward. In current practice, it is very difficult to develop such a long-term spatial strategy that is consistent for a longer period. Instead, ad-hoc urbanism, driven by landownership, short term profits, a housing market under pressure, and a quick development process, prevails, leading to an ever-continuing sprawl of new housing [58–60].

2.2. Current Urban Development

Recent urban development in the Sydney metropolitan area has been characterized by this market-driven pressure and has led to the building of homes of maximized size on the plot they are built on, with black, heat-absorbing roofs, and built according the 4-2-2-concept: four bedrooms, two bathrooms, and a double garage [61]. In the urban design, the location of the station at the center of the area and surrounded by all amenities, offices and multilevel, multifunctional buildings, determines the lay-out of the neighborhood. The dogma to house most people within a radius of 500 m from the station, means the highest densities are found here and housing is planned in a convenient grid (80 × 40 m or 80 × 80 m) further away. Waterways and riparian zones are seen as an obstructive stand in the way of more housing, which are often built in low densities with their back to these green zones. This business-as-usual urbanization model [62], represented by one recent example of urban planning for the Austral-Leppington area in Sydney's west (Figure 2), is based on these urban planning principles:

1. A circle around the station and bus stop determines higher densities (in blue and dark orange);
2. Outside the public transport corridors, other land-use, such as low-density housing and green space, is located far from high densities;
3. The road system is designed to form an efficient grid;
4. Waterways are the spaces that remain after the station, density, and road system are planned. Here, at a distance from the station, low-density housing is planned and is only accessible by car.

The consequence of this is that in many urban plans, the landscape has become the backyard of the city instead of an amenity for enhancing the quality of life and the value of real estate. Moreover, these potential qualities are only available for a limited group of residents: those privileged enough to have their backyard border onto these water and riparian zones.

Figure 2. Recent plan for the Western Sydney area: the plan for Austral & Leppington [62].

Current urban development seems to underestimate the consequences for increasing the vulnerability of its future residents, sidelining the potential of green infrastructure to create a healthy urban environment, both for urban ecology [47–49] and physical [63] and mental [64–67] human wellbeing. Under current market stress this form of urban development will, most likely, continue, resulting in a new phase of urban sprawl in the westerns fringe of the Sydney metropolitan area, most probably in the form of a sea of similar houses in identical neighborhoods (Figure 3). The vast area will be filled up with housing similar to the Austral-Leppington plan. Figure 3 shows the future of such a landscape, in the form of a redesign of the region, copying the current planning principles as a montage.

Figure 3. Copy and pasting of the Austral plan leads to continuous urban sprawl [68].

2.3. NBS Frameworks

To counteract a thoughtless replication of recent urban planning and providing the directives to apply NBS, three major frameworks were used to guide this research.

In the first framework (Figure 4), NBS for climate-proofing, three phases are distinguished [69]. The initial phase, visioning and backcasting, describes the vision for a climate-proof future and the concrete actions to make this vision a reality. In the second stage, the benefits and costs of the vision and its actions are evaluated, and a preferred alternative is chosen. The final step contains the implementation and management of this solution. In three steps, the framework (1) identifies, (2) evaluates, and (3) implements.

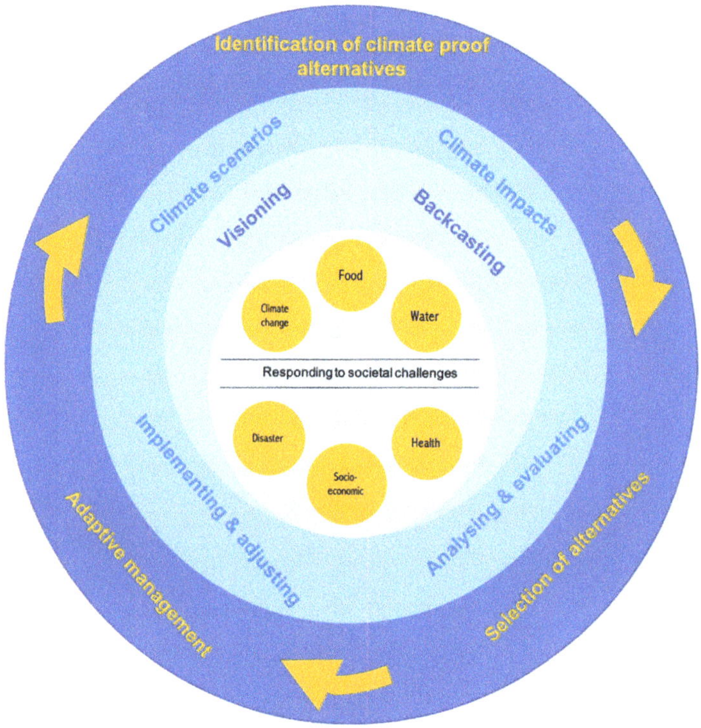

Figure 4. Assessment framework for climate-proof NBS (adapted from: [69]).

The second framework, shown in Figure 5, the co-benefits of NBS [70], consists of three layers. In the first, broadest layer, four system dimensions are defined: the socio-cultural and economic system, biodiversity, ecosystems, and climate. These systems are interlinked with the benefits for human health, environmental performance, synergies to biodiversity, and potential for citizen involvement [35]. The second layer identifies 10 societal challenges faced by cities in light of global environmental change. The third layer, at the heart of the model, integrates these subjects into a holistic approach, enabling cross-cutting solutions to design NBS by understanding the environmental and socio-ecological context, addressing multiple interconnected challenges, implementation across multiple scales, and using multi-actor co-production processes over the long-term [70]. This framework entails thinking at system, subject, and solution levels.

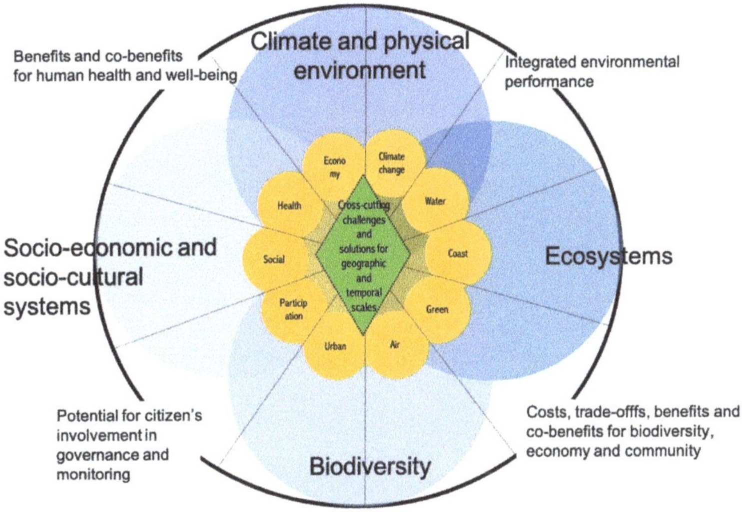

Figure 5. Assessment framework for co-benefits of NBS (adapted from: [70]).

The third framework (Figure 6), NBS-typologies, introduces three types of NBS-solutions. The first type of solutions enables better use and management of existing natural ecosystems. The solutions belonging to type two aim to restore ecosystems, and the third type of solutions aim to create novel ecosystems [71]. In short, these three types can be characterized as managing, restoring or recreating ecosystems for land and water environments when applied to cities or urban regions.

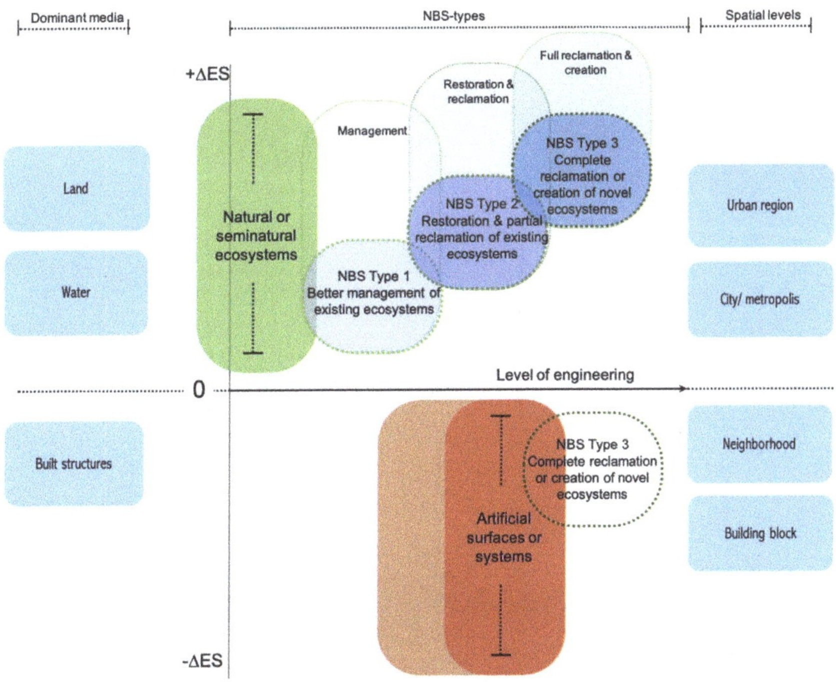

Figure 6. Conceptualization of NBS-types (adapted from: [71], based on [20,72]).

3. Research Problem and Objective

In most urban planning cases, at least in the Australian context [73], current urban development practices start with urban development. While this is obvious, an alternative practice is yet to unfold. As mentioned before, the way cities are planned cause serious problems for human as well as non-human lives. Climate impacts, loss of biodiversity, social and health implications, and depletion of resources, are all, to a larger or lesser extent, caused by urbanization. The market-driven way of exploitation of new neighborhoods, such as in the Sydney metropolitan region, accelerates these problems. In practice, ecosystems, biodiversity, natural water systems, and green space for recreational purposes are seen as additions to the core of planning, which consists of housing and infrastructure. This is a missed opportunity and disciplines, such as landscape architecture, urban planning and design, should use their profession to create resilient cities. The urban fabric is all too often the end to which green infill is added (Figure 7, left).

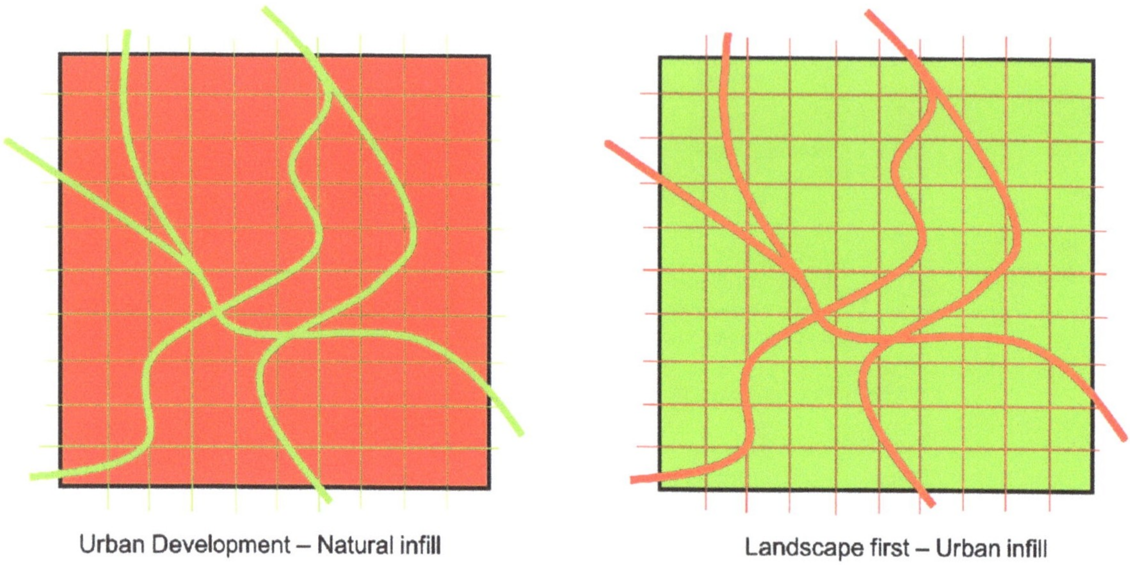

Figure 7. From (**right**) current practice in urban development towards (**left**) landscape-driven planning (by the authors).

Therefore, our research objective is to investigate how NBS can use these solutions to design the landscape system before urban functions are embedded within the features and vulnerabilities of the landscape. This takes landscape urbanism principles, well-known for a long time [74–76], as the basis of the design. The main shift is not so much to use the theory and methodology of landscape urbanism, but to implement it. This would imply a transition from a process that takes housing and the urban program as the start and adding green spaces as an amenity at a later stage (Figure 7, left), towards establishing the qualities of the landscape first, which then determine the way urban infill can be adopted (Figure 7, right).

4. Methodology and Research Design

In the methodology, abstractions of the three frameworks were used to guide design workshops. Taking these frameworks as the point of departure acknowledges the role NBS is expected to play in the design process, as a prime starter of the spatial propositions to follow. Each of the frameworks thus provide input in three consecutive stages of design (Figure 8), starting at the highest abstraction levels, then zooming in to concrete levels of thought. The design workshops are guided to be interactive and iterative to guarantee that the findings of each workshop will act as the foundations for the workshops to follow. The

3. Research Problem and Objective

In most urban planning cases, at least in the Australian context [73], current urban development practices start with urban development. While this is obvious, an alternative practice is yet to unfold. As mentioned before, the way cities are planned cause serious problems for human as well as non-human lives. Climate impacts, loss of biodiversity, social and health implications, and depletion of resources, are all, to a larger or lesser extent, caused by urbanization. The market-driven way of exploitation of new neighborhoods, such as in the Sydney metropolitan region, accelerates these problems. In practice, ecosystems, biodiversity, natural water systems, and green space for recreational purposes are seen as additions to the core of planning, which consists of housing and infrastructure. This is a missed opportunity and disciplines, such as landscape architecture, urban planning and design, should use their profession to create resilient cities. The urban fabric is all too often the end to which green infill is added (Figure 7, left).

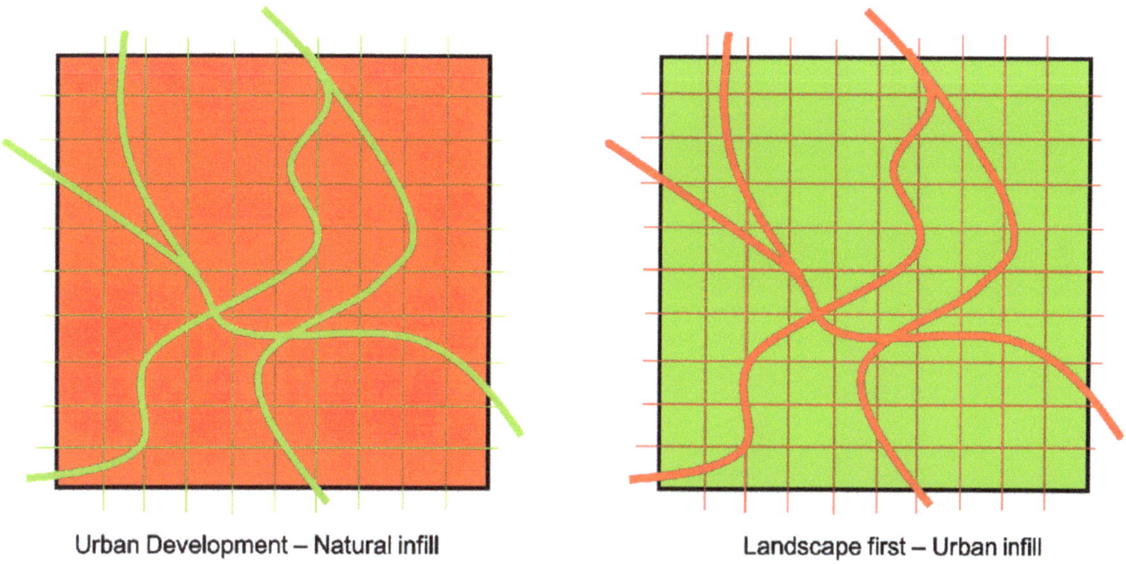

Figure 7. From (**right**) current practice in urban development towards (**left**) landscape-driven planning (by the authors).

Therefore, our research objective is to investigate how NBS can use these solutions to design the landscape system before urban functions are embedded within the features and vulnerabilities of the landscape. This takes landscape urbanism principles, well-known for a long time [74–76], as the basis of the design. The main shift is not so much to use the theory and methodology of landscape urbanism, but to implement it. This would imply a transition from a process that takes housing and the urban program as the start and adding green spaces as an amenity at a later stage (Figure 7, left), towards establishing the qualities of the landscape first, which then determine the way urban infill can be adopted (Figure 7, right).

4. Methodology and Research Design

In the methodology, abstractions of the three frameworks were used to guide design workshops. Taking these frameworks as the point of departure acknowledges the role NBS is expected to play in the design process, as a prime starter of the spatial propositions to follow. Each of the frameworks thus provide input in three consecutive stages of design (Figure 8), starting at the highest abstraction levels, then zooming in to concrete levels of thought. The design workshops are guided to be interactive and iterative to guarantee that the findings of each workshop will act as the foundations for the workshops to follow. The

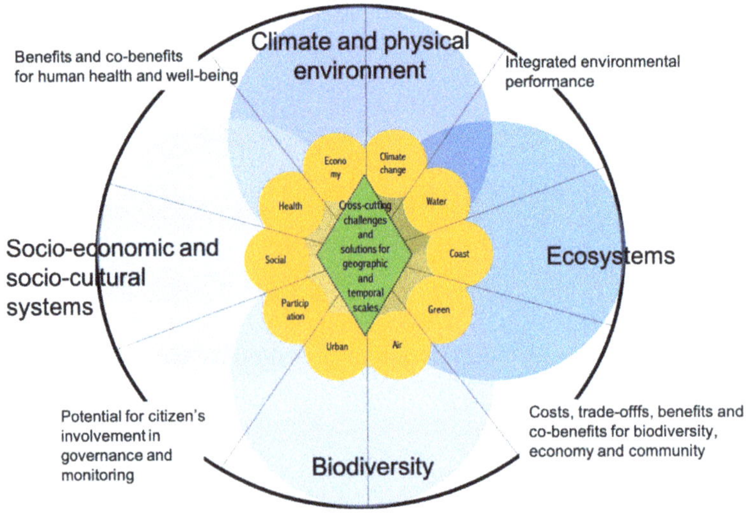

Figure 5. Assessment framework for co-benefits of NBS (adapted from: [70]).

The third framework (Figure 6), NBS-typologies, introduces three types of NBS-solutions. The first type of solutions enables better use and management of existing natural ecosystems. The solutions belonging to type two aim to restore ecosystems, and the third type of solutions aim to create novel ecosystems [71]. In short, these three types can be characterized as managing, restoring or recreating ecosystems for land and water environments when applied to cities or urban regions.

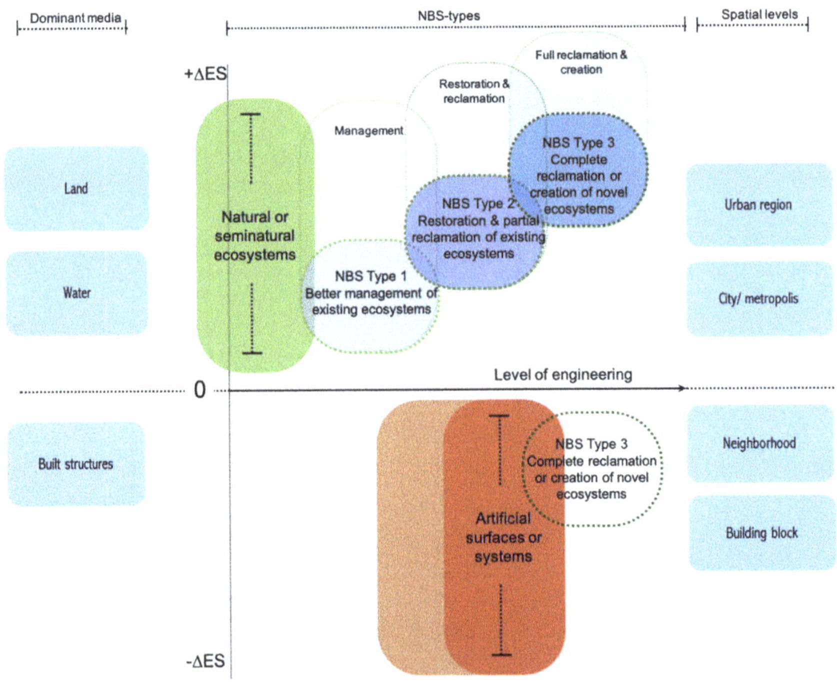

Figure 6. Conceptualization of NBS-types (adapted from: [71], based on [20,72]).

workshops themselves use an action research method [77–79], in which the researchers are participants of the design workshops. Each of the following steps in the research process are shaped using research by design principles [80–85] through which intermediate results can be constantly assessed and valued [86].

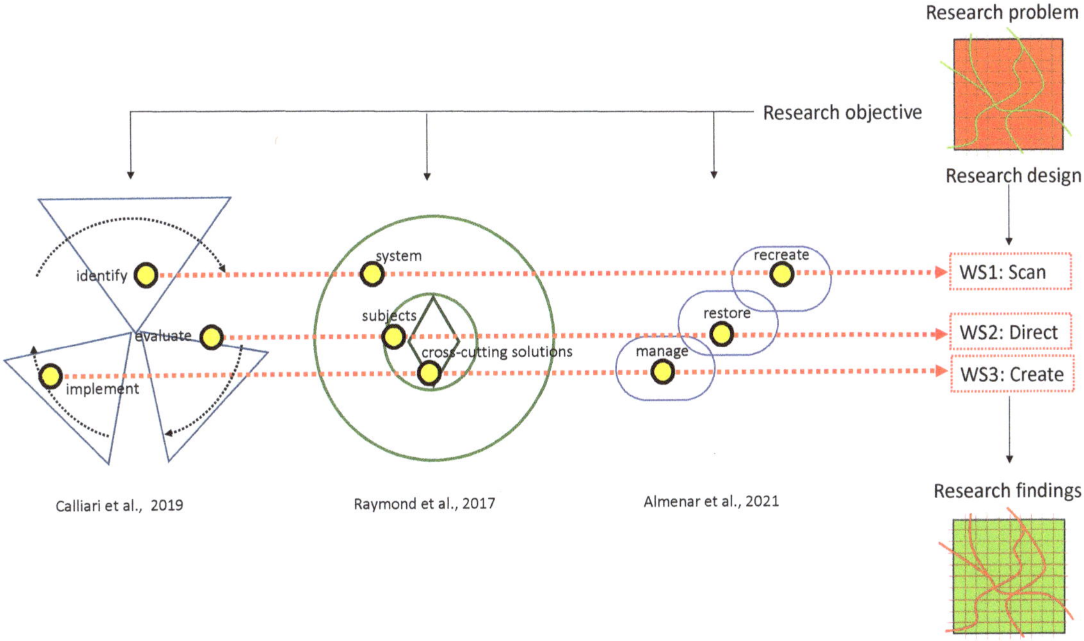

Figure 8. Methodology (by the authors).

In three consecutive design workshops, which took place between April and November 2019 and used the design charrette method [87–90], novel concepts and ideas were explored by allowing individual ideas to emerge and synergize into the overarching objectives and concepts. The design workshops consisted of a combination of work, such as spatial analyses, mapping exercises, collaborative design, conceiving of spatial propositions, and the building of plasticine models, each used in a specific context, timing, and appropriation. Since the objective of this research was to alter the urbanism approach from an urban development point of view to a nature-driven perspective, each of the conducted workshops started from an ecological point of view. In workshop one, the focus lay on scanning opportunities at the highest abstract level (why), in the second workshop, concepts were evaluated and one visionary direction was chosen (what), and, finally, during the third workshop, this vision was designed in greater detail (how).

4.1. Workshop 1: Scanning Opportunities—Identify, System, Recreate

In the first workshop, organized in Parramatta, NSW, Australia, a total of 50 participants took part, including representatives of the regional and state government, the utility sector, local councils, community groups, researchers from four local universities, professional designers and consultants, and organizational change managers. The professional background of the participants was broad and included landscape architects, urban and regional planners, ecologists, agronomists, sustainability experts, water system specialists, environmentalists, policy and public administration professionals, energy experts, students, and local citizens. The researchers took on different roles in the lead up to, during, and after the workshop. They designed the process, facilitated the workshop, participated as designers, conducted the analytical mapping beforehand, reported on, and illustrated the results.

From the NBS frameworks introduced before, the aspects *identify*, *systems* and *recreate* (see Figure 8) were used as the input for the first workshop. The major problems, such as climate change and impacts, degradation of the land, flood risk, social divide, food system, transportation, eco- and water systems, socio-economics, citizen engagement, and their social habitat, were identified and scoping of their future potentials was undertaken in the form of exploratory scenarios. During the workshop, special attention was given to ways novel ecosystems could be recreated, fitting with the historic qualities of the Cumberland Plain [91]. To interlink spatial scales, analytical mapping was conducted before the start of the workshop, which gave layered information on the potential direction for future urban planning.

At this stage of envisioning the future, creative methods, such as the art of the long-term view [92], futuring [93,94], and spatial visioning [95,96], were applied. Focusing on the long-term while making choices in the present requires a method that fits the long- and short-term simultaneously, and, thus, the backtracking method was used [97] to make this process tangible. Additionally, an exploration of involved stakeholders and residents was undertaken, using the method of actor-network mapping [98,99] and assemblage [100–102].

4.2. Workshop 2: Determine Directions—Evaluate, Subject, Restore

The second workshop, also organized and held in Parramatta, was attended by 35 participants. The typology of participants, their professional backgrounds, and the roles of the researchers were very similar to the first workshop. The concepts derived from the NBS frameworks, *evaluate, the subjects*, and *restoration* (see Figure 8), formed an additional set as the input for the sessions. The second workshop started with evaluating the analytical mapping work on the potentials, impacts and problems of the regional ecosystems, and the high-level vision, determined in the first workshop, on how urban developments could be embedded in the landscape. The benefits, associated problems, and implications for the four distinguished systems (socio-economic, ecosystems, biodiversity, and physical) were tested against each of the relevant subjects in this area (green, water, climate, food, participation, health, air, urban social, and economy).

The workshop aimed to find a common (NBS-)direction, framing, and enhancement of the implementation of the vision, connecting the different subjects with each other in a way that is usable as the spatial concept for the final workshop. To achieve this, scenarios were developed, which took the food system as the foundation for exploring future directions. Three spatial propositions were developed: autarkic food provision, a regenerative and reciprocal food system, and high-tech agri-business production. These scenarios were amalgamated into one scenario that formed the basis for a common, urban [103–106] and food-forestry [107,108] direction, capable of thriving alongside the restoration of the regional ecosystem.

4.3. Workshop 3: Creating Ecologies—Implement, Solution, Manage

Workshop three was carried out at the Hawkesbury campus, Richmond, NSW, Australia, and approximately 20 participants took part in this final workshop. The participants were largely of the same background as in the first two workshops, but with a limited number of people due to the intensity of the expected design work. While each type of participant, and their professional background, were still represented, fewer people could participate. The researchers prepared and designed the workshop, facilitated, acted as co-designers, and made sure the results were illustrated and reported.

As a first step, the design work took guidance from workshop two, with food-forestry as the conceptual framework to start the regional design. The NBS-frameworks provided the third set of inputs through the *implementation, cross-cutting solutions*, and *management* (see Figure 8) of ecosystems. This third workshop was concerned with making the visionary outcomes of workshop one and the conceptual direction derived from workshop two into a concrete landscape design for the Western Sydney Parklands. Practical considerations on cross-cutting issues, that work on multiple scales and are multi-actor oriented, were transformed into a concrete spatial proposal for the region, illuminating on the types of food-forests, living areas, and natural spaces were envisaged in the landscape. This third design workshop integrated the what if explorations into a spatial strategy for the long-term, creating a holistic vision of how regional urban development could become a resilient and productive urban landscape.

5. Results

The findings of this research are presented as outputs from the three workshops, reflecting the described methodology. However, many of these findings emerged simultaneously and should be read in parallel.

5.1. Scanning Opportunities

The outcomes of the first workshop were based on the identification of problems and possible visionary futures at the climate, ecological, and social systems level, aiming to recreate the regional ecosystem (type 3 NBS). The main climate change factors impacting the region were taken as the point of departure for the workshop: current maximum temperatures (up to 47 °C) and a maximum potential flood (MPF) [109] level of nine meters above mean water levels, must be seen as the lower limits of change. In this light, it was not enough to restore or manage existing ecosystems and, thus, envisioning a long-term future was undertaken in which the re-creation of the ecological systems was mapped out. In an iterative process, elevation, soil, vegetation, waterways, future flood-risk, and ecological remnants were mapped to understand the nuanced sensitivities of life and the reliability of available water, sunlight, shade, coolness, and other factors that determine the opportunities for the historic Cumberland Plain ecology [91] to re-appear in symbiosis with future human occupation. The landscape is there and is re-established through its creeks and side streams, the contours and potential discharge of rainwater, and the potential natural vegetation [88]. Once this landscape ecology was understood as the basis of regional planning, current steps forward could be defined: estimation of future flood risk (doubling of the current calculated MPF), the remainder of occupiable landscape, the so-called urban fields, and the infrastructural conditions for a nature-based urban development (Figure 9).

As an overarching vision, the landscape was seen as transforming by the natural terraforming forces of the free-moving creek system, shaping a resilient water system, which contains surpluses of water for later use. A transition zone between the Blue Mountains and Nepean and Hawkesbury River basin and the fine network of the South Creek system, with all its side-gullies, was established. Only then, urban infill thinking emerged.

The second major output of this first stage of the research was the mapping of the social network. The participants of the design workshop each defined their role, background, organizational embedding, and relation to every other participant (if apparent), which led to a complex inter-relational actor-network map (Figure 10). In the context of the Western Sydney Parkland development, this was seen as an important means to allow for the involvement of stakeholders and residents alike for two reasons: (1) to connect people from the region to the planning and design process, and (2) to recreate the quality of the social system. The latter was important since the south-western part of the Sydney metropolitan area is seen as lagging behind when it comes to income, education, jobs, crime, and health [42], compared with the wealthier north-west. Together with the understanding of the ecological landscape, this social landscape also played a role in gaining support for NBS, acceptance of the future vision, and permanent involvement of local people to start with during the second design workshop.

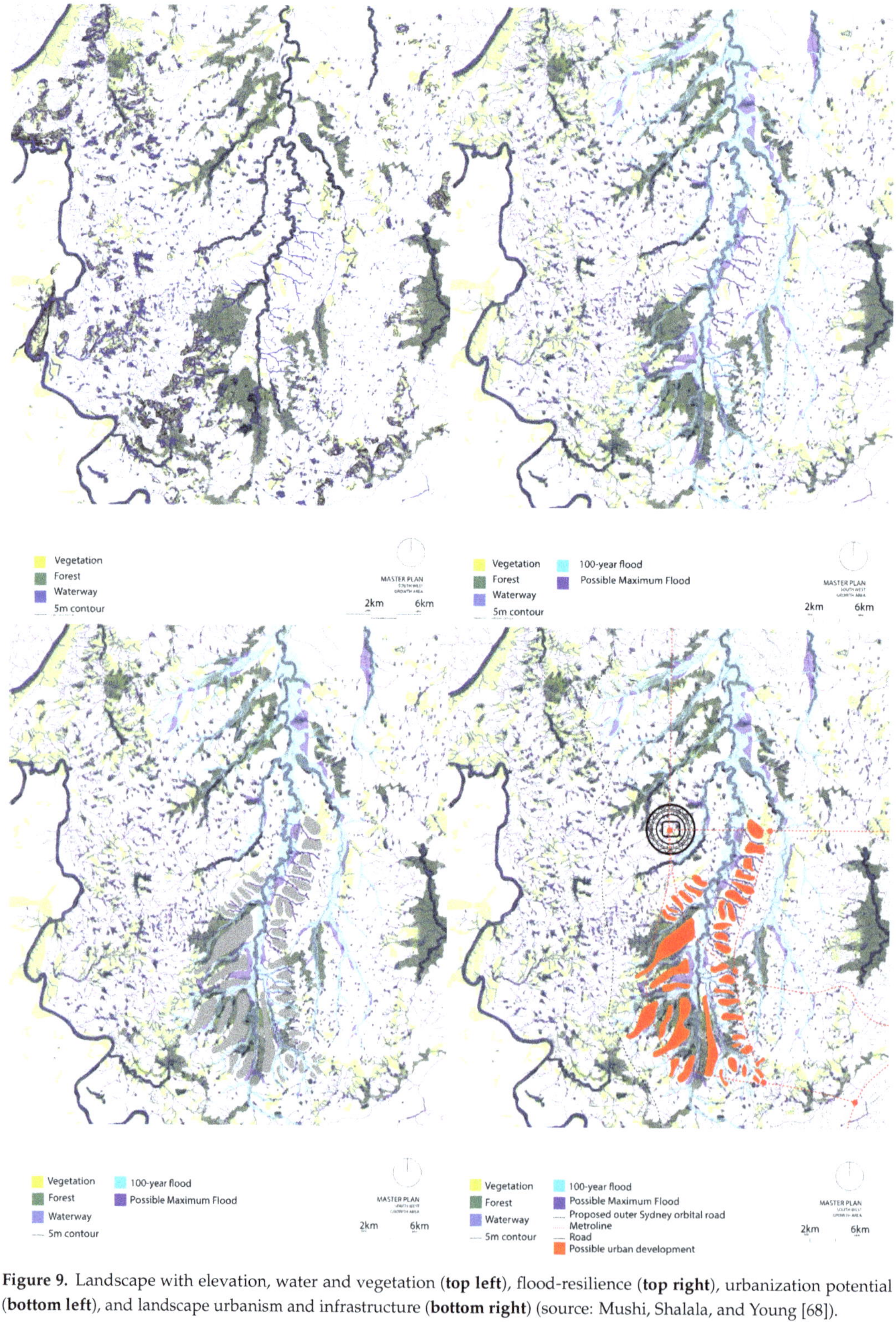

Figure 9. Landscape with elevation, water and vegetation (**top left**), flood-resilience (**top right**), urbanization potential (**bottom left**), and landscape urbanism and infrastructure (**bottom right**) (source: Mushi, Shalala, and Young [68]).

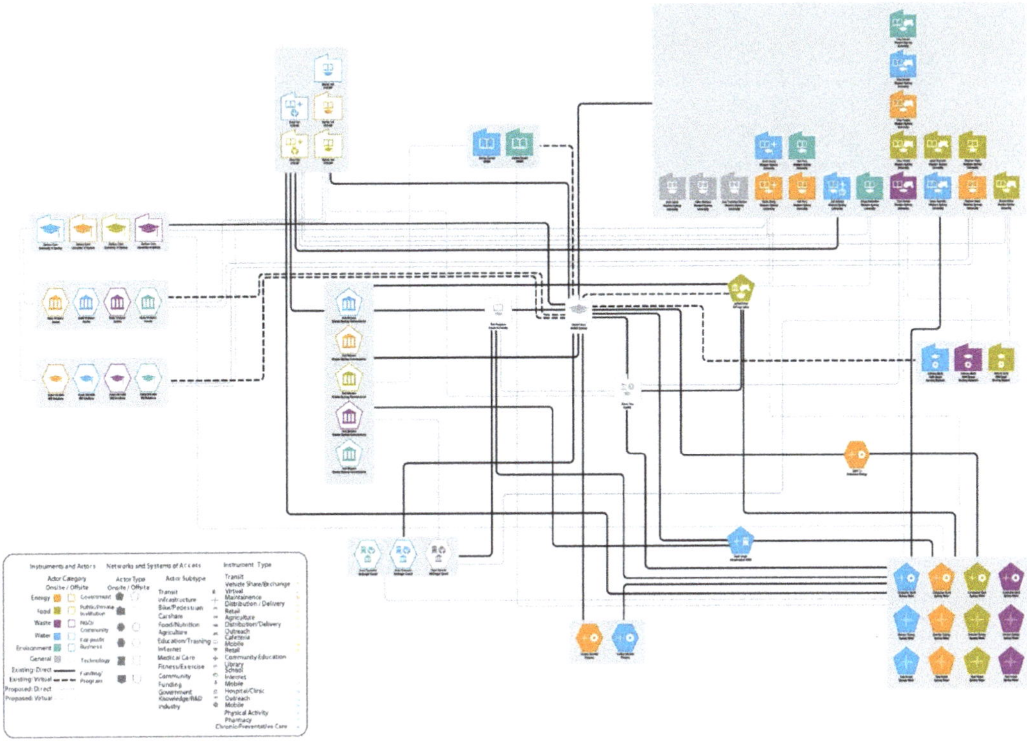

Figure 10. Actor-network mapping of the Western Sydney Parklands citizen and organizations involvement (source: Roggema and Monti; illustrated by Stewart Monti).

5.2. Determine Directions

The actor network that was formed during the first workshop continued working together on the evaluation of the findings during the second workshop. The long-term vision was taken as the subject to evaluate a range of relevant subjects in the Western Parklands: green, water, climate, food, participation, health, air, urban, social, and economy. To balance these subjects of ecology, sociology and economy, a uniting element was found in the choice for the type of food system as this brings in viability, connectedness, biodiversity, and the environment all at the same level. To take the full spectrum of possibilities into account, three scenarios were developed: high-tech, regenerative, and autarkic.

The first scenario (Figure 11) focused on establishing a high-tech agri-business complex. It comprised of high-intensity commercial and industrial developments surrounding the future airport. In this scenario, two unique elements were identified: aerospace and greenhouse agriculture. The South Creek flood plain was reserved and the existing Western Sydney Parklands extended further south. A key focus of this scenario was also a wastewater treatment plant integrated into the floodplain. Wastewater was captured from both the aerospace and greenhouses, along with surrounding residential areas, and fed back into the system.

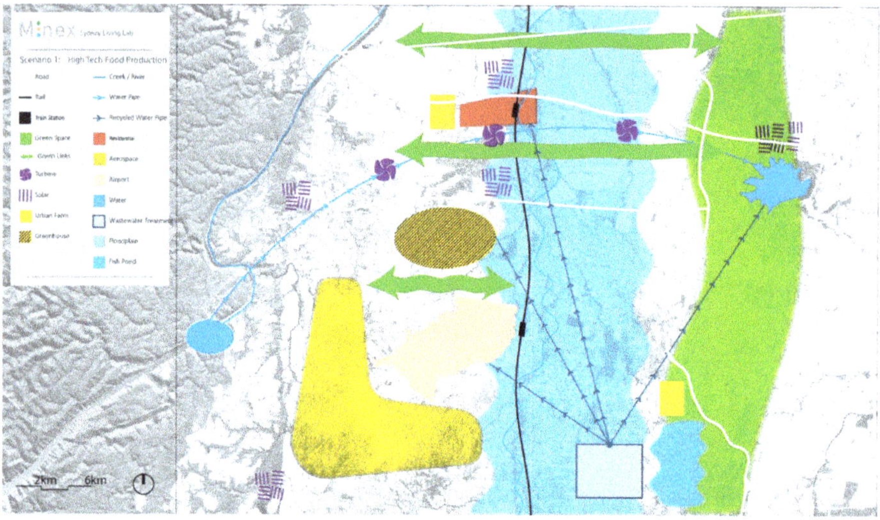

Figure 11. High-tech food system (source: Roggema and Monti; illustrated by Stewart Monti).

The second scenario (Figure 12) was based on the idea that direct linkages of producers to consumers, or smaller retailers, replace the conventional supply chain: a reciprocal food-system. This prosumer concept [110] assumes a significant increase in productivity at the metro-rural scale based on the advantages that come from direct purchasing. Rural villages and connections to urban agriculture in the existing city shift consumer preferences in the market. Ideally this is accompanied by an education program designed to promote eating for health to build market demand. This scenario combines agriculture and energy generation, tying it to residential areas. The daily market was seen as the node connecting these two, providing the means for producers to liaise directly with consumers. The floodplain was kept free from developments and its ecological function was only exploited for small-scale aquaculture. The airport was surrounded by a food-forest, which also provided valuable, publicly accessible green space for nearby residents.

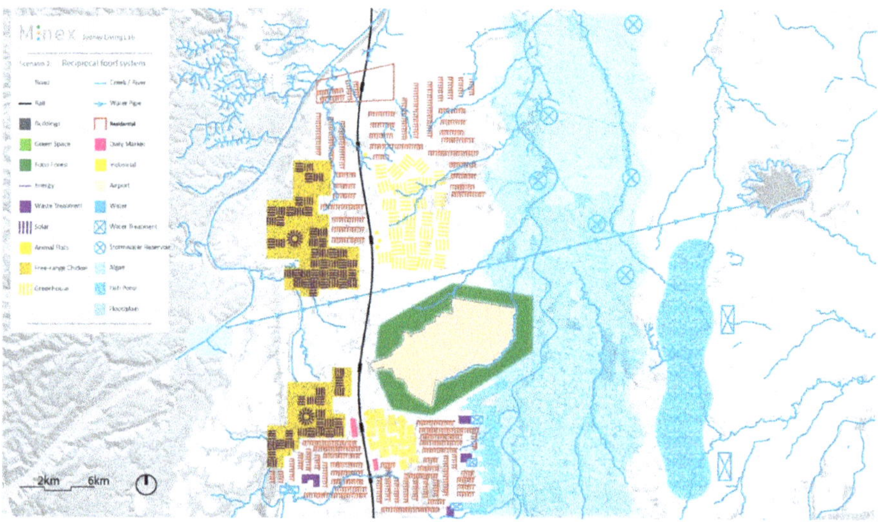

Figure 12. Regenerative food system (source: Roggema and Monti; illustrated by Stewart Monti).

Scenario three (Figure 13) took food production as a public good, mixing agricultural typologies carefully attuned to the characteristics of particular locations. Autarkic and regenerative food production improves soil and water qualities, increases biodiversity and resilience, recycles nutrients, and supports bio-sequestration. This implies new forms of co-governance, providing incentives for local food production. A pair of new *parklands* were introduced, which also run north to south. Agricultural production would take place within these new parklands. On the fringes of these areas, farms were coupled with associated services, residential, energy production and water treatment, to create self-sufficient communities linked by active transport routings.

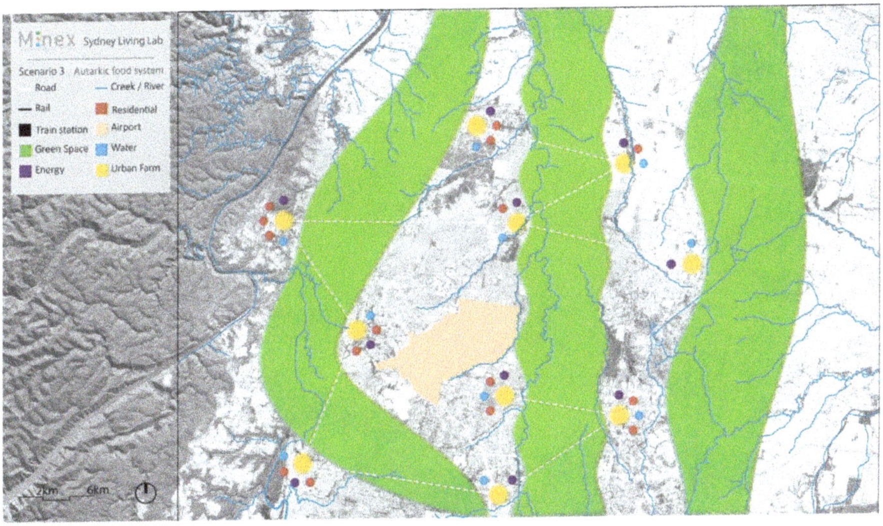

Figure 13. Autarkic food system (source: Roggema and Monti; illustrated by Stewart Monti).

The amalgamation scenario (Figure 14) subsumes aspects of each individual scenario in one tripartite vision. Resilience is the most influential conceptual practice underlining future city-making. It closes food, energy, and water loops, creating a purifying city which commonly manages its resources [111]. It incorporates Aboriginal custodians' knowledge and generates the realization of the so-called OCHRE strategy, which takes Aboriginal values as the foundation for planning and development [112]. Moreover, it introduces localized, regenerative agriculture for remediation which can "start replenishing and treatment at the source then move downstream". It proposes an active community engagement and co-design process shaped as a living lab, applying quadruple helix co-governance. In the amalgamation scenario, there is an important role for the resilience of the South Creek, which was not only recreated as an ecological floodplain area, but was also restored to its historic eco-qualities.

Derived from the amalgamated food-system, a spatial strategy was conceived, which could restore the natural system while meeting the needs of current demands, such as future climate, flood, heat, food, housing, biodiversity, recreation, social interaction, and economic viability. It embraced the vision of a landscape that satisfied the need to create a cooling environment, in which vegetation, water, and shade rhizomatically ramify themselves in smaller branches. The creative working process resulted in the conclusion that a food-forest strategy could bring multiple benefits: operating as a sponge, providing building materials for urban development, increasing biodiversity, and supplying indigenous and local food to the residents of the new city. Planting eight million trees, one for every resident living in the Sydney metropolitan area in 2045, would allow them to start growing, capturing carbon and nitrogen, creating timber for construction purposes, and provide a cooling canopy for later urban infill.

Figure 14. Amalgamated scenario (source: Roggema and Monti: illustrated by Stewart Monti).

The six proposed forest typologies (Figure 15) relate to indigenous forest communities of the Sydney metropolitan area [91] and depend on local factors, such as the soil, humidity, and climate, and require mixed planting (indigenous fruit, special timber, and nuts) or a link to permaculture when dealing with food-forests.

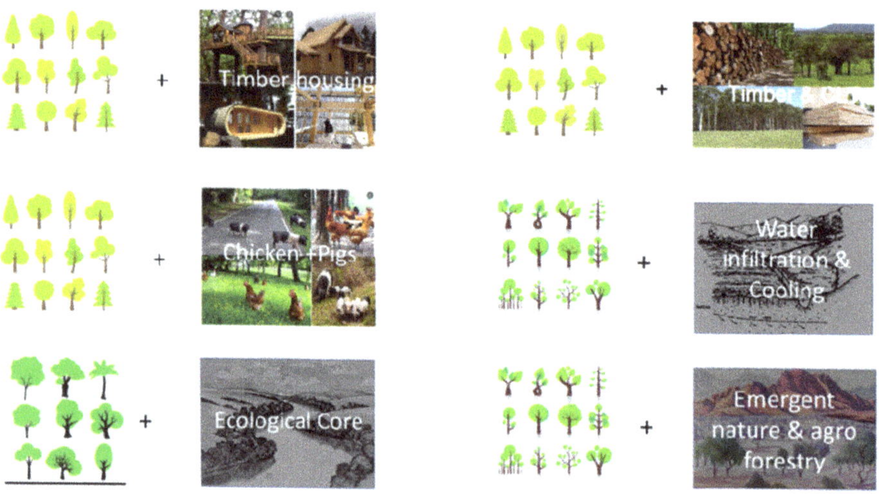

Figure 15. Food-forest typologies (by the authors).

The six typologies are described as follows:

1. A forest in which water can be stored so the environment can be cooled, flattening the potential flood risk. Riparian forest communities are used to flooding and, thus, form a buffer between areas where people live and the (potential) floodplains of the creeks. Apart from the water in the creeks and the evapotranspiration, the shade of the extended tree canopy network results in a much cooler environment. At the same time, these forests form ecological corridors in the region, linking similar habitats.

For these areas near the rivers, the use of alluvial woodland, with common species such as *Eucalyptus amplifolia, E. tereticornis,* and *Angophora floribunda*, and riparian forests, with common species *Eucalyptus botryoides, E. elata, Angophora subvelutina* and *A. floribunda*, were suggested [91];

2. A forest operating as an emerging ecological reserve. Due to deforestation and the pressure of different forms of land use, such as housing development and agriculture, some habitats and linked species can only be found in small pockets of land. Some of these species are rare and endemic. These are relics from historic ecosystems which are slowly disappearing or becoming extinct;
3. A forest used as timber production for building materials. Some areas in the timber forests are meant to be cut (checkerboard pattern) and used for buildings being built at the open patches in the direct vicinity;
4. A forest used as the producer of cross laminated timber (CLT). As some of the timber forests are meant to stay, with ongoing thinning as harvest products for timber and CLT, the set up might be different than traditionally managed timber forests.

For typologies two, three, and four, the Sydney Turpentine-Ironbark Forest, typical of the inner west region, was proposed. This forest type is indigenous to the Sydney metropolitan area and is classified as endangered under the Threatened Species Conservation Act 1995 [113]. This forest community can be very well used in typology two and its common species may serve as a base for typologies three and four. The forest type is dominated by *Syncarpia glomulifera* with *Eucalyptus paniculata* and *E. eugeniodes* occurring less frequently [92]. The *Syncarpia glomulifera,* or turpentine tree, is highly durable, as its timber is used in heavy-traffic flooring and for poles and wharves. It resists marine invertebrates and termites and is one of the most difficult timbers to ignite [114]. It is described as a hardy and adaptable tree, as turpentine tolerates heavy soils and frosts and is suitable for large gardens and parks, where it provides good shade with its dense canopy [115,116].

5. A forest used as a food-growing area (agroforestry). In many of these systems, trees and woody perennials were combined with growing fruits, nuts, and vegetables (annuals). The local opportunity here is to grow native products which are often unknown to consumers. This means making use of indigenous knowledge of plants and food production. Again, this strengthens biodiversity and resilience, and makes people care for their environment;
6. A forest used as a home to free-range pig and chicken farming. It is well known that boars were raised in forests on the island of Corsica in combination with rich productive forests. The boars were fed on chestnuts, corns, and fruits. However, the animals should be part of a well-balanced flow of nutrients and be a sustainable part of the agroforestry system [117].

For typologies five and six, prevailing local forest communities are suggested in combination with permaculture techniques. As this would influence the humidity of the soil over time, more moist-appreciating species could gradually be added to the range of species. With bushfire controlling strategies and the use of different forest typologies, an extended, biodiverse, and multifunctional forest could be created based on traditional knowledge [118]. Although site-specific studies need to be done, the overall effect of growing extended forests in Western Sydney would be that rainfall, especially in dry summers, would increase. Wright et al. found that transpiration helped to drive the seasonal cycle of rainfall in the southern Amazon [119]. These typologies can be used as the design principles in an intelligent combination to direct climate adaptive forms of urbanization.

This forest strategy gave direction to the regional spatial design explorations that were the subject of the third and final design workshop.

5.3. Create Ecologies

The third stage of the research, e.g. the third design workshop, focused on designing the implementation of solutions and looking at the way ecosystems can be managed in an urban context. This built upon the directions from research stages one and two, embracing the long-term vision, putting the landscape at the heart of future regional planning, and directing the regional plan from a food-forest perspective using the six indigenous forest typologies. In this way, NBS were connected to the social and economic systems through enriching urban livability and human health, demands for recreation, food production, circular resource management, and community engagement and co-creation of the future.

Here the results are presented in a series of six mapping stages as an emergence of ecologies (Figure 16a–e).

First, the current water network is captured (Figure 16a). The flow of the main water courses is based on the local topography, determining the ecological gradients and pathways of discharging waterflows through the landscape. The expected flood risks appear in the zones next to the main streams.

Second, alongside the streams, a riparian zone was planned (Figure 16b) in which wet forests would create the ideal conditions for sensitive ecologies, increasing biodiversity and where surpluses of floodwater could be captured, preventing flooding in other parts of the urban landscape.

The third stage entailed the introduction of an ecological grid (Figure 16c) forming the frame for structural ecological connections, linking wet and dry, and nutrient-rich and -poor parts of the landscape. This framework allows for an increased ecological connectivity and exchange, enhancing eco-capacity.

Stage four introduces plantations of timber forests in the inner parts of the eco-framework (Figure 16d) to produce building materials for homes. This timber is grown in areas of 1 km^2, within which housing compounds in high densities were foreseen, with each compound home to about 20,000 people. In the entire Western Sydney Parkland, 50 of these timber cities, in total, could be planned on the higher parts of the region, totaling up to the one million people that are projected to live in this area.

The remaining area within the inner framework (Figure 16e) was designed as a mix of productive agroforestry for fruits and vegetables and some intensive forms of food production, such as aero- and hydroponics.

Finally, under the future flightpath, the food-forest would host free-range chicken and pig farming (Figure 16f) where the animals can, within limits, forage freely under and in between the trees, providing high-quality products from local land. The remaining land was made redundant as a future spatial resource, which could be made productive if circumstances, demands, or developments change. It could be used to grow food or be developed on an ecological basis.

Together, these forest typologies simultaneously build up the landscape and the city. The quality of the urban environment co-evolves with the enrichment and increase of natural values. It contributes to the resilience for sudden climatic changes and impacts that cannot clearly be foreseen. Finally, it provides the urban population with a natural recreation space and a healthy environment to live in or nearby.

(a)

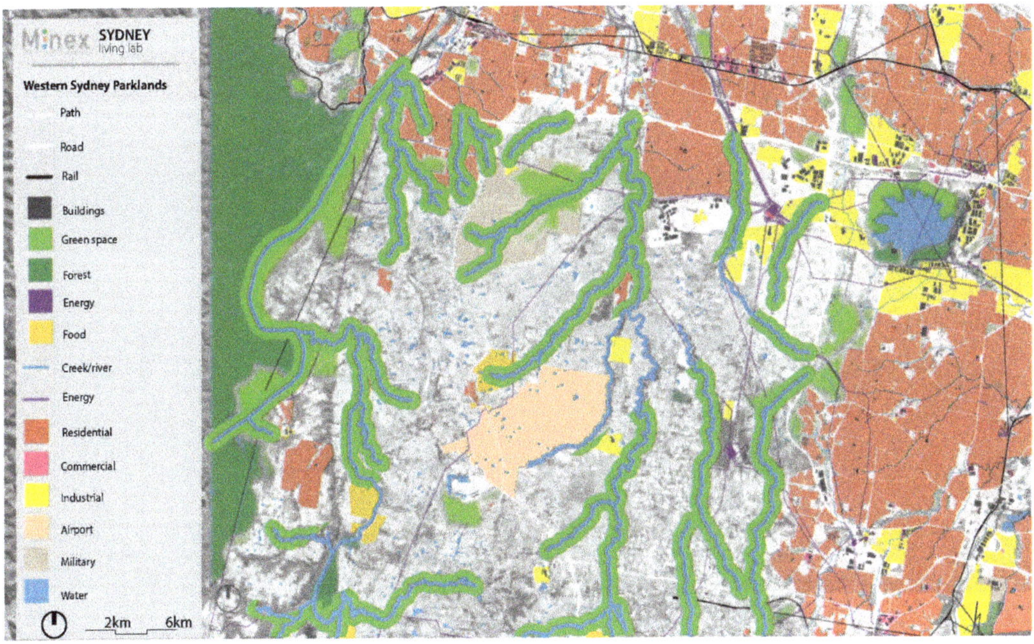

(b)

Figure 16. *Cont.*

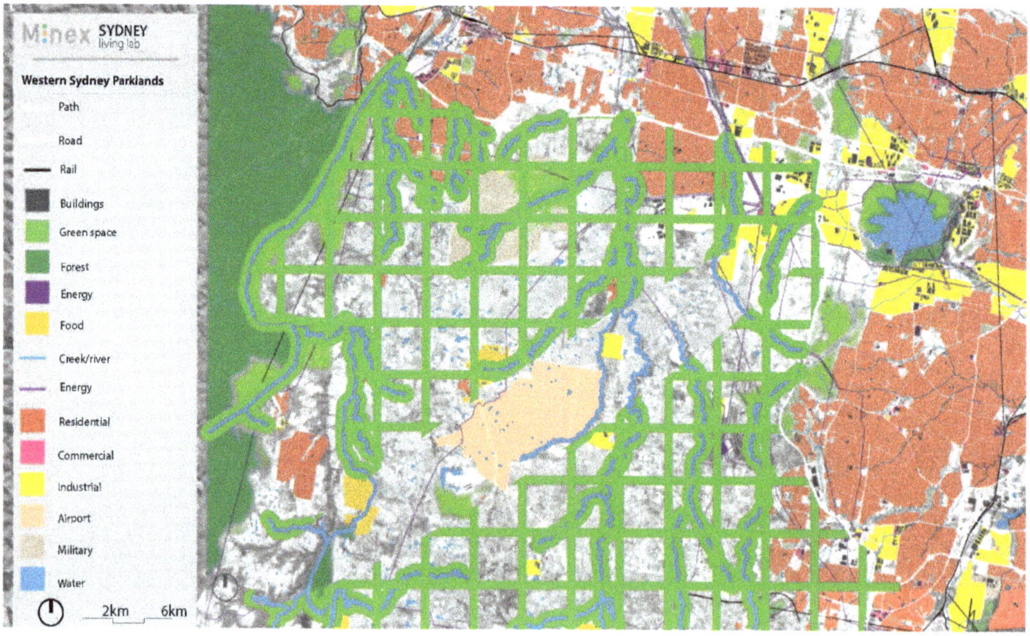

(c)

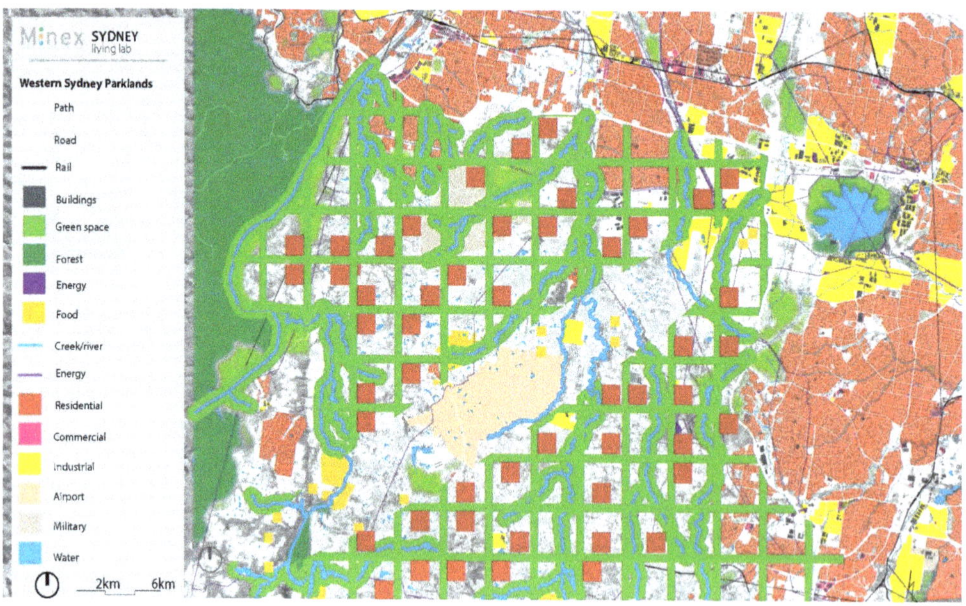

(d)

Figure 16. *Cont.*

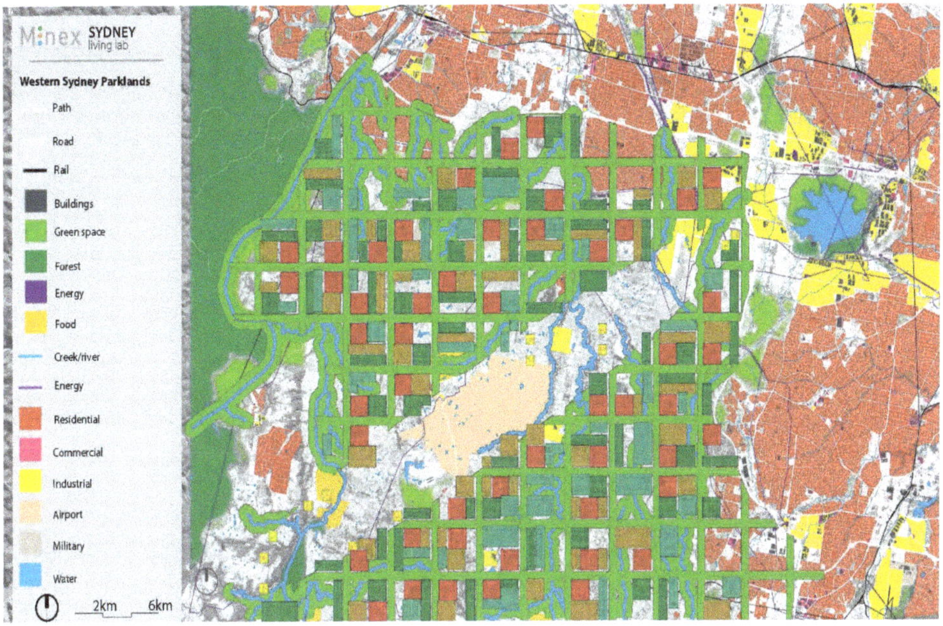

(e)

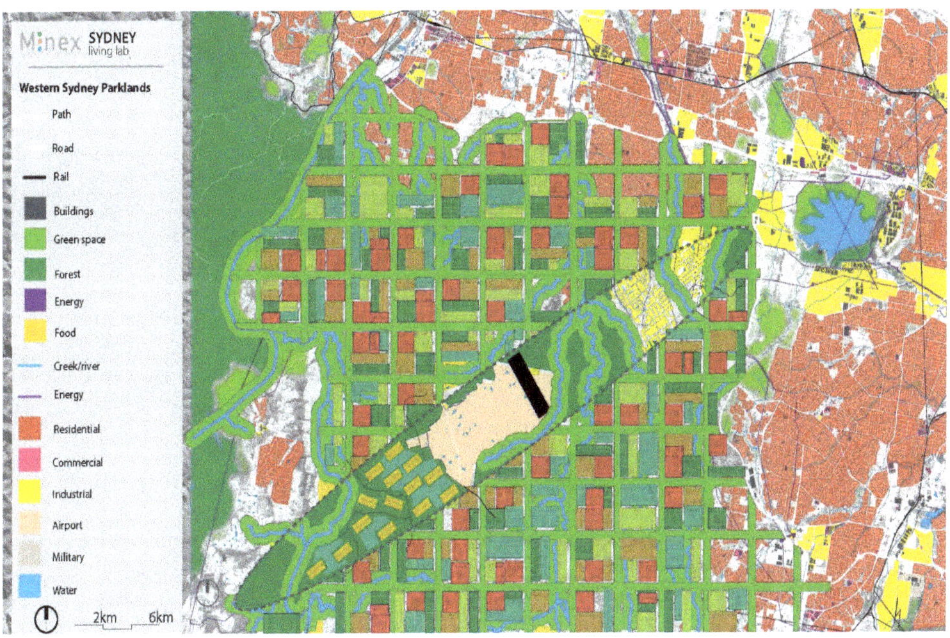

(f)

Figure 16. (**a**) Elevation and water courses (source: by the authors; illustrated by Stewart Monti), (**b**) Riparian ecology zoning of the water system (source: by the authors; illustrated by Stewart Monti), (**c**) Ecological grid connecting water and ecological landscapes (source: by the authors; illustrated by Stewart Monti), (**d**) Housing compounds embedded in the ecological grid (source: by the authors; illustrated by Stewart Monti), (**e**) Agroforestry of fruit, vegetable, and small cattle farming within the compounds of the eco-grid (source: by the authors; illustrated by Stewart Monti), (**f**) Free range cattle under the flightpath (source: by the authors; illustrated by Stewart Monti).

6. Discussion

NBS: hype or novel theory?

NBS is a relatively recent concept, and a range of literature has emerged aiming to theorize on it. A broad spectrum of topics is related to NBS: social, climate, economy, health, green, water, and many more. It is applied to different spatial scales and used in rural, natural landscapes as well as in urban contexts. There is a common feeling about NBS that it is appealing and a potential denominator for creating more resiliency. There are assessment frameworks developed and the concept is used in (urban) planning and concrete projects, some of which are suddenly dubbed nature-based. Therefore, the question that can be asked is whether NBS is hype or a novel theory? For NBS to become an autonomous theory, there should be a unique aspect to it, not just a next (or alternate) label for creating more ecological cities or landscapes. If NBS is only reframing the concept of ecosystem services or thinking in green grids or urban green-blue infrastructure, the danger of copying old concepts, including the reasons for failure, is apparent. Instead, NBS should offer new theoretical insights, for instance regarding the role nature-based takes on in the planning process. This requires further debate about the exact meaning of nature-based in a spatial planning environment and in urban development.

Nature as basis in a dynamic context?

In addition to the above, the implications of taking nature as the basis for solving current and future problems concerning the resilience of cities and landscapes, it can be stated that nature (ecosystems, water systems, the landscape, biodiversity, and all their interrelations) should be the driving force for every development. This goes beyond conserving species or (eco-)systems and aims to understand the functioning of the natural system, improving this through sensitive urban and rural planning and design. The question that deserves more attention is how this understanding of nature can be made applicable in a dynamic planning context.

NBS: detailed assessment or guide for design?

Some argue that by measuring and assessing elements of NBS, these will lead to an easier realization of these solutions. Therefore, several frameworks were proposed, some of which were used in this article. However, the level of detail and the abundance of subjects incorporated into these frameworks increase their complexity and decrease the ease of applying them in a practical planning process. There is a tension between the desire to assess the effect of NBS in detail and the use of more abstract guidance, derived from the frameworks, in conceiving designs that establish and realize NBS in concrete situations. In this article, three frameworks were used to define the direction of thought and, thus, make the NBS frameworks applicable in the design process at three abstraction levels. The discussion could be taken further by exploring an optimal balance between a detailed or an all-encompassing assessment and an abstraction of this and the relative freedom of thought to mold NBS-designs.

This, then, would make it possible to use the NBS-theoretical frameworks in practical design processes. To be useful, the frameworks need to give enough freedom to allow for the searching of novel applications and solutions, while at the same time directing the designers in the process towards basing their solutions on the desired natural qualities, an increase in biodiversity or extensions of ecological areas and corridors.

The interpretation of NBS in a design process is a matter of perspective. Planning NBS could take, at least, three dimensions: (1) when cities and landscapes are designed, integrate more NBS as part of the plan, (2) design cities and landscapes based on nature, and (3) design the desired regional and local ecosystem as one large nature-based solution. These preliminary directions require further elaboration.

NBS: beyond the urban-rural?

The urban-rural dichotomy is dominating the way our cities and landscapes are planned. This paradigm of separating the urban from the rural leads to distinct approaches for rural NBS and NBS in cities. The essence of ecological systems, however, is interconnectivity between the urban and the rural. This implies that NBS should be developed for the entire continuum of the city-urban and fringe-rural, demanding one coherent approach.

NBS as the subject in participatory planning

Finally, the success of NBS could be related to the design process. A participatory planning approach, suggested in this article, emphasizes the role of stakeholders and citizens in the conception of design solutions. This brings expert views and laymen visions together and allows for non-design experts to be directly involved, co-design, and understand proposed NBS. This results in more support for the future vision and less obstruction afterwards. This, as opposed to top-down governmental planning processes or even designer-led visionary approaches, which might be *technically* the best solution, yet are not always the *perceived* best solution.

7. Conclusions

Designing resilient landscapes using abstractions of NBS-frameworks

Current urban development places the urban in the development first, leading to suboptimal or poor outcomes for biodiversity and resilience. NBS is seen as a potential way to create better solutions for both humans and non-humans. In this article, NBS frameworks were used in the design process, enabling the exploration of planning that considers these aspects more thoroughly. The use of three NBS frameworks drove the design process. To achieve a fluid design process, the detailed and all-encompassing nature of the frameworks were abstracted at three levels, forming the input for three consecutive creative design workshops. At the highest level, the long-term visioning took place in which the re-creation of socio-ecological systems could be envisioned in the form of a regional landscape. This vision was then evaluated at the second level of abstraction, subject-by-subject, and led to concrete design directives, in the form of a commonly conceived food-forest strategy, to be implemented at the lowest level of abstraction: cross-cutting solutions, integrating a spectrum of concrete design aspects related to NBS at the regional planning scale.

NBS: connecting the dots through a spatial food-forest strategy

A food-forest strategy, as proposed in this article, offers good opportunities to connect a range of ambitions to enhance ecology, water conservation, timber production, the growth of food, livability, and health. The multiple benefits of forest typologies can structure the urban configuration, is well placed to anticipate future change, increase ecological values, and provide the building materials and space for food production in conjunction with urban spaces. A multifaceted and integrated approach concerning planning, participation, and ecological resilience is highly relevant. It makes sense to use NBS to direct the urban design and planning practice to, not only, increase the size of nature reserves, or improve nature's quality, but to develop a systemic urban future, which is more resilient and healthier to both its natural inhabitants as well as its human ones.

NBS: putting the ecological landscape first

The shifting mindset is to put the ecological landscape first in the planning process, followed by enriching and extending the ecological resilience and qualities, and, ultimately, embedding urban uses and provisions into the plan. By analyzing the current vulnerabilities and systemic characteristics of the ecological and water systems in the landscape, NBS informs and guides urban planning. The plan for the Western Sydney Parklands offers a systemic change, both in the way of working (putting landscape first in the design and planning process), as well in the time horizons considered (using the growth of plants and trees as the timeframe of development). System innovations like these imply systemic changes and enable ecological, social, and technological innovations, as well as the interaction between them. An (urban) design approach in which the landscape is taken as the point of departure brings nature-based planning and design practice to the fore, supporting the implementation of resilient and ecological solutions. When landscape is put first, the starting point of the planning and development of cities begins on a different foot and uses NBS as strategic spatial guiding principles.

Author Contributions: Conceptualization and methodology: R.R.; investigation, R.R., G.K., N.T.; writing—original draft preparation, R.R.; writing—review and editing, R.R., G.K., N.T.; visualization, R.R.; supervision, R.R.; project administration, R.R. All authors have read and agreed to the published version of the manuscript.

Funding: This research received no external funding.

Informed Consent Statement: Not applicable.

Acknowledgments: The authors wish to acknowledge the support of Western Sydney University for the facilitation and support throughout the workshops that founded this study. Moreover, we want to express our gratitude to Stewart Monti for visualizing the results of the workshops, and acknowledge the support in preparing the manuscript of Anouk Roggema.

Conflicts of Interest: The authors declare no conflict of interest.

References

1. WWF. *Living Planet Report 2020—Bending the Curve of Biodiversity Loss*; Almond, R.E.A., Grooten, M., Petersen, T., Eds.; WWF: Gland, Switzerland, 2020.
2. Brink, M. *Country Report for The State of the World's Biodiversity for Food and Agriculture—The Netherlands*; Centre for Genetic Resources, The Netherlands (CGN), Wageningen University and Research Centre (WUR): Wageningen, The Netherlands, 2015; p. 92.
3. PBL. *Halting Biodiversity Loss in the Netherlands: Evaluation of Progress*; Planbureau voor de Leefomgeving Den Haag: Den Haag, The Netherlands, 2008.
4. PBL. *Balans van de Leefomgeving 2014. De Toekomst Is Nú*; Planbureau voor de Leefomgeving Den Haag: Den Haag, The Netherlands, 2014. Available online: https://themasites.pbl.nl/balansvandeleefomgeving/wp-content/uploads/PBL_2014_Balans-van-de-Leefomgeving-2014_1308.pdf (accessed on 31 May 2021).
5. Crutzen, P.J. Geology of mankind. *Nature* **2002**, *415*, 23. [CrossRef]
6. Seto, K.C.; Fragkias, B.G.; Reilly, M.K. A meta-analysis of global urban land expansion. *PLoS ONE* **2011**, *6*, e23777. [CrossRef]
7. Schubert, D. Cities and plans—The past defines the future. *Plan. Perspect.* **2019**, *34*, 3–23. [CrossRef]
8. Molotch, H.; Freudenburg, W.; Paulsen, K. History repeats itself, but how? City character, urban tradition, and the accomplishment of place. *Am. Sociol. Rev.* **2000**, *65*, 791–823. [CrossRef]
9. Sijmons, D. *Het Casco-Concept, een Benaderingswijze voor de Landschapsplanning*; Ministerie van LNV, Directive NBLF: Utrecht, The Netherlands, 1992.
10. Tjallingii, S.P. *Ecopolis: Strategies for Ecologically Sound Urban Development*; Backhuys Publishers: Leiden, The Netherlands, 1995.
11. Tjallingii, S. Planning with water and traffic networks. Carrying structures of the urban landscape. *Res. Urban Ser.* **2015**, *3*, 57–80. [CrossRef]
12. Roggema, R.; van den Dobbelsteen, A.; Biggs, C.; Timmermans, W. Planning for Climate Change or: How Wicked Problems Shape the New Paradigm of Swarm Planning. In Proceedings of the 3rd World Planning Schools Congress, Perth, Australia, 4–8 July 2011.
13. Niemela, J. Ecology and urban planning. *Biodivers. Conserv.* **1999**, *8*, 119–131. [CrossRef]
14. Goddard, M.A.; Dougill, A.J.; Benton, T.G. Scaling up from gardens: Biodiversity conservation in urban environments. *Trends Ecol. Evol.* **2010**, *25*, 90–98. [CrossRef]
15. Haase, D.; Larondelle, N.; Andersson, E.; Artmann, M.; Borgström, S.; Breuste, J.; Gomez-Baggethun, E.; Gren, Å.; Hamstead, Z.; Hansen, R.; et al. A quantitative review of urban ecosystem service assessments: Concepts, models, and implementation. *AMBIO* **2014**, *43*, 413–433. [CrossRef]
16. Kabisch, N.; Qureshi, S.; Haase, D. Human environment interactions in urban green spaces—A systematic review of contemporary issues and prospects for future research. *Environ. Impact Assess. Rev.* **2015**, *50*, 25–34. [CrossRef]
17. European Commission. *Towards an EU Research and Innovation Policy Agenda for Nature-Based Solutions & Re-Naturing Cities. Final Report of the Horizon 2020 Expert Group on 'Nature-Based Solutions and Re-Naturing Cities'*; Publications Office of the European Union: Luxembourg, 2015.
18. Mccormick, K. (Ed.) *Cities, Nature and Innovation: New Directions*; Lund University: Lund, Sweden, 2020.
19. The EU and Nature-Based Solutions. Available online: https://ec.europa.eu/research/environment/index.cfm?pg=nbs (accessed on 12 November 2020).
20. Cohen-Shacham, E.; Janzen, C.; Maginnis, S.; Waters, G. *Nature-Based Solutions to Address Global Societal Challenges*; IUCN: Gland, Switzerland, 2016.

21. Emilsson, T.; Sang, A.O. Impacts of Climate Change on Urban Areas and Nature-Based Solutions for Adaptation. In *Nature-Based Solutions to Climate Change Adaptation in Urban Areas—Linkages between Science, Policy and Practice*; Kabisch, N., Korn, H., Stadler, J., Bonn, A., Eds.; Springer: Cham, Switzerland, 2017; pp. 15–28.
22. Pauleit, S.; Zölch, T.; Hansen, R.; Randrup, T.B.; Konijnendijk van den Bosch, C. Nature-Based Solutions and Climate Change—Four Shades of Green. In *Nature-Based Solutions to Climate Change Adaptation in Urban Areas—Linkages Between Science, Policy and Practice*; Kabisch, N., Korn, H., Stadler, J., Bonn, A., Eds.; Springer: Cham, Switzerland, 2017; pp. 29–50.
23. Depietri, Y.; McPhearson, T. Integrating the Grey, Green, and Blue in Cities: Nature-Based Solutions for Climate Change Adaptation and Risk Reduction. In *Nature-Based Solutions to Climate Change Adaptation in Urban Areas—Linkages Between Science, Policy and Practice*; Kabisch, N., Korn, H., Stadler, J., Bonn, A., Eds.; Springer: Cham, Switzerland, 2017; pp. 91–110.
24. Haase, D. Urban Wetlands and Riparian Forests as a Nature-Based Solution for Climate Change Adaptation in Cities and Their Surroundings. In *Nature-Based Solutions to Climate Change Adaptation in Urban Areas—Linkages between Science, Policy and Practice*; Kabisch, N., Korn, H., Stadler, J., Bonn, A., Eds.; Springer: Cham, Switzerland, 2017; pp. 111–122.
25. Enzi, V.; Cameron, B.; Dezsényi, P.; Gedge, D.; Mann, G.; Pitha, U. Nature-Based Solutions and Buildings—The Power of Surfaces to Help Cities Adapt to Climate Change and to Deliver Biodiversity. In *Nature-Based Solutions to Climate Change Adaptation in Urban Areas—Linkages between Science, Policy and Practice*; Kabisch, N., Korn, H., Stadler, J., Bonn, A., Eds.; Springer: Cham, Switzerland, 2017; pp. 159–185.
26. Braubach, M.; Egorov, A.; Mudu, P.; Wolf, T.; Ward Thompson, C.; Martuzzi, M. Effects of Urban Green Space on Environmental Health, Equity and Resilience. In *Nature-Based Solutions to Climate Change Adaptation in Urban Areas—Linkages between Science, Policy and Practice*; Kabisch, N., Korn, H., Stadler, J., Bonn, A., Eds.; Springer: Cham, Switzerland, 2017; pp. 187–206.
27. Kabisch, N.; Annerstedt van den Bosch, M. Urban Green Spaces and the Potential for Health Improvement and Environmental Justice in a Changing Climate. In *Nature-Based Solutions to Climate Change Adaptation in Urban Areas—Linkages between Science, Policy and Practice*; Kabisch, N., Korn, H., Stadler, J., Bonn, A., Eds.; Springer: Cham, Switzerland, 2017; pp. 207–220.
28. Haase, A. The Contribution of Nature-Based Solutions to Socially Inclusive Urban Development—Some Reflections from a Social-Environmental Perspective. In *Nature-Based Solutions to Climate Change Adaptation in Urban Areas—Linkages between Science, Policy and Practice*; Kabisch, N., Korn, H., Stadler, J., Bonn, A., Eds.; Springer: Cham, Switzerland, 2017; pp. 221–236.
29. Wamsler, C.; Pauleit, S.; Zolch, T.; Schetke, S.; Mascarenhas, A. Mainstreaming Nature-Based Solutions for Climate Change Adaptation in Urban Governance and Planning. In *Nature-Based Solutions to Climate Change Adaptation in Urban Areas—Linkages between Science, Policy and Practice*; Kabisch, N., Korn, H., Stadler, J., Bonn, A., Eds.; Springer: Cham, Switzerland, 2017; pp. 257–274.
30. Droste, N.; Schröter-Schlaack, C.; Hansjürgens, B.; Zimmermann, H. Implementing Nature-Based Solutions in Urban Areas: Financing and Governance Aspects. In *Nature-Based Solutions to Climate Change Adaptation in Urban Areas—Linkages between Science, Policy and Practice*; Kabisch, N., Korn, H., Stadler, J., Bonn, A., Eds.; Springer: Cham, Switzerland, 2017; pp. 307–322.
31. Lafortezza, R.; Chen, J.; van den Bosch, K.C.; Randrup, B.T. Nature-based solutions for resilient landscapes and cities. *Environ. Res.* **2018**, *165*, 431–441. [CrossRef]
32. Frantzeskaki, N. Seven lessons for planning nature-based solutions in cities. *Environ. Sci. Policy* **2019**, *93*, 101–111. [CrossRef]
33. Frantzeskaki, N.; McPhearson, T.; Collier, M.J.; Kendal, D.; Bulkeley, H.; Dumitru, A.; Pintér, L. Nature-Based Solutions for Urban Climate Change Adaptation: Linking Science, Policy, and Practice Communities for Evidence-Based Decision-Making. *BioScience* **2019**, *69*, 455–466. [CrossRef]
34. Randrup, B.T.; Buijs, A.; Konijnendijk, C.C.; Wild, T. Moving beyond the nature-based solutions discourse: Introducing nature-based thinking. *Urban Ecosyst.* **2020**, *23*, 919–926. [CrossRef]
35. Kabisch, N.; Frantzeskaki, N.; Pauleit, S.; Naumann, S.; Davis, M.; Artmann, M.; Bonn, A. Nature-based solutions to climate change mitigation and adaptation in urban areas: Perspectives on indicators, knowledge gaps, barriers, and opportunities for action. *Ecol. Soc.* **2016**, *21*, 39. [CrossRef]
36. Roggema, R. (Ed.) *Contemporary Urban Design Thinking, Vol. 2, Nature Driven Urbanism*; Springer: Dordrecht, The Netherlands, 2019.
37. Roggema, R. From Nature-Based to Nature-Driven: Landscape First for the Design of Moeder Zernike in Groningen. *Sustainability* **2021**, *13*, 2368. [CrossRef]
38. Pinnegar, S.; Randolph, B.; Troy, L. Decoupling growth from growth-dependent planning paradigms: Contesting prevailing urban renewal futures in Sydney, Australia. *Urban Policy Res.* **2020**, *38*, 321–337. [CrossRef]
39. Greater Sydney Commission. *A Metropolis of Three Cities. Greater Sydney Region Plan—Connecting People*; State Government of New South Wales: Sydney, Australia, 2018.
40. Australian Government and Government of NSW. Western Sydney City Deal. Vision. Partnership. Delivery. Sydney, Australia, 2018. Available online: https://www.infrastructure.gov.au/cities/city-deals/western-sydney/files/western-sydney-city-deal.pdf (accessed on 23 September 2019).
41. Benson, D.; Howell, J. *Taken for Granted: The Bushland of Sydney and Its Suburbs*; Kangaroo Press: Sydney, Australia, 1990.
42. Simpson, R.; Roggema, R. How to Design Sydney's Third City? In *Contemporary Urban Design Thinking, Vol. 1, The Australian Approach*; Roggema, R., Ed.; Springer: Dordrecht, The Netherlands, 2018; pp. 39–70.
43. Roggema, R. Landscape First! Nature-Driven Design for Sydney's Third City. In *Contemporary Urban Design Thinking, Vol. 2, Nature Driven Urbanism*; Roggema, R., Ed.; Springer: Dordrecht, The Netherlands, 2019; pp. 81–110.

44. Yan, W.; Roggema, R. Developing a Design-Led Approach for the Food-Energy-Water Nexus in Cities. *Urban Plan.* **2019**, *4*, 123–138. [CrossRef]
45. Gill, S.E.; Handley, J.F.; Ennos, A.R.; Pauleit, S. Adapting cities for climate change: The role of the green infrastructure. *Built Environ.* **2007**, *33*, 115–133. [CrossRef]
46. Kambites, C.; Owen, S. Renewed prospects for green infrastructure planning in the UK. *Plan. Pract. Res.* **2006**, *21*, 483–496. [CrossRef]
47. Wu, J. Urban ecology and sustainability: The state-of-the-science and future directions. *Landsc. Urban Plan.* **2014**, *125*, 209–221. [CrossRef]
48. Rebele, F. Urban ecology and special features of urban ecosystems. *Glob. Ecol. Biogeogr. Lett.* **1994**, *4*, 173–187. [CrossRef]
49. Alberti, M. *Advances in Urban Ecology: Integrating Humans and Ecological Processes in Urban Ecosystems*; No. 574.5268 A4; Springer: New York, NY, USA, 2008.
50. Peck, J.; Theodore, N.; Brenner, N. Neoliberal urbanism: Models, moments, mutations. *Sais Rev. Int. Aff.* **2009**, *29*, 49–66. [CrossRef]
51. Theodore, N.; Peck, J.; Brenner, N. Neoliberal urbanism: Cities and the rule of markets. In *The New Blackwell Companion to the City, 1625*; Bridge, G., Watson, S., Eds.; Wiley: Hoboken, NY, USA, 2011.
52. Lewyn, M. *Government Intervention and Suburban Sprawl: The Case for Market Urbanism*; Springer/Palgrave Macmillan: New York, NY, USA, 2017.
53. Van den Berg, A.E.; Hartig, T.; Staats, H. Preference for nature in urbanized societies: Stress, restoration, and the pursuit of sustainability. *J. Soc. Issues* **2007**, *63*, 79–96. [CrossRef]
54. Bark, R.H.; Osgood, D.E.; Colby, B.G.; Halper, E.B. How do homebuyers value different types of green space? *J. Agric. Resour. Econ.* **2011**, 395–415.
55. Keogh, G.; D'Arcy, E. Property market efficiency: An institutional economics perspective. *Urban Stud.* **1999**, *36*, 2401–2414. [CrossRef]
56. D'Arcy, E.; Keogh, G. Territorial competition and property market process: An exploratory analysis. *Urban Stud.* **1998**, *35*, 1215–1230. [CrossRef]
57. Evans, A.W. The property market: Ninety per cent efficient? *Urban Stud.* **1995**, *32*, 5–29. [CrossRef]
58. Gillham, O. *The Limitless City: A Primer on the Urban Sprawl Debate*; Island Press: Washington, DC, USA, 2002.
59. Harvey, R.O.; Clark, W.A. The nature and economics of urban sprawl. *Land Econ.* **1965**, *41*, 1–9. [CrossRef]
60. Peiser, R.B. Density and urban sprawl. *Land Econ.* **1989**, *65*, 193–204. [CrossRef]
61. Maddox, G. Oran Park, the Suburb that's Grown from a Population of 0 to 7500 in Eight Years. Sydney Morning Herald. 20 April 2018. Available online: https://www.smh.com.au/national/nsw/oran-park-the-suburb-that-s-grown-from-a-population-of-0-to-7500-in-eight-years-20180301-p4z2bb.html (accessed on 21 April 2018).
62. Department of Planning and Environment. Planning for Sydney's South West. Available online: https://www.planning.nsw.gov.au/Plans-for-your-area/Priority-Growth-Areas-and-Precincts/South-West-Growth-Area (accessed on 11 November 2020).
63. Epstein, L.H.; Paluch, R.A.; Roemmich, J.N.; Beecher, M.D. Family-based obesity treatment, then and now: Twenty-five years of pediatric obesity treatment. *Health Psychol.* **2007**, *26*, 381. [CrossRef]
64. Flouri, E.; Papachristou, E.; Midouhas, E. The role of neighborhood greenspace in children's spatial working memory. *Br. J. Educ. Psychol.* **2018**, *89*, 359–373. [CrossRef]
65. Li, D.; Sullivan, W. Impact of views to school landscapes on recovery from stress and mental fatigue. *Landsc. Urban Plan.* **2016**, *148*, 149–158. [CrossRef]
66. Mennis, J.; Mason, M.; Ambrus, A. Urban greenspace is associated with reduced psychological stress among adolescents: A geographic ecological momentary assessment (GEMA) analysis of activity space. *Landsc. Urban Plan.* **2018**, *174*, 1–9. [CrossRef] [PubMed]
67. Thompson, C.; Roe, J.; Aspinall, P.; Mitchell, R.; Clow, A.; Miller, D. More green space is linked to less stress in deprived communities: Evidence from salivary cortisol patterns. *Landsc. Urban Plan.* **2012**, *105*, 221–229. [CrossRef]
68. Mushi, M.; Shalala, R.; Young, J. *Badgerys Urbanism*; UTS School of Architecture: Sydney, Australia, 2017.
69. Calliari, E.; Staccione, A.; Mysiak, J. An assessment framework for climate-proof nature-based solutions. *Sci. Total Environ.* **2019**, *656*, 691–700. [CrossRef] [PubMed]
70. Raymond, C.M.; Frantzeskaki, N.; Kabisch, N.; Berry, P.; Breil, M.; Nita, M.R.; Geneletti, D.; Calfapietra, C. A framework for assessing and implementing the co-benefits of nature-based solutions in urban areas. *Environ. Sci. Policy* **2017**, *77*, 15–24. [CrossRef]
71. Almenar, J.B.; Elliot, T.; Rugani, B.; Philippe, B.; Gutiérrez, T.; Sonnemann, G.; Geneletti, D. Nexus between nature-based solutions, ecosystem services and urban challenges. *Land Use Policy* **2021**, *100*, 104898. [CrossRef]
72. Eggermont, H.; Balian, E.; Azevedo, J.M.N.; Beumer, V.; Brodin, T.; Claudet, J.; Fady, B.; Grube, M.; Keune, H.; Lamargque, P.; et al. Nature-based Solutions: New Influence for Environmental Management and Research in Europe. *Gaia Ecol. Perspect. Sci. Soc.* **2015**, *24*, 243–248. [CrossRef]
73. Roggema, R. Australian Urbanism; State of the Art. In *The Australian Approach. Contemporary Urban Design Thinking*; Roggema, R., Ed.; Springer: Dordrecht, The Netherlands, 2019; Volume 1, pp. 1–10.
74. McHarg, I.L. *Design with Nature*; Natural History Press: New York, NY, USA, 1969.

75. Mostafavi, M.; Doherty, G. (Eds.) *Ecological Urbanism*; Lars Müller Publishers: Baden, Switzerland, 2010.
76. Waldheim, C. *Landscape as Urbanism. A General Theory*; Princeton University Press: Princeton, NJ, USA, 2016.
77. Kindon, S.; Pain, R.; Kesby, M. (Eds.) *Participatory Action Research Approaches and Methods: Connecting People, Participation and Place*; Routledge: London, UK, 2007; Volume 22.
78. McIntyre, A. *Participatory Action Research*; Sage Publications: Thousand Oakes, CA, USA, 2007; Volume 52.
79. Selener, J.D. *Participatory Action Research and Social Change: Approaches and Critique*; Cornell University: Ithaka, NY, USA, 1993.
80. De Jong, T.M. *Kleine Methodologie voor Ontwerpend Onderzoek*; BOOM Uitgevers: Amsterdam, The Netherlands, 1992.
81. Hauberg, J. Research by Design—A research strategy. *Rev. Lusófona Arquit. Educ. Archit. Educ. J.* **2011**, *5*, 46–56.
82. Milburn, L.-A.S.; Brown, R.D. The relationship between research and design in landscape architecture. *Landsc. Urban Plan.* **2003**, *64*, 47–66. [CrossRef]
83. Roggema, R. Research by Design: Proposition for a Methodological Approach. *Urban Sci.* **2016**, *1*, 2. [CrossRef]
84. Rosemann, J. The Conditions of Research by Design in Practice. In Proceedings of the Research by Design: International Conference Faculy of Architecture Delft University of Technology in Co-Operation with the EAASE/AEEA, Delft, The Netherlands, 1–3 November 2000; pp. 63–68.
85. Swann, C. Action Research and the Practice of Design. *Des. Issues* **2002**, *18*, 49–61. [CrossRef]
86. Conversano, I.; del Conte, L.; Mulder, I. Research through Design for accounting values in design. In Proceedings of the 4th Biennial Research through Design Conference, Delft, The Netherlands, 19–22 March 2019; pp. 1–16. [CrossRef]
87. Condon, P.M. *Design Charrettes for Sustainable Communities*; Island Press: Washington, DC, USA, 2008.
88. Lennertz, B.; Lutzenhiser, A. *The Charrette Handbook. The Essential Guide for Accelerated Collaborative Community Planning*; The American Planning Association: Chicago, IL, USA, 2006.
89. Roggema, R.; Martin, J.; Horne, R. Sharing the climate adaptive dream: The benefits of the charrette approach. In Proceedings of the ANZRSAI Conference, Canberra, Australia, 6–9 December 2011.
90. Roggema, R. (Ed.) *The Design Charrette: Ways to Envision Sustainable Futures*; Springer: Dordrecht, The Netherlands, 2013.
91. New South Wales National Parks and Wildlife Service. *Interpretation Guidelines for the Native Vegetation Maps of the Cumberland Plain, Western Sydney*; Final Edition; NPWS: Hurstville, NSW, Australia, 2002.
92. Schwarz, P. *The Art of the Long View. Planning for the Future in San Uncertain World*; Bantam Doubleday Dell Publishing Group Inc.: New York, NY, USA, 1991.
93. Fry, T. *Design Futuring*; University of New South Wales Press: Sydney, Australia, 2009; pp. 71–77.
94. Cornish, E. *Futuring: The Exploration of the Future*; World Future Society: Chicago, IL, USA, 2004.
95. Shipley, R.; Newkirk, R. Vision and visioning in planning: What do these terms really mean? *Environ. Plan. B Plan. Des.* **1999**, *26*, 573–591. [CrossRef]
96. Gaffikin, F.; Sterrett, K. New visions for old cities: The role of visioning in planning. *Plan. Theory Pract.* **2006**, *7*, 159–178. [CrossRef]
97. Roggema, R. Adaptation to climate change: Does spatial planning help? Swarm planning does! *WIT Trans. Ecol. Environ.* **2009**, *127*, 161–172.
98. Callon, M. *Actor-Network Theory—The Market Test. Actor Network Theory and after*; Law, J., Hassard, J., Eds.; Oxford University Press: Oxford, UK, 1999.
99. Bruno Latour, B. *Reassembling the Social: An Introduction to Actor-Network-Theory*; Oxford University Press: New York, NY, USA, 2005; 328p.
100. DeLanda, M. *A New Philosophy of Society: Assemblage Theory and Social Complexity*; Bloomsbury Publishing: London, UK, 2019.
101. Deleuze, G.; Guattari, F. *A Thousand Plateaus*; Athlone Press: London, UK, 1988.
102. DeLanda, M. Deleuzian social ontology and assemblage theory. In *Deleuze and the Social*; Edinburgh University Press: Edinburgh, UK, 2006; pp. 250–266.
103. Nowak, D.J. Institutionalizing urban forestry as a "biotechnology" to improve environmental quality. *Urban For. Urban Green.* **2006**, *5*, 93–100. [CrossRef]
104. Miller, R.W.; Hauer, R.J.; Werner, L.P. *Urban Forestry: Planning and Managing Urban Greenspaces*; Waveland Press: Long Grove, IL, USA, 2015.
105. Konijnendijk, C.C.; Ricard, R.M.; Kenney, A.; Randrup, T.B. Defining urban forestry—A comparative perspective of North America and Europe. *Urban For. Urban Green.* **2006**, *4*, 93–103. [CrossRef]
106. Park, H. A Model of Food Forestry and its Monitoring Framework in the Context of Ecological Restoration. Ph.D. Thesis, Pusan National University, Busan, Korea, 2016.
107. Park, H.; Kramer, M.; Rhemtulla, J.M. Urban Food Forestry: Current state and future perspectives. *Urban For. Urban Green.* **2019**, *45*, 126482.
108. Clark, K.H.; Nicholas, K.A. Introducing urban food forestry: A multifunctional approach to increase food security and provide ecosystem services. *Landsc. Ecol.* **2013**, *28*, 1649–1669. [CrossRef]
109. Bewsher Consulting PTY LTD. *South Creek Floodplain Risk Management Study and Plan for the Liverpool Local Government Area*; Final Report; Liverpool Council: Liverpool, Australia, 2004.
110. McLean, L.; Roggema, R. Planning for a Prosumer Future: Central Park case study. *Urban Plan.* **2019**, *4*, 172–186. [CrossRef]

111. Ostrom, E.E.; Dietz, T.E.; Dolšak, N.E.; Stern, P.C.; Stonich, S.E.; Weber, E.U. *The Drama of the Commons*; National Academy Press: Washington, DC, USA, 2002.
112. GANSW. *Connecting with Country*; Draft; GANSW: Sydney, Australia, 2020.
113. NSW. *Act of Parliament 1995, Threatened Species Conservation Act 1995, NSW Act of Parliament*; Online Edition; State of New South Wales: Sydney, Australia, 1995.
114. Boland, D.J.; Brooker, M.I.H.; Chippendale, G.M.; McDonald, M.W. *Forest Trees of Australia*; CSIRO Publishing: Collingwood, VIC, Australia, 2006; p. 628.
115. Syncarpia Glomulifera. Available online: https://en.wikipedia.org/wiki/Syncarpia_glomulifera (accessed on 21 February 2021).
116. Elliot, R.W.; Jones, D.L.; Blake, T. *Encyclopaedia of Australian Plants Suitable for Cultivation: Volume 9 Sp-Z*; Lothian Press: Port Melbourne, VIC, Australia, 2010; p. 159.
117. Holmgren, D. *Essence of Permaculture*; Holmgren Design Services: Hepburn, Australia, 2007.
118. Steffensen, V. *Fire Country, How Indigenous Fire Management Could Help Save Australia*; Hardie Grant Publishing: Sydney, Australia, 2020.
119. Wright, J.S.; Fu, R.; Worden, J.R.; Chakraborty, S.; Clinton, N.E.; Risi, C.; Sun, Y.; Yin, L. Rainforest-initiated wet season onset. *Proc. Natl. Acad. Sci. USA* **2017**, *114*, 8481–8486. [CrossRef] [PubMed]

land

MDPI

Article

Urban Land Use Efficiency under Resource-Based Economic Transformation—A Case Study of Shanxi Province

Xin Janet Ge [1,*] and Xiaoxia Liu [2]

1 School of Built Environment, University of Technology Sydney, Ultimo, NSW 2007, Australia
2 Faculty of Public Management, Shanxi University of Finance and Economics, Taiyuan 030006, China; liuxx@sxufe.edu.cn
* Correspondence: xinjanet.ge@uts.edu.au; Tel.: +61-2-9514-8074

Abstract: Shanxi, one of China's provinces, has been approved by the State Council as the only state-level comprehensive reform zone for resource-based economic transformation in 2010. Consequently, the implementation of National Resource-based Cities Sustainable Development Planning (2013–2020) and The State Council on Central and Western Regions Undertaking of Industrial Transformation Guide were also introduced. As a result, many agricultural lands were urbanized. The question is whether the transformed land was used efficiently. Existing research is limited regarding the impact of the government-backed transformation of the resource-based economy, industrial restructuring, and urbanization on land use efficiency. This research investigates urban land use efficiency under the government-backed resource-based economy transformation using the Bootstrap-DEA and Bootstrap-Malmquist methods. The land use efficiency and land productivity indexes were produced. Based on the empirical study of 11 prefectural cities, the results suggest that the level of economic development and industrial upgrading are the main determinants of land use efficiency. The total land productivity index declined after the economic reform was initiated. The findings imply that the government must enhance monitoring and auditing during policy implementation and evaluate the policy effects after for further improvement. With the scarcity of land resources and urban expansion in many cities worldwide, this research also provides an approach to determining the main determinants of land use efficiency that could guide our understanding of the impact of the future built environment.

Keywords: land use efficiency; land productivity index; technological change index; technical efficiency change index; Bootstrap-DEA; Bootstrap-Malmquist; China

Citation: Ge, X.J.; Liu, X. Urban Land Use Efficiency under Resource-Based Economic Transformation—A Case Study of Shanxi Province. *Land* **2021**, *10*, 850. https://doi.org/10.3390/land10080850

Academic Editors: Sarel Cilliers, Elizelle Juanee Cilliers and Luca Salvati

Received: 26 May 2021
Accepted: 10 August 2021
Published: 14 August 2021

1. Introduction

Rapid urbanization and industrialization, supported by an extensive land use pattern, have resulted in land urbanization occurring significantly faster than population urbanization in China. From 2005 to 2015, the area of urban built-up areas in China increased from 32,221 km^2 to 52,102.31 km^2, with an annual growth rate of 5.47%. Meanwhile, the urban resident population increased from 562.1 million to 7711.6 million, with an annual growth rate of 3.38% [1]. Moreover, the inefficient use has coexisted with urban sprawl, which has exacerbated the tensions of the human–land relationship [2,3]. Concurrently, as in many other countries, countless negative effects have emerged in the development mode of urban expansion, including but not limited to the reduction of high-quality farmland and deterioration of the environment and ecosystems [4–6], as well as an increase in potential risks relating to food security [7].

Before the economic reform in China, resource-based cities played a significant role in their contribution to the nation's wealth and local employment. Resource-based cities are cities dominated by extractive industries and heavy industry under a low-tech and extensive economic development model. They depend primarily on the exploiting and processing of natural resources, such as minerals, energy, or virgin forests [8]. In contrast, the economic activities of non-resource-based cities rely on textiles, electrical and electronics, transport equipment, and manufacturing. There are 262 resource-based cities in China according to the Plan of Sustainable Development for Resource-based Cities (2013–2020) [9], where 25 percent of the resource-based cities' developments are derived from coal mining.

Previously, the central government has dominated the tax rules and distribution of resources. The State collects a major portion of taxes from the state-owned resource industry and sets an unfavorable price for the resource industry. Because of diminished revenue from unfavorable selling prices and tax collection, the local council was seriously deprived of funds for any development, leaving the local community exposed to potential risks from the damaged ecological system and vulnerability for its heavy reliance on employment by the resource-related industry [10]. Together with other centrally-designed strategic plans for the nation's industry structure, the unfavorable pricing system led to an unbalanced and distorted industry distribution in China. Coastal cities and other energy-consuming heavy industries are the beneficiaries of such a low coal price policy. Studies have shown that coastal cities and eastern China are the regions with relatively high GDP outputs and higher land use efficiencies [10–13].

The use of natural resources is not sustainable as they are non-renewable. Resource-based cities have encountered more prominent land use problems, as land use is more extensive, and the ecological environment has deteriorated significantly [12]. The root cause of this result is a single energy-intensive, high-emission, and low-tech industrial structure formed by a long-term reliance on resource development [14]. Against this backdrop, the Chinese Government has recognized the importance of developing sustainable resource-based cities. As of 2000, the State started taking initiatives and measures, backed by a series of documentations and billions in financial aid, aiming to improve resource-based cities. The measurements include optimizing the industrial structure by cultivating replacement industries, such as biomedicine and smart manufacturing; recovering the destroyed land by mining into "ecological land", such as parkland, woodland, grassland, lake, and marsh water surface to improve the living environment; and promoting the urbanization of mined-out and subsided areas through immigration relocation or centralized resettlement. However, the earlier preliminary studies have shown that economic transformation is effective in addressing the conflict of land use and economic development, resulting in a protected and restored ecological environment [10]; the impacts of economic transformation policies on land use efficiency need further testing.

Land use efficiency is an economic concept. From an economic perspective, land supply is inelastic. Improving land use economic efficiency can be interpreted as either the incensement of economic output while maintaining the same land input for economic activities or keeping the same economic output with a lowered land input for economic activities. On the other hand, the rapid progress of urbanization on a global level has seen land use efficiency become an even more pressing worldwide issue for all developed, developing, and underdeveloped countries due to the scarcity of land. From an international perspective, a wide range of policies and regulations has been developed to improve urban land use efficiency and manage urban sprawl. Developed countries, represented by the United States and Europe [15–17], have adopted government intervention measures. Some examples are zoning ordinances and urban planning, smart growth by increasing the development density and curbing urban sprawl [16], and spatial planning towards compact urban forms to manage urban growth effectively [17]. They have also established tradeable land development rights to improve land use efficiency [18]. Similarly, a series of policies have been introduced and implemented for managing land use in recent years in China, including cultivated land protection policies [19,20], land use planning, and the allocation of a construction land quota. More toolkits have been introduced to enhance land use efficiency in urban built-up areas, including industrial upgrading policies and urban renewal patterns. Some examples are the reconstruction of "three-old" (old town, old factory house, old village house) in Guangzhou and urban renewal in Shenzhen. However, urban land use efficiency is not only a response to land use management policies and measures, it is also the result of the joint effects of resource endowment, economic development level, industrial structure, and population. The constant land depletion could cause land damage problems, such as collapses, landslides, mudslides, ground subsidence, ground fissures, open pits, and mining pits, and land occupation problems, such as waste rock (soil, slag) yards, coal gangue piles, tailings ponds and industrial squares. The above-mentioned problems cause damage to the topography and landscape and have a serious impact on the land's function. It is thus important to understand the impact of land use efficiency and the integration of economic, environmental and societal impacts that land use changes have on sustainability.

In recent years, the evaluation of urban land use efficiency and the identification of driving factors responsible for urban land use change has attracted more attention from scholars of various countries. Zitti et al. [21] have examined the direction and intensity of metropolitan growth in post-war Athens, analyzing development drivers using regression models based on 13 indicators. Valerio et al. [22] have studied the pace and dynamics of urban expansion in the Metropolitan City of Rome. Abdullahi et al. [23] have evaluated and forecasted the residential development of Kajang City in Malaysia based on compact development. The investigation of efficiency and drivers of urban land use in China appeared in different research scales—such as national, city circle, provincial, or single cities—and have focused on the construction of an evaluation index system, evaluation method, and analysis of efficiency difference [2,24–26]. An evaluation index system is usually constructed using the input-output indicators of the economy, society, and environment. The key indices include investment intensity, labor, financial expenses, urban land release, and economic output. The data envelopment analysis (DEA), a nonparametric frontier approach, has been widely used to model and analyze land use efficiency because it has the advantage of consolidating all relevant indicators into one overall indicator to support a more comprehensive performance comparison. Recently, DEA has been extended to the stochastic production function with improved measurement accuracy, the Malmquist index analysis, and Bootstrap-DEA [27]. The indicators and methods of previous research provide a good foundation for the following research. However, little research has been conducted on urban land use efficiency throughout the resource-based economic transformation. Further, there is a lack of information concerning a targeted evaluation index system. Therefore, developing an index system and tool for measuring urban land use efficiency in a resource-based economy transformation is worthwhile.

This paper studies the impact of urban land use efficiency on the resource-based economy under the transformation period. The research further explores the main influencing factors of a resource-based city in the context of economic transformation. Cities in the Shanxi Province, China, are selected as a case study since most have developed a single industrial structure dominated by coal mining and coal chemical industries. Coal mining has resulted in undesirable land use consequences, such as mined-out subsidence areas, soil contamination, land occupation, and industrial and mining wasteland.

In November 2010, Shanxi Province was approved by the State Council as the only state-level comprehensive reform zone for resource-based economy transformation in the country. Under the premise of protecting and restoring the ecological environment, Shanxi Province was allowed to conduct a series of reforms in various fields, including systems and mechanisms, to promote emerging industries, organic clusters and networks, and enhance economic vitality. In 2017, the State Council further emphasized the economic transformation reform, putting forward suggestions for deepening Shanxi Province's transformation reform. The aim was to explore a new road of transformation, upgrade innovation-driven development, and form a scalable and replicable experience. This research is significant for the government's understanding of the effects of implementing economic transformation policy and the factors that affect land use efficiency. The research findings can be used for guiding the planning of land for urban development.

We began developing the input–output indicators according to the goals of resource-economy transformation using the Bootstrap-DEA and Bootstrap-Malmquist index for measuring urban land use efficiency before and after the resource transformation reform. Further, we analyzed dominant factors affecting the changes in urban land use efficiency.

The remainder of the paper is organized as follows. Section 2 introduces the methodology and data consisting of the input-output indicators selected, Bootstrap-DEA, and the Bootstrap-Malmquist index for measuring urban land use efficiency. Section 3 uses the approach described in Section 2 to conduct an empirical study on the change of urban land use efficiency in Shanxi Province from 2006 to 2015. In Section 4, a regression model using panel data is employed to analyze the main determinants that may influence urban land use efficiency. Section 5 concludes this study.

2. Methodology and Data

Urban land use efficiency is an important indicator for measuring the efficient allocation, operating status, and management level of urban input resources. The theoretical framework of land use efficiency concerns an application in economic analysis for comparing the alternatives of recourse allocation doing business and, an analysis of environmental and ecological systems to ensure the protection and the renewability of valuable land resources [28–30]. The efficiency is usually explained by the changes in effect compare the land use related activities to the changes in resources consumed to produce the effect. The study of land use efficiency aims for decision-making to optimize business operations and improve the management of land units.

Despite qualitative methods and the cost–benefit analysis method, input–output analysis using the Malmquist index and Bootstrap-DEA [27] is widely used to assess, analyze, and model procedures related to land use. The efficiency is measured by examining whether a maximum output could be achieved by a given mix of inputs (i.e., technical efficiency). The input–output analysis also addresses the issue when it is impossible to increase the output without increasing inputs (i.e., technological efficiency).

We selected a combination of Bootstrap-DEA, Bootstrap-Malmquist indices, and a panel data model with Hausman test regression analysis to derive the scores of land use efficiencies and a land productivity index, which was then further refined into a technological change index and technical efficiency change index. This was done to facilitate a robust and comprehensive analysis of land use efficiency and its relationship with some external explanatory variables, such as the level of economic development, government spending, industrial upgrading, industrial reconstruction and urban population.

2.1. Data Envelopment Analysis

The data envelopment analysis (DEA) is a linear programming methodology that measures the efficiency of multiple decision-making units (DMUs) when the production process presents a structure of multiple inputs and outputs [31], which can deal with multiple inputs and outputs simultaneously and does not require any prior information on the theoretical framework of the production function [32]. In this study, the input-oriented data envelope analysis (DEA) is used to measure the urban land use efficiency of each city at a certain period since the control input is more feasible in this field. Suppose that each city is a decision-making unit (DMU)—each $DMU_j(j = 1, 2, \cdots, n)$, demands m inputs $x_{ij}^t(i = 1, 2, \cdots, m)$, to generate s outputs $y_{rj}^t(r = 1, 2, \cdots, s)$ at t period. The piecewise linear production technology can be defined as follows to determine the kth of DMU at the time t:

$$\begin{cases} minh_k = \theta - \varepsilon\left(\sum\limits_{i=1}^{m} s_i^- + \sum\limits_{r=1}^{s} s_r^+\right) \\ s.t.\ \sum\limits_{j=1}^{n} \lambda_j\, x_{ij} - \theta x_{ik} + s_i^- = 0, i = 1, \cdots, m \\ \sum\limits_{j=1}^{n} y_{rj}\lambda_j - s_r^+ = y_{rk}, r = 1, \cdots, s \\ \lambda_j, s_i^-, s_r^+ \geq 0, j = 1, \cdots, n, i = 1, \cdots, m, r = 1, \cdots, s \end{cases} \tag{1}$$

where λ_j denotes the non-negative multiplicative vector of inputs and outputs; t (t = 2006, … , 2015) denotes the year s_i^-; *and* s_r^+ denotes the slack variables of inputs and outputs, respectively. Further, θ denotes the efficiency of DMU estimation, and ε denotes non-Archimedes infinite small amount in the model, which is 10^{-4}.

When $\theta = 1$, the DMU is at a perfect and efficient position on the production possibility curve. When $\theta < 1$, the DMU is inefficient, and it is necessary to modify the input-output structure.

2.2. Malmquist Indices

We applied the Malmquist productivity index to construct an urban land use efficiency model for the application purposes to study the changes of urban land use efficiency over time. The Malmquist productivity index, first pioneered by Caves et al. [33], relies on distance functions. Färe et al. [34] further developed it by considering technical inefficiency in productivity measurement. The Malmquist index derived from DEA has been widely employed to evaluate productivity growth assessment and its decomposition into technological change and technical efficiency change [35] in many fields. Some examples of the empirical studies were financing and banking [36], farming [37], IT systems [38], environmental studies [39], and other areas. The method is relatively easy to compute and does not require formational form assumptions and price information. One of the pitfalls is that it can only serve binary comparisons since it does not verify the transitivity property [40]. In recent years, its application for evaluating land use efficiency can be found in literature [2,24–26,40]. In this paper, the input-oriented DEA-CCR (the *CCR model* is based on the assumption that the constant return to scale exists at the efficient frontiers) and Malmquist productivity indices are adopted for the sample of cities used here.

2.3. Malmquist Indices Model for the Efficiency of Land Use

Taking Shanxi Province as a decision-making system, each of the 11 prefectural-level cities is a decision-making unit, represented by $DMU_j (j = 1, 2, \cdots, 11)$. In each DMU_j at time t, there are m inputs $x_{ij}^t (i = 1, 2, \cdots, m)$ and s outputs $y_{rj}^t (r = 1, 2, \cdots, s)$. From t_1 to t_2, the Malmquist indices formula in concerning the input-output is described as follows:

$$M_I\left(x_j^{t_2}, y_j^{t_2}, x_j^{t_1}, y_j^{t_1}\right) = \sqrt{\frac{D_c^{t_1}\left(x_j^{t_2}, y_j^{t_2}\right)}{D_c^{t_1}\left(x_j^{t_1}, y_j^{t_1}\right)} \times \frac{D_c^{t_2}\left(x_j^{t_2}, y_j^{t_2}\right)}{D_c^{t_2}\left(x_j^{t_1}, y_j^{t_1}\right)}} \tag{2}$$

where in (2), $D_c^{t_1}\left(x_j^{t_2}, y_j^{t_2}\right)$ indicates the Shepard distance function between the combination of input-output factor to the frontier at time t_1 to t_2, $D_c^{t_1}\left(x_j^{t_1}, y_j^{t_1}\right)$, $D_c^{t_2}\left(x_j^{t_2}, y_j^{t_2}\right)$, $D_c^{t_2}\left(x_j^{t_1}, y_j^{t_1}\right)$ are defined in the same way under the constant scale of the decision-making unit DMU_j; and $M_I\left(x_j^{t_2}, y_j^{t_2}, x_j^{t_1}, y_j^{t_1}\right)$ measures changes in productivity for DMU_j from t_1 to t_2 $(t_1 < t_2)$.

According to Worthington [36], productivity improvement over time can be measured by technical efficiency improvements (ΔE) or technological improvement (ΔT). Technical efficiency is the ability to use a minimal amount of input to make a given output level, whereas technological changes measure the ability to combine inputs and outputs optimally. As a result, the performance of changes in productivity can be further broken down into two parts—changes in efficiency (ΔE) and technological change (ΔT)—which can be described in Formula (3):

$$M_I\left(x_j^{t_2}, y_j^{t_2}, x_j^{t_1}, y_j^{t_1}\right) = \underbrace{\frac{D_c^{t_2}\left(x_j^{t_2}, y_j^{t_2}\right)}{D_c^{t_1}\left(x_j^{t_1}, y_j^{t_1}\right)}}_{\Delta E} \underbrace{\sqrt{\frac{D_c^{t_1}\left(x_j^{t_2}, y_j^{t_2}\right)}{D_c^{t_2}\left(x_j^{t_2}, y_j^{t_2}\right)} \times \frac{D_c^{t_1}\left(x_j^{t_1}, y_j^{t_1}\right)}{D_c^{t_2}\left(x_j^{t_1}, y_j^{t_1}\right)}}}_{\Delta T} \tag{3}$$

or

$$\text{Changes in productivity } (DMU_j) = \text{changes in technical efficiency } (\Delta E) * \text{technological change } (\Delta T) \quad (4)$$

where in (3) and (4), ΔE denotes the extent of efficiency change to the frontier from t_1 to t_2, representing the change in technical efficiency. Further, ΔT denotes the extent of technological change to the frontier from t_1 to t_2, reflecting the technological change. This research method has been applied to many applications, including the assessment of the productive performance of countries [41], financial institutions [42,43], manufacturing industries [44], and the like.

2.4. Bootstrap-Malmquist Method for the Correction of Urban Land Use Efficiency

Both the traditional DEA and Malmquist models are based on the observed finite samples. The estimated results are subject to sampling variation sensitivity [27,45]. With this in mind, the Bootstrap method is an improved resampling technique; a replacement sample data pool is created by stimulating the formation process of the original sample. This is used for a serious repeatable preformation of statistic calculations, such as the significance test, the original sample empirical distribution, and corrections to the original means and deviation [46].

Based on the principle and method of the Bootstrap-Malmquist method proposed by Simar and Wilson [47], the Bootstrap-Malmquist technique for measuring the efficiency of land is established as follows:

(a) In the original data pool for DMU_j, the Malmquist indices estimate $\hat{M}_j\left(x_j^{t_2}, y_j^{t_2}, x_j^{t_1}, y_j^{t_1}\right), j = 1, 2, \cdots, n$;

(b) Based on Simar and Wilson [47], the proposed binary Kernel density and Reflection method, the Bootstrap resampling technique is used to obtain a combination of input-output replacement samples P^* from the decision-making unit:

$$P^* = \left\{x_j^{*t}, y_j^{*t} \middle| j = 1, 2, \cdots, n; t = t_{1,}t_2\right\}. \quad (5)$$

(c) The multivariate linear programming model for solving the pseudo-sample distance function is shown in (6):

$$\left[\hat{D}_j^{t_1}\left(x_j^{t_2}, y_j^{t_2}\right)\right]^{-1} = max\theta,$$

where

$$\begin{aligned} &\theta y_{rj}^{t_2} \le \sum_{j=1}^{11} \lambda_j^{t_1} y_{rj}^{t_1}, r = 1, 2, \cdots, s \\ \text{s.t. } &\sum_{j=1}^{11} \lambda_j^{t_1} x_{ij}^{t_1} \le x_{ij}^{t_2}, i = 1, 2, \cdots, m \\ &\lambda_j^{t_1} \ge 0, j = 1, 2, \cdots, n \end{aligned} \quad (6)$$

Based on resamples P^*, we must solve (6) deepening Shanxi Province's transformation reform, from distance function $D_j^{t_1}\left(x_j^{t_1}, y_j^{t_1}\right)$, and derive Bootstrap estimates $\hat{D}_j^{*t_1}\left(x_j^{t_2}, y_j^{t_2}\right)$. We must change the corresponding factors in (6) at the same period to reconstruct the model and repeat these procedures to solve the equation. The remaining distance function Bootstrap estimates are derived $\hat{D}_j^{*t_2}\left(x_j^{t_2}, y_j^{t_2}\right), \hat{D}_j^{*t_2}\left(x_j^{t_1}, y_j^{t_1}\right), \hat{D}_j^{*t_1}\left(x_j^{t_2}, y_j^{t_2}\right)$.

(d) The Malmquist indices Bootstrap estimate $\hat{M}_j^*$ is calculated by applying Bootstrap estimates.

(e) Repeating steps (b–d) B times (B = 2000) results in B times of estimate $\left\{\hat{M}_{jb}^*\right\}_{b=1}^{B}$. According to research by [48], if 2000 sampling times of B are obtained, they should ensure a guaranteed coverage of confidence intervals.

A smooth distribution of Bootstrap-Malmquist indices was obtained by adopting the above steps, which were used to replace the original data distribution. Therefore, corrections for deviations to the original Malmquist index can be made using $\left\{\hat{M}^{*}_{jb}\right\}_{b=1}^{B}$ to construct a confidence area at a 5 percent and 10 percent significant level, respectively, followed by a further examination of productivity significance. With the same principles applied, values for technical efficiency change and technological changes can be derived using the same confidence intervals.

2.5. Selection of Variables and Indicators, and the Source of Data

In this paper, 11 prefecture-level cities in Shanxi Province were selected as units for decision-making and analysis including Taiyuan, Datong, Yangquan, Changzhi, Jincheng, Shuozhou, Jinzhong, Yuncheng, Zhangzhou, Linyi, and Luliang. From 2006 to 2015, the year 2010 was chosen as the reference point, at which the State Government initiated the comprehensive reform.

For the state-initiated economic reform policies with financial aids for the resource-based cities, it set major tasks, such as the establishment of long-term incentive measures for sustainable development; the cultivation of alternative industries; the need to solve social problems with a focus on employment; the enhancement of environmental treatment, and ecological system protection; and the reinforcement of the resources survey and mineral rights management.

Therefore, the following input–output variables that measure urban land use efficiency were selected in the model to assess the effectiveness of government reform initiatives, concerning optimizing the ecosystem, industrial upgrading, and industrial restructuring.

Land, capital, and labor were nominated as input variables in non-agricultural industries that impact urban economic benefits and the change of ecologic environment. These input variables can be quantified individually as areas of urban construction land, fixed-capital stock of secondary and tertiary industries, and the number of people employed in secondary and tertiary industries. The added value of secondary and tertiary industries, and the green coverage rate of built-up areas, were chosen as output indicators, measuring urban economic output and ecologic output. Table 1 depicts the details.

Table 1. Variables of land use efficiency model.

Variables Type	Variables	Measurement	Expected Sign
Input Variables	Capital	2nd & tertiary industries fixed asset (10,000 yuan)	Positive
	Land	Urban construction land area (square kilometer)	Positive
	Labor	Number of employments on 2nd & tertiary industries	Positive
	Research expenses	R&D research funds (10,000 yuan)	Positive
	Energy consumption	Coal consumption (Ton/10,000 yuan)	Negative
Output Variables	Economic output	Value-added of 2nd and tertiary industry (10,000 yuan)	Positive
	Quality of Environment	Green area coverage rate (%)	Positive
		Days completed for Green area coverage (%)	Positive

The urban construction land area and the green area coverage rate of the built-up areas were sourced from the China Urban Construction Statistical Yearbooks from 2007 to 2016. Others were sourced from the Shanxi Statistical Yearbooks from 2007 to 2016, covering fixed assets investment, employed persons, and added value. All relate to the secondary and tertiary industries, with an additional fixed-asset investment price index and price reduction index. Supplementary data in case of incomplete data were sourced from local city yearbooks or statistical bulletins.

Data Processing

The fixed capital stock of the secondary and tertiary industries is estimated by the perpetual inventory method. The calculation formula is

$$k_t^j = k_{t-1}^j \times (1 - \delta) + I_t^j / p_t \quad (7)$$

where k_t^j, k_{t-1}^j are the fixed capital stocks of j city at the years t and $t-1$, respectively. δ is the depreciation rate concerning the recommended value calculated by Zhang et al., [49], which is set at 9.6%. I_t^j is the investment value of fixed assets of the secondary and tertiary industries for each individual city at the time of t year (price at the year of t). p_t is a calculated fixed-asset investment price index with a baseline starting from 2005. The determination of the baseline concerning the fixed-capital stock in 2005 can be referred to as the inter-provincial fixed-capital stock calculated by Zhang et al. [49]. This derives the total fixed-capital stock of the whole society in Shanxi Province in 2000 (the current year's price level). Furthermore, the total social fixed-assets investment in 2001–2005 was used to apply the perpetual inventory method. The total fixed capital stock of the whole society in Shanxi Province in 2005 (remaining the constant price level of 2000) was calculated and subsequently restored to the year 2005 price level. Based on this, the weighing of the added value of the second and tertiary industries in Shanxi Province in 2005 was multiplied by the fixed-capital stock of the whole society, resulting in the derivation of the fixed-capital stock of the secondary and tertiary industries in Shanxi Province. Further derivation of each chosen city's added value of the secondary and tertiary industries is then calculated by its weighting in the province. Similar weighting principles can be repeated to derive the fixed capital stock in the secondary and tertiary industries in the base period for each city. The secondary and tertiary industries' value added was reduced to the constant price level of 2005 using the price reduction index.

3. Empirical Results and Discussions

The percentage increase of input variables, such as the area of urban construction land, fixed-capital stock for the secondary and tertiary industries, and the number of people employed in the secondary and tertiary industries, are given in Figure 1. The percentage increase of output variables, such as the value added by the secondary and tertiary industries and the percentage of green coverage of the built-up areas, are shown in Figure 2.

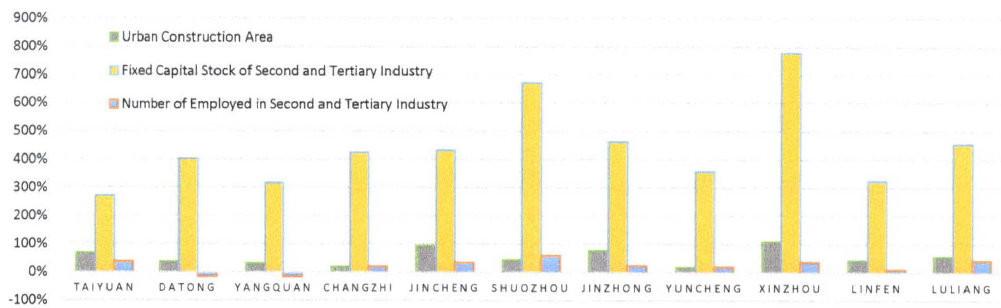

Figure 1. The percentage increase of input variables (i.e., urban construction area, the fixed capital stock of secondary and tertiary industry, and the number of people employed in the secondary and tertiary industries from 2006 to 2015).

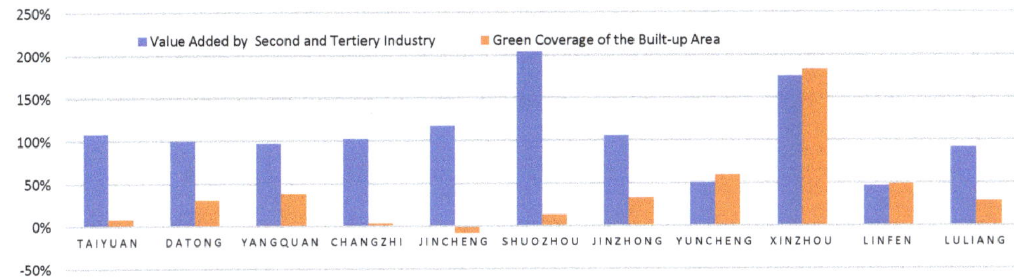

Figure 2. The percentage increase of output variables, such as the value added by the secondary and tertiary industries and green coverage of built-up areas, between 2006 and 2015.

It is clear that there is a 300–800% increase in investments into the fixed-capital stock for the secondary and tertiary industries among the cities. The area for urban construction increased from 20% to 100% in 2015 from 2006. Meanwhile, the number of people employed by the secondary and tertiary industries show variations among cities, with 500,000 more people employed in 2015 above the 1.4 million in 2006 for Taiyuan. Figure 2 shows a clear increase from 50 to 200% for value added, with a reduction of 7%, and various increases of up to 180% for the green coverage of built-up areas.

3.1. Differences in Scores of Land Use Efficiency between Cities

Based on the DEA-CCR and Bootstrap-Malmquist models, MATLAB's DEA analysis tool was used to calculate the Bootstrap-DEA scores of urban land use efficiency, the values of Malmquist indices, the value of Bootstrap-Malmquist indices, the decomposed values of Bootstrap technical efficiency change, and the technological change. The Bootstrap-DEA results appeal different ranges from 0.688 to 0.868, as shown in Table 2, by studying the urban land use efficiency of 11 cities in Shanxi Province between 2006 and 2015. The higher the score, the higher the economic outputs and more green area coverage, thus increased land use efficiency.

Table 2. The scores of land use efficiencies (Bootstrap-DEA) for 11 prefecture-level cities in Shanxi Province 2006–2015.

City	2006	2007	2008	2009	2010	2011	2012	2013	2014	2015	Mean	S.D.
Taiyuan	0.847	0.803	0.824	0.714	0.798	0.813	0.816	0.892	0.952	0.914	0.837	0.068
Datong	0.865	0.782	0.795	0.607	0.609	0.592	0.564	0.610	0.672	0.789	0.688	0.108
Yangquan	0.764	0.643	0.855	0.823	0.866	0.897	0.870	0.880	0.899	0.918	0.841	0.082
Changzhi	0.905	0.847	0.906	0.769	0.815	0.888	0.829	0.855	0.867	0.865	0.855	0.043
Jincheng	0.947	0.760	0.804	0.669	0.756	0.771	0.777	0.819	0.836	0.901	0.804	0.079
Shuozhou	0.922	0.881	0.907	0.841	0.837	0.829	0.834	0.854	0.875	0.905	0.869	0.034
Jinzhong	0.879	0.741	0.742	0.618	0.673	0.657	0.643	0.679	0.707	0.763	0.710	0.076
Yuncheng	0.957	0.828	0.766	0.638	0.553	0.579	0.553	0.623	0.739	0.835	0.707	0.139
Xinzhou	0.887	0.892	0.928	0.712	0.597	0.579	0.541	0.653	0.646	0.802	0.724	0.143
Linfen	0.938	0.884	0.910	0.767	0.727	0.751	0.727	0.738	0.769	0.844	0.805	0.081
Luliang	0.934	0.880	0.897	0.835	0.831	0.835	0.828	0.860	0.868	0.916	0.868	0.038

Further studies were conducted to evaluate the impact of comprehensive reform on the changes in urban land use efficiency for the chosen 11 cities (refer to Table 3). The estimated results indicate that the estimations of most cities are statistically significant as presented for the significance level at 10% or 5%.

The comprehensive reform was introduced in 2005. Therefore, comparisons of the calculated average scores of land use efficiencies for each of these cities were made between the two periods (i.e., before [2006–2010] and after [2011–2015]). The widened range of the average scores of land use efficiencies among the cities for the two periods is revealed in Figure 3. Four cities (Taiyuan, Yangquan, Changzhi, and Jincheng) have improved their land productivities during 2011–2015. Other cities have reduced their average scores of land use efficiencies after the comprehensive reform.

Table 3. The year-on-year land productivity change index for prefectural level cities in Shanxi Province 2006–2015 (Malmquist).

City	2006/2007	2007/2008	2008/2009	2009/2010	2000/2011	2011/2012	2012/2013	2013/2014	2014/2015
Taiyuan	1.028 **	0.949 **	0.919 **	0.983 **	0.978 **	0.932 **	0.853	0.894	0.927 **
Datong	0.962	0.945 **	0.816	0.900 **	0.951 **	0.894 **	0.838	0.877	0.951
Yangquan	0.916	1.006	0.860 **	0.845 **	0.868 **	0.867 **	0.879 *	0.868 **	0.885 **
Changzhi	0.998 **	1.035 **	0.939	0.981 **	1.042 **	0.875 **	0.807 *	0.828 **	0.783 **
Jincheng	0.894	1.003 **	0.890	0.939 **	0.984 **	0.922 **	0.822	0.824 **	0.841
Shuozhou	1.146 **	0.837 **	1.054 **	0.947 **	0.994 **	0.931 **	0.914 **	0.895 **	0.819 **
Jinzhong	0.897	0.968 **	0.933	1.011 **	0.986 **	0.810 **	0.829 **	0.839 **	0.834
Yuncheng	0.918	0.920 **	0.891	0.885	1.184	0.892 **	0.880 *	0.927 *	0.875
Xinzhou	1.057 **	1.017 **	0.796	0.849	0.966 **	0.873 **	1.069 *	0.922 **	1.003
Linfen	1.008	1.044 **	0.867	1.022 **	1.069 **	0.877 **	0.845 **	0.849 **	0.872 *
Luliang	1.083 *	0.993 **	1.009 *	1.103 **	1.086 **	0.941 **	0.853 **	0.847 **	0.839

Note: * Significance level at 10%; ** Significance level at 5%.

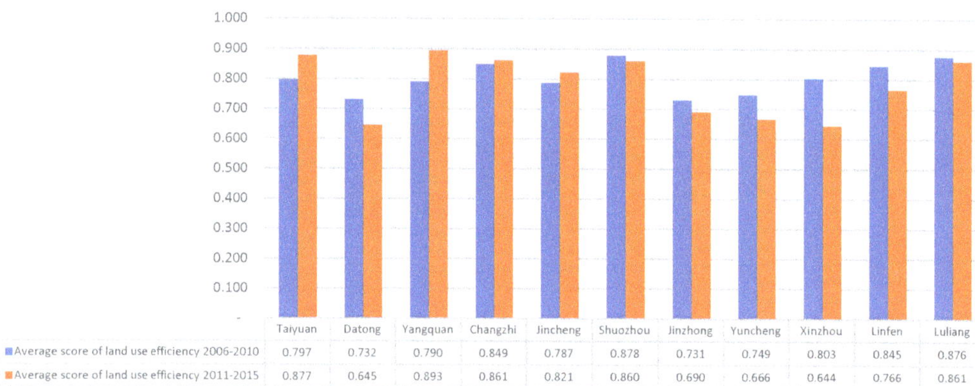

Figure 3. Comparisons between the average scores of land use efficiencies before and after the comprehensive economic transformation.

Note that cities with higher improved average scores of land use efficiencies, such as Taiyuan and Yangquan, generally have a larger increase of value added in the meantime, with a relatively smaller increase of investment in the fixed stock. This effect is vice versa for cities such as Xinzhou, Linfen, and Yunchong, where a drop in land use efficiency is presented.

Figure 4 shows the trend of overall (average) score of land use efficiencies for the total of studied cities in Shanxi Province during 2006–2015. The overall land efficiency scores are marked with a declining trend from 2005 to 2010, then an uptrend after 2010. This finding coincides with Pu et al. [50] who studied industrial land use efficiency in Hunan province. Over the studied period, economic events occurred, such as fluctuations in coal prices, coal demands, and the introduction of economic reform. Nevertheless, the combined factors have inevitably impacted the level of land use efficiency in the area. Further regression analysis on what factors have had a significant impact on the score of land use efficiency is to be demonstrated later in the paper.

Figure 4. The overall (average) land use efficiency score trend for the total of studied cities in Shanxi Province from 2006–2015.

3.2. Identifying Trend Changes in Land Use Efficiency

The results demonstrated in Section 3.1 show that the estimated value of urban land productivity in the chosen cities during the observation period is significant in most cases. However, the number of cities and years in which the Malmquist index value is greater than 1 is relatively few, indicating the scarcity of cities and years in which the urban land productivity has increased from the previous year. In fact, the overall level of urban land productivity has declined, especially after the comprehensive reform.

Further decomposed values of Bootstrap technical efficiency change and the technological change derived from the urban land productivity are shown in Tables 4 and 5 to identify the reasons for the declined urban land productivity.

Table 4. The year-on-year technical efficiency change index for prefectural cities in Shanxi Province 2006–2015 (Bootstrap-Malmquist).

City	2006/2007	2007/2008	2008/2009	2009/2010	2000/2011	2011/2012	2012/2013	2013/2014	2014/2015
Taiyuan	0.911	1.026	0.821	1.112	1.051	0.989	1.105	1.096	0.998
Datong	0.870	1.022	0.729	1.012	1.002	0.951	1.091	1.094	1.202
Yangquan	0.817	1.340	0.939	1.090	1.054	0.957	1.028	1.004	1.021
Changzhi	0.888	1.082	0.822	1.062	1.108	0.925	1.034	1.012	1.020
Jincheng	0.755	1.069	0.824	1.089	1.043	0.985	1.063	1.013	1.108
Shuozhou	0.953	1.016	0.928	0.997	1.006	0.988	1.025	1.014	1.039
Jinzhong	0.798	1.014	0.815	1.088	1.025	0.929	1.067	1.029	1.104
Yuncheng	0.822	0.932	0.816	0.858	1.075	0.951	1.117	1.171	1.157
Xinzhou	0.958	1.058	0.740	0.829	0.988	0.943	1.230	0.986	1.257
Linfen	0.912	1.022	0.808	0.971	1.028	0.948	1.033	1.029	1.117
Luliang	0.951	1.018	0.929	0.990	1.006	0.989	1.020	1.017	1.045

Table 5. The year-on-year technological change index for prefectural level cities in Shanxi Province 2006–2015 (Bootstrap-Malmquist).

City	2006/2007	2007/2008	2008/2009	2009/2010	2000/2011	2011/2012	2012/2013	2013/2014	2014/2015
Taiyuan	1.063	0.942	1.033	0.895	0.969	0.917	0.787	0.830	0.927
Datong	1.042	0.942	1.033	0.895	0.969	0.917	0.788	0.815	0.814
Yangquan	1.055	0.762	0.853	0.795	0.846	0.867	0.879	0.868	0.885
Changzhi	1.055	0.974	1.054	0.930	0.979	0.916	0.798	0.818	0.794
Jincheng	1.100	0.953	1.005	0.868	0.969	0.918	0.792	0.816	0.784
Shuozhou	1.146	0.837	1.054	0.947	0.993	0.930	0.914	0.895	0.819
Jinzhong	1.046	0.973	1.057	0.935	0.966	0.869	0.798	0.817	0.782
Yuncheng	1.036	1.005	1.009	1.037	1.101	0.928	0.809	0.800	0.786
Xinzhou	1.031	0.981	0.983	1.034	1.000	0.901	0.886	0.946	0.836
Linfen	1.056	1.039	0.993	1.050	1.062	0.912	0.832	0.833	0.810
Luliang	1.083	0.993	1.009	1.103	1.086	0.941	0.853	0.847	0.839

The results of Table 4 indicate there is no clear pattern before 2011 concerning the values of technical efficiency change of urban land use in Shanxi Province as the pattern fluctuates with alternative upward and downward trends. However, the pattern becomes clear after 2012, with an overall rising trend. Therefore, based on the characteristics of industrial structure dominated by coal production in Shanxi Province, comparison studies were carried out between the technical efficiency changes of urban land use with the coal price factor in the same period. It is found that before 2011, the fluctuation of the technical efficiency change of urban construction land highly coincided with the fluctuated trend of coal prices. This finding suggests that the heavy-weighted coal-based industrial structure during this period could be the significant factor determining the technical efficiency of urban land use. However, in 2011, comprehensive reform on the transformation of the resource-based economy was initiated in Shanxi Province, adversely impacting the coal price. The average price of standard coal dropped from 1260 Yuan/1000 kg at the end of 2011 to 780 Yuan/1000 kg in 2015 [51]. Another important comprehensive reform commenced from 2013–2015, focusing on prompting the transformation of the coal-based economy to an energy-efficient economy. Notably, since 2011, urban land technical efficiency changes in various cities show an overall upward trend; this is especially true from 2013–2015, in which the energy-efficient economy reform was promoted. This finding suggests that the reform from a coal-based economy to an energy-efficient economy has enhanced the improvements of technical efficiency in land use.

The results in Table 5 illustrate that the technological changes in urban land use in Shanxi Province fluctuate with ascending and descending trends before 2011. However, thereafter, a clear declining trend emerges as the values of technological changes drop year by year for all the studied 11 prefectural-level cities, without exceptions. In the meantime, background studies were conducted to analyze the economic status of ShangXi Province. From 2006 to 2011, major programs focusing on the restructuring and integration of the coal industry were in operation [52]. However, there was no material change concerning the predominantly coal-based industrial structure. The fact that change of ownership due to restructuring and integration could harm the industry's stability, with likely consequences of deterring investors from investing in the industry was apparent. Despite the favorable market condition of coal price, it was evident that preference in the industry was to expand production capacity over technology advancement. Hence, technological improvement was not significant.

From 2012 to 2015, market conditions for coal prices became unfavorable. A series of reforms took place in Shanxi Province, including industrial restructuring, a reduction in raw coal production capacity, extensions to the coal industry chain via focusing on the deep processing of coal, coal–electricity integration, and the development of coal gas [53]. The reform aimed to enhance the improvement of production technology and industrial upgrading. The initial results were positive. However, a minimal impact was expected due to the replacement industries (e.g., tourism and modern logistics) still being in their infancy. At the same time, new strategically important industries, such as modern manufacturing coupled with information technology and industrialization, were still in the planned initial construction stage, with no production ability. Further, the frontiers of the input-output portfolio did not move forward over this time.

In summation, under the background of comprehensive reform, as mentioned above, there are signs of relatively improved technical efficiency of urban land throughout the observation. However, it is still at the early stages of the learning curve, and it is expected that a longer period will be required for the potential impact to occur for the technological progress. This is in agreement with conclusions made by previous relevant research [10] that the economic reforms are not costless. This is especially true when it bears considerable costs at its initial stage [54]. Therefore, further studies are needed to appropriately appraise the extent of the impact that will improve land use efficiency by industrial upgrading and industrial restructuring reform. The sample size employed 11 cities in Shanxi as a case study, which could be a limitation. Further research should collect more data to verify the results of this research.

Figure 5 presents the trends of an overall (average) year-on-year change of the Malmquist productivity index (MPI), the technological change index and the technical efficiency change index for the total studied cities during 2006–2015. It shows a negative correlation between the score of land use efficiency (Figure 4) and technological change (Figure 5) after 2011. However, it displays a positive correlation between the score of land use efficiency and technical efficiency change between 2005 and 2015. The fact that the fluctuation of total land productivity coincides positively with the pattern of technological change suggests the importance of technological advancement, although the area will benefit from assistance provided by government funds and transfers of payment [10]. Still, there is a need to examine the efficiency and effectiveness of funds utilization closely. In particular, it is recommended that future focus should be directed towards improving technologies.

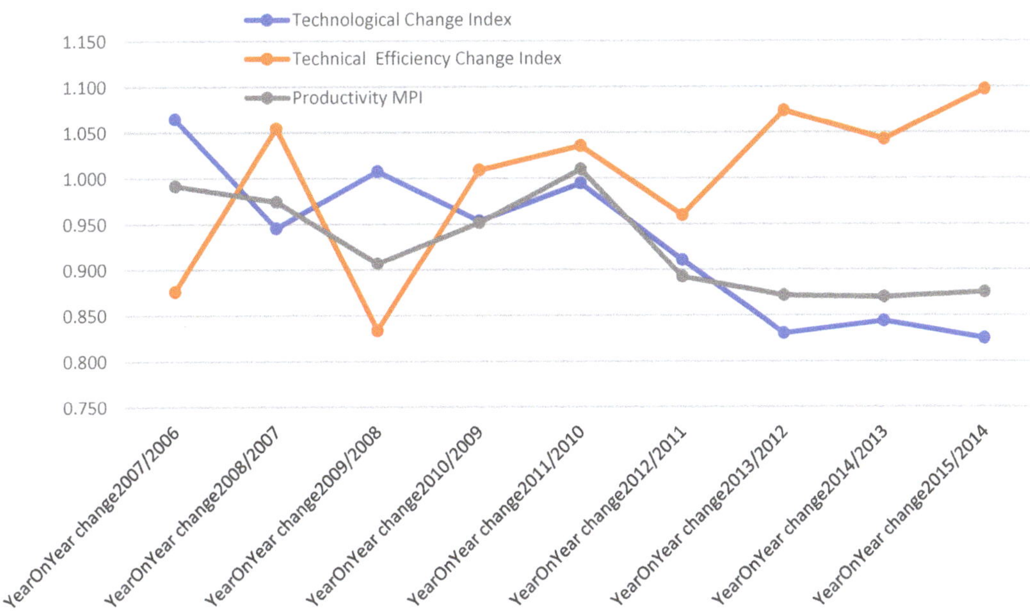

Figure 5. The trends of overall (average) year-on-year change of the Malmquist productivity index, technological change index, and technical efficiency change index for the total studied cities in Shanxi Province from 2006 to 2015.

4. Analysis of Factors Affecting Land Use Efficiency Change

Relevant research shows that urban land use efficiency is influenced by many factors, such as the level of urban economic development status, industrial structure, level of urbanization, and government management [24]. The econometric regression analysis method was used to identify the main factors affecting the change of urban land use efficiency in Shanxi Province.

4.1. The Explanatory Variables, Source of Data, and Construction of Panel Data Model

The Bootstrap scores of the land use efficiency in each city were treated as explanatory variables, which *LE* represent. Considering the transitional characteristics of the resource-based economic transformation in Shanxi Province, there were five major aspects selected for close examination: the level of economic development, industrial upgrading, industrial restructuring, the level of urbanization, and potential government influence on economic transformation. The explanatory variables were quantified, including the level of economic development (GDPP), represented by average GDP per head of the prefectural city (10,000 Yuan/person); level of industrial upgrading (CP), represented by coal consumption per unit of GDP (1000 kg of standard coal/10,000 Yuan in GDP), a reverse index to be modified into a positive indicator by using the reciprocal method; industrial restructuring (IS), represented by the weighing of the added value generated by the tertiary industry to the regional GDP; the urbanization level (UPL), represented by the proportion of the urban population to the number of residents living in the prefectural-level city; and government influence (GOV), represented by the ratio of budgeted government expenditure to GDP for the fiscal year. The raw data of the above variables were sourced from the Shanxi Statistical Yearbook (2006–2016).

A regression analysis model was constructed using panel data from 2006 to 2015. First, the Hausman test was conducted to test the hypothesis of the random effect model, with test results rejecting the random-effects model. Further efforts were made by establishing the fixed effect model, subsequently followed by performing the redundant fixed effect test. It was found that these test results accept the individual fixed-effects model. The model was constructed as follows:

$$LE_{it} = C_i + C + \beta_1 GDPP_{it} + \beta_2 CP_{it} + \beta_3 IS_{it} + \beta_4 UPL_{it} + \beta_5 GOV_{it} + \varepsilon_{it} \quad (8)$$

where LE_{it} indicates the economic efficiency of the construction land for the city i in the year t. Further, the main explanatory variables $GDPP_{it}$, CP_{it}, IS_{it}, GOV_{it}, UPL_{it}, and GOV_{it} indicate the level of economic development, industrial upgrading, industrial restructuring, urbanization, and potential government influence, respectively, for the city i in the year t. C is the common intercept term and C_i is the individual difference; in the model, the intercepted item $C + C_i$ for each city shows its irrelevance to the change of time. It is challenging to observe and reach a conclusion on how and to what extent the selected variants can impact land use efficiency due to the complexity and individuality of each city. ε_{it} is a random error term.

4.2. Results and Analysis

Figure 6 shows the percentage increase of explanatory variables, such as GDPP, coal consumption CP, IS, GOV, and UPL, from 2006 to 2015.

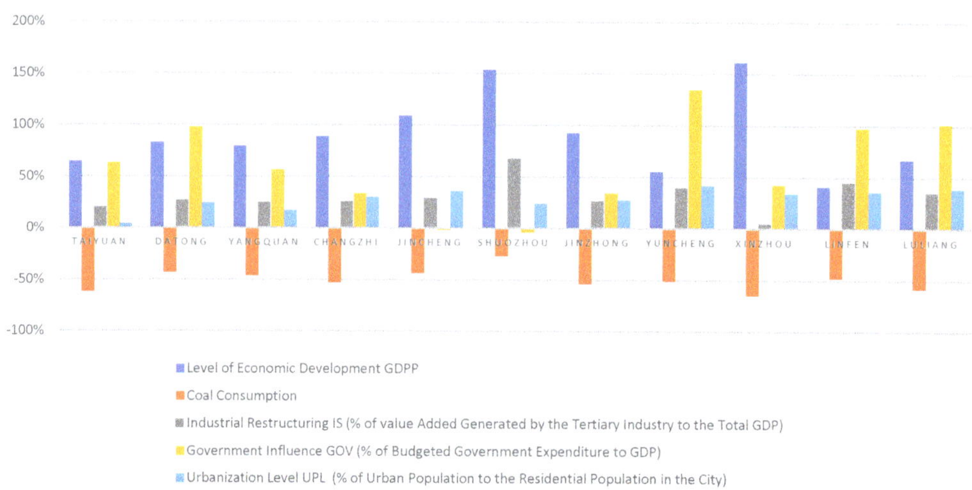

Figure 6. The percentage increase of explanatory variables such as GDPP, coal consumption CP, IS, GOV, and UPL, from 2006 to 2015.

As for GDPP, it had an increase of 50–150% for all cities, with Taiyuan achieving the highest GDPP of 48,000 Yuan per capital and Xinzhou have the lowest of 16,000 Yuan per capital in 2015. On the other hand, a reduction of coal consumption of between 7% and 65% for all cities was observed. Linfen had the highest consumption of coal at 4200 kg per 10,000 Yuan GDP generated in 2006, with 2600 kg for Taiyuan being the lowest. Both cities improved by 2015, with 2200 kg for Linfen and just 1000 kg for Taiyuan. In regard to the change of value added by the tertiary industry, there was an increase of 20% (Taiyuan) to 60% (Shuozhou). The percentage of value added by the tertiary industry to GDP varied among those studied cities, with the highest being 61% for Taiyuan in 2015, which increased from 50% in 2006. In contrast, Luliang had the lowest percentage of 26% and 37% in 2006 and 2015, respectively. As for the government influence factor, the percentage of government spending to GDP had increased up to 130% for Xinzhou and 50% for Taiyuan, with little variation for Jincheng and Shuozhou. The percentage increase of the urbanization level varied between 9% (Taiyuan) and 43% (Yuencheng) among the 11 cities, with Taiyuan at 81% and 84% for 2006 and 2015, respectively—the highest level for an urbanized city. Luliang was among the lowest urbanized cities, with 33% and 46% for 2006 and 2015, respectively.

The estimates using EVIEWS9.0 software for the random-effects model (model I) and the individual fixed-effects model (model II) are shown in Table 6. The Hausman test results reject the random effect model, indicating the existence of an individual or time difference. Further tests followed on the constructed individual fixed-effects model (model II). The redundant fixed-effects test showed the following: statistic value for F at 6.75, which is significant at the 1% significance level. Therefore, test results from the individual fixed-effects model (model II) suggest a reasonable fit with the regression analysis that the F statistics are significant, and the hypothesis is found to be true.

Table 6. Panel data model on regression analysis of explanatory variables with land efficiency score.

Explanatory Variables	Model I	Model II
C	0.704 *** (8.189)	0.616 *** (3.129)
GDPP	0.055 ** (2.575)	0.056 ** (2.111)
CP	0.168 *** (2.726)	0.301 *** (3.353)
IS	−0.001 (−0.568)	0.0046 * (1.9117)
UPL	−0.002 (−0.991)	−0.007 (−1.599)
GOV	0.0017 (1.032)	0.0047 ** (2.553)
R^2	0.09	0.534
F value	Not Significant	7.2 ***
HausmanTest	Random Effect Declined	-

Note: ***, **, * Indicates significance level of 1%, 5%, and 10%, respectively. The *t* value is shown in brackets.

According to the regression results of model II using the Hausman estimator, the determinants for the changes in urban land use efficiency in Shanxi Province are analyzed as follows.

First, the economic development level coefficient expressed as GDPP at 0.056 was positive and significant (at 5%), indicating that the level of economic development improves urban land use efficiency. It is suggested by Chen et al. [11], Sun et al. [12], and Yang et al. [13] that the urban economic agglomeration effect and the scale of urban economy effect could have played an important role in improving land use efficiency. Their studies found that Western cities tend to have relatively low land use efficiency and discovered higher land use efficiency for eastern areas, aligning with the distribution of economic development in China.

Second, the output per unit energy consumption coefficient expressed as CP at 0.301 was positive and significant (at the 1% level). The upper grading of basic industrial structure in Shanxi Province, especially with the predominance of coal and heavy industry structure and the comprehensive reform focusing on introducing energy-efficient technology and general improvement of industrial technology, has inevitably contributed to the increase in land use efficiency.

Third, the industrial restructuring (IS) coefficient was derived with a positive value of 0.0046 at the significant level of 10%; nevertheless, the low coefficient value indicates that the impact from the service industry was minimal for land use efficiency. Research results on the formulation of the third industry within the province suggest that traditional services—such as wholesale and retail, transportation and warehousing, real estate, and the like—are still the major players, although the weighting of the financial industry is rising year by year. Nevertheless, the weighting added by recent technological development in the service industry, such as information transmission, software, and IT services, is still comparatively low, with its highest contribution towards the tertiary industry achieved at 6.8% in 2015. Therefore, for contribution improvement by the third industry, the focus should be directed towards increasing the weighting of the latest technology developments in the tertiary industry. This holds the key to enhanced land use efficiency [11,13].

Fourth, the urban population level (UPL) coefficient was found to be negative at a non-significant value of 0.007, indicating that the level of population urbanization in the province has yet to form a population agglomeration effect. Additionally, its impact at its current stage is negligible on urban land use efficiency. Further investigation of the urban population density in Shanxi Province demonstrates the existence of differentiation in the population density change. The cities with a decreasing urban population density during the observation period are Datong, Yangquan, Yuncheng, and Linyi. This is especially true for the case of Datong; its urban population density as the number of people per km^2 was reduced to 9492 in 2015 from 13,134 in 2006. The contributing reason could be a rapid expansion of land urbanization by enlarged urban areas in some cities, which lag behind in population urbanization. This could lead to a decline in urban land use efficiency.

Last, the government influence (GOV) coefficient was positive at 0.0047 with a significance level of 5%. The Provincial Government has adopted financial interventions to accommodate industrial agglomeration, such as capital investment in improving urban infrastructure construction, financial subsidy policies to support industrial development, industrial upgrades and technological transformation, and establishing a specially-designated park construction zone. This proves to have positively impacted land use efficiency.

5. Conclusions

This study has provided a checkpoint for the policy assessment and guidance for government policy-makers to further enhance the competitiveness and sustainability of the province. The province should make continuous improvements to energy-efficient technology, with further scope for the improvement of industrial structure, such as introducing policies to encourage the establishment of industries with high added value (e.g., high technology industry and financial services). In addition, the province should be prudent in managing the process of population urbanization and industry urbanization. Furthermore, the functional attributes of land use should be assessed by the urban planners for creating reasonable, harmony, and sustainable city that integrates residential, commercial, institutional, and industrial functions.

This paper studied the 11 prefectural cities in Shanxi Province and established an index system to evaluate land use efficiency from an input–output perspective. Based on the statistical data from 2006 to 2015, the changes in urban land use efficiency and total factor productivity in Shanxi Province were measured before and after the comprehensive reform on the transformation of the resource-based economy by applying the Bootstrap-DEA and Bootstrap-Malmquist index method. The panel-data analysis model was also constructed for exploring mechanisms and the dominant factors affecting the change of land use efficiency. The key conclusions are summarized below.

First, during the examining period, the urban land use efficiency of Shanxi Province, on the whole, formed a U-shaped change trend, with the bottom line observed in 2009–2011 and an upward trend after 2012. However, the changes in land use efficiency among cities are quite different. After the comprehensive reform, a significant increase in urban land use efficiency has been observed for Taiyuan and Yangquan, whereas the decline in Zhangzhou, Datong, Linfen, and Yuncheng has been significant.

Second, the dynamic changing process of total factor productivity of urban land does not appear promising. Before the comprehensive reform, productivity improvement had only been noticed for a few cities in a particular year. In particular, a descending trend was prominent for the overall level of total factor productivity after the comprehensive reform, which can mainly be attributed to the lack of technological advancements.

Third, from the perspective of the attributing factors, economic development and industrial upgrading significantly positively impacted urban land use efficiency. Industrial structure and government influence on economic transformation have an enabling influence, while the level of the urban population had a negative and less notable effect. This explains why Taiyuan has the highest score of land use efficiency as a result of being the capital city of the province, with the highest GDPP and value added by the tertiary industry, and the least by coal consumption.

This research is useful for understanding, learning, and evaluating the utilization efficiency of allocated resources. Moreover, it holds potential significance for the State to establish and adjust supporting policies in the future for resources-based cities, especially when the State has further planned to deepen the province's economic transformation.

With the challenges of the scarcity of resources, climate change, and an increasing population, the improvement of technical efficiency driven by the reform and upgrading of traditionally resource-based industries is insufficient to help improve of the total factor productivity of urban land. In fact, the fundamentals for the improvement of land use efficiency and total factor productivity are to foster and develop the modern tertiary industry and strategically emerging alternative industries and improve the quality of the tertiary industry. However, in the resource-based cities, with limited capital resources, combined with other factors, such as the lack of essential technological innovation, it is even more important to introduce governmental policies to support implementing alternative strategic industries. From the perspective of land use, the blind expansion of the urban land use scales not in synchronization with industry transformation will only lead to inefficient land use. It is necessary to synchronize with transformation projects, develop land stock, integrate existing industrial park zones, and improve the level of urban land saving and intensive use to develop a future sustainable built environment.

Author Contributions: Conceptualization, X.J.G. and X.L.; literature review, X.J.G. and X.L.; methodology and data curation, X.L.; resources, X.J.G.; visualization, X.J.G. and X.L.; writing—original draft preparation, X.L.; writing—re-write, review and editing, X.J.G. All authors have read and agreed to the published version of the manuscript.

Funding: This research received no external funding.

Institutional Review Board Statement: Not applicable.

Informed Consent Statement: Not applicable.

Data Availability Statement: The data presented in this study are available on request from the corresponding author.

Conflicts of Interest: The authors declare no conflict of interest.

References

1. *China Urban Construction Statistics Yearbook, Various Issues, 2006–2016*; China Plans to Press: Beijing, China, 2016.
2. Chen, Y.; Chen, Z.; Xu, G.; Tian, Z. Built-up land efficiency in urban China: Insights from the general land use plan (2006–2020). *Habitat Int.* **2016**, *51*, 31–38. [CrossRef]
3. Choy, L.; Lai, Y.; Lok, W. Economic performance of industrial development on collective land in the urbanization process in China: Empirical evidence from Shenzhen. *Habitat Int.* **2013**, *40*, 184–193. [CrossRef]
4. Alberti, M. The Effects of Urban Patterns on Ecosystem Function. *Int. Reg. Sci. Rev.* **2005**, *28*, 168–192. [CrossRef]
5. Delbecq, B.A.; Kuethe, T.H.; Borchers, A.M. Identifying the extent of the urban fringe and its impact on agricultural land values. *Land Econ.* **2014**, *90*, 587–600. [CrossRef]
6. Guastella, G.; Pareglio, S.; Scodeck, P. A spatial econometric analysis of land use efficiency in large and small municipalities. *Land Use Policy* **2017**, *63*, 288–297. [CrossRef]
7. William, J.M.; Andrés, V. Interactions between food security and land use in the context of global change. *Land* **2018**, *7*, 53. [CrossRef]
8. Liu, Y. Introduction to land use and rural sustainability in China. *Land Use Policy* **2018**, *74*, 1–4. [CrossRef]
9. The State Council. *The Plan for the Sustainable Development of Resource-Based Cities in China (2013–2020) (In Chinese)*; Government Document: Beijing, China. 2013. Available online: http://www.gov.cn/zwgk/2013-12/03/content_2540070.htm (accessed on 15 March 2019).
10. Li, H.J.; Long, R.Y.; Chen, H. Economic transition policies in Chinese resource-based cities: An overview of government efforts. *Energy Policy* **2013**, *55*, 251–260. [CrossRef]
11. Chen, W.; He, R.; Wu, Q. A Novel efficiency measure model for industrial land use based on subvector data envelope analysis and spatial analysis method. *Complexity* **2017**, *2017*, 1–11. [CrossRef]
12. Sun, W.; Li, Y.; Wang, D.; Fan, J. The efficiencies and their changes of China's resources-based cities employing DEA and malmquist index models. *J. Geogr. Sci.* **2012**, *22*, 509–520. [CrossRef]
13. Yang, X.D.; Wu, Y.X.; Dang, H. Urban land use efficiency and coordination in China. *Sustainability* **2017**, *9*, 410. [CrossRef]
14. Dong, S.C.; Li, Z.H.; Li, B.; Xue, M. The economic transformation and strategic exploration of resource-based cities in China. China population. *Resour. Environ.* **2007**, *17*, 12–17.
15. Burton, E. The Compact City: Just or Just Compact? A Preliminary Analysis. *Urban Stud.* **2000**, *37*, 1969–2006. [CrossRef]
16. Gabriel, S.A.; Faria, J.A.; Moglen, G.E. A multi-objective optimization approach to smart growth in land development. *Socio-Econ. Plan. Sci.* **2006**, *40*, 212–248. [CrossRef]
17. Rudolf, S.C.; Kienast, F.; Hersperger, A.M. Planning for compact urban forms: Local growth-management approaches and their evolution over time. *J. Environ. Plan. Manag.* **2018**, *61*, 474–492. [CrossRef]
18. Proeger, T.; Meub, L.; Bizer, K. The role of communication on an experimental market for tradable development rights. *Land Use Policy* **2017**, *68*, 614–624. [CrossRef]
19. Ding, C. Land policy reform in China: Assessment and prospects. *Land Use Policy* **2003**, *20*, 109–120. [CrossRef]
20. Meng, Y.; Zhang, F.R.; An, P.L.; Dong, M.L.; Wang, Z.Y.; Zhao, T. Industrial land-use efficiency and planning in Shunyi, Beijing. *Landsc. Urban Plan.* **2008**, *85*, 40–48. [CrossRef]
21. Zitti, M.; Ferrara, C.; Perini, L.; Salvati, L. Long-term urban growth and land use efficiency in southern Europe: Implications for sustainable land management. *Sustainability* **2015**, *7*, 3359–3385. [CrossRef]
22. Valerio, Q.; Anna, B.; Francesco, C. Monitoring land take by point sampling: Pace and dynamics of urban expansion in the metropolitan city of Rome. *Landsc. Urban Plan.* **2015**, *143*, 126–133. [CrossRef]
23. Abdullahi, S.; Pradhan, B.; Mojaddadi, H. City Compactness: Assessing the Influence of the Growth of Residential Land Use. *J. Urban Technol.* **2018**, *25*, 21–46. [CrossRef]
24. Chen, W.; Wu, Q. Economic efficiency of urban construction land in the yangtze river delta and its influencing factors. *Econ. Geogr.* **2014**, *34*, 142–149.
25. Liang, T.; Zhao, Q.L.; Chen, C. Spatial differentiation characteristics and optimization path analysis of urban land use efficiency in China—Based on empirical study of 287 cities above prefectural level. *China Land Sci.* **2013**, *7*, 48–54.
26. Tu, F.; Yu, X.; Ruan, J. Industrial land use efficiency under government intervention: Evidence from hangzhou, China. *Habitat Int.* **2014**, *43*, 1–10. [CrossRef]
27. Simar, L.; Wilson, P.W. Statistical inference in nonparametric frontier models: The state of the art. *J. Product. Anal.* **2000**, *13*, 49–78. [CrossRef]
28. Atstaja, D.Z.; Dimante, D.Z.; Brivers, I.; Malzubris, J.; Keneta, M.; Tambovceva, T.; Sina, I.; Liviņa, A.; Ievins, J.; Grasis, J.; et al. *Environment and Economics*; Monographic ed.; Academic Press of University of Latvia: Riga, Latvia, 2011. (In Latvian)
29. Costanza, R.; Perrings, C.; Cleveland, C.J. *The Development of Ecological Economics*; Edward Elgar Publishing Company: Cheltenham, UK, 1997.
30. van den Bergh, J.; Barendregt, A.; Gilbert, A. *Spatial Ecological-Economic Analysis for Wetland Management: Modelling and Scenario Evaluation of Land-Use*; Cambridge University Press: Cambridge, UK, 2004; pp. 36–49.
31. Charnes, A.; Cooper, W.W.; Rhodes, E. Measuring the efficiency of decision making units. *Eur. J. Oper. Res.* **1978**, *2*, 429–444. [CrossRef]

32. Choi, Y.; Wang, N. the economic efficiency of urban land use with a sequential slack-based model in Korea. *Sustainability* **2017**, *9*, 79. [CrossRef]
33. Caves, D.W.; Christensen, L.R.; Diewert, W.E. The economic theory of index numbers and the measurement of input, output, and productivity. *Econom. J. Econom. Soc.* **1982**, *50*, 1393–1414. [CrossRef]
34. Färe, R.; Grosskopf, S.; Lovell, C.A.K. Indirect productivity measurement. *J. Product. Anal.* **1992**, *2*, 283–298. [CrossRef]
35. Färe, R.; Grosskopf, S.; Norris, M.; Zhang, Z. Productivity growth, technical progress, and efficiency change in industrialized countries. *Am. Econ. Rev.* **1994**, *84*, 66–83.
36. Worthington, A.C. Technical efficiency and technological change in Australian Building Societies. *Abacus* **2000**, *36*, 180–197. [CrossRef]
37. Odeck, J. Statistical precision of DEA and malmquist indices: A bootstrap application to Norwegian grain producers. *Omega* **2009**, *37*, 1007–1017. [CrossRef]
38. Ko, J.H.; Park, S.H.; Kim, D.C. Efficiency analysis of project management offices for large-scale information system projects: Insights for construction megaprojects. *Constr. Econ. Build.* **2015**, *15*, 34–47. [CrossRef]
39. Zhou, P.; Ang, B.W.; Han, J.Y. Total factor carbon emission performance: A Malmquist index analysis. *Energy Econ.* **2010**, *32*, 194–201. [CrossRef]
40. Dakpo, K.H.; Desjeux, Y.; Jeanneaux, P.; Latruffe, L. Productivity, technical efficiency and technological change in French agriculture during 2002–2015: A färe-primont index decomposition using group frontiers and meta-frontier. *Appl. Econ.* **2019**, *51*, 1166–1182. [CrossRef]
41. Wu, Y.R. Is China's economic growth sustainable? A productivity analysis. *China Econ. Rev.* **2000**, *11*, 278–296. [CrossRef]
42. Berg, S.A.; Førsund, F.R.; Jansen, E.S. Malmquist indices of productivity growth during the deregulation of Norwegian banking, 1980–1989. *Scand. J. Econ.* **1992**, S211–S228. [CrossRef]
43. Fukuyama, H. Measuring efficiency and productivity growth in Japanese banking: A nonparametric frontier approach. *Appl. Financ. Econ.* **1995**, *5*, 95–107. [CrossRef]
44. Jajri, I.; Ismail, R. *Technical Efficiency, Technological Change and Total Factor Productivity Growth in Malaysian Manufacturing Sector*; MPRA Paper, Munich Personal RePEc Archive, Paper No. 1966, posted 2; University of Malaya: Kuala Lumpur, Malaysia, 2007; pp. 1–19.
45. Daraio, C.; Simar, L. *Advanced Robust and Nonparametric Methods in Efficiency Analysis: Methodology and Applications*; Springer: Berlin/Heidelberg, Germany, 2007.
46. Mo, H.D. Re-sampling technique—Bootstrap method. *J. Yangzhou Univ.* **2001**, *4*, 1–6.
47. Simar, L.; Wilson, P.W. Estimating and bootstrapping Malmquist indices. *Eur. J. Oper. Res.* **1999**, *115*, 459–471. [CrossRef]
48. Hall, P. On the number of bootstrap simulations required to construct a confidence interval. *Ann. Stat.* **1986**, *14*, 1453–1462. [CrossRef]
49. Zhang, J.; Wu, G.Y.; Zhang, J.P. Estimation of China's inter-provincial material capital stock: 1952–2000. *Econ. Res.* **2004**, *10*, 35–44.
50. Pu, W.F.; Zhang, A.L.; Wen, L.J. Can China's resource-saving and environmentally friendly society really improve the efficiency of industrial land use? *Land* **2021**, *10*, 751. [CrossRef]
51. Beijing Zhiyan Kexin Consulting. The Analysis of Coal Price Trend in 2017 and Impact Study in China (in Chinese), Industry Research Report. 2017. Available online: https://www.sohu.com/a/162801835_775892 (accessed on 22 May 2019).
52. Geng, H.G.; Zhang, K.; Zhang, H.W. Research on sustainable development of resource-based small industrial and mining cities—A case study of Yangquanqu Town, Xiaoyi, Shanxi Province, China. *Procedia Eng.* **2011**, *21*, 633–640.
53. Zheng, Y.T. Focusing on the implementation of key tasks and promote the construction of the comprehensive reform pilot zone—Read the shanxi comprehensive reform experiment "implementation plan (2013–2015)" and "2013 action plan". *Advance* **2013**, *9*, 4–6.
54. Yan, P.F.; Wang, B. Technical efficiency, technological progress and productivity growth: Empirical analysis based on DEA. *Econ. Res.* **2004**, *12*, 88–96.

land

Article

Exploring the Impact of Urban Built Environment on Public Emotions Based on Social Media Data: A Case Study of Wuhan

Yuanyuan Ma [1], Yunzi Yang [1] and Hongzan Jiao [1,2,*]

[1] Department of Urban Planning, School of Urban Design, Wuhan University, Wuhan 430072, China; yuanyuanma@whu.edu.cn (Y.M.); yangyunzi@whu.edu.cn (Y.Y.)
[2] Engineering Research Center of Human Settlements and Environment of Hubei Province, Wuhan 430072, China
* Correspondence: Jiaohongzan@whu.edu.cn; Tel.: +86-27-6877-3062

Abstract: In the era of public participation in government, public emotions and expectations are important considerations influencing urban construction, planning, and management. A desirable urban environment can make people feel at ease and comfortable and contribute to promoting positive public emotions. However, in the process of rapid urban development, the high-density and overloaded urban built environment has triggered people's mental tension and anxiety and has contributed to negative emotions. Thus, this study aimed to explore the spatial distribution of public emotions and urban built environments in cities and to thoroughly investigate the correlation between urban built environments and public emotions. Considering the lack of dynamic elements analysis and emotions spatial analysis in previous studies, this study takes Wuhan City as an example, uses social media big data as the basis for text emotion analysis, introduces dynamic traffic elements, and establishes a multidimensional urban built environment measurement index system from five aspects: land use, spatial form, road and traffic, green space and open space, and daily life service facilities. Subsequently, the spatial distribution characteristics of public sentiment and urban built environment elements in Wuhan were analyzed. Finally, a geographically weighted regression method was used to analyze the degree of influence of different urban built environment elements on public emotions. The results showed that public emotions in Wuhan are not homogeneously distributed in terms of score and space and that there are significant differences. The urban built environment has a significant influence on public emotions. Higher land use mix, higher road network density, higher number of public transportation facilities, higher number of public open spaces, lower traffic congestion, and impact of freight transportation play important roles in promoting positive emotions. Therefore, in the process of urban construction, planners and decision makers should purposefully improve the quality of the built environment. Measures can include improving the mix of land functions, alleviating traffic congestion, avoiding the negative effects of freight traffic, rationally constructing green and open spaces, and improving various living facilities. This can help contribute toward improving urban functions and urban environments, and promote the construction of a people-oriented healthy city.

Keywords: public emotions; urban built environment; social media data; dynamic traffic influencing factors; geographically weighted regression

Citation: Ma, Y.; Yang, Y.; Jiao, H. Exploring the Impact of Urban Built Environment on Public Emotions Based on Social Media Data: A Case Study of Wuhan. *Land* **2021**, *10*, 986. https://doi.org/10.3390/land10090986

Academic Editors: Sarel Cilliers and Elizelle Juanee Cilliers

Received: 31 July 2021
Accepted: 15 September 2021
Published: 17 September 2021

1. Introduction

A city is the result of a complex combination of various characteristics and is a space where people live their daily lives [1,2]. As a spatial carrier of material exchange, a city is where production, consumption, sharing, and experience are combined [3]. Studies have shown that the urban built environment influences people's perception of a city and evokes different feelings and emotions [4]. Moreover, the renewal and planning of a built environment has a long-term impact on the future development of cities [5,6].

Nowadays, people-oriented urban construction has become a global trend, and assessing a built environment from the public's perspective has become an important element of urban renewal and urban planning [5,7–9]. Emotion is a complex multidimensional trait that reflects human personality and behavioral characteristics [10–12]. Emotional stress is influenced by the degree of habitability. A prolonged exposure to urban environments with poor habitability contributes to excessive emotional stress, generates negative emotions, and triggers mental illness [13–15]. It was suggested that space has the power to elicit emotions as early as the 1950s [16]. A comprehensive investigation of the relationship between urban built environments and people's emotions is important for scientific urban renewal and urban planning and the construction of people-centered urban spaces.

However, the reality is that the lack of useful information about the relationship between multidimensional urban built environments and emotions has hindered the progress toward understanding which elements in cities influence public emotions and how to improve the urban built environment to build humane urban spaces. Taking Wuhan City as an example, this paper proposes a multidimensional urban built environment measurement index method incorporating dynamic traffic elements, combined with textual emotion analysis based on social media big data, and introduces a geographically weighted regression (GWR) analysis for a multifaceted interdisciplinary investigation. We pay attention to the relationship between urban built environments and public emotions, explored from a macroperspective to provide suggestions for urban planning and development. The methodology used in this study includes four steps: (1) emotion analysis of social media data texts in the study area; (2) exploration of the spatial distribution characteristics of emotions in the study area; (3) establishment of an index system to explore the spatial structure and characteristics of the urban built environment in the study area; (4) exploration of the correlation between elements of the urban built environment and emotions based on GWR. Here, this study changes the previous research approach of exploring the relationship between urban built environments and emotions from a single aspect. A multidimensional measure of the impact of urban built environments on people's emotions and a more comprehensive analysis of the relationship between cognition and emotions can provide a new perspective for studying urban public emotions, thus bridging the gap in existing literature.

The rest of this paper is organized as follows. After a brief literature review in Section 2, we describe the natural and social backgrounds of the study area, data sources, and preprocessing in Section 3. Section 4 presents the methodotlogy of the study. The results are reported in Section 5. Section 6 presents the discussion and conclusions of the study.

2. Literature Review

In this section, we briefly review some relevant research about understanding the interrelationship between urban built environments and emotions using emerging big data and natural language processing techniques.

Cities are the main sites where people conduct their activities, and the behavior of people in cities is closely related to their emotions [17]. The interrelationship between human cognition and emotion has become an important research topic in biology and medicine, including the regulation of negative emotions by cognition [18,19] and emotion recognition and emotion expression in sick adolescents [20]. Since the 21st century, urbanization has brought about many urban problems, such as traffic congestion, environmental pollution, and lack of supporting facilities [21–23]. The high-intensity, overloaded urban built environment has triggered physiological and psychological problems and negatively affected public emotions [24]. The impact of urban built environments on human emotions has become a hot topic in urban planning research. Relevant studies have shown that built environments include both static and dynamic elements of space [25,26]. Several scholars have studied the impact of static elements in the built environment on people's emotional health, including architectural forms, public facility support, parks and green spaces, and land use [27–29]. In recent years, studies have found that dynamic elements in cities also have important effects on public emotion, including the interrelationship between emotion traffic, and the effect of congestion on emotion [30,31]. However, these studies tend to focus only on the effects of single-level elements in the urban built environment on emotion, rather than integrating static elements with dynamic ones. Therefore, this study combines dynamic and static elements to evaluate urban built environments and more comprehensively explore the impact of such environments on people's emotions.

Most of the early studies collected people's behaviors, opinions, and emotions through traditional social questionnaires, which had the disadvantages of having a small sample size and a long acquisition time [32,33]. In recent years, some scholars have used wearable physiological sensors to investigate the relationship between cognition and emotion, and this method can more accurately perceive emotional changes [34]. However, similar to conventional methods, physiological sensors have drawbacks such as small sample size and difficulty in obtaining data, which are unsuitable for large scale studies. With the development of Internet technology, various types of big data have been widely used in academic research. These big data have the advantages of having large sample sizes and easier access [35]. Among them, social media data is one of the most common big data. People use various social media platforms to share their activities, opinions, and emotions to further showcase the urban environment and reveal the hidden characteristics of urban spaces [36,37], as well as providing information on movements [38,39]. The development of social media data provides a new opportunity to systematically study the interrelationship between urban built environment and emotions, bridging the information gap of traditional surveys. In recent studies, some scholars have attempted to use social media data for urban and mental health-related studies, further revealing the research potential of social media data in the field of human behavior and urban spatial perception [14,40,41]. With the advancement of technology and social platforms, social media provides more diverse data (e.g., text, audio, and images), and scholars have explored multiple aspects in information extraction based on text and image data in social networks [42–44]. The application of new computer technologies such as deep learning and natural language processing provides a good basis for further mining the emotional and perceptual information in social media data [45,46]. Therefore, this study uses natural language processing methods to retrieve public emotion from social media data, combines static and dynamic elements to comprehensively evaluate the urban built environment, and thoroughly explores the interrelationship between urban built environments and public emotions.

3. Materials

3.1. Study Area

The study area is concentrated in Wuhan (113°41′–115°05′ E, 29°58′–22°31′ N). Wuhan is the capital city of Hubei Province and the central city of central China, located in the eastern part of the Jianghan Plain and the middle reaches of the Yangtze River, where the Yangtze River, the third largest river in the world, and its largest tributary, the Han River, meet in the city, forming the towns of Wuchang, Hankou, and Hanyang (Figure 1). Wuhan is an advanced metropolis with a long history, developed transportation, and a large population. It has a complex and diverse urban spatial environment that is representative of the rapidly developing cities of central China.

The Wuhan Urban Development Zone is the main gathering area of urban functions and the key expansion area of urban space in Wuhan, with a resident population of 8,474,600, accounting for 84.5% of the city's resident population, and a total area of 3261 km^2. In the Wuhan Urban Master Plan (2010–2020), the Wuhan Urban Development Area is divided into one main urban area and six new city clusters with a spatial structure of "1 + 6". Given that the urban development within the Wuhan Urban Development Zone is mature, the urban built-up area and urban population are mainly distributed in this area. Therefore, the urban development zone of Wuhan City is taken as the main study area in this work.

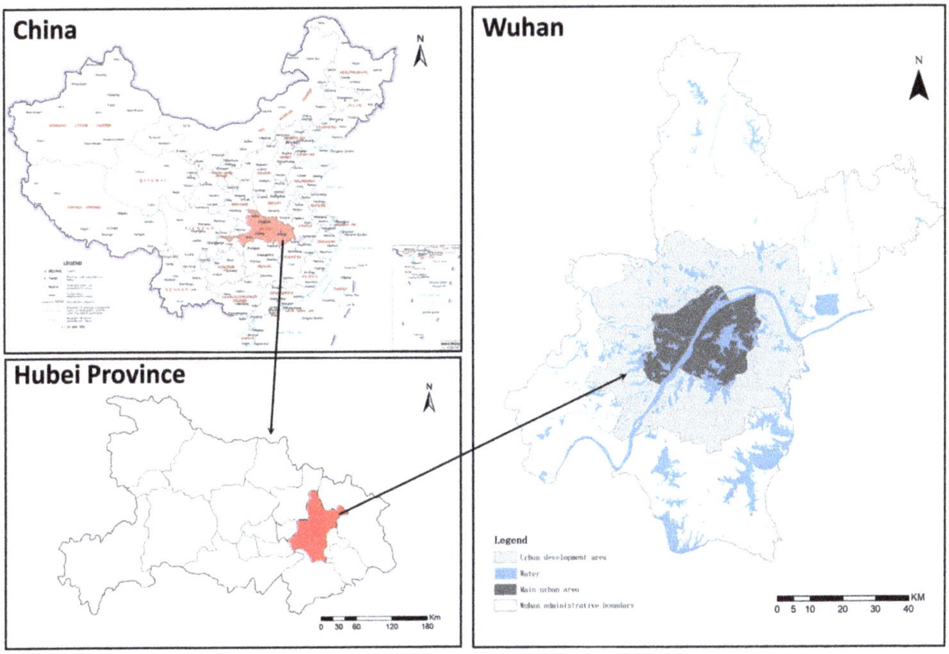

Figure 1. Location of study area.

3.2. Data Sources

3.2.1. Data Unit

The traffic analysis zone (TAZ) is the basic data unit of traffic planning and plays an important role in traffic planning. Compared with other types of urban spatial divisions, each traffic zone has more similar characteristics, such as population density, economic density, and land function, which can help measure the urban built environment in each zone more accurately and fairly [47]. Therefore, this study considers 7940 traffic districts within the Wuhan urban development area to finely characterize the urban built environment and people's emotions in the area (Figure 2).

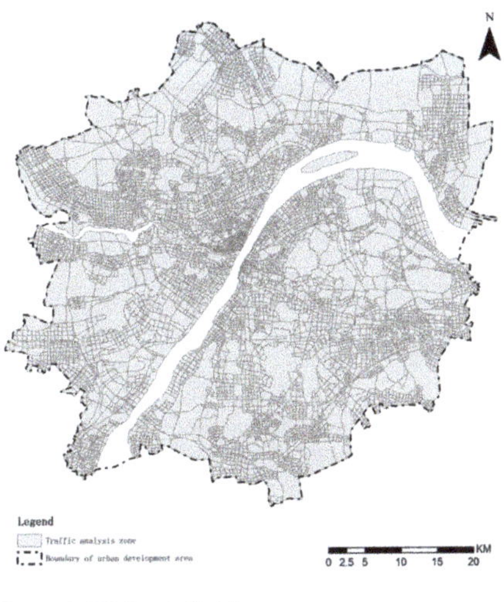

Figure 2. TAZ subdivision.

3.2.2. Weibo and POI Data

Sina Weibo is a social media platform operated by Sina Corporation to publicly share, disseminate, and receive information with free, open, and shared characteristics [48]. Users can log onto Weibo through their website or mobile app and share their current geographic location and upload text, images, and videos to express their moods and opinions. Approximately 57% of Sina Weibo users are in China, and approximately 87% of Sina Weibo activities occur in China, while Sina Weibo is one of the most visited websites in mainland China. This study obtained information on users' check-in data points through the official open API interface of Sina Weibo, including the geographical location (i.e., latitude and longitude), text content posted, and time of posting when users check-in. Table 1 shows some of the original data. The data cover dates ranging from 1 January 2018 to 31 December 2018 in Wuhan area, with a total of 1,138,655 check-ins. Point of interest (POI) data were obtained through the Baidu Map API interface in 2018 by administrative area and time, with a total of 168,424 entries in 13 categories.

Table 1. Part of the original data display.

ID	Longitude	Latitude	Text	Number
0	113.85714	30.254629	246 line scenery map appreciation, the original village station sign is still ...	20 February 2018
1	113.83689	30.65835	Today's driving is not as fast as walking.	27 January 2018
2	113.82184	30.660141	Another thief took public resources for private	30 January 2018
3	114.10257	30.346842	Wuhan's snow is falling more and more, the highway is closed! I want to ...	26 January 2018
4	113.75531	31.019424	Keep a good attitude, that very important person will come, just sooner ...	14 February2018
5	114.04178	30.70154	Take a break. Let's go on	23 March 2018
6	114.10257	30.346842	The excitement on the way out! Now, when others think about how to ...	27 January 2018

3.2.3. Road Traffic Data

Traffic data has a more pronounced cyclical nature than social media data. Studies have shown that urban traffic is usually repeated on a monthly cycle [49,50]. Therefore, traffic data for one month was gathered as a dynamic element.

Traffic dynamics is a new concept derived from the traditional traffic flow theory, which is simply the current state and future trend of traffic flow [51]. In this study, the traffic congestion at the center point of each road segment is used to represent the traffic congestion of this road segment, and the hourly traffic congestion on the main roads in Wuhan was obtained through the open API interface of Baidu Maps to measure the overall road traffic congestion in the study area. A total of 15,432,210 data were collected from 1 March 2018 to 31 March 2018. Table 2 shows some of the original data. The "Statue" field in the table represents the congestion of the current road in Baidu Map, where "1" means the road is unobstructed and "5" means serious congestion. The cost of time through the road is used to represent the congestion score, which is recorded as the "value" field. The calculation process is as follows: congestion = (1 − expedite) + congested × 2 + blocked × 5.

Table 2. Part of the traffic dynamics data display.

Longitude	Latitude	Status	Expedite	Congested	Blocked	Value
113.85714	30.254629	1	50.00%	0.00%	50.00%	3
113.83689	30.65835	1	0.00%	100.00%	0.00%	3
113.82184	30.660141	1	33.33%	33.33%	33.33%	2.998
114.10257	30.346842	1	40.00%	40.00%	20.00%	2.4
113.75531	31.019424	1	25.00%	75.00%	0.00%	2.25
114.04178	30.70154	1	20.00%	20.00%	20.00%	2.2
114.10257	30.346842	1	50.00%	33.33%	16.67%	2.0001

The freight corridor flows were obtained from the truck GPS monitoring data provided by Wuhan Truck Management Company. The daily average truck flows of the major freight corridors in Wuhan were monitored from 1 March 2018 to 31 March 2018, and a total of 1,072,815 data were obtained.

3.3. Data Pre-Processing

In the early stage of this study, the original Weibo data needed to be screened to ensure their validity and also to improve the accuracy of the results. First, since Weibo data contains not just subjective emotions, it is necessary to remove data that do not reflect subjective opinions and contain only blank text such as images and videos. Second, the ArcGIS10.7 tool was used to filter and delete Weibo data from surrounding areas whose search range exceeds the range of the Wuhan Urban Development Zone. Third, using JIEBA and term frequency–inverse document frequency (TF-IDF) methods to extract the subject words of each data, the data was filtered based on the subject words to get the topics related to the urban built environment [52]. Finally, we obtained 883,440 check-in data entries. Similarly, the POI data were cleaned using geographic coordinates (latitude and longitude) to obtain 101,954 POIs located within the urban development zone of Wuhan. The POIs were classified into seven categories based on attributes: transportation facilities, park square areas, scenic spots, leisure and entertainment, shopping services, restaurant services, and medical service facilities.

The Weibo check-in data and POIs were imported into ArcGIS 10.7, and after pre-processing, such as coordinate correction, a distribution of the check-in points and POI points in the study area was generated (Figure 3).

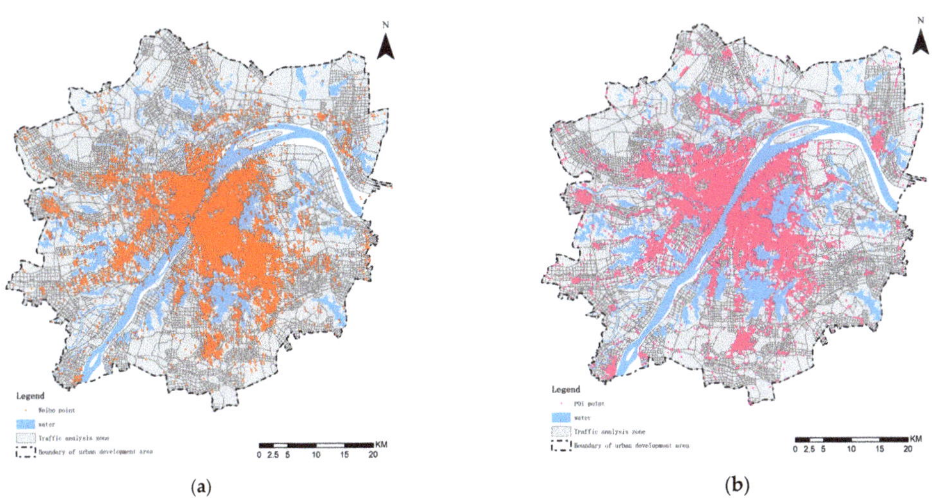

Figure 3. Visualization results of Weibo and POI points: (**a**) Weibo sign-in data distribution display; (**b**) POI data distribution display.

To express the average performance of the dynamic urban road traffic congestion in March 2018, the daily traffic dynamics data generated during the same time of year were averaged, and weights were assigned on an hourly basis to obtain the average urban road traffic congestion score. The freight corridor data were processed in the same manner. Figure 4 shows the results, where the congestion conditions of low-level roads are not shown because of display effect considerations.

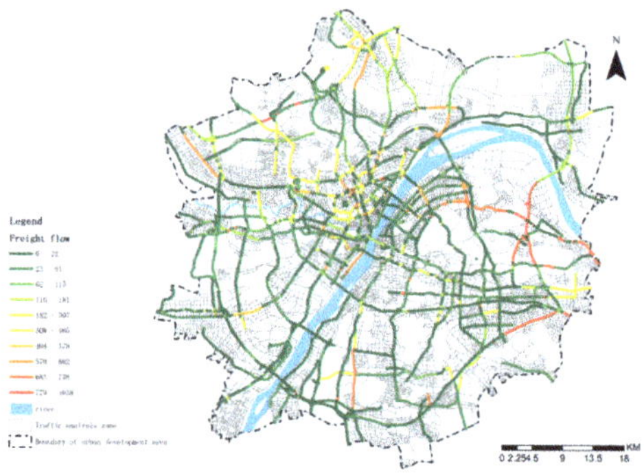

Figure 4. Traffic corridor flow display.

4. Methods

4.1. Lexicon-Based Emotion Analysis of Weibo Text

4.1.1. Splitting of the Weibo Text

Participle is to cut the Weibo text into individual words according to Chinese grammar rules. In the process of element selection, different word participle effects directly affect the emotion scores of words, thus affecting the results of data analysis [53]. Weibo text data have the characteristics of short text content, a high degree of colloquialism, an uncritical language structure, and common use of Internet words, which makes the emotion analysis method for Weibo different from conventional text emotion analyses [8]. A total of 245 professional terms and 483 Internet terms were added. Moreover, the more complex forms of Chinese expressions increased the difficulty of performing an emotion analysis of Weibo data.

In this study, the JIEBA word participle was used, which is an open source word separation method with simple installation and high accuracy. It is suitable for analyzing large amounts of text and is widely used by noncomputer professionals [54]. The JIEBA participle analysis has three modes, in which the precision mode word participle can segment the utterance precisely with no overlapping words, which helps solve the ambiguity problem to a great extent. Therefore, this study selects the precision mode of the JIEBA word participle and invokes the cut method to realize the participle of Weibo text. At the same time, since there are some drawbacks in the lexicon in terms of specialized terms and emerging random Internet terms, this study manually added specialized terms related to the urban built environment and some Internet terms to further improve the accuracy of the word participle. After the word participle, we used the deactivation word list of HIT and the deactivation word database of Machine Intelligence Laboratory of Sichuan University to remove the irrelevant deactivation words and obtain the processed Weibo text.

4.1.2. Integration of Emotional Lexicon

Loading of the Emotion Lexicon

The rule-based emotion lexicon approach is an intuitive and effective method for emotion analysis [55]. The emotion lexicon is a collection of words or phrases labeled with emotion intensity, which are of any lexical nature and can be nouns, verbs, or adjectives. Emotion words are typically polarized into positive and negative emotion words, and the scores of emotion words can be quantified. For example, the emotion of the word "like" is positive, whereas the emotion of the word "dislike" is negative.

This study uses the How Net emotion lexicon as the basis for emotion analysis [56]. How Net emotion lexicon is widely used for emotion analysis in social situations. Liu and Li proposed to establish a believable vocabulary on semantic knowledge named How Net, and then obtained the sentiment polarity of words through comparison with the similarity between the words [57]. Zhu succeeded in judging the semantic orientation of Chinese online reviews based on the How Net lexicon [58]. Casas-Garriga proposed the emotion word polarity calculation method based on the semantic similarity and semantic correlation of How Net lexicon [59], in which the positive emotion word levelness was 1 and the negative emotion word levelness was −1.

Complement of Negative Vocabulary and Degree Adverbs

The presence of negative vocabulary tends to change the emotional tendency of an utterance, and the degree adverbs in a sentence have an important influence on the intensity of emotion [60]. This study adds and integrates the commonly used negative vocabulary and adverbs of degree in How Net to obtain a more comprehensive negation word database and adverbs of degree database.

The database of negative vocabulary was obtained by supplementing the How Net negative vocabulary; the common negative words are listed in Table 3. In addition, the number of negatives preceding the emotion word reverses the emotion polarity. In case of odd numbers, the intensity of the sentiment word is multiplied by −1, and in the case of even numbers, it is multiplied by 1.

Table 3. Negative vocabulary displays.

Weight	Some Negative Vocabulary
−1	no, not, can't, not much, don't have to, didn't, no, don't, none, non, don't, in vain, hugh, white, empty, in vain, in no way . . .

The How Net list of degree adverbs was supplemented to obtain 219 degree adverbs. The degree adverbs were divided into six levels, representing different strengths and weaknesses of emotional tendencies. Each level was assigned a different weight value according to the gradient descent formula, and the gradient descent formula was expressed as follows.

$$T_{k+1} = T_1\left(\frac{\sqrt{2}}{2}\right)^k \quad k = 1,\ 2,\ 3,\ 4,\ 5 \tag{1}$$

where T_1 is the weight value of the first level "extremely, most"; the constant $\frac{\sqrt{2}}{2}$ is the gradient descent rate. The grading of the degree adverbs used in this study is shown in Table 4.

Table 4. Negative vocabulary displays.

Degree	Weight	Some Adverbs of Degree	Number
1	3	Very, extremely, fully, absolutely, most	69
2	2.1	super, over, excessive, more than, bias, extra	30
3	1.5	quite a lot, especially, extraordinarily, greatly	42
4	1.06	More, more and more, also, further	37
5	0.75	slightly, a little, somewhat	29
6	0.53	not a little, not very, not much, relatively	12

4.1.3. Algorithmic Rule Construction for Weibo Emotion

This study is based on the emotion classification method using How Net lexicon, supplemented by negative vocabulary and degree adverbs, and integrated to obtain the emotion lexicon. The emotion value of the Weibo text is calculated for the emotion words, expressions and symbols, negative vocabulary, and degree words in the text after the word participle process. The specific method is as follows: Python is used to read the emotion lexicon, negative vocabulary database, and degree adverbs database. Each text is traversed, divided into words and compared with the integrated emotion lexicon, with a score of 1 if it is a positive word and −1 if it is a negative word. If the degree adverbs appear in the sentence, the score is multiplied by the weight indicator. For the presence of negative words, an even score is multiplied by 1 and an odd score is multiplied by −1. Subsequently, the emotion scores are calculated. To better demonstrate the distribution of the emotion scores, the scores are normalized and distributed in the [−1,1] interval. Figure 5 shows the specific algorithmic model.

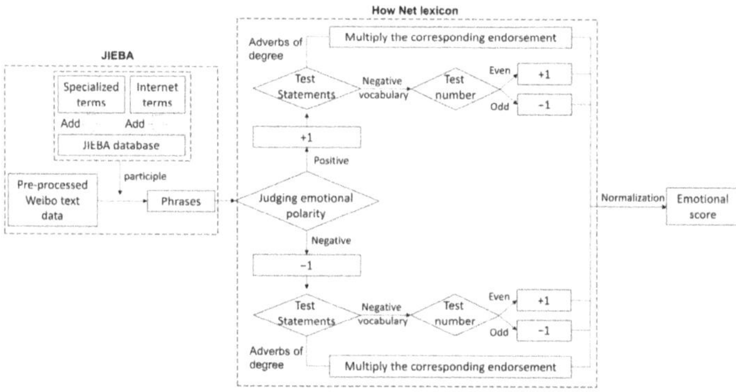

Figure 5. Emotional score algorithm.

Based on the algorithm model, the accuracy of the emotion analysis of the Weibo text based on the sentiment lexicon was tested to be approximately 81.25%.

4.2. Construction of Urban Built Environment Index System

Based on previous research on the relationship between emotions, urban space and behavioral activities, and built environment, in this study 15 built environment-related elements were selected from five aspects, namely land use, spatial form, road and traffic, green and open spaces, and daily life service facilities, to construct a system of urban built environment indicators (Table 5) [5,27–29,61].

Table 5. Urban built environment measurement index system.

Tier 1 Indicators	Tier 2 Indicators	Units
Land use	Mixed degree of land use	–
Spatial form	Floor area ratio	%
	Building density	–
Road and traffic	Road density	km/km^2
	Bus stop density	pcs/km^2
	Distance to nearest subway station	km
	Transportation facilities POI density	pcs/km^2
	Freight traffic impact	–
	Traffic congestion	–
Green and open spaces	Park square POI density	pcs/km^2
	Scenic spots POI density	pcs/km^2
Daily life service facilities	Restaurant POI density	pcs/km^2
	Leisure and entertainment POI density	pcs/km^2
	Shopping service POI density	pcs/km^2
	Medical services POI density	pcs/km^2

The mix of parcels for various uses creates a diverse urban built environment. This study uses POI data, which represents the most fine-grained land use, to measure the mixed degree of land use in cities [62]. Mixed degree of land use is quantified by using the spatial entropy of the POI, with higher entropy values indicating more functional types and a higher mixed degree. The spatial entropy calculation formula is expressed as follows.

$$P_n = C_n / A_n \quad (2)$$

$$S_s = -\sum_n^M P_n \times \log P_n \quad n = 1,\ 2,\ 3, \ldots, \text{M} \quad (3)$$

where C_n refers to the number of a certain type of POI in the n-th TAZ, A_i is the total number of POIs in the n-th TAZ, and S_s is the spatial information entropy.

4.3. Geographically Weighted Regression

The GWR model, proposed by Fortheringham et al., is a refined model for a local spatial correlation analysis based on the conventional general linear regression analysis and parametric analysis [63]. The model embeds the geographic location of sample data into regression parameters and calculates the local regressions of the observations in the data and similar observations in its vicinity to overcome the shortcomings of general linear regression models. In recent years, the GWR has been widely used in the field of urban planning to study the factors influencing urban transportation, urban public space, and urban vitality [64,65]. The coefficients of the variables in this model can be visualized in an identifiable manner to provide better support for analyzing Weibo check-in behavior. Considering the spatial heterogeneity of sentiment distribution, this study uses the GWR model to describe the spatial relationship between emotion values and the built-up urban environment in a more refined way. The model is defined as follows.

$$y_i = \beta_0 + \sum_{k=1}^{p} \beta_k(u_i, v_i)x_{ik} + \varepsilon_i \quad i = 1,2,3,\ldots,\text{N} \quad (4)$$

where the dependent variable y_i is the average emotion of each traffic cell, β_0 is the intercept value and is a constant, (u_i, v_i) is the geographic coordinate of the sampling point, $\beta_k(u_i, v_i)$ is the characteristic elasticity coefficient at sampling point i, and ε_i is the random error.

The regression coefficients of the GWR were estimated using a locally weighted least-squares method based on the decay of the distance influenced by the observations around the spatial location of each TAZ i. The estimated parameters can be expressed as follows.

$$\hat{\beta}(u_i, v_i) = \left[x^T w(u_i, v_i) x\right]^{-1} x^T w(u_i, v_i) Y \quad (5)$$

In this study, the most commonly used Gaussian kernel function was employed as the estimated weight function to represent the relationship between the weights w and distances d. The matrix representation is as follows.

$$w_{ij} = \exp\left(-\left(d_{ij}/b\right)^2\right)$$

where b represents the bandwidth, a non-negative decay parameter as a function of the distance between the weights. The higher the bandwidth, the slower the influence decay of the weights as the distance between locations i. and j increases. The choice of bandwidth has an important influence on the GWR results. The corrected Akaike information criterion (AICc) method was chosen in this study to select the appropriate bandwidth.

5. Results

5.1. Distribution of Emotions

5.1.1. Score Distribution of Emotions

To study the characteristics of public emotion distribution within the urban development zone of Wuhan, this study scored the pre-processed Weibo texts individually. The scores were taken as absolute values and the processed results were tallied according to two categories to further analyze the intensity distribution of positive and negative emotions. A positive emotion score indicates that the emotions in the text content are positive, and the higher the score, the greater the positivity of the emotions; a negative emotion score indicates that the emotions in the text content are negative and pessimistic, and the higher the score, the greater the negativity of the emotions [8,14]. Table 6 presents the statistical data related to the emotion score distribution. The statistical results show that there are significantly more positive emotions than negative emotions in the study area, and compared to positive emotions, negative emotions have a more pronounced polarization phenomenon. The highest numbers are in the strong emotion interval of 0.8–1, and the distribution of the second strongest emotion interval of 0.6–0.8 and the weak emotion interval of 0–0.2 are more in number, with evident changes in the emotional intensity.

Table 6. Intensity distribution of emotion scores.

Value	0–0.2	0.2–0.4	0.4–0.6	0.6–0.8	0.8–1	Sum
Positive	69,438	60,536	89,502	117,797	236,944	574,217
Negative	70,890	41,225	46,762	50,853	99,493	309,223

A further decomposition of the emotion scores was performed (Figure 6). The results showed that positive emotions accounted for a significantly higher proportion in the overall emotion than negative emotions, and the balance of the distribution of the negative emotions was slightly higher than that of the positive emotions. The positive emotion scores were mainly distributed in the intervals of 0.9–1, 0.8–0.9, and 0.6–0.7, indicating a strong emotional tendency. The negative emotion scores were mainly distributed in the intervals of −1–−0.9 and −0.2–−0.1, with evident differences in emotional tendencies.

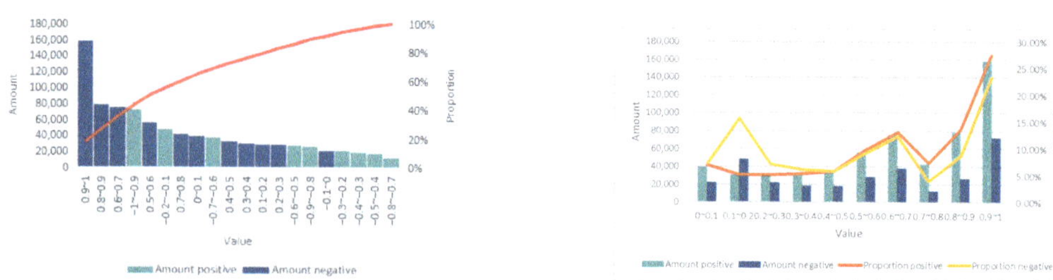

(a) (b)

Figure 6. Number and percentage distribution of emotion scores: (**a**) Overall characteristics of the number of emotions distribution display; (**b**) Internal emotional intervals and polarity differences distribution display.

5.1.2. Spatial Distribution of Emotions

To more fairly represent the spatial distribution of emotions within the study area to further explore the spatial distribution characteristics of the emotions instead of individual emotion points, this study used the geographic coordinate information of Weibo data to synthetically calculate the average of the internal emotion point scores of each TAZ, and comprehensively analyze the spatial distribution of emotion score from the average, the top, and lowest quartile. Figure 7 shows the results.

From the overall spatial distribution of the emotions, the scores of the main city center are significantly higher than those of the surrounding new cities in all three aspects. The emotion scores of the Wuchang area east of the Yangtze River are significantly higher than those of the Hankou and Hanyang areas in the west, showing distribution characteristics of strong positive emotions and weak negative emotions. The distribution of the emotions in new cities shows more evident clustering characteristics, and there is a circle-decreasing effect.

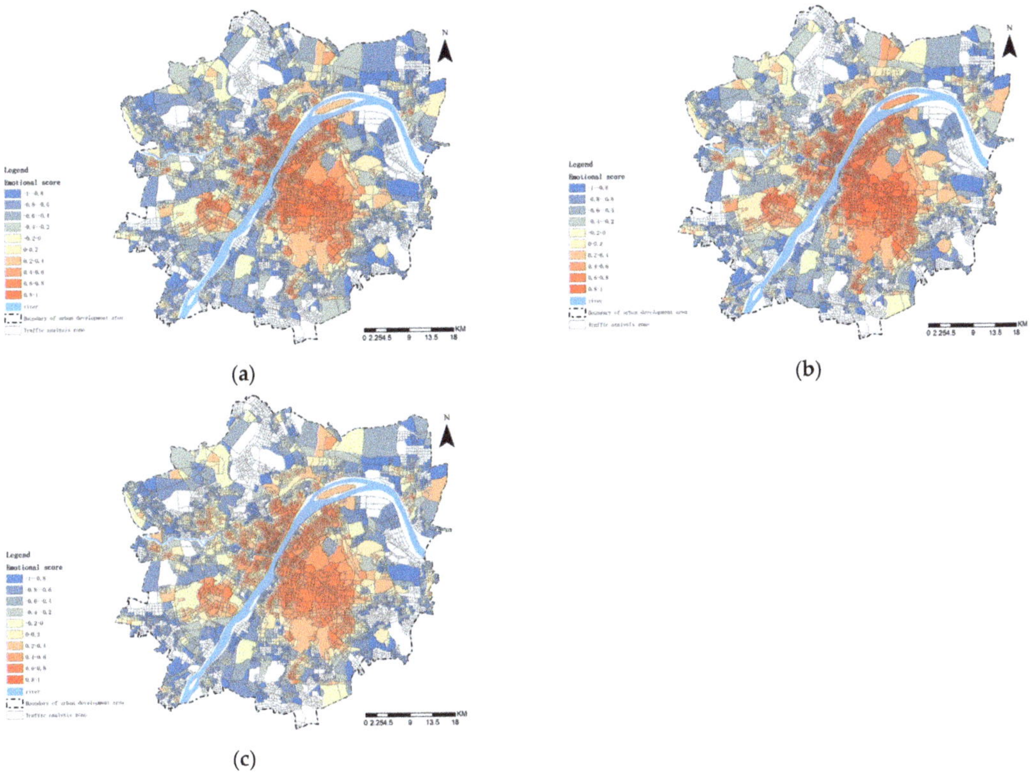

Figure 7. Spatial distribution of emotions. (**a**) Average score; (**b**) Top quartile score; (**c**) Lowest quartile score.

Looking at the emotion classification in more detail, positive emotions are mainly distributed along the Yangtze and Hanshui Rivers in the main urban area, with a ribbon distribution feature. Moreover, a small number of positive areas exist outside the main urban area. The proportion of positive emotions in the Wuchang area is the highest, followed by the Hankou area and finally the Hanyang area. The distribution of positive emotion scores is more homogeneous in the Hankou area and varies most within the Wuchang area. Negative emotions are mainly distributed in six new cities in the periphery, and there are evident circle differences; the closer to the main urban area, the less significant the negative emotions, mainly −0.4–−0.2 points. In addition, the closer to the positive emotion gathering point within the new city, the weaker the negative emotions.

5.2. *Distribution Characteristics of the Urban Built Environment*

In this study, 15 built environment-related elements were selected from five aspects: land use, spatial form, road and traffic, green space and open space, and daily life service facilities, covering both static and dynamic elements, to build a built environment measurement index system. To reflect the spatial distribution pattern of the urban built environment in different dimensions more intuitively, a study was conducted from two aspects: static elements and dynamic elements.

5.2.1. Spatial Distribution of Static Elements

Urban land use reflects the layout of urban functions, and studies have demonstrated that land use has a significant impact on human activities [66]. A mixed degree of land use is a response to the diversity and complexity of land use properties within a particular area in the city, and the higher the mix, the richer the built environment.

The POI entropy index was calculated for each traffic cell to reflect its land use mix (Figure 8). The results showed that the overall land use mix is high in the study area and has significant distributional differences. The closer to the central city, the higher the mixing degree and the more complex the urban functions. The land use mixing degree in Hankou and Wuchang areas is higher than that in Hanyang area, and both have evident functional aggregation areas.

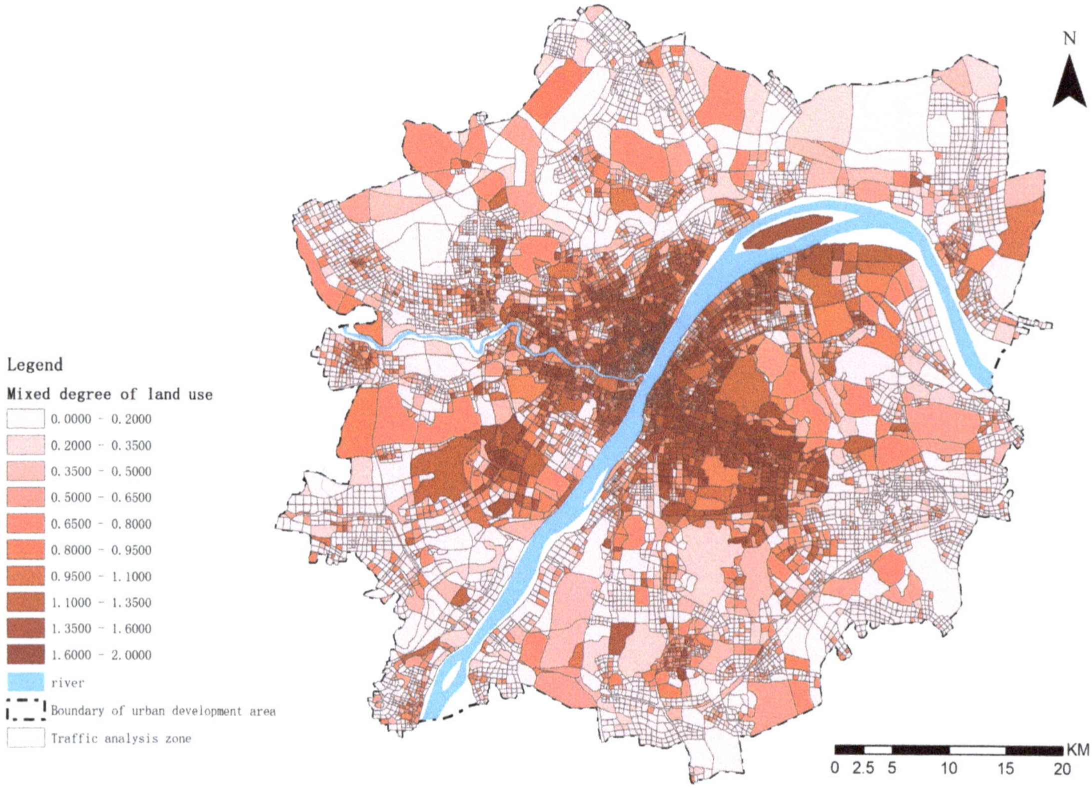

Figure 8. Mixed degree of land use.

The spatial form of the urban built environment within the TAZ unit was measured comprehensively by calculating two indicators: the average floor area ratio, and building density (Figure 9). The results show that the floor area ratio and building density are generally high in the study area, reflecting a high degree of urban construction and dense buildings. Specifically, there is a consistency in the spatial distribution of the floor area ratio and the building density, both showing the characteristics of high in the central urban area and low in the external areas, indicating an evident circle-decreasing effect. The average floor area ratio in the central urban area is typically in the range of 1.615%–4.755%, and the building density is typically in the range of 13.472%–33.800%.

The static road and traffic in the study area was measured in terms of the road density, bus stop density, distance to the nearest subway station, and transportation facilities POI density (Figure 10). The overall urban road and traffic is highly developed, with a dense road network and extensive coverage of bus stops. The construction of subway and transportation facilities is mainly in the densely populated areas of the main urban areas, with evident distribution differences.

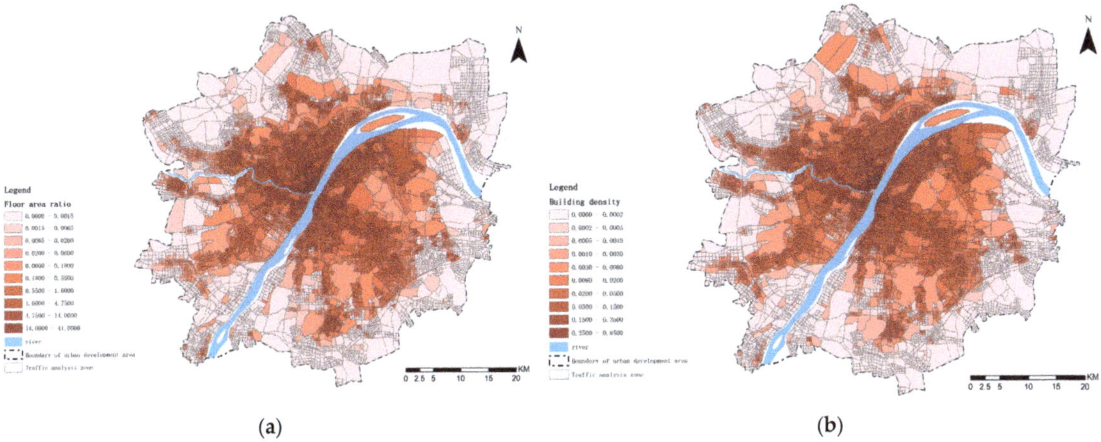

Figure 9. Spatial form: (**a**) Floor area ratio; (**b**) Building density.

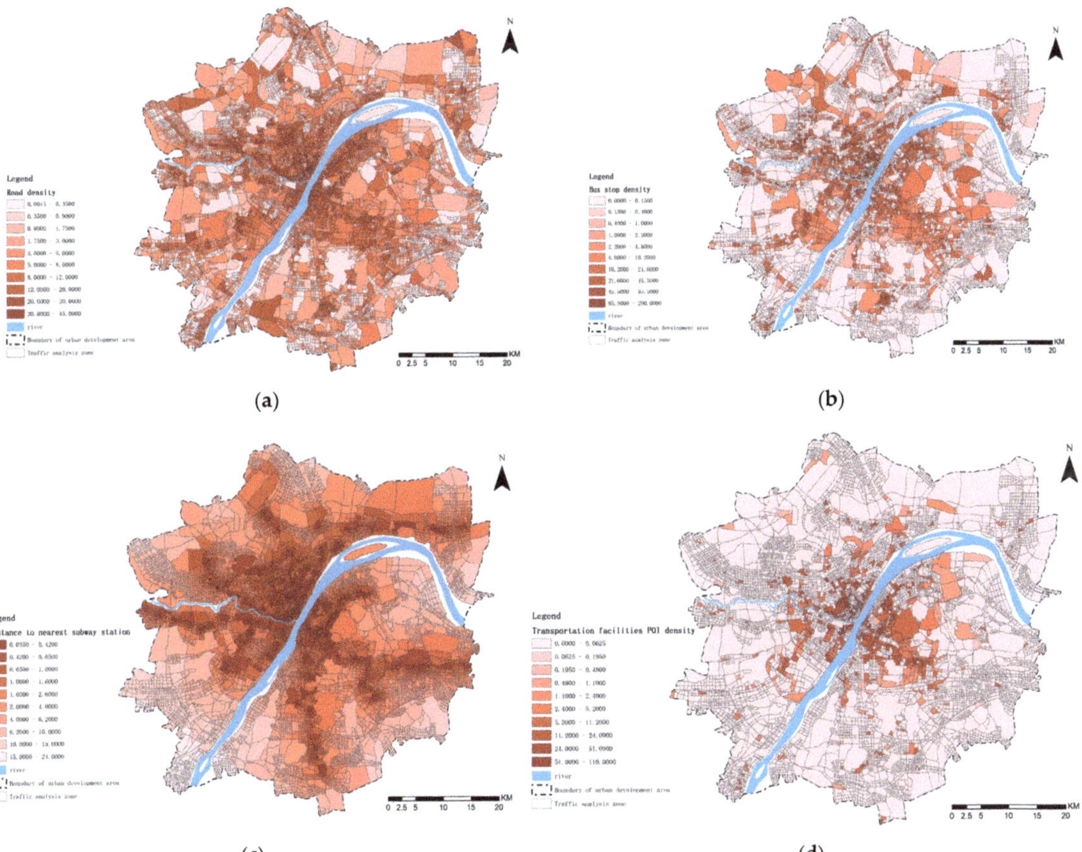

Figure 10. Road and traffic: (**a**) Road density; (**b**) Bus stop density; (**c**) Distance to nearest subway station; (**d**) Transportation facilities POI density.

Public green spaces and open spaces play an important role in the emotional health of the public [14]. Figure 11 shows the results of green and open spaces distributions. Green and open spaces are significantly lacking in the study area. This is mainly reflected in the generally low distribution density of park green spaces and scenic spots in the study area, with few high-distribution densities in the Wuchang area.

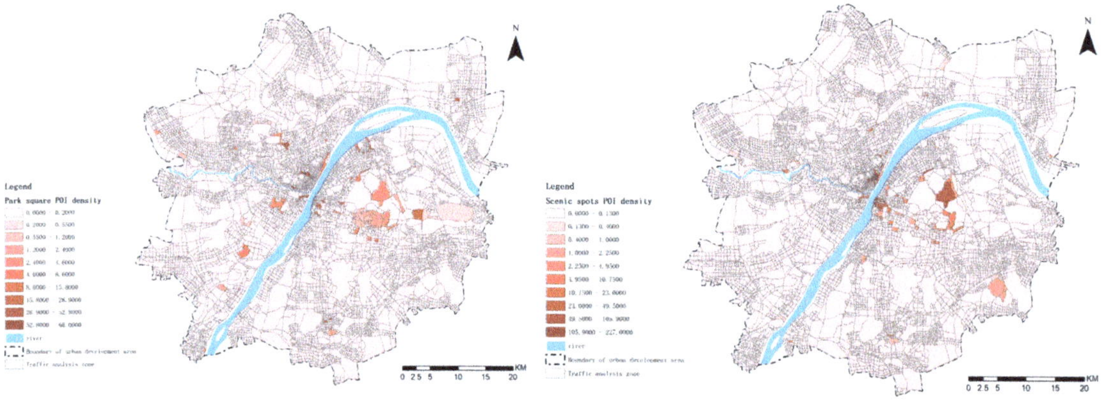

(**a**) (**b**)

Figure 11. Green and open spaces: (**a**) Park square POI density; (**b**) Scenic spots POI density.

The spatial distribution of daily living places was reflected in the distributions of four types of POIs: restaurants, shopping services, leisure and entertainment, and medical service facilities (Figure 12). The results showed that most of the daily life service facilities in the study area are widely distributed, sufficient in number, and rich in type in the TAZ. The distribution of daily living service facilities shows a highly similar cluster distribution, with high density in the central area and low density in the peripheral areas. Among them, the density of dining and shopping facilities is significantly higher than that of medical and leisure facilities.

5.2.2. Spatial Distribution of Dynamic Elements

Urban road traffic is a dynamic process, and the physical aspects, such as the road density, distribution of traffic stations, and service facilities, do not fully reflect the real situation of urban roads and traffic. In this study, two dynamic traffic elements, such as the traffic dynamics and freight corridor flow, are added to further measure the urban built environment of the study area.

Traffic dynamics can reflect the general state of urban road traffic congestion (Figure 13a). The results showed that the traffic in the study area is generally smooth, with slight congestion in some areas and significant congestion in the main urban area, particularly in the central area along the river. The degree of congestion in the north–south direction is significantly higher than that in the east–west direction, and the congestion in the Wuchang and Hankou areas is more serious.

Freight corridors are important corridors for urban road transportation, but they are often neglected. This study further analyzed the degree of impact of freight corridors on the surrounding area of roads (Figure 13b). The results showed that the impact of freight traffic is stronger in the main urban area because of its dense population and high road network density.

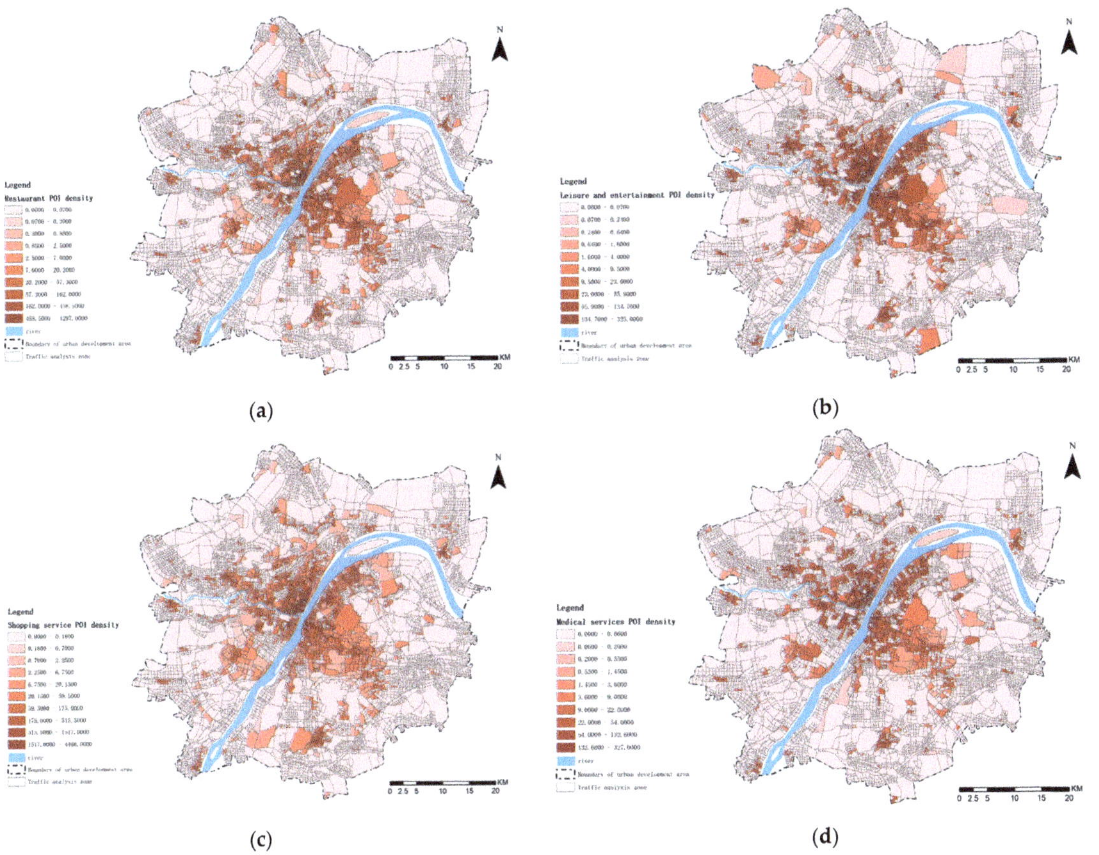

Figure 12. Daily life service facilities: (a) Restaurant POI density; (b) Leisure and entertainment POI density; (c) Shopping service POI density; (d) Medical services POI density.

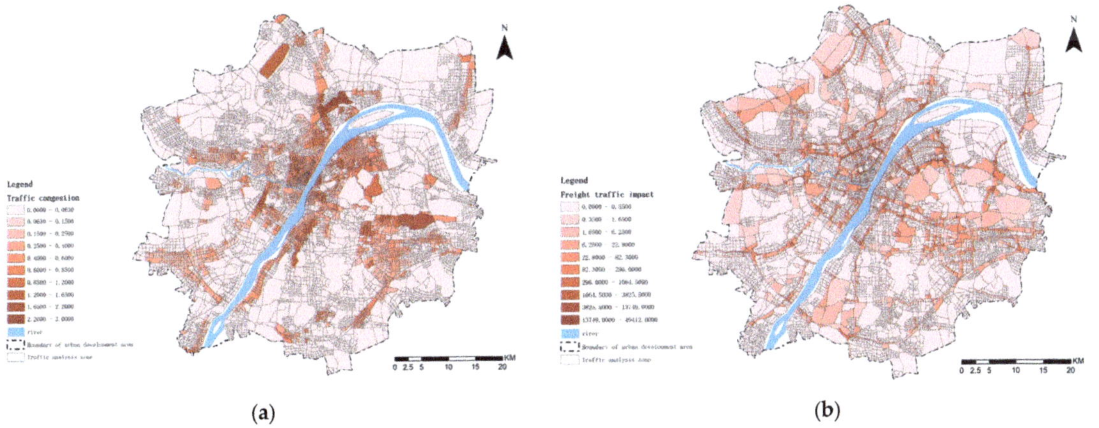

Figure 13. Road and traffic: (a) Traffic congestion; (b) Freight traffic impact.

5.3. Characteristics of the Built Environment under Different Emotional Polarities

To better investigate the characteristics of built environments under different emotional polarities, the mean and standard deviation of the built environment characteristics under positive and negative emotions were calculated and compared with the overall situation in the study area. Table 7 presents the results.

Table 7. Statistics of built environment element values under different emotional polarities.

Urban Built Environment		Total		Positive		Negative	
		Average Value	**Standard Deviation**	**Average Value**	**Standard Deviation**	**Average Value**	**Standard Deviation**
Land use	Mixed degree of land use	0.5475	0.5822	1.0763	0.5204	0.3523	0.4720
Spatial form	Floor area ratio	2.4440	3.7361	5.3191	4.6709	1.3828	2.6230
	Building density	0.0609	0.0849	0.1201	0.0976	0.0390	0.0676
Road and traffic	Road density	6.8872	6.5022	10.9595	6.9866	5.3841	5.6112
	Bus stop density	6.6934	13.9940	12.0664	18.6165	4.7102	11.2072
	Distance to nearest subway station	3.4621	4.1325	1.1058	1.1331	4.3319	4.4835
	Transportation facilities POI density	2.3355	7.7844	5.9778	12.2998	0.9911	4.5187
	Freight traffic impact	48.8836	389.3905	63.7367	302.2095	43.4011	416.8560
	Traffic congestion	0.2319	0.3717	0.3991	0.4236	0.1701	0.3298
Green and open spaces	Park square POI density	0.3240	3.0059	0.7911	4.7848	0.1516	1.9518
	Scenic spots POI density	0.9026	8.6549	2.8130	15.8469	0.1974	2.8316
Daily life service facilities	Restaurant POI density	18.7648	64.1207	45.6255	86.2631	8.8501	50.1771
	Leisure and entertainment POI density	14.2956	36.3408	35.6877	52.7102	6.3995	23.4804
	Shopping service POI density	53.9652	170.9552	130.2214	258.9595	24.4489	1,106,538
	Medical services POI density	8.0928	21.6962	18.8236	32.1794	4.1319	14.2851

The results showed that the positive areas have a higher mix of urban functions, denser distribution of public facilities, and shorter distances to subway stations. At the same time, the built environment varies more within the positive areas. The mean value of the distance to the nearest metro station is negative area > overall situation > positive area, while the mean value of the other built environment elements is positive area > overall situation > negative area. The standard deviation of the built environment elements in the positive area is generally positive area > overall situation > negative area.

5.4. Correlation between Built Environment and Emotions

5.4.1. Numerical Distribution of Correlation Coefficients

A GWR model has been introduced to investigate the coupling relationship between urban built environment and emotion, in which the emotion score is taken as the dependent variable and the urban built environment elements are taken as independent variables. After passing the covariance test that none of them are collinear, the GWR is calculated. Table 8 presents the results. Based on the parameters listed in the table, the R^2 after model fitting is 0.655, and the adjusted R^2 is 0.640, indicating that the model fits well. The multiple regression equation fitted by the independent variables can explain 64.0% of the variation.

Table 8. Geographically weighted regression parameters.

Varname	Variable
Bandwidth	7501.522
Residual Squares	101.261
Effective Number	244.445
Sigma	0.131
AICc	−7382.546
R^2	0.655

The regression coefficients of the independent variables responded to the degree of influence of the independent variables on the dependent variable. The mean, minimum, maximum, and positive and negative percentages of the coefficients of each independent variable were counted. Table 9 shows the results. The greater the mixed degree of land use, floor area ratio, road density, bus stop density, scenic spots POI density, and medical facility POI density, the more conducive it is to the positive emotions. The greater the distance from the nearest subway stop, the more likely it is to lead to negative emotions.

Table 9. Distribution of the regression coefficients of independent variables.

Explanatory	Average	Minimum	Maximum	% of Positive	% of Negative
Mixed degree of land use	0.04706	−0.21335	0.207051	98.84%	1.16%
Floor area ratio	−0.19383	−0.04334	0.398989	59.29%	40.71%
Building density	−0.048451	−4.86299	3.956941	48.56%	51.44%
Road density	0.001778	−0.00785	0.009447	68.59%	31.41%
Bus stop density	0.000687	−0.00272	0.006895	66.55%	33.45%
Distance to nearest subway station	−0.29818	−0.13458	0.010312	4.07%	95.93%
Transportation facilities POI density	−0.000313	−0.15775	0.21269	48.46%	51.54%
Freight traffic impact	−0.000015	−0.00094	0.000202	42.49%	57.51%
Traffic congestion	−0.040992	−0.11581	0.190944	21.56%	78.44%
Park square POI density	0.002059	−0.13386	0.60973	54.93%	45.07%
Scenic spots POI density	0.00629	−0.25801	0.112994	71.02%	28.98%
Restaurant POI density	0.000106	−0.0013	0.002887	48.80%	51.20%
Leisure and entertainment POI density	0.00193	−0.00735	0.003355	55.86%	44.14%
Shopping service POI density	0.000125	−0.00111	0.002496	69.36%	30.64%
Medical services POI density	0.000243	−0.0061	0.004681	73.38%	26.62%

5.4.2. Spatial Distribution of GWR Correlation Coefficients

The regression coefficients of the independent variables provide a visual representation of the spatial distribution and differences in the effects of the built environment elements on the emotional polarity.

The sensitivity of emotional polarity to the independent variables varies across the different TAZs in the study area, showing significant spatial differences. Figure 14 shows the distribution of the estimated coefficients of the GWR of the respective variables. Among them, the east side area of Wuhan is more likely to be influenced in a more positive direction than the west side. This distribution is also reflected in the north–south direction, with the north side being influenced to a greater extent. In addition, it is significantly reflected in several factors such as the mixed degree of land use, road density, bus stop density, and the degree of freight traffic impact (Figure 14d,e,h).

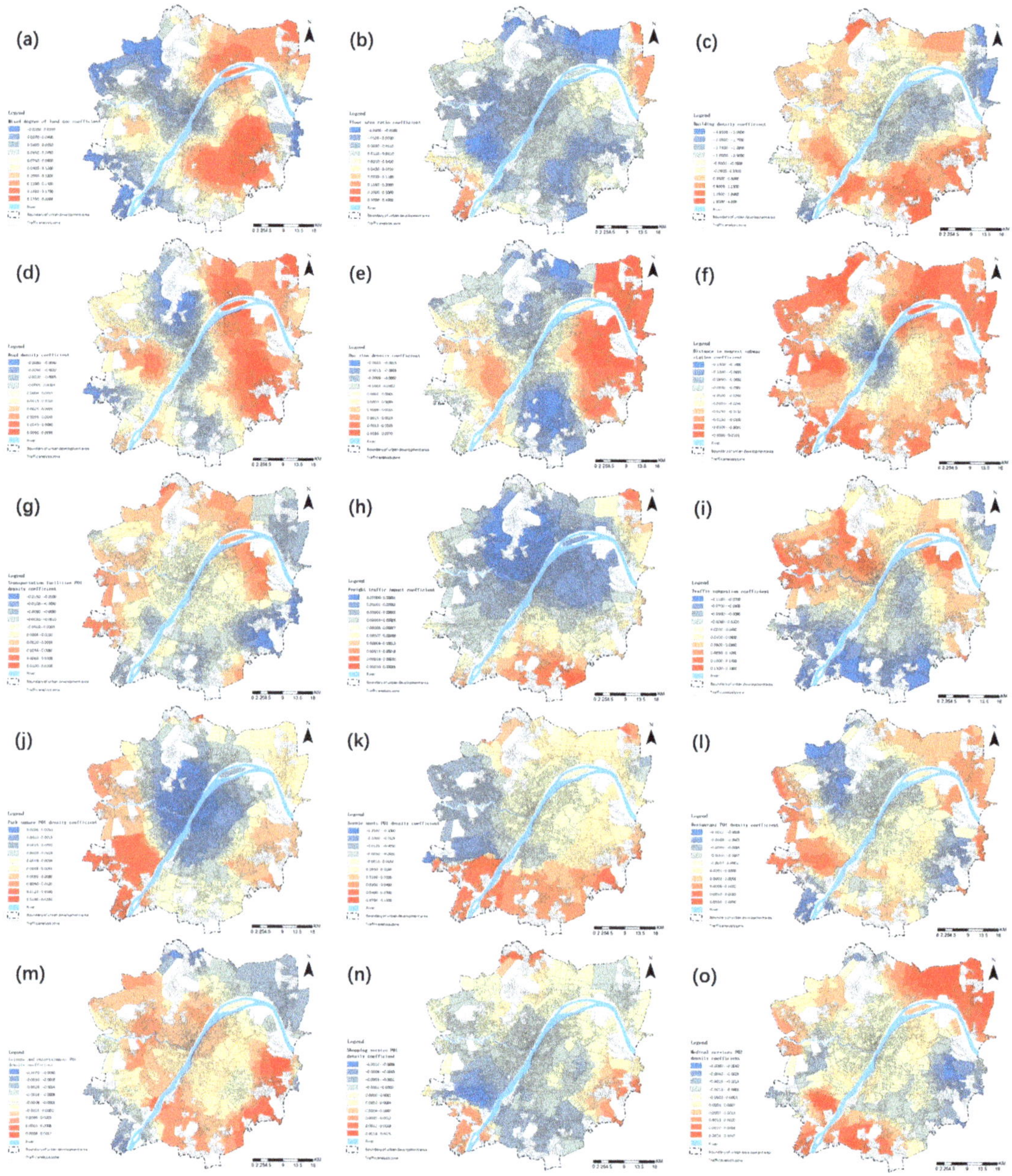

Figure 14. GWR coefficients: (**a**–**o**) Representing the GWR coefficients between different built environment elements and emotions.

Specifically, in terms of the built environment factors, the correlations of the different built environment elements with emotions are spatially distinct and uneven. In terms of the mixed degree of land use, the correlation is highest in the Wuchang area, particularly in the eastern side, and lower in the Hankou and Hanyang areas, on both sides of the Hanshui basin (Figure 14a). The correlation of building density and the floor area ratio is significantly lower in the main urban area than in the outer areas, with Wuchang having a significantly lower value than Hankou and Hanyang (Figure 14b,c). The correlation between the road density, bus stop density, and emotions has evident consistency, showing a clear trend of circle distribution, with low correlation at the center and a high correlation between the east and west sides (Figure 14d,e). There is a significant negative correlation between the distance to the nearest subway station and emotions. The closer the distance to the subway station, the more positive the emotion (Figure 14f). The density of traffic facilities has a significantly higher impact on emotion in the outer areas than in the main city, with the west side significantly stronger than the east side (Figure 14g). The correlation of the degree of freight traffic impact is more evident in the outer areas, which is closely related to the industrial layout status and the distribution of traffic corridors (Figure 14h). There is an evident spatial divergence in the degree of urban road traffic congestion in the main urban area and the outer areas, which is reflected by a weak correlation in the main urban area and a strong correlation in the outer areas. This is related to the generally congested traffic conditions within the city (Figure 14i). The trend in the correlation between the density of POI distribution and emotion of public open space and daily activity space has evident consistency, such as park square space, scenic spots, restaurants, shopping services, leisure and entertainment, and medical service facilities. The higher the density of the distribution of facilities, the more positive the emotion. This indicates that the configuration of public open space and daily life service facilities is conducive to evoking positive emotions (Figure 14j–o).

The above correlation analysis shows that urban built environments and emotions have an evident correlation. The distribution of this correlation has evident spatial consistency and spatial divergence, which is strongly influenced by geographical location.

6. Discussion and Conclusions

Research on analyzing emotions and urban built environments through social media data is emerging in the context of other cities around the world. Zhu et al. have explored emotional differences in urban green spaces in 34 parks in three northern Chinese cities [14]. Ma et al. have explored the emotion distribution of social media texts around the waterfront of Wuhan [8]. However, existing studies have mainly focused on single elements such as green areas and waterfronts in urban areas, paid little attention to the overall built environment in cities, and used short research periods or few research data sets. This study takes Wuhan City as an example and explores the spatial distribution of urban public emotions and urban built environment from the perspective of public emotions, providing a new perspective for the future enhancement of urban built environment and people-centered urban construction, and proposing recommendations for the renewal, planning, and development of humanized urban spaces and built environments. The research results for Wuhan prove the feasibility of this approach. Wuhan is an advanced metropolis with a long history, developed transportation, and a large population. It has a complex and diverse urban spatial environment that is representative of the rapidly developing cities of central China. These representative characteristics ensure that the research results obtained for Wuhan can be extended to similar cities.

6.1. Results Discussion

This study proposes a new approach and perspective that combines text emotion analysis of social media big data and multidimensional urban built environment incorporating dynamic elements, and explores the spatial distribution of public emotions and urban built environments in cities. A GWR model and spatial analysis are introduced to explore the relationship between emotions and the built environment and to provide targeted suggestions for improving the quality of the built environment from the perspective of emotions. The results of the study indicate that the scores and spatial distribution of public sentiment within the Wuhan urban development zone have significant differences, and the built-up urban environment has a significant impact on public emotion.

Specifically, on the one hand, the distribution of positive and negative emotions in Wuhan has evident differences in terms of the scores and spatial heterogeneity. The number and scores of positive emotions are significantly higher than those of negative emotions. Positive emotions are mainly distributed in the main urban areas and other areas where functions are gathered, while negative emotions are mainly distributed in the outer areas. In comparison, the emotion value of each TAZ is the average of all values in the region and represents the overall situation rather than an individual case. There is a clear divergence between the main urban area and the new urban area in the study area. Within the main urban area, Wuchang is the political, educational, and cultural center with a better natural environment. Hankou is mainly the commercial and trading center of Wuhan city, and parts of Hankou and Hanyang have undergone urban renewal, and their urban environment and landscape are better. Therefore, it has a high score within the main urban area. New urban areas are in the development stage, with more industrial areas and poorer overall environment, which is more likely to cause negative emotions. On the other hand, the impact of urban built environments on the public also has a large variability. Among the 15 indicators, those with the greatest impact on public sentiment values are mixed degree of land use, road density, traffic congestion, and freight traffic impact. Hence, the quality of Wuhan's future urban built environment can be improved by enhancing land use mix, reasonably increasing the density and accessibility of the road network, increasing the number of public transportation facilities, improving traffic congestion, weakening the impact of freight traffic, and increasing public open space within an appropriate range. Further enriching the functions of the urban built environment, enhancing urban vitality, mitigating the negative impacts of urban traffic, and reasonably developing and constructing urban green and public spaces are important for promoting positive public emotions and mitigating negative ones.

6.2. Values and Innovations

Four innovative and unique research contents and methods have been reported in this paper: first, a multidisciplinary interdisciplinary inquiry method that combines GIS, natural language processing, and psychology was developed. Various methods, such as text emotions, GWR, and multifactor evaluation, were incorporated into the study. Second, social media big data were used to respond to public sentiment in a more relevant manner. Unlike conventional purposeful survey methods, social media data can provide continuous information on users' opinions and attitudes toward cities, further revealing their hidden characteristics. The use of social media data makes an important contribution to urban planning and decision making and to deeply interpreting the role of urban space and built environment from a human perspective, which helps further improve the relationship between people and cities and enhance urban well-being. Third, the introduction of dynamic traffic environment elements provided a comprehensive measure of urban built environment from a multidimensional perspective, which further revealed the environmental elements affecting people's emotions in urban spaces in a more comprehensive manner. Fourth, unlike existing studies, the specific built environment elements significantly related to emotions were clarified in this study to provide direct feedback for urban renewal and urban planning.

6.3. Limitations

This study, like any other, has some limitations. First, iconic words and symbols were used to rate Weibo texts. Because of the diverse forms of Chinese expressions, our method does not fully and accurately reflect users' emotions. Methods such as machine learning can be incorporated in subsequent studies to improve the interpretation of emotions. Second, the age distribution of people in Weibo data has limitations. Despite the rapid development of mobile networks, the number of Weibo users is limited, particularly in terms of the age distribution. Children and older people were not considered in the study given the lack of related Weibo data. Hence, further attention should be paid to the coupling between the emotions of different age groups and the urban built environment by combining traditional questionnaire methods in subsequent studies. Third, our consideration of the elements that influence emotion is still not comprehensive enough. The influence of urban environmental and social factors, such as air quality and crime rates, is ignored. Future research can further investigate the elements influencing emotions in a comprehensive manner.

Author Contributions: Conceptualization, H.J. and Y.M.; methodology, H.J. and Y.M.; software, Y.M.; validation, H.J., Y.M. and Y.Y.; formal analysis, Y.M.; investigation, H.J.; resources, H.J.; data curation, Y.M. and Y.Y.; writing—original draft preparation, Y.M.; writing—review and editing, H.J. and Y.Y.; visualization, Y.M. and Y.Y.; supervision, H.J.; project administration, H.J.; funding acquisition, H.J. All authors have read and agreed to the published version of the manuscript.

Funding: This research was funded by the National Key Research and Development Program "Research and Development of Emergency Response and Collaborative Command System with Holographic Perception of Traffic Network Disaster "(2020YFC1512002)".

Data Availability Statement: Not applicable.

Conflicts of Interest: The authors declare no conflict of interest.

References

1. Benabbou, R.; Lee, H. Exploring the evolution of urban emotions in the City of Seoul using social media information. *Int. J. Knowl. Based Dev.* **2019**, *10*, 232–248. [CrossRef]
2. Herrera-Yague, C.; Schneider, C.M.; Couronne, T.; Smoreda, Z.; Benito, R.M.; Zufiria, P.J.; Gonzalez, M.C. The anatomy of urban social networks and its implications in the searchability problem. *Sci. Rep.* **2015**, *5*, 10265. [CrossRef]
3. Yang, Z.H.; Liu, P.G. Expert system of the urban community safety plan based on ArcIMS data frame. *Ind. Saf. Environ. Prot.* **2008**, *7*, 550–555.
4. Jing, Q. Urban space and emotion. *Urban Plan. Forum* **2004**, 3, 6–8.
5. Hijazi, I.H.; Koenig, R.; Schneider, S.; Li, X.; Bielik, M.; Schmit, G.N.J.; Donath, D. Geostatistical analysis for the study of relationships between the emotional responses of urban walkers to urban spaces. *Int. J. E-Plan. Res.* **2016**, *5*, 1–19. [CrossRef]
6. Kumlu, K.B.Y.; Tudes, S. Effect of physical urban environment on sustainable urban development. In *Lecture Notes in Civil Engineering, Proceedings of 3rd International Sustainable Buildings Symposium, Dubai, United Arab Emirates, 15–17 March 2017*; Firat, S., Kinuthia, J., AbuTair, A., Eds.; Springer: Berlin, Germany, 2018; Volume 6, pp. 501–512.
7. Martínez, L. Health differences in an unequal city. *Cities* **2021**, *108*, 102976. [CrossRef]
8. Ma, Y.; Ling, C.; Wu, J. Exploring the spatial distribution characteristics of emotions of Weibo users in Wuhan waterfront based on gender differences using social media texts. *ISPRS Int. J. Geo-Inf.* **2020**, *9*, 465. [CrossRef]
9. Qiu, J.H.; Liu, J.P. Research on urban image building based on the people-oriented. In Proceedings of the International Conference on Engineering and Business Management 2010, Chengdu China, 25–27 March 2010; pp. 3006–3009.
10. Lee, S. The relationship between positive negative emotion, emotion-focused coping style and emotional eating behavior. *J. Table Food Coord.* **2020**, *15*, 61–80. [CrossRef]
11. Strack, F.; Pauli, P.; Weyers, P. Editorial: Emotion and behavior. *Front. Psychol.* **2016**, *7*, 313. [CrossRef]
12. Lindquist, K.A.; Barrett, L.F.; Bliss-Moreau, E.; Russell, J.A. Language and the perception of emotion. *Emotion* **2006**, *6*, 125–138. [CrossRef]
13. Ward, J.S.; Duncan, J.S.; Jarden, A.; Stewart, T. The impact of children's exposure to greenspace on physical activity, cognitive development, emotional wellbeing, and ability to appraise risk. *Health Place* **2016**, *40*, 44–50. [CrossRef] [PubMed]
14. Zhu, X.; Gao, M.; Zhang, R.; Zhang, B. Quantifying emotional differences in urban green spaces extracted from photos on social networking sites: A study of 34 parks in three cities in northern China. *Urban For. Urban Green.* **2021**, *62*, 127133. [CrossRef]
15. Calvillo Cortés, A.B.; Falcón Morales, L.E. Emotions and the urban lighting environment: A cross-cultural comparison. *SAGE Open* **2016**, *6*, 1. [CrossRef]
16. Byrne, F.L. Libby prison—A study in emotions. *J. South. Hist.* **1958**, *24*, 430–444. [CrossRef]
17. Weijs-Perree, M.; Dane, G.; van den Berg, P. Analyzing the relationships between citizens' emotions and their momentary satisfaction in urban public spaces. *Sustainability* **2020**, *12*, 7921. [CrossRef]
18. Kobylinska, D.; Karwowska, D. How automatic activation of emotion regulation influences experiencing negative emotions. *Front. Psychol.* **2015**, *6*, 66–71. [CrossRef]
19. Vanderlind, W.M.; Millgram, Y.; Baskin-Sommers, A.R.; Clark, M.S.; Joormann, J. Understanding positive emotion deficits in depression: From emotion preferences to emotion regulation. *Clin. Psychol. Rev.* **2020**, *76*, 101826. [CrossRef] [PubMed]
20. Burli, M.; Schmid, M.; In-Albon, T. Emotion regulation and interpersonal relationships in adolescents with NSSI—Emotion recognition and emotion expression. *Nervenheilkunde* **2013**, *32*, 24–29.
21. Wang, G.-L.; Zhang, W. Effects of urban expansion and changes in urban characteristics on PM2.5 pollution in China. *Huan Jing Ke Xue* **2019**, *40*, 3447–3456. [CrossRef]

22. Liu, B.; Cheng, K.; Ma, Z. Research on the impacts of urbanization on logistics carbon emission changes. *China Popul. Resour. Environ.* **2016**, *26*, 54–60.
23. Ji, X.; Zhang, Y.; Chen, F. Deprivation-based transportation equity evaluation model for rapid urbanization areas. *J. Transp. Syst. Eng. Inf. Technol.* **2012**, *12*, 7–13.
24. Adli, M.; Schondorf, J. Does the city make us ill? The effect of urban stress on emotions, behavior, and mental health. *Bundesgesundheitsblatt Gesundh. Gesundh.* **2020**, *63*, 979–986. [CrossRef]
25. Zhu, H.; Gao, Q. Review on "emotional turn" and emotional geographies in recent western geography. *Geogr. Res.* **2015**, *34*, 1394–1406.
26. Resch, B.; Summa, A.; Sagl, G.; Zeile, P.; Exner, J.P. Urban emotions-geo-semantic emotion extraction from technical sensors, human sensors and crowdsourced data. In *Progress in Location-Based Services 2014*; Lecture Notes in Geoinformation and Cartography; Gartner, G., Huang, H., Eds.; Springer: Berlin, Germany, 2015; pp. 199–212.
27. Li, X.; Hijazi, I.; Koenig, R.; Lv, Z.H.; Zhong, C.; Schmitt, G. Assessing essential qualities of urban space with emotional and visual data based on GIS technique. *ISPRS Int. J. Geo-Inf.* **2016**, *5*, 218. [CrossRef]
28. Pizzato, G.Z.; Guimaraes, L.B.D. Emotional attributes of urban furniture. In *Advances in Intelligent Systems and Computing, Proceedings of the 20th Congress Of the International Ergonomics Association, Florence, Italy, 26–30 August 2018*; Bagnara, S., Tartaglia, R., Albolino, S., Alexander, T., Fujita, Y., Eds.; Springer: Berlin, Germany, 2019; Volume 824, pp. 2087–2097.
29. Qu, Z.Q.; Lu, Y.M.; Jiang, Z.Q.; Bassett, E.; Tan, T. A psychological approach to "public perception" of land-use planning: A case Study OF Jiangsu Province, China. *Sustainability* **2018**, *10*, 3056. [CrossRef]
30. Wang, S.R.; Cao, D.L.; Lin, D.Z.; Chao, F. Traffic condition analysis based on users emotion tendency of microblog. In *Advances in Computational Intelligence Systems*; Advances in Intelligent Systems and Computing; Chao, F., Schockaert, S., Zhang, Q., Eds.; Springer: Berlin, Germany, 2018; Volume 650, pp. 299–311.
31. Hu, T.Y.; Xie, X.F.; Li, J. Negative or positive? The effect of emotion and mood on risky driving. *Transp. Res. Part F Traffic Psychol. Behav.* **2013**, *16*, 29–40. [CrossRef]
32. Niedenthal, P.M.; Rychlowska, M.; Wood, A.; Zhao, F. Heterogeneity of long-history migration predicts smiling, laughter and positive emotion across the globe and within the United States. *PLoS ONE* **2018**, *13*, e0197651. [CrossRef]
33. Morello, E.; Ratti, C. A digital image of the city: 3D isovists in Lynch's urban analysis. *Environ. Plan. B Plan. Des.* **2009**, *36*, 837–853. [CrossRef]
34. Zhang, J.H.; Chen, M.; Zhao, S.K.; Hu, S.Q.; Shi, Z.G.; Cao, Y. ReliefF-based EEG sensor selection methods for emotion recognition. *Sensors* **2016**, *16*, 1558. [CrossRef]
35. Kandt, J.; Batty, M. Smart cities, big data and urban policy: Towards urban analytics for the long run. *Cities* **2021**, *109*, 102992. [CrossRef]
36. Yang, L.; Wu, L.; Liu, Y.; Kang, C. Quantifying tourist behavior patterns by travel motifs and geo-tagged photos from flickr. *ISPRS Int. J. Geo-Inf.* **2017**, *6*, 345. [CrossRef]
37. Frias-Martinez, V.; Frias-Martinez, E. Spectral clustering for sensing urban land use using Twitter activity. *Eng. Appl. Artif. Intell.* **2014**, *35*, 237–245. [CrossRef]
38. Rizal, A.R.; Doherty, B.; Haeusler, M.H. Enabeling low cost human presence tracking. In Proceedings of the 21st Conference on Computer-Aided Architectural Design Research in Asia, Melbourne, Australia, 30 March–2 April 2016; pp. 45–54.
39. Krueger, R.; Hang, Q.; Ivanov, N.; Mahtal, S.; Thom, D.; Pfister, H.; Ertl, T. Bird's-eye—Large-scale visual analytics of city Dynamics using social location data. *Comput. Graph. Forum* **2019**, *38*, 595–607. [CrossRef]
40. Chen, T.T.; Hui, E.C.M.; Wu, J.M.; Lang, W.; Li, X. Identifying urban spatial structure and urban vibrancy in highly dense cities using georeferenced social media data. *Habitat Int.* **2019**, *89*, 102005. [CrossRef]
41. Ullah, H.; Wan, W.G.; Haidery, S.A.; Khan, N.U.; Ebrahimpour, Z.; Muzahid, A.A.M. Spatiotemporal patterns of visitors in urban green parks by mining social media big data based upon WHO reports. *IEEE Access* **2020**, *8*, 39197–39211. [CrossRef]
42. Ye, C.; Zhang, F.; Mu, L.; Gao, Y.; Liu, Y. Urban function recognition by integrating social media and street-level imagery. *Environ. Plan. B-Urban Anal. City Sci.* **2020**, *48*, 1430–1444. [CrossRef]
43. Liu, J.X.; Bi, H.; Wang, M.L. Using multi-source data to assess livability in Hong Kong at the community-based level: A combined subjective-objective approach. *Geogr. Sustain.* **2020**, *1*, 284–294. [CrossRef]
44. Feng, Y.; Sester, M. Extraction of pluvial flood relevant volunteered geographic information (VGI) by deep learning from user generated texts and photos. *ISPRS Int. J. Geo-Inf.* **2018**, *7*, 39. [CrossRef]
45. Huang, L.; Li, S.S.; Zhou, G.D. Emotion corpus construction on microblog text. In *Lecture Notes in Artificial Intelligence, Proceedings of the 16th Chinese Lexical Semantics Workshop, Bejing China, 9–11 May 2015*; Lu, Q., Gao, H.H., Eds.; Sprigner: Berlin, Germnay, 2015; Volume 9332, pp. 204–212.
46. Roberts, H.; Resch, B.; Sadler, J.; Chapman, L.; Petutschnig, A.; Zimmer, S. Investigating the emotional responses of individuals to urban green space using Twitter data: A critical comparison of three different methods of sentiment analysis. *Urban Plan.* **2018**, *3*, 21–33. [CrossRef]
47. Ding, C.; Wang, Y.; Yang, J.; Liu, C.; Lin, Y. Spatial heterogeneous impact of built environment on household auto ownership levels: Evidence from analysis at traffic analysis zone scales. *Transp. Lett.* **2016**, *8*, 26–34. [CrossRef]
48. Wu, J.; Chen, X.R.; Chen, S.L. Temporal characteristics of waterfronts in Wuhan city and people's behavioral preferences based on social media data. *Sustainability* **2019**, *11*, 6308. [CrossRef]

49. Zhang, S.; Huang, Z.X. Fractal analysis of the relation between the observation scale and the prediction cycle in short-term traffic flow prediction. *Int. J. Intell. Transp. Syst. Res.* **2019**, *17*, 1–8. [CrossRef]
50. Ai, Y.F.; Bai, Z.M.; Su, H.; Zhong, N.; Sun, Y.H.; Zhao, J.D.; IEEE. Traffic flow prediction based on expressway operating vehicle data. In Proceedings of the 11th International Conference on Intelligent Computation Technology and Automation, Changsha, China, 22–23 September 2018; pp. 322–326.
51. Fu, F.J.; Lou, Q.F.; Wang, B.; Wang, D.H. An identification method of traffic state inequality for urban roads. In *Applied Mechanics and Materials*; Bao, T., Ed.; Trans Tech Publications, Ltd.: Bäch, Switzerland, 2014; Volume 505–506, pp. 979–984.
52. Curiskis, S.A.; Drake, B.; Osborn, T.R.; Kennedy, P.J. An evaluation of document clustering and topic modelling in two online social networks: Twitter and Reddit. *Inf. Process. Manag.* **2020**, *57*, 102034. [CrossRef]
53. Wang, J.Y.; Ji, F.J.; Liu, B.Y.; Wang, N.; Yin, H.; Zhang, F.F. Research on mass news classification algorithm based on spark. In Proceedings of the 2020 International Conference on Big Data & Artificial Intelligence & Software Engineering, Bangkok, Thailand, 30 October–1 November 2020; pp. 408–414.
54. Ding, Y.; Li, R.; Zhen, Y.; Li, W. Research on semi-automatic construction method of high-speed train. *Mach. Des. Res.* **2020**, *36*, 185–189.
55. Bandhakavi, A.; Wiratunga, N.; Padmanabhan, D.; Massie, S. Lexicon based feature extraction for emotion text classification. *Pattern Recognit. Lett.* **2017**, *93*, 133–142. [CrossRef]
56. Fu, X.H.; Liu, G.; Guo, Y.Y.; Wang, Z.Q. Multi-aspect sentiment analysis for Chinese online social reviews based on topic modeling and HowNet lexicon. *Knowl. Based Syst.* **2013**, *37*, 186–195. [CrossRef]
57. Liu, Q.; Li, S. Word semantic similarity computation based on Hownet. In Proceedings of the 3th Chinese Lexical Semantic Workshop (CLSW2002), Taipei, Taiwan, 2–4 July 2002.
58. Zhu, Y.; Min, J.; Zhou, Y.; Huang, J.; Wu, L. Semantic orientation computing based on HowNet. *J. Chin. Inform. Process.* **2006**, *20*, 14–20.
59. Casas-Garriga, G. Discovering unbounded episodes in sequential data. In *Knowledge Discovery in Databases: PKDD 2003, Proceedings of the European Conference on Principles of Data Mining and Knowledge Discovery, Cavtat-Dubrovnik, Croatia, 22–26 September 2003*; Lecture Notes in Artificial Intelligence; Lavrac, N., Gamberger, D., Todorovski, L., Blockeel, H., Eds.; Sprigner: Berlin, Germany, 2003; Volume 2838, pp. 83–94.
60. Niu, Y.; Pan, M.; Wei, O.; Cai, X. Emotion analysis of Chinese microblogs using lexicon-based approach. *Comput. Sci.* **2014**, *41*, 253.
61. He, K.K.; Chen, Y. Urban Traffic Congestion Based on System Dynamics: Taking Wuhan City as an Example. In Proceedings of the 9th International Conference on Internet and Distributed Computing Systems (IDCS), Wuhan, China, 28–30 September 2016; pp. 372–380.
62. Yue, Y.; Zhuang, Y.; Yeh, A.G.O.; Xie, J.Y.; Ma, C.L.; Li, Q.Q. Measurements of POI-based mixed use and their relationships with neighbourhood vibrancy. *Int. J. Geogr. Inf. Sci.* **2017**, *31*, 658–675. [CrossRef]
63. Fotheringham, A.S.; Brunsdon, C. Local forms of spatial analysis. *Geogr. Anal.* **1999**, *31*, 340–358. [CrossRef]
64. Nilsson, P. Natural amenities in urban space—A geographically weighted regression approach. *Landsc. Urban Plan.* **2014**, *121*, 45–54. [CrossRef]
65. Zhang, X.X.; Huang, B.; Zhu, S.Z. Spatiotemporal influence of urban environment on taxi ridership using geographically and temporally weighted regression. *ISPRS Int. J. Geo-Inf.* **2019**, *8*, 23. [CrossRef]
66. Zhao, X.F.; Ji, Y.N.; IOP. The urban public space betterment and land use sustainability under the human behavior. *IOP Conf. Ser. Earth Environ. Sci.* **2018**, *113*, 012143. [CrossRef]

land

Article

House Price Forecasting from Investment Perspectives

Song Shi [1,*], Vince Mangioni [1], Xin Janet Ge [1], Shanaka Herath [1], Fethi Rabhi [2] and Rachida Ouysse [3]

[1] School of the Built Environment, University of Technology Sydney, Sydney, NSW 2007, Australia; Vincent.Mangioni@uts.edu.au (V.M.); Xinjanet.ge@uts.edu.au (X.J.G.); Shanaka.herath@uts.edu.au (S.H.)
[2] School of Computer Science and Engineering, University of New South Wales, Sydney, NSW 2052, Australia; f.rabhi@unsw.edu.au
[3] School of Economics, University of New South Wales, Sydney, NSW 2052, Australia; rouysse@unsw.edu.au
* Correspondence: song.shi@uts.edu.au

Abstract: Housing market dynamics have primarily shifted from consumption- to investment-driven in many countries, including Australia. Building on investment theory, we investigated market dynamics by placing investment demand at the center using the error correction model (ECM). We found that house prices, rents, and interest rates are cointegrated in the long run under the present value investment framework. Other economic factors such as population growth, unemployment, migration, construction activities, and bank lending were also important determinants of the housing market dynamics. Our forecasting results show that the Sydney housing market will continue to grow with no significant price decline in the foreseeable future.

Keywords: housing market dynamics; price forecast; error correction model (ECM); Sydney housing market

Citation: Shi, S.; Mangioni, V.; Ge, X.J.; Herath, S.; Rabhi, F.; Ouysse, R. House Price Forecasting from Investment Perspectives. *Land* **2021**, *10*, 1009. https://doi.org/10.3390/land10101009

Academic Editor: Sarel Cilliers

Received: 20 August 2021
Accepted: 23 September 2021
Published: 26 September 2021

1. Introduction

A prosperous and healthy housing market is critical for the modern economy. Yet, there are crucial research gaps, and, therefore, there is a need to better understand housing market dynamics. One of the most critical observations we made in past decades is that housing market dynamics have primarily shifted from consumption- to investment-driven in many countries, including Australia. Investment activities (including speculation on capital gains) have been the center of housing demand. More importantly, this investment-driven housing demand is believed to continue in the foreseeable future. With firm guidance from investment theory, this paper examines housing market dynamics through a lens that places investment demand at the center of housing market dynamics using an augmented error correction model (ECM). Since the proposed ECM model contains information set on both current and past values of house prices, the results are useful in a one-step-ahead price forecast which is important in real estate markets.

The ECM method has been widely applied to house price forecasting in many countries such as the US [1], Netherlands [2], Euro areas [3], New Zealand [4], and Australia [5]. As pointed out by Wheaton, Chervachidze, and Nechayev [6], the method can handle both the stationarity and endogeneity problems that plague the time-series data for local housing price analysis. However, the empirical application of an ECM requires researchers to select possible "determinants" of the local housing market and distinguish their long- or short-term relationship in the market. Traditional reduced-form studies of housing market supply and demand struggle to provide firm guidance from theory on the selection of housing determinants, which means that the ECM specifications are often ad hoc in practice. The novelty of our approach is to specify an investment ECM model based on the long-run relationship among house prices, rents, and interest rates and a variety of short-term impacts from other economic variables. To the best of our knowledge, this is the first paper distinguishing between the investment and consumption determinants in an ECM specification of house price analysis.

We found that the key market drivers for housing demand are prices, rents, and interest rates. There is a long-run relationship between housing prices, rents, and interest rates. House prices are also influenced by other local and nationwide economic variables such as population growth, unemployment rates, migration numbers, construction costs, bank mortgage lending, and spatial correlation among Australian capital cities. These economic variables directly affect house prices in the short-term and indirectly affect house prices through rents and interest rates in the long term.

Housing is both a consumption good and an investment asset. For many Australians, the great Australian dream is to own their family homes, leading to a better, more secure, and prosperous life. Nowadays, the rising housing prices in many Australian regions, especially in large capital cities such as Sydney and Melbourne, are making it increasingly difficult for many to achieve this dream. Over the last 40 years, housing prices in Sydney have increased more than thirty times; in contrast, household incomes have merely increased by about ten times. Although the Sydney population has almost doubled during the same period and household formations are becoming smaller, these do not fully explain why Australian house prices have priced out so many hard-working ordinary Australians. Yanotti and Wright [7] studied the determining factors of purchasing investment properties through analyzing mortgage finance data and identified where the investment stock is physically located. They found a mismatched investment of residential property and rent-al demand in Australia.

Traditional house price models derived from reduced-form structural supply and demand equations are valid but have a misplaced investment demand. In 1989, Mankiw and Weil [8] published their famous paper titled "The Baby Boom, The Baby Bust, and the Housing Market". By examining significant demographic changes in the US housing market, they concluded that real housing prices would fall substantially in the future. Thirty years later, global house prices, including in the US and Australia, have increased considerably rather than decreased. Why do their conclusions, based on demographic changes, have proved to be inaccurate? The answer largely depends on how we understand and interpret housing market dynamics.

In housing finance, making a deposit and then repaying the mortgage principal over 20 to 30 years represents an opportunity cost of ownership that is valued by the present value of future benefits. Shiller [9] argued that house prices should be equal to the current discounted value of future rents. This present value framework borrowed from finance literature applies to both investment and consumption purposes. For owner-occupied houses, there is an implicit rent called a user cost of housing paid by the owner [10–16]. Since the accurate estimation of a user cost can be difficult or even problematic [17], researchers tend to use market rents as a proxy for user cost in practice. Compared to the estimation of user cost, market rents are more readily observable in a marketplace. Shi, Jou, and Tripe [4] applied the present value model to investigate how the real rental rate and interest rate affect the real housing price. Using a similar asset pricing approach, Bourassa, Hoesli, and Oikarinen [18] found that a price-to-rent ratio measure performs well in housing market dynamics analysis.

To test the validity of our proposed investment modeling, we undertook a cointegration and error correction analysis for fifteen years of house prices in the Greater Sydney Area between 2004 and 2018. We first used data from 2004 to 2013 to develop the model and then tested the model's forecast performance in an out-of-sample test during 2014 and 2018. The results showed an average forecast error of 2.2%, or between AUD 19,000 and AUD 23,000, compared to the average house price in the Greater Sydney Area of AUD 863,000 during the testing period. Moreover, the model showed a superior forecast performance compared to the alternative models. We also forecasted house prices in the Greater Sydney Area for the foreseeable future, that is, from 2019 to 2030. The results indicated that house prices should bottom out in 2019 and would continue to rise in the future. The chance for a significant downward market price adjustment is less likely. Note that our price forecasting was based on the data up to 2018, i.e., the best available information at the time of writing this paper. Although we cannot forecast future events like a pandemic, their economic impacts on the housing market are easy to incorporate in our models via the changes of interest rates and other economic variables in the specified dynamic system.

In the next section, we review the theoretical base for the modeling. We then outline the econometric tools used, describe the data, discuss the results, and provide the conclusions.

2. Theoretical Base for Modeling

The theoretical basis follows the present-value model, which is commonly used to estimate the value of stocks in the share market to estimate the values for housing. Under the proposed present-value theory, housing is valued as a financial asset where price relates to the expected future cash flows discounted to the present by using an expected discount rate. For owner-occupied housing, there is an implicit rent (called a user cost or economic rent) paid by the homeowners; for investment properties, there is market rent paid by the renters. Thus, housing price P at time t can be written as follows:

$$P_t = E_t\left[\sum_{i=1}^{n} \frac{D_{t+i}}{(1+R)^i}\right] + E_t\left[\frac{P_{t+n}}{(1+R)^n}\right] \tag{1}$$

where D_t is the dividend or cash flow at time t and R is the discount rate. E denotes the expected value.

In the finance literature, the first term is often called the fundamental value, and the second term is the price bubble. When n is sufficiently large, the second term converges to zero. The model implies that the current asset price is simply the sum of all expected present values of future cash flows, discounted at a constant rate. The model has been widely applied in finance to value shares.

In the property market, a simplified version of the above present model, called a capitalization approach, is used to value income-producing properties. However, while the assumption of a constant expected income D and discount rate R is analytically convenient, they contradict the evidence that both the expected income and the investor's expected rate of return vary over time. Campbell and Shiller [19,20] suggested a log-linear present-value model with time-varying expected returns, where a log asset price at time t is written as follows:

$$p_t = \frac{k}{1-\rho} + E_t\left[\sum_{j=0}^{n} \rho^j \left[(1-\rho)d_{t+1+j} - r_{t+1+j}\right]\right] + E_t\left[\rho^j p_{t+j}\right] \tag{2}$$

where $\rho = 1/\left(1+\exp\left(\overline{d-p}\right)\right)$, $\overline{(d-p)}$ is the average log dividend-price ratio, $k = -\log(\rho) - (1-\rho)\log\left(\frac{1}{\rho} - 1\right)$, and $r_{t+1} = \log(P_{t+1} + D_{t+1}) - \log(P_t)$.

When the time horizon n increases to infinity, the third term, which is the discounted expected value of the asset price, shrinks to zero. Accordingly, the current asset price can be presented as follows:

$$p_t = \frac{k}{1-\rho} + E_t\left[\sum_{j=0}^{\infty} \rho^j \left[(1-\rho)d_{t+1+j} - r_{t+1+j}\right]\right] \tag{3}$$

This equation can be rewritten in terms of the log dividend-price ratio, which is:

$$d_t - p_t = -\frac{k}{1-\rho} + E_t\left[\sum_{j=0}^{\infty} \rho^j \left[\Delta d_{t+1+j} + r_{t+1+j}\right]\right] \tag{4}$$

The above Equation (4) is referred to by Campbell and Shiller [19,20] as the dividend-ratio model. This equation implies that the log dividend-price ratio should be stationary, provided the changes in log dividends and the expected stock return are stationary. Where the log dividend-price ratio is nonstationary, it is very likely that the expected stock return is nonstationary (highly persistent), even when the above present-value model is valid.

3. Econometric Tools for Model Development

The Augmented Dickey–Fuller (ADF) unit root test is applied to test for the stationarity of time-series variables. As expected, they are integrated order one I(1) processes[1]. Because house prices, rents, and interest rates are I(1) variables, they are further tested for cointegration under the proposed present-value framework.

Cointegration tests are widely used in time-series econometrics. In empirical tests, cointegration and unit root tests between stock prices and dividends give mixed findings depending on the time period studied. Using the annual US stock market data from 1871 to 1986, Campbell and Shiller [21] found that stock prices and dividends are not cointegrated. The deviation between prices and dividends is quite persistent. On the other hand, Diba and Grossman [22] indicated a possible cointegration relationship between stock prices and dividends for the US stock market. In terms of the housing market, Gallin [23] found the log rent-price ratio is stationary by using aggregated quarterly data for the US housing market. When Brooks, Katsaris, McGough, and Tsolacos [24] examined the monthly prices of UK equity-traded property stocks from 1986 to 1998, they found that prices and rents are not cointegrated over the sample period.

The error correction model (ECM) is the next application step in this study under the proposed present-value framework. The ECM model is based on the assumptions that house prices, rents, and interest rates are cointegrated and that house prices are affected by both long-run cointegration and short-run dynamics among prices, rents, and interest rates. Since house prices are affected by other factors and tend to be seasonal, key micro- and macro-economic variables and seasonal quarterly dummy variables are included as external control variables in the ECM model. The identified key economic variables include population and immigration growth, building costs, building supply, the amount of mortgage lending, and housing price growth in Australian capital cities. They are selected for the proposed ECM model according to their statistical significance in the modeling. To be specific, the Johansen cointegration test and the ECM model were written as follows:

$$x_t = \alpha_0 + \alpha_1 x_{t-1} + \beta_1 y_{t-1} + \gamma_1 z_{t-1} + \delta_1 \mu_{t-1} + E_t + S_t + \varepsilon_t \tag{5}$$

where x, y, and z are log price, rent, and interest rate in the first differences, α_0 is a constant, μ is the error correction term, E is a vector of key micro and macroeconomic variables such as population, employment, immigration, bank lending, construction activities, etc. (see Table 1 for exact variables used in Equation (5)), S is a vector for seasonal dummies, and ε is white noise.

For simplicity, we excluded any potential structure breaks or regime shifts in the above models. This is justified as our main objective is forecasting rather than econometric modeling. The literature showed that forecasts are not substantially affected by the presence of structural breaks [25]. Elliott and Muller [26,27] further showed that gains from modeling a structural break might be offset by imprecisely estimated break dates and post-break parameters. Therefore, ignoring a break may lead to more accurate forecasts [28].

Table 1. Definition of variables.

Variable	Definition	Sources
SYD_HP_MEDIAN_H	Median house prices in the Greater Sydney Area	SIRCA
SYD_RENT_H	Median house rents in the Greater Sydney Area	SIRCA
SYD_HP_MEDIAN_U	Median unit prices in the Greater Sydney Area	SIRCA
SYD_RENT_U	Median unit rents in the Greater Sydney Area	SIRCA
AUS_INT_90D	Australia 90-day bill rate	Datastream
AUS_INT_10YRF	Australian ten-year government bond yield	Datastream
AUS_LEN_DF	Bank mortgage lending for all dwelling finance in Australia	ABS
AUS_CONF	Building construction cost index of Australia	ABS
NSW_PPGF	New South Wales population growth	ABS
SYD_SUP_DF	Total number of dwelling (all types) supply in the Greater Sydney Area	NSW Planning & Environment
SYD_NOMF	Net migration numbers in the Greater Sydney Area	ABS
ECC_PCF	Residential Property Price Index percentage change from the corresponding quarter of the previous year-weighted average of eight capital cities	ABS
	Variable Transformation	
D(LOG(SYD_HP_MEDIAN_H))	Changes in median house prices	
D(LOG(SYD_RENT_H))	Changes in median house rents	
D(LOG(SYD_HP_MEDIAN_U))	Changes in median unit prices	
D(LOG(SYD_RENT_U))	Changes in median unit rents	
D(LOG(AUS_LEN_DF))	Growth rate of bank mortgage lending	
D(LOG(AUS_CONF))	Changes in building construction costs	
D(LOG(NSW_PPGF))	Changes in population growth	
D(LOG(SYD_NOMF))	Changes in net migration numbers	

Note: D denotes the first difference and LOG represents the natural logarithm.

4. Variable Selection and Description

This analysis focused on the Greater Sydney Area, which covers 35 LGAs as defined by the Australian Bureau of Statistics (ABS)[2]. The sample period spanned from Q3 2004 to Q4 2018. The reason the analysis started in 2004 is mainly due to data availability. For example, there is no reliable rental time-series data prior to 2004. All time-series data are in quarterly frequencies and collected from various data sources. In total, 50 variables were considered in the proposed ECM model, covering a broad range of national, regional, and local economies. Some key variables and data sources are listed below and detailed in Table 1[3].

(1) Australian Bureau of Statistics (ABS). This includes, for example, time-series data on regional population and immigration growth, nationwide bank mortgage lending, building construction cost, and price changes in other capital cities
(2) Securities Industry Research Center of Asia-Pacific (SIRCA). This includes, for example, time-series data on median prices and rents for houses and units in the Greater Sydney Area. We assume that the mix and quality of quarterly residential property sales for houses and units (i.e., the number of bedrooms, bathrooms, car parks, lot size, land tenure, a mix of high- and low-value properties, and transaction types, etc.) are relatively stable over time. The impact of forced sales on reported median price

indices as discussed in Renigier-Biłozor, Walacik, Źróbek, and d'Amato [29] is small in this study.

(3) DataStream. This includes, for example, market data on Australia's interest rates and government bond yields

(4) NSW Planning & Environment. This includes, for example, time-series data related to local housing and land supply.

All the variables and forecasting results are expressed here in nominal terms, i.e., they are the actual value at the time of the analysis without adjusting for inflation. Depending on the purpose of their research, economists choose to use either real or nominal terms in their analyses. However, there is no consensus in the literature on what inflation indices should be used to deflate the variables. To avoid introducing unnecessary measurement errors in deflating variables for the analysis and later converting prices back to nominal values for the forecasting, nominal values are adopted in this market dynamic analysis.

5. Model Development and Testing Results

5.1. Model Development

Table 2 shows the ECM modeling results for the Greater Sydney Area between 2004 and 2018. Column (1) contains the results for houses, and column (2) is for units. The adjusted R-squared is 0.843 for houses, which means that a variation of 84.3% in quarterly house price changes has been explained by the ECM house price model. For units, the adjusted R-squared is 0.812. Thus, both house and unit price changes are well modeled over the sample period.

Under the present-value framework, asset prices, rents, and interest rates are expected to be cointegrated in the long run. The cointegration results in Table 2 show prices, rents, and interest rates are indeed cointegrated[4]. Both house and unit price changes are positively related to rent changes but negatively related to interest rate changes. For every one percent increase in rents, it increases house prices by about 0.262%, while for every one percent increase in interest rates, it decreases house prices by 0.582%. In contrast, unit prices are more sensitive to rent and less sensitive to interest rate changes. For every one percent increase in rents, it increases unit prices by 0.5%, while for every one percent increase in interest rates, it decreases unit prices by about 0.412%. One possible explanation for this is that houses are more expensive than units. Thus, house owners are more sensitive to interest rate changes than unit owners. Another explanation is that units are more likely to be held for investment purposes. Investors focus more on income instead of interest rate changes because interest costs are tax-deductible for property investment.

The long-term relationship, as indicated by the cointegration term CointEq1, is negative (−0.048 for houses and −0.046 for units) and statistically significant, which means the speed of adjustment from a short-run towards a long-run equilibrium in the Sydney housing market is about 4.6–4.8 percent each quarter. The results indicate a very slow price adjustment process in the housing market.

Short-term price dynamics also affect price changes. Table 2 shows that the current period price change is statistically significant in relation to the last period price change and the price change in other Australian capital cities (apart from Sydney). One percentage point increase in the last period price change increases the current period price change by 0.536% and 0.639% for houses and units, respectively. Although other economic factors are not statistically significant in the price model, they are important in rent and interest rate models. For example, rents are found to be negatively related to the last period's net migration figure in New South Wales and the new dwelling supply in Sydney. Interest rates, however, are found to relate positively to the last period's government bond yields, construction costs, net migration figure, and dwelling supply. Interest rates are also negatively related to Australian mortgage lending[5].

Table 2. Model development—Error correction model estimates, Q3 2004–Q4 2018.

	(1)	(2)
	House Price Changes	**Unit Price Changes**
CointEq1	−0.048 ***	−0.046 ***
	(0.014)	(0.015)
D(LOG(SYD_HP_MEDIAN(-1)))	0.536 ***	0.639 ***
	(0.071)	(0.076)
D(LOG(SYD_RENT(-1)))	0.009	−0.007
	(0.107)	(0.054)
D(LOG(AUS_INT_90D(-1)))	0.017	−0.007
	(0.021)	(0.014)
Constant	−0.002	−0.000
	(0.003)	(0.003)
D(LOG(AUS_INT_10YRF))	−0.003	−0.001
	(0.015)	(0.011)
D(LOG(AUS_LEN_DF))	0.007	0.011
	(0.034)	(0.025)
D(LOG(AUS_CONF))	−0.291	−0.210
	(0.269)	(0.215)
D(LOG(NSW_PPGF))		−0.003
		(0.004)
D(LOG(SYD_SUP_DF))	0.003	−0.006
	(0.008)	(0.007)
D(LOG(SYD_NOMF))	0.042	0.052 *
	(0.036)	(0.028)
ECC_PCF	0.002 ***	0.001 ***
	(0.000)	(0.000)
Seasonal dummy	Yes	Yes
Adj. R-squared	0.843	0.812
Sum sq. resids	0.003	0.006
S.E. equation	0.009	0.012
F-statistic	24.567	2.966
Log likelihood	200.977	184.466
Akaike AIC	−6.447	−5.878
Schwarz SC	−5.95	−5.38
Mean dependent	0.012	0.008
S.D. dependent	0.022	0.014
The cointegration equation results		
LOG(SYD_HP_MEDIAN_H(-1))	1.000	1.000
LOG(SYD_RENT_H(-1))	−0.262 (0.249)	−0.500 (0.117)
LOG(AUS_INT_90D(-1))	0.582 (0.077) ***	0.412 (0.043) ***
Constant	−12.475	−10.609

Notes: Descriptions of the explanatory variables are provided in Table 1. Standard errors are shown in ( ). *, **, and *** denote significance levels at the 10%, 5%, and 1% level, respectively.

Overall, both house and unit price equations are well supported by the ECM modeling under the present-value framework. Our results show that prices are mostly driven by the short-run dynamics in the market, and the price adjustment process to its long-run equilibrium is very slow. It takes about 20 quarters for the housing market to adjust towards its long-run equilibrium. The results imply a self-fulfilling phenomenon in the Sydney housing market, which is consistent with the bubble literature on the Australian and New Zealand housing markets [5,30,31].

5.2. *Out-of-Sample Testing, Underlying Variable Assumptions, and Measures of Forecasting Accuracy*

To validate the proposed ECM model forecasting performance, we used an in-sample (training) data period from 2004 to 2013 to estimate the model and a pseudo-out of-sample period between 2014 and 2018 to test its forecasting performance. In other words, we used the first 10 years of data to build the model and compared its forecasted prices to the actual prices in the next five years from 2014 to 2018. The ECM models developed for the periods between 2004 and 2013 are included in Appendix A.

Since other economic variables are included as exogenous inputs in the proposed ECM model, we needed to make some assumptions about those variables in the out-of-sample testing period before forecasting prices. One approach was to take them as observable. In this case, forecasting errors between the forecasted and actual prices would have to be due to the ECM model itself rather than any estimation errors introduced from the other forecasted variables. However, this take-as-given approach is not realistic. In reality, we simply do not know those economic variables in advance. Thus, we introduced a conditional forecasting method of using time-series techniques to forecast those underlying economic variables. These advanced time-series forecasting techniques include auto-regressive integrated moving average (ARIMA) and the auto-regressive (AR) methods. Four statistical measures were used to evaluate a forecast performance and the statistical difference between the forecasted prices and actual prices. They are the Root Mean Square Error (RMSE), Mean Absolute Error (MAE), Mean Absolute Percentage Error (MAPE), and Theil inequality coefficient. While the first three statistics are based on quadratic loss and average errors, the Theil index measures the proximity between the forecasts examining the ability of the forecast series to match the mean and the variance of the actual series. The model forecast evaluation results are shown in Table 3.

Table 3. Out-of-sample testing, Q1 2014–Q4 2018.

Variable	obs.	RMSE	MAE	MAPE	Theil
Panel A: Actual					
Median house prices	20	22,578	19,003	2.207	0.013
Median unit prices	20	31,923	23,047	3.141	0.024
Panel B: Statistical models					
		ARIMA model			
Median house prices	20	53,771	45,449	5.368	0.032
Median unit prices	20	24,593	20,107	2.299	0.014
		AR(1) model			
Median house prices	20	59,000	49,203	5.776	0.035
Median unit prices	20	18,484	14,111	2.068	0.014
		AR(4) model			
Median house prices	20	70,685	59,990	7.172	0.042
Median unit prices	20	21,640	18,154	2.729	0.016
		AR(8) model			
Median house prices	20	64,694	54,936	6.534	0.038
Median unit prices	20	21,370	18,047	2.688	0.016
		AR(16) model			
Median house prices	20	46,217	39,780	4.712	0.027
Median unit prices	20	25,551	19,379	2.836	0.019
		AR(20) model			
Median house prices	20	37,305	31,007	3.677	0.022
Median unit prices	20	30,753	23,981	3.397	0.023

Notes: RMSE denotes the Root Mean Square Error; MAE is for the Mean Absolute Error; MAPE is for the Mean Absolute Percentage Error; and Theil is for the Theil Inequality Coefficient.

Panel A of Table 3 shows the model forecast evaluation results for houses when underlying variables are taken as given. Between the forecasted and actual quarterly house prices, the RMSE is AUD 22,578, the MAE is AUD 19,003, the MAPE is 2.207%, and the Theil for inequity measure is 0.013. The average house price in the Greater Sydney Area is about AUD 863,000 during the forecasting period between 2014 and 2018. Thus, the forecasted quarterly house prices closely track the actual house prices, which is demonstrated in Figure 1.

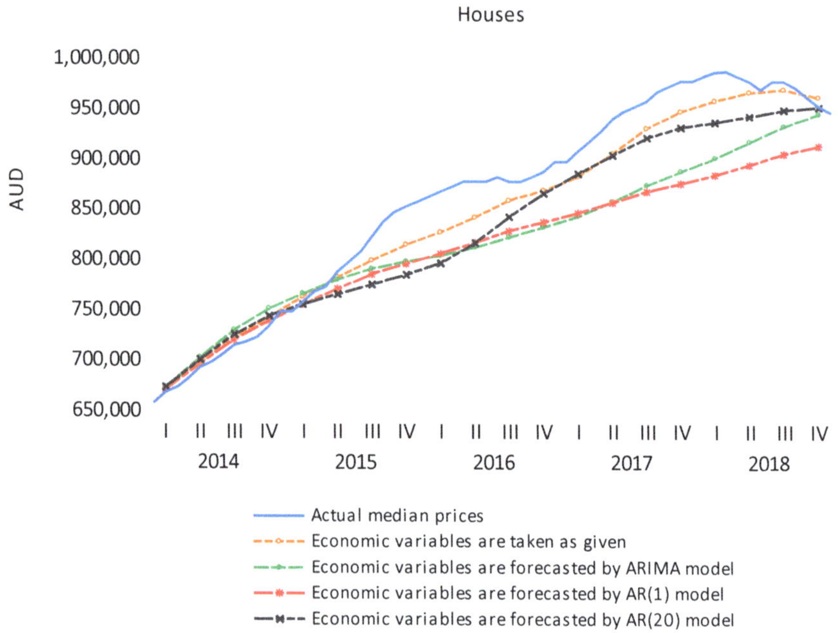

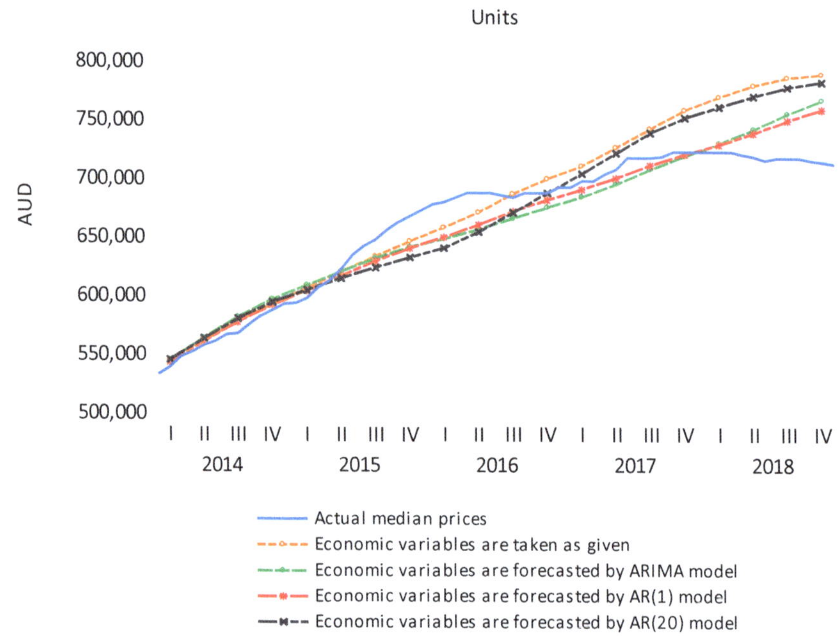

Figure 1. Out-of-Sample forecast, Q1 2014–Q4 2018.

Panel B of Table 3 shows the results of conditional forecast when the underlying economic variables are unknown and must be forecasted. Two popular and widely-applied time-series forecasting techniques (ARIMA and AR models) were considered in this analysis. The results show that using an AR(1) process in the underlying economic variable forecast produced the best forecasting results for units, while an AR(20) process produced the best forecasting results for houses. The ARIMA model produced somewhat forecasting results in between an AR(1) and AR(20) process. The results showed that the forecasting performance of developed models is conditional on the assumed knowledge of underlying variables. As AR(1) and AR(20) processes tend to set the boundary of forecasting, we chose to rely on the more sophisticated ARIMA model and selected AR(1) and AR(20) processes for future price forecasting in the next section. The results provide some insights into forecasting for decision-making under different underlying variable assumptions.

5.3. Forecasting Results

Table 4 presents the forecast median house and unit prices for the Greater Sydney Area from 2019 to 2030. These prices are forecast-based on the model developed using the data between 2004 and 2018 described in Section 4. The underlying economic variables are estimated using the ARIMA and selected AR models. In other words, these are fitted values to each one of the macroeconomic variables, E_t.

Table 4. Forecasted median prices, Q1 2019–Q4 2028.

	Houses			Units		
Time Period	**ARIMA**	**AR(1)**	**AR(20)**	**ARIMA**	**AR(1)**	**AR(20)**
2019Q1	931,067	935,696	930,709	708,278	711,058	707,385
2019Q2	911,194	919,994	910,228	703,633	709,259	702,693
2019Q3	899,668	910,522	898,726	701,262	709,549	701,167
2019Q4	899,592	902,946	898,575	704,042	709,965	705,222
2020Q1	911,688	897,870	910,480	710,599	709,785	712,396
2020Q2	937,604	896,699	935,054	725,269	712,940	727,899
2020Q3	969,586	898,580	963,080	742,813	717,425	744,982
2020Q4	1,000,454	899,203	988,440	761,172	721,061	762,201
2021Q1	1,026,564	899,777	1,010,434	776,647	723,352	776,808
2021Q2	1,051,763	902,546	1,035,673	793,422	728,505	794,303
2021Q3	1,072,154	907,377	1,061,955	807,664	734,725	810,899
2021Q4	1,084,812	910,416	1,085,048	818,901	739,903	827,012
2022Q1	1,090,764	913,157	1,103,916	825,515	743,606	838,728
2022Q2	1,097,922	918,003	1,122,849	833,694	750,172	852,274
2022Q3	1,104,906	924,863	1,141,454	841,290	757,768	864,616
2022Q4	1,110,291	929,794	1,158,450	848,505	764,205	877,111
2023Q1	1,115,409	934,287	1,176,081	854,240	769,020	887,132
2023Q2	1,127,265	940,781	1,201,009	864,403	776,693	902,642
2023Q3	1,143,098	949,182	1,229,239	876,219	785,338	919,601
2023Q4	1,159,630	955,449	1,255,977	888,902	792,691	937,265
2024Q1	1,176,506	961,117	1,280,240	900,605	798,278	951,949
2024Q2	1,199,588	968,711	1,307,132	916,464	806,760	970,516
2024Q3	1,224,766	978,157	1,329,884	933,008	816,191	986,871
2024Q4	1,247,488	985,306	1,341,254	948,795	824,226	999,107
2025Q1	1,267,094	991,752	1,340,653	961,684	830,379	1,003,737
2025Q2	1,289,900	1,000,113	1,336,449	977,148	839,503	1,009,010
2025Q3	1,311,892	1,010,324	1,330,797	991,885	849,581	1,011,236
2025Q4	1,328,916	1,018,109	1,324,326	1,004,629	858,177	1,012,546
2026Q1	1,341,300	1,025,120	1,321,241	1,013,773	864,786	1,012,437
2026Q2	1,356,635	1,034,069	1,329,277	1,025,616	874,466	1,019,592
2026Q3	1,371,711	1,044,895	1,343,665	1,037,221	885,119	1,029,243
2026Q4	1,382,851	1,053,181	1,359,345	1,047,595	894,211	1,040,725
2027Q1	1,390,980	1,060,639	1,374,828	1,055,374	901,217	1,050,439
2027Q2	1,404,237	1,070,076	1,396,359	1,067,094	911,409	1,066,013
2027Q3	1,419,395	1,081,436	1,420,529	1,079,755	922,604	1,082,358
2027Q4	1,432,427	1,090,149	1,441,346	1,092,156	932,161	1,098,464

Table 4. *Cont.*

	Houses			Units		
Time Period	ARIMA	AR(1)	AR(20)	ARIMA	AR(1)	AR(20)
2028Q1	1,444,141	1,097,988	1,459,068	1,102,630	939,534	1,110,928
2028Q2	1,462,484	1,107,862	1,480,598	1,117,673	950,221	1,128,060
2028Q3	1,483,687	1,119,714	1,503,627	1,133,929	961,946	1,144,919
2028Q4	1,503,083	1,128,815	1,524,299	1,149,803	971,957	1,161,090
2029Q1	1,521,148	1,137,001	1,544,480	1,163,369	979,686	1,174,553
2029Q2	1,545,758	1,147,287	1,572,678	1,181,283	990,865	1,194,720
2029Q3	1,572,689	1,159,614	1,606,490	1,199,862	1,003,123	1,216,369
2029Q4	1,596,568	1,169,086	1,640,703	1,217,315	1,013,591	1,238,525

Notes: the prices are forecasted based on various statistical assumptions of exogenous variables in the error correction model (ECM). Prices are in nominal terms.

For both houses and units, the best-predicted price pattern was conditional on a high order serial correlation in the underlying variable assumptions such as an AR(20) process, while the worst predicted price path was found if the underlying economic variables follow a low order serial correlation such as an AR(1) process. The results from the ARIMA variable assumptions were in between. Figure 2 shows the future price movements based on different underlying variable assumptions.

It shows that housing prices are more likely to follow a high-growth scenario, as indicated by the ARIMA underlying variable assumptions; the market will bottom out in 2019–2020 and continue to grow in the future. Under this scenario, the median price will reach AUD 1,596,568 for houses and AUD 1,217,315 for units in 2030, presenting an annual compounding growth rate of 5%. This forecast is supported by the alternative variable assumptions following an AR(20) process. Even in a low-growth scenario, as indicated in the AR(1) variable assumptions, there will be no significant price decline in the foreseeable future; rather, prices will mostly be flat or in a slow-growth mode with an average annual growth rate of 2% in the next decade.

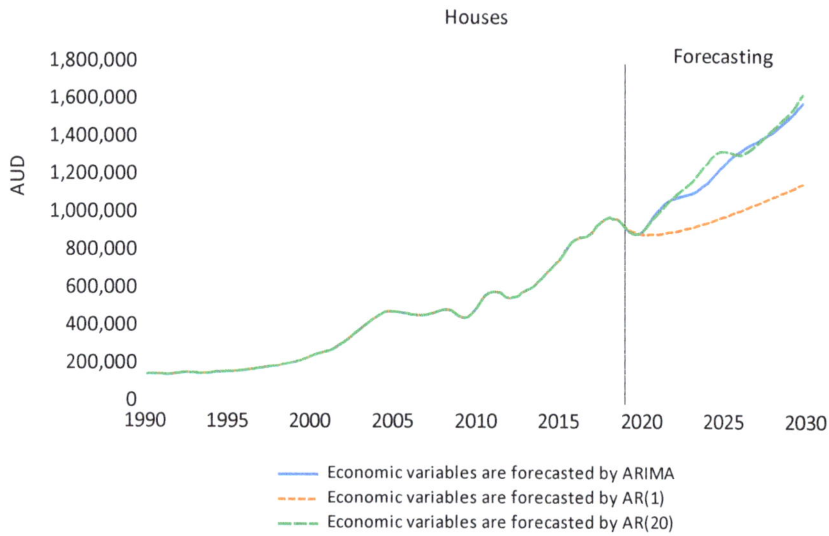

Figure 2. *Cont.*

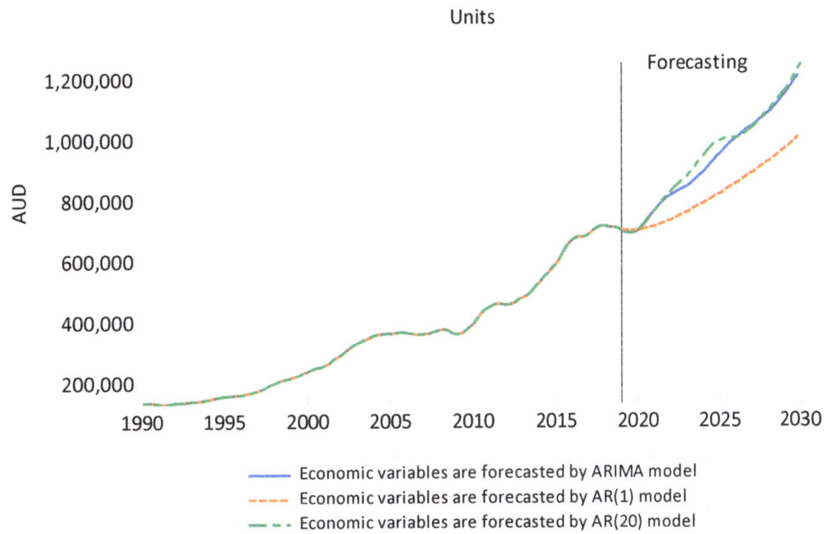

Figure 2. Forecasted Median Prices, 2019Q1–2029Q4.

5.4. Robustness Check

5.4.1. Unconditional Forecast

The forecasting performance of the developed ECM model depends on the forecasting of underlying economic factors. This raises some concerns about the usefulness of the proposed ECM model. Alternatively, we used the unconditional forecast of the standard AR(1) process and ECM model without other economic factors to validate the results. The advantage of using an AR(1) process for house price forecasting is that it is simple and does not require any other variables in the modeling. As the standard ECM model only depends on prices, rents, and interest rates, there is no need to forecast other economic variables in this situation. Using the data from 2004 to 2013, we developed the standard AR(1) and ECM models and placed them in 5-year out-of-sample testing between 2014 and 2018 for performance comparison. The results are presented in Appendix B.

Our results show that the forecasting performance of a standard ECM model without underlying economic variables is less accurate than a standard AR(1) forecast, despite the ECM model having a higher adjusted R-squared (0.78) than the AR(1) process (0.62) in the model development period between 2004 and 2013. The findings were in line with the forecasting literature, stating that models, which fit the historical data well, do not necessarily perform better than others in forecasting. It is not surprising that our proposed conditional ECM model is superior to the unconditional forecast models such as the standard AR(1) and ECM models. Our findings stress the importance of including other economic variables in the ECM model development and, thereafter, forecasting.

5.4.2. Forecasted Median Rents

Our proposed ECM model can also be used to predict future rents. To check the validity of our price model, we further checked whether the predicted rents were in line with market expectations. The predicted rental price movements are presented in Appendix C. Our results show that the rental markets for both houses and units will continue to creep up in the foreseeable future. For houses, median rents are forecasted to increase from AUD 550 in 2019 to about AUD 760 per week in 2030. For units, the rental prices are forecasted to increase from AUD 530 in 2019 to AUD 840 per week in 2030, all in nominal prices. Our model shows a faster rental growth for units than houses. This could be due to the demand shift in the housing market as owning becomes unfordable for many people, especially for young people who choose to rent rather than own. As a result, the rental demands in the unit market grow at a faster pace than that of the housing market.

5.4.3. Forecasted Interest Rates

Interest rates (90-day bill rates) are a key variable in the proposed ECM forecasting model. Similar to the rent, we checked what the "predicted" interest rates look like. The predicted interest rate graph is included in Appendix D. Our model forecasts a continually downward interest rate curve that falls to near zero in the future, and virtually there is no difference between rates predicted in the house and unit markets. This forecast is in line with the recent interest rate cuts and the forecast by the Reserve Bank of Australia (RBA) that interest rates could stay low 'for a long, long time' [32]. Therefore, our model is robust in relation to the interest rate check. Note that interest rates, rents, and house prices are endogenously determined in the ECM model, which means that we can only take robustness checks on their predicted values but cannot conduct stress tests related to the increase or decrease in these variables in the future period.

6. Conclusions

Housing market dynamics have shifted from traditional consumption-motivated to investment-driven in many developed countries, including Australia, over past decades. Traditional reduced-form studies of housing market supply and demand are ad hoc without firm guidance from theory and struggle to confront the challenges brought about by the dramatically increasing importance of the housing market for both policymakers and the public in their decision-making. Using sophisticated econometric tools available for modern time-series analysis and forecasting, we explored the housing market dynamics by placing the investment demand at the center guided by the present value investment theory. We found that the principal variables in the housing market dynamics (long-run relationship) are prices, rents, and interest rates. Meanwhile, other influencing variables, such as bank mortgage lending, building construction costs, population growth, dwelling supply, and net migration, affected the market in the short run via rent and interest rates. We further demonstrated that the proposed investment model has a superior forecasting performance compared to alternative models. The results showed that the Sydney housing market is more likely to follow a high-growth scenario in the foreseeable future with an average compounding growth rate of 5% per annum.

As the interest rate is a national variable set by the Reserve Bank of Australia (RBA), an important policy implication from this research is if the State Government wants to stabilize housing prices to confront housing affordability, rent control for residential properties is a sensible way to do it. Alternative tools and policies could include increasing the building supply, restricting migration, and putting in place various kinds of purchase restriction controls. From the perspective of the Central Government, the most effective way to slow the housing market is via macro-prudential tools such as placing a loan-to-value ratio restriction on housing purchases.

Importantly, we need to point out that our forecast results are based on the time-series data up to 2018. The model parameters could be changed when new information is available. One advantage of our ECM forecast is that the model can be automated for updates in an IT platform designed to manage house price forecasts. Of course, the accuracy of our forecast depends on our understanding of the housing market dynamics and econometric tools adopted in this study.

Author Contributions: Conceptualization, S.S.; methodology, S.S.; software, S.S.; validation, S.S. and R.O.; formal analysis, S.S.; investigation, S.S., V.M., X.J.G., S.H., F.R. and R.O.; resources, S.S. and V.M.; data curation, S.S.; writing—original draft preparation, S.S.; writing—review and editing, S.S., V.M., X.J.G., S.H., F.R. and R.O.; project administration, S.S. and V.M.; funding acquisition, V.M., S.S., X.J.G., S.H. and F.R. All authors have read and agreed to the published version of the manuscript.

Funding: NSW Landcom.

Acknowledgments: We thank Yang Zhang for their assistance with this research. We acknowledge Landcom, the NSW Government's land and property development organization, for funding this project.

Conflicts of Interest: The authors declare no conflict of interest.

Appendix A

Table A1. Error Correction Model Estimates for the Period between Q3 2004 and Q4 2013.

	(1)	(2)
	House Price Changes	**Unit Price Changes**
CointEq1	−0.084 ***	−0.074 **
	(0.025)	(0.030)
D(LOG(SYD_HP_MEDIAN(-1)))	0.577 ***	0.674 ***
	(0.090)	(0.105)
D(LOG(SYD_RENT(-1)))	0.104	−0.016
	(0.148)	(0.057)
D(LOG(AUS_INT_90D(-1)))	0.039	−0.008
	(0.033)	(0.021)
Constant	−0.003	0.000
	(0.005)	(0.004)
D(LOG(AUS_INT_10YRF))	−0.017	−0.015
	(0.028)	(0.019)
D(LOG(AUS_LEN_DF))	−0.006	−0.002
	(0.043)	(0.029)
D(LOG(AUS_CONF))	−0.143	−0.073
	(0.3540)	(0.278)
D(LOG(NSW_PPGF))		−0.006
		(0.005)
D(LOG(SYD_SUP_DF))	0.011	−0.000
	(0.012)	(0.009)
D(LOG(SYD_NOMF))	0.072	0.076 **
	(0.045)	(0.033)
ECC_PCF	0.001 *	0.000
	(0.000)	(0.000)
Seasonal dummy	Yes	Yes
Adj. R-squared	0.827	0.806
Sum sq. resids	0.002	0.001
S.E. equation	0.010	0.007
F-statistic	14.650	11.954
Log likelihood	131.030	146.459
Akaike AIC	−6.16	−6.919
Schwarz SC	−5.556	−6.272
Mean dependent	0.008	0.009
S.D. dependent	0.023	0.015
The cointegration equation results		
LOG(SYD_HP_MEDIAN_H(-1))	1.000	1.000
LOG(SYD_RENT_H(-1))	−0.373 (0.202) *	−0.550 (0.079) ***
LOG(AUS_INT_90D(-1))	0.427 (0.102) ***	0.280 (0.042) ***
Constant	−11.588	−10.113

Notes: Descriptions of the explanatory variables are provided in Table 1. Standard errors are shown in (). *, **, and *** denote significance levels at the 10%, 5%, and 1% level, respectively.

Appendix B

Table A2. Forecasting Performance by Alternative AR(1) and ECM Models.

Models	obs.	RMSE	MAE	MAPE	Theil
	Panel A: Houses				
AR(1)	20	84,743	74,407	8.179	0.051
ECM without economic variables	20	91,834	78,256	9.626	0.055
	Panel B: Units				
AR(1)	20	27,524	22,353	3.256	0.021
ECM without economic variables	20	46,490	39,064	6.140	0.036

Note: We first developed a price model based on the standard AR(1) and ECM techniques using the available data between 2004 and 2013; we then used the developed models to forecast future house and unit prices and compare them to the actual prices. The AR(1) model depends only on the lagged value of prices, while the ECM model depends only on the lagged value of prices, rents, and interest rates. No other economic variables are included in the AR(1) or ECM modeling.

Appendix C

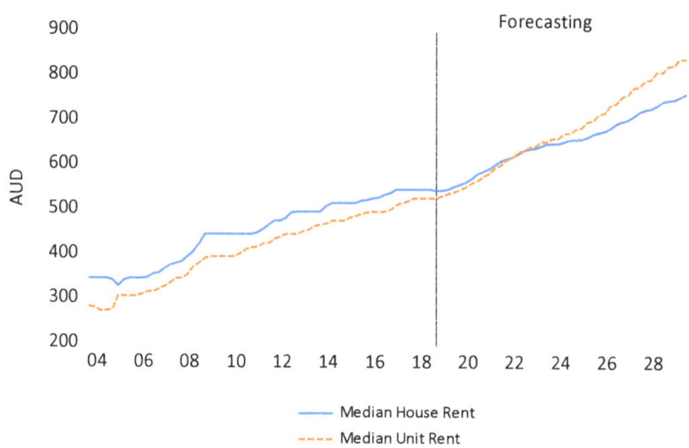

Figure A1. Forecasted Median Rents.

Appendix D

Figure A2. Forecasted Interest Rates.

Notes

1 A time series process with a unit root (a random walk).

2 The statistical area of the Greater Sydney Area is maintained by the Australian Bureau of Statistics. For 35 LGAs and their geographic locations and boundaries, please go to the Australian Bureau of Statistics website at: https://dbr.abs.gov.au/index.html (accessed on 25 July 2021).

3 The authors went through a wide range of data collection in this study. As not all variables are available or statistically significant in our model, we only report the key variables used in the ECM model, as shown in Table 1. Please contact the corresponding author for a complete list of variables collected in this study.

4 Results of Johansen's cointegration test are available on request from the corresponding author.

5 Results are available on request by contacting the corresponding author.

References

1. Malpezzi, S. A simple error correction model of house prices. *J. Hous. Econ.* **1999**, *8*, 27–62. [CrossRef]
2. Francke, M.K.; Vujic, S.; Vos, G.A. *Evaluation of House Price Models Using an ECM Approach: The Case of the Netherlands*; Methodological Paper No. 2009-05; Ortec Finance Research Center: Rotterdam, The Netherlands, 2009.
3. Gattini, L.; Hiebert, P. *Forecasting and Assessing Euro Area House Prices through the Lens of Key Fundamentals*; Working Paper Series No. 1249; European Central Bank: Frankfurt, Germany, 2010.
4. Shi, S.; Jou, J.B.; Tripe, D. Can interest rates really control house prices? Effectiveness and implications for macro prudential policy. *J. Bank. Financ.* **2014**, *47*, 15–28.
5. Shi, S.; Valadkhani, A.; Smyth, R.; Vahid, F. Dating the timeline of house price bubbles in Australian capital cities. *Econ. Rec.* **2016**, *92*, 590–605. [CrossRef]
6. Wheaton, W.; Chervachidze, S.; Nechayev, G. *Error Correction Models of MSA Housing "Supply" Elasticities: Implications for Price Recovery*; Working Paper Series No. 07-15; Center for Real Estate, Massachusetts Institute of Technology: Cambridge, MA, USA, 2014.
7. Yanotti, M.B.; Wright, D. Residential property in Australia: Mismatched investment and rental demand. *Hous. Stud.* **2021**. [CrossRef]
8. Mankiw, N.G.; Weil, D.N. The baby boom, the baby bust, and the housing market. *Reg. Sci. Urban Econ.* **1989**, *19*, 235–258. [CrossRef]
9. Shiller, R. Long-term perspective on the current boom in home prices. *Econ. Voice* **2006**, 3, 1–11. [CrossRef]
10. Engsted, T.; Nielsen, B. Testing for rational bubbles in a coexplosive vector autoregression. *Econ. J.* **2012**, *15*, 226–254. [CrossRef]
11. Engsted, T.; Hviid, S.J.; Pedersen, T.Q. Explosive bubbles in house prices? Evidence from the OECD countries. *J. Int. Financ. Mark. Inst. Money* **2016**, *40*, 14–25. [CrossRef]
12. Gyourko, J.; Sinai, T. The spatial distribution of housing-related ordinary income tax benefits. *Real Estate Econ.* **2003**, *31*, 527–575. [CrossRef]
13. Hendershott, P.H.; Slemrod, J. Taxes and the user cost of capital for owner-occupied housing. *J. Am. Real Estate Urban Econ. Assoc.* **1983**, *10*, 375–393. [CrossRef]
14. Himmelberg, C.; Mayer, C.; Sinai, T. Assessing high house prices: Bubbles, fundamentals, and misperceptions. *J. Econ. Perspect.* **2005**, *19*, 67–92. [CrossRef]
15. Meese, R.; Wallace, N. Testing the present value relation for housing prices: Should I leave my house in San Francisco? *J. Urban Econ.* **1994**, *35*, 245–266. [CrossRef]
16. Poterba, J. Tax subsidies to owner-occupied housing: An asset market approach. *Q. J. Econ.* **1984**, *99*, 729–752. [CrossRef]
17. Case, K.E.; Shiller, R.J. The efficiency of the market for single family homes. *Am. Econ. Rev.* **1989**, *79*, 125–137.
18. Bourassa, S.C.; Hoesli, M.; Oikarinen, E. Measuring house price bubbles. *Real Estate Econ.* **2016**, *47*, 534–563. [CrossRef]
19. Campbell, J.Y.; Shiller, R.J. The dividend-price ratio and expectations of future dividends and discount factors. *Rev. Financ. Stud.* **1988**, *1*, 195–227. [CrossRef]
20. Campbell, J.Y.; Shiller, R.J. Stock prices, earnings, and expected dividends. *J. Financ.* **1988**, *43*, 661–676. [CrossRef]
21. Campbell, J.Y.; Shiller, R.J. Cointegration and tests of present value models. *J. Political Econ.* **1987**, *95*, 1062–1088. [CrossRef]
22. Diba, B.T.; Grossman, H.I. Explosive rational bubbles in stock prices? *Am. Econ. Rev.* **1988**, *78*, 520–530.
23. Gallin, J. The long-run relationship between house prices and rents. *Real Estate Econ.* **2008**, *36*, 635–658. [CrossRef]
24. Brooks, C.; Katsaris, A.; McGough, T.; Tsolacos, S. Testing for bubbles in real estate price cycles. *J. Prop. Res.* **2001**, *18*, 341–346. [CrossRef]
25. Stock, J.H.; Watson, M.W. Evidence on structural instability in macroeconomic time series relations. *J. Bus. Econ. Stat.* **1996**, *14*, 11–30.
26. Elliott, G.; Muller, U.K. Confidence sets for the date of a single break in linear time series regressions. *J. Econ.* **2007**, *141*, 1196–1218. [CrossRef]
27. Elliott, G.; Muller, U.K. Pre and post break parameter inference. *J. Econ.* **2014**, *180*, 141–157. [CrossRef]
28. Pesaran, M.H.; Pick, A.; Timmermann, A. Variable selection, estimation and inference for multi-period forecasting problems. *J. Econ.* **2011**, *164*, 173–187. [CrossRef]

29. Renigier-Biłozor, M.; Walacik, M.; Źróbek, S.; d'Amato, M. Forced sale discount on property market–How to assess it? *Land Use Policy* **2018**, *78*, 104–115. [CrossRef]
30. Berry, M.; Dalton, T. Housing prices and policy dilemmas: A peculiarly Australian problem? *Urban Policy Res.* **2004**, *22*, 69–91. [CrossRef]
31. Shi, S.; Kabir, M.H. Catch animal spirits in auction: Evidence from New Zealand property market. *Real Estate Econ.* **2018**, *46*, 59–84. [CrossRef]
32. Commins, P. Lowe Warns Rates Could Stay Low 'for a Long, Long Time'. The Australian Financial Review. Available online: https://www.afr.com/policy/economy/lowe-warns-rates-could-stay-low-for-a-long-long-time-20190925-p52uun (accessed on 25 September 2019).

land

MDPI

Article

The Heterogeneous Influence of Infrastructure Construction on China's Urban Green and Smart Development—The Threshold Effect of Urban Scale

Lingyan Xu [1,2,*], Dandan Wang [1] and Jianguo Du [1,2]

1 School of Management, Jiangsu University, 301 Xuefu Road, Zhenjiang 212013, China; 2211910022@stmail.ujs.edu.cn (D.W.); djg@ujs.edu.cn (J.D.)
2 Research Center for Green Development and Environmental Governance, Jiangsu University, Zhenjiang 212013, China
* Correspondence: xulingyan333@163.com

Abstract: The construction of green and smart cities is an important approach to enhancing the level of high-quality development and modern governance, in which infrastructure construction is the antecedent condition. From the perspective of green total factor productivity (GTFP), this paper adopts the SBM–GML (Slack-Based Model and Global Malmquist–Luenberger) index to measure the urban green and smart development level (GSDL) considering smart input–output factors. Based on the panel data of China's 223 prefecture-level cities from 2005 to 2018, the dynamic impacts, temporal, and spatial differences of energy, transportation, and telecommunication infrastructure construction on the urban GSDL are discussed, and the threshold effects of urban scale are tested. The following conclusions are drawn: (1) On the whole, energy infrastructure inhibits the urban GSDL, while transportation and telecommunication infrastructures significantly promote it. There are distinct spatial and temporal characteristics among the impacts of these three infrastructures on the urban GSDL, in which the facilitating role of transportation and telecommunication infrastructures are further enhanced during the period of 2013–2018. Furthermore, the impacts of these three infrastructures on the urban GSDL all show "U" shape in terms of non-linearity. (2) Economic development level and industrial structure have significant positive effects on the urban GSDL, whereas human capital only has positive effect in the northeast and southwest regions, and government scale shows no positive impact yet. (3) There is a single threshold for the impact of urban scale on these three infrastructures, among which the impacts of energy and transportation infrastructures on the urban GSDL remain consistent before and after the threshold, while the impact of telecommunication infrastructure on the urban GSDL varies from having no significance to being positive when crossing the threshold. Thus, capital investment for infrastructure construction should be further allocated reasonably, the positive potential of human capital should be fully released, and the urban scale should be appropriately controlled in the future.

Keywords: infrastructure construction; urban green and smart development level; urban scale; threshold effect

Citation: Xu, L.; Wang, D.; Du, J. The Heterogeneous Influence of Infrastructure Construction on China's Urban Green and Smart Development—The Threshold Effect of Urban Scale. *Land* **2021**, *10*, 1015. https://doi.org/10.3390/land10101015

Academic Editors: Elizelle Juanee Cilliers and Sarel Cilliers

Received: 30 August 2021
Accepted: 24 September 2021
Published: 27 September 2021

1. Introduction

1.1. Motivation

Urbanization is an important driving force for economic growth [1], which is closely related to capital and also has great impact on the environment [2]. It is shown that the urbanization rate of China has exceeded 60% by the end of 2020 [3]. Although urbanization has promoted industrialization and economic growth remarkably, it has also resulted in serious deterioration of ecology, frequent extreme weather, and other problems, which have been accumulated over the decades [4]. As such, the traditional extensive urban development mode needs to be changed urgently. In this context, the construction of green and smart cities has been successively distributed in various prefecture-level cities with national policies implemented in China [5,6]. The construction of green and smart cities is based on the sustainable and low-carbon development and utilization of the internet, cloud computing, artificial intelligence, and other smart technologies to improve the level of urban production, social management, and public services [7] and eventually optimize the industrial structure, extensively form green production and lifestyles, and promote the urban economic development [8,9]. Faced with the development opportunities of urban green and smart construction, China's cities actively deploy infrastructure construction related to energy, transportation, telecommunication to provide fundamental support in enhancing low-carbon emission reduction, transportation structure, comprehensive transportation network efficiency, and information resource sharing. As such, the urban GSDL would be enhanced [10,11]. However, redundant infrastructure construction is also accompanied by high energy consumption, and the adoption of infrastructures for commodity trade and production will also generate new pollution [12], which will be a bottleneck of the urban GSDL. Therefore, this paper explores the complex relationship between infrastructure construction and the urban GSDL by taking China as an example.

1.2. Literature Review and Contribution

1.2.1. Measurement of the Urban GSDL

Urban development promotes capital accumulation, which in return improves urban development by investing in infrastructure construction [13], such as Barcelona [14]. Furthermore, industrial development and infrastructure construction respond to the needs of capital expansion [14]. In this frame, the urban development modeled by capitalism has brought some distortion development models, which bring about the result that the land price becomes a determinant factor of the economic and political issues [15]. Continuous capitalist pursuit of profit-seeking has led to new speculative real estate booms, such as new office buildings and high-end residential buildings associated with infrastructure and cultural and entertainment spaces [14,16], which are less related to public needs and urban requirements [17]. As Mora and Camerin noted, urban regeneration and reconstruction projects were carried out for rent demand to achieve the goal of capital accumulation [18], which gradually damaged the architectural and urban development framework [3].

On the contrary, the construction of green and smart cities is not only an effective approach to solve urban problems but also an important decision to realize sustainable development [19,20]. Ding and Wang [21] and Du et al. [22] conducted a comprehensive evaluation index from multidimensional perspectives to evaluate the urban GSDL. Others analyzed the urban GSDL by combining PCA–GRA, entropy power, and cloud model [23]. In addition, some literature adopted GTFP (green total factor productivity) from the perspective of input–output efficiency to analyze the green productivity and innovative efficiency of China's provinces and cities [24,25], and the spatial and temporal disparities between regions [26], which provided a new method for estimating the urban GSDL. Existing research also showed that the quality and efficiency of urban development were not only affected by the degree of green development but also related to urban smart elements. Urban smart elements could also promote the efficiency of urban governance and public services through technological innovation and facilitate resource utilization, thus promoting the GTFP and empowering high-quality development [27,28]. Jiang et al. [7] showed that the construction of smart cities provided vital support for urban green development, which promoted the urban GTFP by accelerating technological innovation. Xia and Xu [29] estimated GTFP with non-parametric methods and found that the GDP growth rate was unrelated to the GTFP, while smart city construction could significantly facilitate the green utilization efficiency of urban land [30]. These conclusions suggest that green and smart factors should be considered when we estimate the quality of urban development.

1.2.2. Relationship between Infrastructure Construction and the Urban GSDL

Infrastructure construction is considered as a prerequisite for urban economy, and the development of urban economy also accelerates infrastructure construction. Historically, urban planning and development has been deeply rooted in the process of infrastructure construction, which provides ontological foundations and operation space for cities [31]. However, infrastructure construction, which was the main way of urban regeneration for capitalist cities in the past, often ignored public needs during the process [16,17]. Since the contradiction between environmental and economic development are becoming more and more prominent, a multi-agent participation mode of social capital was adopted in infrastructure construction under the regulation of government; thus, the public welfare and service nature of infrastructure construction was achieved [32].

Literature focuses more on the relationship between infrastructure construction and urban renewal, including the impacts of infrastructure construction on the urban green development or smart development, while green and smart elements are barely taken as a unity to explore their relationship. Scholars mainly expounded the relationship between infrastructure construction and green development or smart development from the following two aspects.

On the one hand, from the perspective of the driving force for urban development, infrastructure construction is the hardware foundation to accelerate the marketability of resources and elements, which is of benefit to integrating innovators [6], accelerating economic growth [33], and enhancing the resource allocation efficiency [34]. Wei and Chen [35] noted that the transportation infrastructure played an important role in the construction of energy-saving and low-carbon cities, which was closely related to the construction of green and smart development cities [36]. Other research showed that infrastructure construction made the largest contribution to GTFP when compared with the scheme, technology, and other factors [37,38], which had a more significant effect in relatively backward areas [39]. Furthermore, infrastructure construction could not only directly affect the economy and environment but also indirectly affect the urban smart development, living quality, and low-carbon development through technological innovation [40], international trade [41], industrial, and talents agglomeration [42,43]. For example, transportation infrastructure could reduce trade costs and improve the access to the market, which strengthened the driving force of technological innovation and productivity improvement through resource importation and regional spillover effect [44,45], among which economic agglomeration and market accessibility played a mediating role [46]. Additionally, the telecommunication development was conducive to breaking regional market restrictions and reducing coordination costs of enterprises, which was crucial to regional integrated market construction and beneficial to improving production efficiency through scale and intensive economy [47]. Fully equipped telecommunication infrastructure could accelerate the evolution of production structure through talent gathering and optimization of resources, thus improving the scale and network effect of economic development and urban productivity level [48]. Moreover, due to the marketability of production elements, infrastructure construction may also promote or inhibit the development of adjacent areas [49]. For example, the transportation infrastructure would enhance urban accessibility and promote the agglomeration of innovation elements, strengthen knowledge and technology spillover effect, and, accordingly, the innovation structure among regions could be influenced [50,51]. Meanwhile, developed telecommunication infrastructure would facilitate technological innovation and knowledge-based economy, thus forming spatial preference and agglomeration characteristics by influencing the prospect of adjacent areas through spatial spillover effect [52,53].

On the other hand, infrastructure construction requires a large input of energy and other resources, which will cause consumption and pollution. Moreover, excessive infrastructure construction will also destroy environmental carrying capacity, which is harmful to the sustainable development of the economy and society; therefore, the urban GSDL will be restricted. Wang [54] adopted the GML index and threshold regression model to find that traditional energy consumption would have a negative impact on GTFP with both too much and too low degree, which depended on whether the technology level of energy transformation could meet the production requirements or not [55]. Kong et al. [56] also found that the negative effect of energy infrastructure construction on the environment was more prominent, and the large-scale production caused by the elements aggregation was prone to result in resource monopoly, which resulted in the resource allocation distortion and inefficient utilization, and then vertically deepened the negative externality of energy. Besides, by reducing logistics costs and promoting the marketability of elements, transportation infrastructure also accelerated the flow of resources to big cities, resulting in a "siphon effect" and deepening the degree of regional differentiation [57]. Furthermore, the research even demonstrated such phenomena as "short-term effects" were superior to "long-term effects" [58]. Sun et al. [59] noted that urban and rail transportation showed positive effects on reducing air pollution, while the construction of rail transportation had a negative short-term impact on air quality. In addition, the effect of infrastructure construction on GSDL would also be restricted by other factors, such as economic development level, industrial structure, energy structure, and resource endowment [60]. Particularly, the internal difference such as urban scale could also influence urban construction [61]. Jain and Tiwari [62] showed that improving bus and bicycle infrastructures could minimize equivalent CO_2 emissions, while different strategies should be taken based on the urban scale.

In conclusion, the GTFP is an important criterion to evaluate the quality of economic development. The existing research has provided many important ideas and research methods, while green and smart elements are barely integrated into a unified analytical-framework to measure the urban GSDL from the perspective of total factor productivity (TFP), which could not reflect the two-wheel driving force of green and smart for urban sustainable development. Research has paid more attention to the influence of infrastructure construction on regional innovation, economic development, TFP, and GTFP, while few focused on the heterogeneous effects of infrastructure construction on the urban GSDL from the perspective of their linear and nonlinear relationships. Furthermore, the threshold effect of urban characteristics also needs to be deepened. Therefore, based on the panel data of 223 prefecture-level cities from 2005 to 2018, this paper firstly adopts SBM–GML index to evaluate the urban GSDL by considering the green and smart factors based on the GTFP accounting framework. Secondly, the linear and nonlinear effects of infrastructure construction on the urban GSDL are analyzed. Thirdly, heterogeneous effects of infrastructure construction on urban GSDL are explored from regional and temporal perspectives. Finally, the threshold effects of urban scale are explained.

2. Materials and Methods

2.1. Measure of GSDL

2.1.1. Measure Model Construction

The SBM model is an improved DEA (data envelopment analysis) model, which is developed to evaluate the efficiency of decision units with multiple input and output indicators and eliminate the estimation bias caused by the difference in radial and angle selection. Combined with the GML index, the SBM model could better describe the variation of productivity [63]. The form of non-angular and non-radial SBM model containing the unexpected output is as follows (Formula (1)):

$$
\begin{aligned}
&\min\rho = \frac{1-\frac{1}{m}\sum_{i=1}^{m} s_i^-/x_{i0}}{1+\frac{1}{s_1+s_2}\left(\sum_{i=1}^{s_1} s_r^g/y_{r0}^g+\sum_{i=1}^{s_2} s_r^b/z_{r0}^g\right)} \\
&s.t.\ X\lambda + s_i^- = x_k,\ Y^g\lambda - s_r^g = y_0^g,\ Z^b\lambda + s_r^b = z_0^b \\
&\lambda,\ s_i^-,\ s_r^g,\ s_r^b \geq 0
\end{aligned}
\quad (1)
$$

In Formula (1), ρ is the ratio of actual input–output relative to the average narrowing and expansion of technological frontier; m, s_1, s_2 denote the quantity of input, expected and unexpected output respectively, and s^-, s^g, s^b are the corresponding relaxation variables.

The GML index from t to $t+1$ is defined, and the specific Formula (2) is as follows:

$$GML^{t,t+1}(x^t, y^t, b^t, x^{t+1}, y^{t+1}, b^{t+1}) = \frac{1+D^G(x^t,y^t,b^t)}{1+D^G(x^{t+1},y^{t+1},b^{t+1})} \times \frac{1+D^t(x^t,y^t,b^t)}{1+D^{t+1}(x^{t+1},y^{t+1},b^{t+1})} \times \frac{1+D^G(x^t,y^t,b^t)}{1+D^t(x^t,y^t,b^t)} \times \frac{1+D^{t+1}(x^{t+1},y^{t+1},b^{t+1})}{1+D^G(x^{t+1},y^{t+1},b^{t+1})} \quad (2)$$

In Formula (2), $GML^{t,t+1}$ represents the GSDL variations of units in period t to t +1. When $GML^{t,t+1} > 1$, it indicates that GSDL has been improved in the current period when compared with that in previous period. The calculation Formula (3) is as follows:

$$GSDL_{2006} = GSDL_{2005} * GML_{2006} \quad (3)$$

2.1.2. Indicators of GSDL Measurement

According to Du et al. [22] and Lu et al. [64], this paper adopts the SBM–GML index to measure the urban GSDL with the following indicators. The input indicators include the fixed capital stock, labor, energy, and the fiscal expenditure of science, technology, and education. The output indicators are comprised of four aspects including economic, green, smart, and undesired environmental elements, among which regional GDP indicates economic output, international internet users and patent application quantity indicate smart output, the harmless disposal rate of domes-tic garbage, sewage treatment rate and greenery coverage of urban area indicate green output, discharge of industrial wastewater, industrial smoke, and dust emissions and SO2 emissions indicate undesired environmental output. The specific indicators are shown in Table 1:

Table 1. Indicators of GSDL.

Indicator	Variable	Unit	Computation Method
Input indicators	Fixed capital stock	100 million yuan	Perpetual inventory method
	Labor	10 thousand people	The number of urban employees at the end of year
	Electricity consumption	10 thousand kilowatts	Total electricity consumption
	Education and technology expenditure	10 thousand yuan	Financial expenditure on science, technology, and education
Output indicators	Regional GDP	100 million yuan	Regional GDP of the year
	International internet users	10 thousand people	The number of urban international Internet users
	Patent application quantity	Part	The number of urban patent application
	Harmless disposal rate of domestic garbage	%	Percentage of the disposal of harmless garbage
	Sewage treatment rate	%	Percentage of sewage disposed
	Greenery coverage of urban area	%	Greening coverage rate in built-up areas of the city
	Discharge of industrial wastewater	10 thousand tons	Industrial wastewater discharge volume of the city
	Industrial smoke and dust emissions	Tons	Industrial smoke and dust emissions' volume of the city
	Industrial SO_2 emissions	Tons	Industrial SO_2 emissions' volume of the city

In addition, the capital stock is measured by the perpetual inventory method according to Zhang et al. [65]. Regional GDP is represented by the real GDP, which takes 2003 as the base period. The entropy value method with the time variable is adopted to calculate the expected pro-environmental output and unexpected pollution output.

2.2. *Construction of Empirical Model*

The construction of energy, transportation, and telecommunication infrastructure has different functions and adoptions, which may result in differentiated impacts on the urban GSDL. Therefore, this paper first tests the linear and nonlinear effects of energy, transportation, and telecommunication infrastructure construction on the urban GSDL. Then, the regional and temporal heterogeneity effects of infrastructures are further analyzed. Finally, the threshold characteristics of urban scale are explored.

2.2.1. Panel Regression Model

According to the research objectives, the static dynamic panel regression model is first constructed as follow (Formula (4)):

$$GSDL_{it} = \beta_0 + \sum_{n=1}^{3} \alpha_n inf_{nt} + \sum_{i=1}^{4} \lambda_i Control_{it} + \varepsilon_{it} \quad (4)$$

In Formula (4), inf_n (n = 1, 2, 3) represents the infrastructure construction of energy, transportation, and telecommunication, respectively. $Control_{it}$ denotes economic development level, industrial structure, human capital level, and government size, respectively. The sum of α_n and β_1 are indicated as the regression coefficient, and ε_{it} is the random error. Based on Formula (4), the first-order lag term of the explained variable is added to explore the long-term influence between the explanatory variable and the explained variable, and the dynamic panel regression model is obtained as follow (Formula (5)):

$$GSDL_{it} = \beta_0 + \beta_1 GSDL_{t-1} + \sum_{n=1}^{3} \alpha_n inf_{nt} + \sum_{i=1}^{4} \lambda_i Control_{it} + \varepsilon_{it} \quad (5)$$

In Formula (5), $GSDL_{t-1}$ is the first-order lag term of the explained variable ($L.GSDL_t$). In order to explore the nonlinear influence relationship between the core explanatory variable and the explained variable, the square term of the explanatory variable is introduced to Formula (5) to construct a nonlinear panel regression model which is as follow (Formula (6)):

$$GSDL_{it} = \beta_0 + \beta_1 GSDL_{t-1} + \sum_{n=1}^{3} \alpha_n inf_{nt} + \sum_{n=1}^{3} \alpha_n {inf^2}_{nt} + \sum_{i=1}^{4} \lambda_i Control_{it} + \varepsilon_{it} \quad (6)$$

In Formula (6), $infn^2$ (n = 1, 2, 3) represents the square term of energy, transportation, and telecommunication infrastructure construction, respectively.

2.2.2. Threshold Regression Model

Previously, the dynamic panel regression model is constructed to test the influence of core explanatory variables and control variables on the urban GSDL. However, it is unable to describe the structural breakpoint of urban scale. Thus, this paper further explores the threshold effect of urban scale by adopting threshold regression model, which is constructed according to Hansen's panel data regression theory as follow (Formula (7)):

$$GSDL_{it} = u_i + \beta_1 X_{it} \cdot 1(q_{it} \le \lambda) + \beta_2 X_{it} \cdot 1(q_{it} > \lambda) + \varepsilon_{it} \quad (7)$$

In Formula (7), GSDL denotes the urban GSDL, X_{it} is the explanatory variable, q_{it} is the threshold variable, γ is the threshold value to be estimated, and ε_{it} is the random disturbance term. On this basis, a double threshold regression model is constructed as follow (Formula (8)):

$$GSDL_{it} = u_i + \beta_1 X_{it} \cdot 1(q_{it} \le \gamma_1) + \beta_2 X_{it} \cdot 1(\gamma_1 < q_{it} \le \gamma_2) + \beta_3 X_{it} \cdot 1(q_{it} > \gamma_2) \quad (8)$$

In Formula (8), the threshold value meets the requirement of $\gamma_1 < \gamma_2$.

2.3. *Variable Description*

Explained variable: The urban GSDL, which is calculated by SBM–GML.

Explanatory variables: Referring to Yeaple and Golub [66], infrastructures are comprised of energy, transportation, and telecommunication. In this paper, energy infrastructure (lnENER) is measured by the total amount of gas supply, transportation infrastructure (lnTRANS) is expressed by the length of urban highway, and telecommunication infrastructure (lnTELE) is represented by telecommunications revenue.

Threshold variables: Urban scale is the embodiment of regional economic development, which would aggregate resource elements and influence resource utilization efficiency [67]. As such, it is supposed to restrict the impact of infrastructure construction on the urban GSDL. The urban total population at the end of the year is applied to represent the urban scale.

Other control variables: Based on the existing research, this paper adopts economic development level (lnPGDP), industrial structure (STR), human capital level (HC), and government size (GOVER) to control the influence of external factors on the urban GSDL. Specifically, cities with high economic development levels will have greater support for technological innovation and more investment for pro-environmental issues and infrastructure construction. As consequence, the urban GSDL would be enhanced. Furthermore, industrial emissions are the main source of pollution, which directly restrict urban green development; therefore, it is necessary to enhance the urban GSDL by innovation and upgrading the industrial structure [68]. Then, the proportion of the tertiary industry is adopted to evaluate the influence of infrastructure construction on the urban GSDL. Additionally, due to the optimization of human capital elements is conducive to accelerating innovation and technology accumulation [69], this paper adopts the number of college students to represent human capital level. Furthermore, the government scale is measured by the ratio of government fiscal expenditure to GDP.

2.4. Data Source

The data of many urban patent applications before 2005 are missing and their caliber is not consistent; as well, the data in 2019 was incomplete. This paper selected samples from 2005 to 2018 to ensure the credibility of the research. Considering the integrity and continuity of panel data, the samples of cities with more missing data were excluded, and then 223 cities were selected as research samples, among which some missing data were supplemented by the linear interpolation method. As such, the panel data of 223 cities in China from 2005 to 2018 are obtained. The data of variables are mainly from "China Urban Statistics Yearbook", "Statistical Yearbook", and" Science and Technology Yearbooks" from 2006 to 2019 and "Statistical Bulletin" from 2005 to 2018 of provinces and cities.

3. Results

3.1. Linear Regression Analysis

Before the regression analysis, the variance inflation factor test is carried out to avoid estimation bias caused by collinearity. According to the literature, if the VIF of the test is much less than 10, it indicates that there is no colinear problem among variables. In order to overcome the influence of extreme data, all data are truncated by 1% before and after. The regression results are shown in Table 2, where column (1) represents the regression without the control variables, and column (2) to (5) represents the results after the step-by-step addition of control variables, respectively.

According to the results in Table 2, the first-order lag term of the urban GSDL is significantly positive at the 1% level, even adding the control variables, which indicates that the urban GSDL varies dynamically and continuously with self-strengthening and path-dependence effects. While the effect of lnENER on the urban GSDL has fluctuated from positive to negative. The reason is that the traditional energy consumption is still dominant, which would restrict the urban GSDL, even though China's energy consumption structure has been optimized in recent years. The results also show that the positive effect of lnTRANS on the urban GSDL is continuously strengthened and significant at the 1% level, as well as the positive effect of lnTELE is significantly positive at the 5% level. This is mainly due to the role of traffic infrastructure in improving the flow velocity of elements and resources shared among regions. Telecommunication infrastructure can facilitate resource integration and technological innovation, which will inject the power source for urban development.

Table 2. Dynamic linear regression.

	(1)	(2)	(3)	(4)	(5)
	Urban GSDL	Urban GSDL	Urban GSDL	Urban GSDL	Urban GSDL
L.GSDL	0.6251 *** (0.0149)	0.5861 *** (0.0149)	0.5571 *** (0.0155)	0.5561 *** (0.0155)	0.5557 *** (0.0155)
lnENER	0.0533 *** (0.0178)	−0.0693 *** (0.0201)	−0.0830 *** (0.0201)	−0.0832 *** (0.0201)	−0.0838 *** (0.0201)
lnTRANS	0.1088 *** (0.0466)	0.1604 *** (0.0199)	0.1604 *** (0.0495)	0.1663 *** (0.0496)	0.1655 *** (0.0496)
lnTELE	0.0783 *** (0.0145)	0.0232 ** (0.0148)	0.0233 ** (0.0147)	0.0229 *** (0.0147)	0.0219 ** (0.0147)
lnPGDP		0.2409 *** (0.0200)	0.2331 *** (0.0199)	0.2233 *** (0.0206)	0.2191 *** (0.0211)
STR			0.1397 *** (0.0227)	0.1350 *** (0.0228)	0.1345 *** (0.0228)
HC				0.0021 * (0.0012)	0.0022 * (0.0012)
GOVER					0. 0719 (0.0738)
CONS	−5.0399 *** (0.3568)	−2.9332 *** (0.3892)	−2.7627 *** (0.3875)	−2.7434 *** (0.3875)	−2.6855 *** (0.3920)

Note: The values in parentheses are standard deviations. *, **, *** indicate significance at 10%, 5%, and 1% levels, respectively.

From the view of control variables, the lnPGDP and STR can strongly promote the urban GSDL at the level of 1%. This indicates that China's policy of developing economy and adjusting the urban industrial structure is valid, which has effectively ensured the growth of urban GSDL. However, the effect of HC on the urban GSDL is only significant at the level of 10%, which shows limited influence and is not consistent with the expectation. This may be because the human capital level may be inconsistent with the urban GSDL, such as the contradiction between the structure and low equilibrium of talents possibly leading to the dilemma of "excessive competition" or "curse of talents" [70]. Moreover, the impact of GOVER on the urban GSDL is insignificant, indicating that the urban GSDL would be increased by optimizing the structure and utilization efficiency of fiscal expenditure rather than by simply expanding fiscal expenditure.

3.2. Nonlinear Regression Analysis

Based on the linear regression tested above, the explanatory variables of energy, transportation, and telecommunication infrastructures' square term are further adopted into the dynamic regression equation to test the nonlinear impact of infrastructures on the urban GSDL. The results are shown in Table 3.

Columns (1) to (3) represent the nonlinear relationship of energy, transportation, and telecommunication infrastructure construction with urban GSDL, respectively. Table 3 shows that the coefficients of $\ln ENER^2$, $\ln TRANS^2$, and $\ln TELE^2$ are all positive significantly at the 1% level, which indicates that the long-term impact of infrastructure construction on the urban GSDL has a U-shaped nonlinear effect with the trend of first inhibiting and then promoting. The reason for this is that infrastructure construction requires long-term, large invest and resource consumption, which is usually at a low level in the early stage and then reaches the inflection point after a large amount of resource input to achieve scale effect and agglomeration effect. With the improvement of infrastructure construction, network connection among regions, energy utilization, and transport efficiency, communication sharing would be enhanced, and the cost of industrial structure transformation and economic model transformation would be reduced. Eventually, innovation and productivity will be increased, and the urban GSDL will be promoted accordingly.

Table 3. Nonlinear regression.

	(1)	(2)	(3)
	Urban GSDL	**Urban GSDL**	**Urban GSDL**
L.GSDL	0.5709 *** (0.0150)	0.5540 *** (0.0163)	0.5553 *** (0.0155)
lnENER	−1.6591 *** (0.1882)		
$lnENER^2$	0.0589 *** (0.0072)		
lnTRANS		−1.3411 *** (0.3899)	
$lnTRANS^2$		0.0649 *** (0.0220)	
lnTELE			−0.4372 ** (0.1893)
$lnTELE^2$			0.0184 ** (0.0076)
lnPGDP	0.3601 *** (0.0207)	0.3147 *** (0.0203)	0.2556 *** (0.0181)
STR	0.3722 *** (0.0259)	0.3789 *** (0.0257)	0.3849 *** (0.0261)
HC	0.0003 (0.0013)	0.0029 ** (0.0013)	0.0020 (0.0013)
GOVER	0.1335 (0.0873)	0.1007 (0.0881)	0.0875 (0.0889)
CONS	8.4869 *** (1.1943)	4.1780 ** (1.7588)	0.5696 (1.1481)

Note: The values in parentheses are standard deviations. **, *** indicate significance at 5% and 1% levels, respectively.

3.3. Heterogeneity Regression Analysis

3.3.1. Regional Heterogeneity Analysis

Due to the differences in developing speed and quality, the impacts of infrastructure construction on different regions of urban GSDL are heterogeneous. Therefore, this paper refers to the existing classification criteria and divides research samples into six regions to explore the regional differences. The specific results are shown in Table 4.

As shown in Table 4, the relationship between the first-order lag GDSL and the urban GSDL is significantly positive at the 1% level in all regions, which indicates that the urban GSDL variations are consistent in time and path-dependent among regions. The lnENER has no significant effect on the urban GSDL in central, southern, and southwest China while showing obvious inhibitory effects in other regions. The possible reason is that the resource curse phenomenon brought by the resource-based cities are mostly in north China, the large proportion of manufacturing industry in the northeast of China, and the imbalance of energy supply and demand caused by "light abandonment" and "wind abandonment" are mainly in the western region. All these problems exert a negative impact on the improvement of urban GSDL. The lnTRANS significantly improves urban GSDL in east, southwest, and northwest China at the 1% level. By contrast, the lnTELE only promotes urban GSDL in east, central, and southern China at the 5% level. This is mainly because the east and center of China are highly developed regions in China, such as Shanghai, Hangzhou, and Shenzhen, which have comparatively advanced technologies, a high level of innovation, and the strong radiation capacity to form strong motivation for the development of urban GSDL.

Table 4. Regional heterogeneity regression.

	North China	Northeast China	East China	Central South	Southwest China	Northwest China
	Urban GSDL	Urban GSDL	Urban GSDL	Urban GSDL	Urban GSDL	Urban GSDL
L.GSDL	0.5385 *** (0.0485)	0.5081 *** (0.0536)	0.5255 *** (0.0288)	0.5628 *** (0.0296)	0.5153 *** (0.0511)	0.5952 *** (0.0553)
lnENER	−0.1223 * (0.0656)	−0.3265 *** (0.0814)	−0.0877 ** (0.0431)	0.0119 (0.0370)	−0.1007 (0.0634)	−0.1085 ** (0.0468)
lnTRANS	0.1929 (0.2006)	0.1247 (0.2193)	0.2212 *** (0.0768)	0.0878 (0.1026)	0.3315 *** (0.1265)	0.3249 *** (0.1552)
lnTELE	−0.0018 (0.0623)	0.0487 (0.0589)	0.0331 ** (0.0278)	0.0192 ** (0.0221)	0.0604 (0.0447)	0.0605 (0.0481)
lnPGDP	0.2679 *** (0.0900)	0.2350 *** (0.0797)	0.1883 ** (0.0404)	0.1893 *** (0.0370)	0.1090 *** (0.0353)	0.1100 * (0.0654)
STR	−0.0003 (0.0023)	0.2590 *** (0.0773)	0.2475 *** (0.0588)	0.0598 (0.0451)	0.2112 *** (0.1063)	0.1721 ** (0.0767)
HC	−0.0009 (0.0036)	0.0180 ** (0.0070)	0.0022 (0.0025)	0.0038 (0.0021)	0.0119 *** (0.0046)	−0.0004 (0.0024)
GOVER	0.2828 (0.8695)	0.3113 (0.6601)	0.1394 (0.0610)	0.1046 (0.2405)	−0.2219 (0.1697)	0.2303 (0.4910)

Note: The values in parentheses are standard deviations. *, **, *** indicate significance at 10%, 5%, and 1% levels, respectively.

The influence of control variables demonstrates regional differences, as well. The lnPGDP always plays a significant role in promoting the urban GSDL, indicating that the current green economic development mode and smart city construction could promote the improvement of the ecological environment and urban operation efficiency in China. Overall, the coefficients and significance of STR are in line with expectations, which dramatically promote the urban GSDL, indicating that the function of industrial structure adjustment in China is confirmed. The impact of HC on the urban GSDL only passes the test at the 5% significance level in the northeast and southwest China. According to the estimated coefficients, the urban GSDL will increase by 1.8% and 1.2% when HC increases by 1% in northeast and southwest China, respectively. It could be attributed that adjacent regions' human capital encourages technological innovation, which requires more energy input, resulting in more pollution emissions. The "rebound effect" induced by technology offsets the improvement effect of human capital on the urban GSDL [71]. Additionally, the role of GOVER is not significant, which may be related to the structure of fiscal expenditure. The excessive government intervention is not conducive to guiding market elements flow to the productive departments, consequently hindering the innovation of urban green production technology and the construction of smart infrastructure.

3.3.2. Period Heterogeneity Analysis

The construction of low-carbon cities and smart cities launched in 2010 and 2012, respectively, which intensified infrastructure construction consisting of energy, transportation, and telecommunication. Therefore, this paper takes 2012 as the time demarcation point to study the different impacts of infrastructure construction during 2005–2012 and 2013–2018. Specific results are shown in Table 5.

According to the regression results in Table 5, the lnENER significantly inhibits urban GSDL at the level of 1%, while the impact of lnTRANS and lnTELE on the urban GSDL is not significant from 2005 to 2012. In contrast, the negative impact of lnENER on urban GSDL is greatly reduced, while the positive impacts of lnTRANS and lnTELE are dramatically enhanced from 2013 to 2018. Furthermore, the lnTRANS and lnTELE increase by 1%, and the urban GSDL would be enhanced by 26.19% and 6.02%, respectively. The possible reason for this is that the economic crisis in 2008 brought a series of sequelae to China. In order to stimulate economic vitality, the 4 trillion Yuan expansion plan "Ten Major Industrial Revitalization Plan" was launched, which might result in the imbalance of China's economic structure [72]. Therefore, the effective role of infrastructure in promoting capital, trade, and marginal productivity in this period was limited, and the efficiency of the scale economy was low. Afterward, the low carbon cities and smart cities construction policies were launched successively in 2010 and 2012, which further strengthened the construction of green development and enhanced the technical level and investment efficiency. As such, the innovation compensation and scale economy effect of infrastructure construction on urban development were well enhanced, and the allocation efficiency of resource elements was effectively consolidated, which all promoted the development of urban GSDL.

Table 5. Time heterogeneity regression results.

Time	2005–2012 Urban GSDL	2013–2018 Urban GSDL
L.GSDL	0.3281 *** (0.0242)	0.3845 *** (0.0276)
lnENER	−0.1152 *** (0.0340)	−0.0590 (0.0375)
lnTRANS	0.0935 (0.0681)	0.2619 ** (0.1104)
lnTELE	0.0011 (0.0188)	0.0602 ** (0.0275)
lnPGDP	0.0946 ** (0.0371)	0.1867 *** (0.0392)
STR	0.1080 (0.0728)	0.1485 *** (0.0313)
HC	0.0066 *** (0.0021)	0.0042 * (0.0026)
GOVER	0.1776 (0.1453)	−0.0811 (0.1190)
CONS	−0.0023 (0.5975)	−3.8485 *** (1.1300)

Note: The values in parentheses are standard deviations. *, **, *** indicate significance at 10%, 5%, and 1% levels, respectively.

3.4. Threshold Regression Analysis

The threshold test can not only explore the relationship among variables but also depict breakpoint of the relationship. China has huge developing differences among regions, and this urban scale is adopted for further analysis of threshold effect on the urban GSDL, which can reflect urban characteristics to some extent. The threshold regression results are shown in Table 6.

According to the results in Table 6, the p values of the single threshold are all less than 0.1, while the p values of the double threshold are all more than 0.1, indicating that only the single threshold effect is significant. Furthermore, the lnENER and lnTRANS show a significantly negative impact on the urban GSDL both before and after crossing the threshold, while their inhibitory effect weakens after crossing the threshold. This may be due to the expansion of the urban scale, which is usually accompanied by the increase in energy consumption and waste discharge. When the urban scale is smaller than 588, the urban GSDL will decrease by 19.40% if the lnTRANS increases by 1%. While the urban scale is beyond 588, the inhibitory effect of lnTRANS on the urban GSDL would be weakened even still significant. When the urban scale is smaller than 588, the impact of lnTELE on the urban GSDL is not significant; however, the urban GSDL will increase by 5.04% for every 1% increase of lnTELE when the urban scale is greater than 588. This is because large cities are more likely to form low-lying areas where resource elements are prone to be gathered; thus, more employment opportunities would be generated, the matching efficiency of information resources under the digital economy would be improved, and, finally, the urban GSDL would be promoted.

Table 6. Threshold effect test.

	Single Threshold			Double Threshold		
	Urban GSDL	**Urban GSDL**	**Urban GSDL**	**Urban GSDL**	**Urban GSDL**	**Urban GSDL**
H_0 H_1		No threshold Has single threshold			Has single threshold Has double threshold	
Threshold value	588	588	588	198.43 588	198.43 588	198.43 588
F statistics	137.21 ***	136.84 ***	136.69 ***	125.44 ***	125.17 ***	124.01 ***
p value	0.023	0.026	0.036	0.2740	0.2870	0.3800
Th−0	−1.1414 *** (0.0235)	−0.1940 *** (0.0343)	0.0238 (0.0176)	−0.1697 *** (0.0242)	−0.2584 *** (0.0367)	−0.0143 (0.0120)
Th−1	−0.1162 *** (0.0235)	−0.1194 *** (0.0232)	0.049 *** (0.0178)	−0.1375 *** (0.0234)	−0.2018 *** (0.0342)	0.0249 (0.0176)
Th−2				−0.1131 *** (0.0234)	−0.1678 *** (0.0343)	0.0504 *** (0.0178)
Conclusion	reject	reject	reject	accept	accept	accept
lnENER		−0.1361 *** (0.0235)	−0.1360 *** (0.0235)		−0.1378 *** (0.0234)	−0.1364 *** (0.0234)
lnTRANS	−0.1852 *** (0.0342)		−0.1851 *** (0.0342)	−0.2015 *** (0.0343)		−0.2113 *** (0.0347)
lnTELE	0.0299 * (0.0176)	0.0296 * (0.0176)		0.0250 (0.0175)	0.0249 (0.0175)	
lnPGDP	0.3783 *** (0.0241)	0.3800 *** (0.0241)	0.3794 *** (0.0241)	0.3858 *** (0.0241)	0.3873 *** (0.0241)	0.3857 *** (0.0241)
STR	0.3927 *** (0.0257)	0.3929 *** (0.0257)	0.3928 *** (0.0257)	0.3880 *** (0.0256)	0.3886 *** (0.0256)	0.3877 *** (0.0256)
HC	0.0026 ** (0.0013)	0.0026 ** (0.0013)	0.0026 ** (0.0013)	0.0026 ** (0.0013)	0.0025 * (0.0013)	0.0030 ** (0.0013)
GOVER	0.1120 (0.0875)	0.1107 (0.0875)	0.1125 (0.0875)	0.1116 (0.0871)	0.1102 (0.0872)	0.1168 (0.0873)

Note: The values in parentheses are standard deviations. *, **, *** indicate significance at 10%, 5%, and 1% levels, respectively.

4. Discussion

Against the background of green and smart city construction in China, it is urgent for cities to seize the opportunity and break through the shackles of the black economic development model to enhance their urban GSDL and social governance quality. Therefore, cities actively deploy energy, transportation, telecommunication, and other infrastructure construction to provide circulation conditions and sharing platforms for the urban GSDL. Based on this background, this paper develops the GTFP by adopting smart input–output factors to evaluate the urban GSDL of 223 samples in China from 2005 to 2018 and explores the linear and nonlinear effects of energy, transportation, and telecommunication infrastructure construction on the urban GSDL. The heterogeneous effects of spatial and temporal are also analyzed, and the threshold characteristics are identified from the perspective of urban scale.

Firstly, in the part of empirical analysis, this paper considers the linear and nonlinear relationships of infrastructures and adds the first-order lag term of explained variable and the square term of explanatory variables to make the research results more consistent with reality. The results show that the lnENER inhibits the urban GSDL in general, due to the incomplete transformation of energy consumption structure and the insufficient ability of energy industry technology [59]. The lnTRANS and lnTELE have significantly positive effect on the urban GSDL, which is attributed to the improvement of elements circulation and expansion of resource sharing brought by transportation infrastructure construction, as well as the technical innovation and information sharing promoted by telecommunication [73]. Secondly, this paper verifies the heterogeneous effects of infrastructure construction on the urban GSDL from spatial and temporal characteristics, which could further the existing research and provide policy reference for the urban government. Additionally, as the outbreak of COVID−19 becomes the biggest challenge of this century by far, a safer and disaster-resilient public transport is required, which can also meet the needs of private vehicle-owning individuals after the unprecedented disease [74]. In light of the background and these requirements, infrastructures are supposed to be constructed appropriately advanced. Thirdly, the threshold regression effect provides a new perspective for the government to take urban scale and other urban characteristics into consideration during the construction of green and smart cities. Urbanization has accelerated the movement of people to cities, with nearly half the world's population now living in urban settlements. However, rapid urban growth is mostly accompanied by environmental degradation and traffic congestion, which outstrip urban service capacity [75]. Thus, the urban scale should be reasonably controlled to maximize the positive effect of infrastructure construction.

In the future, how to effectively solve the endogenous problem of the empirical model and the appropriate instrumental variables need to be further deepened. Additionally, the influence mechanism of the spatial spillover effect is also conducive to further expanding the depth of research. Finally, it is also meaningful to consider the long-term impact of civilian infrastructure on urban development, such as medical care, culture, and education.

5. Conclusions

Based on the above elaboration, this paper evaluated the urban GSDL of 223 cities in China from 2005 to 2018. Then, the dynamic impacts and temporal and spatial differences of energy, transportation, and telecommunication infrastructure construction on the urban GSDL were discussed, and the threshold effects of urban scale were tested. The main research conclusions are as follows:

(1) Transportation and telecommunication infrastructures play significant roles in promoting the urban GSDL, and the effects are further strengthened during 2013–2018, while energy infrastructure shows an insignificant effect on the urban GSDL. From the perspective of nonlinear relationships, the impact of infrastructures on the urban GSDL shows a U shape.

(2) There are regional differences in the influence of the control variables on the urban GSDL. Among them, the economic development level and industrial structure promote the urban GSDL in general, while the positive relationship between the human capital level and the urban GSDL is only significant in the northeast and southwest regions. Additionally, the government scale shows an insignificant positive impact on the urban GSDL.

(3) From the perspective of the threshold effect, there is only a single threshold effect of urban scale on the infrastructures. The impacts of energy and transportation infrastructures on the urban GSDL remain consistent before and after the threshold, while the impact of telecommunication infrastructure on the urban GSDL varies from having no significance to being positive when crossing the threshold.

In view of the conclusions, this paper puts forward the following policy recommendations. First, reasonable allocation of capital investment in infrastructure construction is crucial to achieving the dislocation of regional development. Different measures should be taken according to urban conditions to promote the ordered and high-quality quantification of infrastructure construction and form the integrated development power through the construction of energy, transportation, and telecommunication infrastructure. Second, it is obligatory to fully unleash the positive potential of elements and strengthen coordination among regions. On the one hand, industrial structure adjustment, human capital level, and urban scale should be integrated for the urban GDSL. On the other hand, the administrative barriers among regions should be broken, and those backward regions such as the central and western regions and northeast China should be encouraged to cooperate with the eastern regions through entrusted management and investment cooperation to realize complementary advantages and mutual benefits. Finally, the urban scale should be reasonably controlled, and cities can learn from the superblock model of Barcelona, which reclaims public space for people, reduces motorized transport, and promotes sustainable mobility and active lifestyles, and consequently achieve green development and mitigate the effects of climate change [76]. Governments, enterprises, and scientists should make more efforts to break through the technical difficulties of adopting renewable energy and releasing the side-effects of energy demand caused by urban population and spatial expansion. Moderate investment is also required to ensure the sustainability of new infrastructure investment and thus to avoid the low efficiency of financial investment caused by large-scale government investment.

Author Contributions: Conceptualization, L.X. and J.D.; methodology, L.X. and D.W.; software, D.W.; validation, L.X. and J.D.; formal analysis, L.X. and D.W.; data checking, D.W.; writing—original draft preparation, D.W.; writing—review and editing, L.X. and D.W.; visualization, D.W.; supervision, L.X. and J.D.; funding acquisition, L.X. All authors have read and agreed to the published version of the manuscript.

Funding: This research is funded by the National Nature Science Foundation of China (72174076, 71704066, 72174054, 71704068, and 91846203), National Social Science Foundation of China (20BGL191),Special research project of think-tanks in Grand Canal Cultural Belt Construction Research Institute (DYH21YB08), Youth Talents Cultivation Program of Jiangsu University (411160001), 2021 Jiangsu Graduate Scientific Research Innovation Plan project (KYCX21_3316). The APC was funded by the National Nature Science Foundation of China (72174076).

Institutional Review Board Statement: Not applicable.

Informed Consent Statement: Not applicable.

Data Availability Statement: The data used and/or analyzed during the current study are available from the corresponding author on reasonable request.

Acknowledgments: The authors would like to thank the anonymous reviewers for their valuable comments.

Conflicts of Interest: The authors declare no conflict of interest.

References

1. Dong, L.-Y.; Shang, J.; Ali, R.; Rehman, R.-U. The coupling coordinated relationship between new-type urbanization, eco-environment and its driving mechanism: A case of Guanzhong, China. *Front. Environ. Sci.* **2021**, *9*, 128. [CrossRef]
2. McNeill, D. Mapping the European urban left: The Barcelona experience. *Antipode* **2003**, *35*, 74–94. [CrossRef]
3. He, Y. Remodeling and regeneration: Evolution logic, dynamic mechanism and action framework of urban renewal. *Mod. Econ. Discuss.* **2021**, *6*, 94–100.
4. Han, J. Can urban sprawl be the cause of environmental deterioration? Based on the provincial panel data in China. *Environ. Res.* **2020**, *189*, 109954. [CrossRef] [PubMed]
5. Xin, B.; Qu, Y. Effects of smart city policies on green total factor productivity: Evidence from a quasi-natural experiment in China. *Int. J. Environ. Res. Public Health* **2019**, *16*, 2396. [CrossRef] [PubMed]
6. Qian, Y.; Liu, J.; Cheng, Z.; Forrest, J.-Y.-L. Does the smart city policy promote the green growth of the urban economy? Evidence from China. *Environ. Sci. Pollut. Res.* **2021**, *163*, 1–15.
7. Jiang, H.; Jiang, P.; Wang, D.; Wu, J. Can smart city construction facilitate green total factor productivity? A quasi-natural experiment based on China's pilot smart city. *Sustain. Cities Soc.* **2021**, *69*, 102809. [CrossRef]
8. Li, X.; Fong, P.; Dai, S. Towards sustainable smart cities: An empirical comparative assessment and development pattern optimization in China. *J. Clean. Prod.* **2019**, *215*, 730–743. [CrossRef]

9. Shamsuzzoha, A.; Niemi, J.; Piya, S.; Rutledge, K. Smart city for sustainable environment: A comparison of participatory strategies from Helsinki, Singapore and London. *Cities* **2021**, *114*, 103194. [CrossRef]
10. Nondo, C. Is there a relationship between information and communication technologies infrastructure, electricity consumption and total factor productivity? *Int. J. Energy Econ. Policy* **2018**, *8*, 207–218.
11. Zhang, S.-D.; Yan, Y.; Li, B. Urban infrastructure, talent gathering and innovation. *Soft Sci.* **2020**, *12*, 1–12.
12. Wu, Z.-D.; Ye, Q.-L. Water pollution loads and shifting within China's inter-province trade. *J. Clean. Prod.* **2020**, *259*, 120879. [CrossRef]
13. Pacione, M. Development pressure and the production of the built environment in the urban fringe. *Scott. Geogr. Mag.* **2008**, *107*, 162–169. [CrossRef]
14. Camerin, F. From "Ribera Plan" to "Diagonal Mar", passing through 1992 "Vila Olímpica". How urban renewal took place as urban regeneration in Poblenou district (Barcelona). *Land Use Policy* **2019**, *89*, 104226. [CrossRef]
15. Savini, F.; Aalbers, M.-B. The de-contextualisation of land use planning through financialisation: Urban redevelopment in Milan. *Eur. Urban Reg. Stud.* **2016**, *23*, 878–894. [CrossRef]
16. Katz, C. Vagabond capitalism and the necessity of social reproduction. *Antipode* **2001**, 33, 709–728. [CrossRef]
17. Sklair, L. *The Icon Project: Architecture, Cities, and Capitalist Globalization*; Oxford University Press: New York, NY, USA, 2017; pp. 26–49.
18. Mora, A.-Á.; Camerin, F. La herencia del "urban renewal" en los procesos actuales de regeneración urbana: El recorrido renovación-regeneración a debate. *Ciudad Y Territ. Estud. Territ.* **2019**, *51*, 5–26.
19. Chen, J. Innovation ecosystem for green smart city building in China. *Front. Eng. Manag.* **2015**, *2*, 325–330. [CrossRef]
20. Albino, V.; Berardi, U.; Dangelico, R.-M. Smart cities: Definitions, dimensions, performance, and initiatives. *J. Urban Technol.* **2015**, *22*, 3–21. [CrossRef]
21. Ding, H.-F.; Wang, Y.-H. The evaluation and analysis of the Chinese urban economic intelligence index. *Stat. Decis. Mak.* **2016**, *13*, 16–20.
22. Du, J.-G.; Wang, Y.; Zhao, A.-W. Study on the effect of smart city construction on urban development and green development. *Soft Sci.* **2020**, *34*, 59–64.
23. Chen, L.; Zhang, H.-X. Evaluation of green smart city in China based on enropy weight-cloud model. *J. Syst. Simul.* **2019**, *31*, 136–144.
24. Zhang, J.-S.; Tan, W. Study on the green total factor productivity in main cities of China. *Zb. Rad. Ekon. Fak. Rijeci Cas. Za Ekon. Teor. Praksu* **2016**, *34*, 215–234.
25. Liu, G.; Wang, B.; Zhang, N. A coin has two sides: Which one is driving China's green TFP growth? *Econ. Syst.* **2016**, *40*, 481–498. [CrossRef]
26. Huang, X.-L.; Han, X.-F.; Ge, P.-F. Spatial temporal evolution and influence mechanism of green total factor productivity in "Belt and Road" countries. *Econ. Manag.* **2017**, *39*, 6–19.
27. Mitra, A.; Sharma, C. Infrastructure, information & communication technology and firms' productive performance of the Indian manufacturing. *J. Policy Modeling* **2016**, *38*, 353–371.
28. Ji, X.; Yao, Y.-X.; Long, X.-L. What causes pm2.5 pollution? Cross-economy empirical analysis from socioeconomic perspective. *Energy Policy* **2018**, *119*, 458–472. [CrossRef]
29. Xia, F.; Xu, J.-T. Green total factor productivity: A re-examination of quality of growth for provinces in China. *China Econ. Rev.* **2020**, *62*, 101454. [CrossRef]
30. Wang, A.; Lin, W.; Liu, B.; Wang, H.; Xu, H. Does smart city construction improve the green utilization efficiency of urban land? *Land* **2021**, *10*, 657. [CrossRef]
31. Neuman, M.; Smith, S. City planning and infrastructure: Once and future partners. *J. Plan. Hist.* **2010**, *9*, 21–42. [CrossRef]
32. Schomaker, R.-M. Public-private governance regimes in the global sphere. *Public Organ. Rev.* **2017**, *17*, 121–138. [CrossRef]
33. Xu, Z.; Das, D.-K.; Guo, W.; Wei, W.-D. Does power grid infrastructure stimulate regional economic growth? *Energy Policy* **2021**, *155*, 112296. [CrossRef]
34. Yang, K.-Z. New logic of the coordinated development of Beijing, Tianjin and Hebei: Local quality-driven development. *Econ. Manag.* **2019**, *33*, 1–3.
35. Wei, T.; Chen, S.-Q. Dynamic energy and carbon footprints of urban transportation infrastructures: Differentiating between existing and newly-built assets. *Appl. Energy* **2020**, *277*, 115554. [CrossRef]
36. Zhang, M.; Liu, X.-X.; Ding, Y.-T. Assessing the influence of urban transportation infrastructure construction on haze pollution in China: A case study of Beijing-Tianjin-Hebei region. *Environ. Impact Assess. Rev.* **2021**, *87*, 106547. [CrossRef]
37. Bresson, G.; Etienne, J.-M.; Mohnen, P. How important is innovation? A Bayesian factor-augmented productivity model based on panel data. *Macroecon. Dyn.* **2016**, *20*, 1987–2009. [CrossRef]
38. Piselli, P.; Bronzini, R. Determinants of long-run regional prouctivity: The role of R&D, human capital and public infrastructure. *Reg. Sci. Urban Econ.* **2009**, *39*, 187–199.
39. Kodongo, O.; Ojah, K. Does infrastructure really explain economic growth in Sub-Saharan Africa? *Rev. Dev. Financ.* **2016**, *6*, 105–125. [CrossRef]
40. Agenor, P.-R.; Alpaslan, B. Infrastructure and industrial development with endogenous skill acquisition. *Bull. Econ. Res.* **2014**, *70*, 313–334. [CrossRef]
41. Wang, R.-Z.; Ou, G.-L.; Wang, Q.-P. Can trade liberalization increase total-factor productivity? Based on regulatory effects of traffic infrastructure. *J. Beijing Jiao Tong Univ. (Soc. Sci. Ed.)* **2019**, *18*, 38–47.

42. Huang, Q.-H.; Shi, P.-H.; Hu, J.-F. Industrial agglomeration and high-quality economic development: Examples of 107 prefecture-level cities along the Yangtze River Economic Belt. *Reform* **2020**, *1*, 87–99.
43. Hulten, C.-R.; Bennathan, E.; Srinivasan, S. Infrastructure, externalities, and economic development: A study of the Indian manufacturing industry. *World Bank Econ. Rev.* **2006**, *20*, 291–308. [CrossRef]
44. Sun, X.-H.; Liu, X.-L.; Xu, S. The agglomeration effect of transportation infrastructure and service industry—Comes from the multi-layer linear analysis at the provincial and municipal levels. *Adm. Rev.* **2017**, *29*, 214–224.
45. Abe, K.; Wilson, J.-S. Investing in port infrastructure to lower trade costs in East Asia. *East Asian Econ. Rev.* **2011**, *15*, 3–32.
46. Cao, Y.-Q.; Yang, Y.-L.; Xiang, H. Study on the impact of transportation infrastructure on total factor productivity in the service industry—Based on productive capital stock data. *Explor. Econ. Probl.* **2021**, *4*, 37–50.
47. Zhao, W.; Deng, F.-H.; Huo, W.-D. Trade effect of internet infrastructure in countries along the "Belt and Road"—Mediation effect analysis based on trade cost and total factor productivity. *J. Chongqing Univ. (Soc. Sci. Ed.)* **2020**, *26*, 19–33.
48. Pradhan, R.-P.; Bagchi, T.-P. Effect of transportation infrastructure on economic growth in India: The VECM approach. *Res. Transp. Econ.* **2013**, *38*, 139–148. [CrossRef]
49. Wang, C.; Lim, M.-K.; Zhang, X.-Y.; Zhao, L.-F.; Lee, P.-T.-W. Railway and road infrastructure in the belt and road initiative countries: Estimating the impact of transport infrastructure on economic growth. *Transp. Res. Part A Policy Pract.* **2020**, *134*, 288–307. [CrossRef]
50. Salas-Olmedo, M.-H.; García, P.; Gutiérrez, J. Accessibility and transport infrastructure improvement assessment: The role of borders and multilateral resistance. *Transp. Res. Part A* **2015**, *82*, 110–129. [CrossRef]
51. Yang, S.-Y.; Li, Z. Influence of the opening of high-speed railway on the regional innovation pattern and its mechanism of action. *South. Econ.* **2020**, *5*, 49–64.
52. Liu, J.-G.; Zhang, W.-Z. Spatial spillover association effect of regional total factor productivity in China. *Geogr. Sci.* **2014**, *34*, 522–530.
53. Keller, W. Geographic localization of international technology diffusion. *Am. Econ. Rev.* **2002**, *92*, 120–142. [CrossRef]
54. Wang, Y.-L. Research on the relationship between green energy use, carbon emissions and economic growth in Henan province. *Front. Energy Res.* **2021**, *9*, 701551. [CrossRef]
55. Wang, M.; Feng, C. Decoupling economic growth from carbon dioxide emissions in China's metal industrial sectors: A technological and efficiency perspective. *Sci. Total Environ.* **2019**, *691*, 1173–1181. [CrossRef] [PubMed]
56. Kong, J.-J.; Zhang, C.; Han, C.-F. Research on the interactive growth strategies of infrastructure systems and natural ecology. *China's Popul. Resour. Environ.* **2018**, *28*, 44–53.
57. Puga, D. European regional policies in light of recent location theories. *J. Econ. Geogr.* **2002**, *2*, 113–150. [CrossRef]
58. Xie, J. Infrastructure construction and China regional total factor productivity—Is based on the spatial measurement analysis of 285 prefecture-level cities. *Sci. Decis.-Mak.* **2018**, *4*, 71–94.
59. Sun, C.-W.; Zhang, W.-Y.; Luo, Y.; Xu, Y.-H. The improvement and substitution effect of transportation infrastructure on air quality: An empirical evidence from China's rail transit construction. *Energy Policy* **2019**, *129*, 949–957. [CrossRef]
60. Liu, H.-J.; Yang, Q. Spatial difference and impact factors of TFP growth in China under resource environment constraints. *Manag. Sci.* **2014**, *27*, 133–144.
61. Kumo, W.-L. Infrastructure investment and economic growth in South Africa: A granger causality analysis. *Afr. Dev. Bank Group Work. Pap. Ser.* **2012**, *160*, 505–532.
62. Jain, D.; Tiwari, G. How the present would have looked like? Impact of non-motorized transport and public transport infrastructure on travel behavior, energy consumption and CO_2 emissions—Delhi, Pune and Patna. *Sustain. Cities Soc.* **2016**, *22*, 1–10. [CrossRef]
63. Yang, X.; Li, X.-P.; Zhong, C.-P. Study on the evolution trend and influencing factors of industrial biased technology progress in China. *Quant. Econ. Tech. Econ. Res.* **2019**, *36*, 101–119.
64. Lu, P.; Liu, J.-H.; Wang, Y.-X.; Ruan, L. Can industrial agglomeration improve regional green total factor productivity in China? An empirical analysis based on spatial econometric. *Growth Chang.* **2021**, *52*, 1011–1039. [CrossRef]
65. Zhang, J.; Wu, G.-Y.; Zhang, J.-P. Estimation of Chinese inter-provincial material capital stock: 1952–2000. *Econ. Res.* **2004**, *10*, 35–44.
66. Yeaple, S.-R.; Golub, S.-S. International productivity differences, infrastructure, and comparative advantage. *Rev. Int. Econ.* **2007**, *15*, 223–242. [CrossRef]
67. Teng, Z.-W. Spatial separation and drivers of green total factor productivity in China's service industry. *Quant. Res.* **2020**, *37*, 23–41.
68. Zhang, W.-D.; Ding, H.; Shi, D.-Q. Influence of smart city construction on total factor productivity—Based on quasi-natural experiments. *Tech. Econ.* **2018**, *37*, 107–114.
69. Florida, R. The rise of the creative class. *Wash. Mon.* **2002**, *35*, 593–596.
70. Yin, Q.; Li, Z.-W. How innovation overflow affects inter-provincial environmental total factor productivity—Based on DEA-ESDA method. *Sci. Technol. Prog. Countermeas.* **2019**, *36*, 45–54.
71. Zhang, W.; Hu, Y. Effect of innovative human capital on green total factor productivity in Yangtze River Delta—Empirical analysis of space Dubin model. *China Popul. Resour. Environ.* **2020**, *30*, 106–120.
72. Fang, F.-Q.; Tian, G.; Xiao, H. Influence of infrastructure on China economic growth and mechanism research on expansion–Based on barro growth model. *Econ. Theory Econ. Manag.* **2020**, *12*, 13–27.

73. Zhou, W.-W.; Li, X.-P.; Li, J. Space overflow effect of infrastructure construction on whole factor productivity—Based on 271 prefecture-level panels under the background of "One Belt and One Road". *Econ. Probl. Explor.* **2020**, *6*, 64–76.
74. Thombre, A.; Agarwal, A. A paradigm shift in urban mobility: Policy insights from travel before and after COVID-19 to seize the opportunity. *Transp. Policy* **2021**, *110*, 335–353. [CrossRef]
75. Moore, M.; Gould, P.; Keary, B.-S. Global urbanization and impact on health. *Int. J. Hyg. Environ. Health* **2003**, *206*, 269–278. [CrossRef] [PubMed]
76. Mueller, N.; Rojas-Rueda, D.; Khreis, H.; Cirach, M.; Andrés, D.; Ballester, J.; Bartoll, X.; Daher, C.; Deluca, A.; Echave, C.; et al. Changing the urban design of cities for health: The superblock model. *Environ. Int.* **2019**, *134*, 105132. [CrossRef] [PubMed]

land

Article

Elevating the Value of Urban Location: A Consumer Preference-Based Approach to Valuing Local Amenity Provision

Shanaka Herath

Faculty of Design, Architecture & Building, University of Technology Sydney, Ultimo, NSW 2007, Australia; Shanaka.Herath@uts.edu.au; Tel.: +61-(02)-9514-8077

Abstract: Estimating the non-market monetary values of urban amenities has become commonplace in urban planning research, particularly following Rosen's seminal article on hedonic theory in 1974. As a revealed preference method, the hedonic approach decouples the market price of a house into price components that are attributable to housing characteristics. Despite the potential contribution of this theory in a planning context, three main limitations exist in the conventional applications: (1) variable measurement issues, (2) model misspecification, and (3) the problematic common use of global regression. These flaws problematically skew our understanding of the urban structure and spatial distribution of amenities, leading to misinformed policy interventions and poor amenity planning decisions. In this article, we propose a coherent conceptual framework that addresses measurement, specification, and scale challenges to generate consistent economic estimates of local amenities. Finally, we argue that, by paying greater attention to the spatial equity of amenity values, governments can provide greater equality of opportunities in cities.

Keywords: urban amenity; house price; hedonic theory; urban planning; local regression

Citation: Herath, S. Elevating the Value of Urban Location: A Consumer Preference-Based Approach to Valuing Local Amenity Provision. *Land* **2021**, *10*, 1226. https://doi.org/10.3390/land10111226

Academic Editors: Sarel Cilliers and Elizelle Juanee Cilliers

Received: 16 October 2021
Accepted: 6 November 2021
Published: 11 November 2021

1. Introduction

The world is rapidly becoming urbanized, and it is forecast that by 2050, 68% of people on earth will live in towns and metropolitan areas [1]. In both the developing and developed world, city growth is putting unprecedented pressure on governments already struggling to accommodate the demand for housing and to service residents with essential amenities. These major challenges of urban expansion face all nations. In honoring their spatial planning responsibilities, governments must continuously monitor the adequacy of urban amenities, particularly in rapidly evolving metropolitan areas. The tracking of amenity availability and amenity values can help planners identify the localities in need of particular facilities and neighborhoods vulnerable to over-development.

The typical response to planning challenges has been to anticipate and plan for future population growth. Thus, current planning methods primarily involve demographic analysis and forecasting. This approach of planning for the future population takes account of the expected future demand; however, it overlooks urban amenities in local areas and the level of demand they attract. In contrast, any reliable information about the current demand for amenities would be a good indicator of residents' preferences and the value attributable to the amenities in a particular city.

To address the acutely inequitable patterns of urban development [2], policymakers and practitioners need sophisticated tools to measure and monitor existing as well as emerging spatial patterns of demand for housing and access to amenities. Such tools may greatly assist our understanding of the impact of policy, particularly at the local level. Typically, the provision of amenities and services transpires at the state or council level but with little systematic attention to the local impact of those services and amenities [3]. Thus, without appropriate tools and monitoring, governments may fail to ensure that adequate levels of amenities are available to residents in different parts of a city.

This article puts forward a novel framework to assess the values that residents are willing to ascribe to (and hence pay for) amenities in their local areas. By establishing the theoretical interconnections between house prices, the amenity preferences of residents, and urban planning, and building on innovative econometric techniques, we develop a framework consisting of new analytical models capable of reliably identifying the values of high-demand amenities in cities and, therefore, can serve as planning tools. While addressing a significant theoretical and empirical gap, the proposed framework could also inform planning policy by improving the capacity to identify localities well served with amenities and services, thus providing a basis for policy action to enhance spatial equality in cities. This article thereby contributes to a long tradition of planning scholarship, arguing that developing new knowledge about the quality of life is critical for effective urban planning [4].

This research sets out to advance the approaches to understanding the local demand for and the values of amenities in urban locations. In terms of hedonic theory, house prices embed the *values* that residents assign to nearby amenities. In this research, we construct the analytical framework on the premise that these values can be calibrated through modeling and fed into planning decisions on the optimal location of amenities for the local population. Dissecting amenity values in this way constitutes a *consumer preference-based approach* to urban amenity provision. The framework can shed light on the 'locational value' of specific localities by analyzing the local differences of amenities through the lens of house prices.

Specifically, the framework enables research questions such as the following to be addressed: 'What are the amenities most preferred by residents in different local areas in the city?' 'Do transport nodes closer to homes have a positive or a negative impact?' 'How does the positive effect of better schools vary geographically?' 'Is there a clear positive impact of proximity to urban parks?' As the model incorporates the structural attributes of houses and the characteristics of the local neighborhood, it also enables preferences for specific structural and neighborhood characteristics to be analyzed.

Section 2 of this article explores the theoretical interconnections between house prices, the amenity preferences of residents and urban planning. Section 3 discusses the development of a coherent analytical framework that addresses measurement, specification, and scale challenges. The article concludes in Section 4 with a review of the main findings and policy implications.

2. Theoretical Interconnections between House Prices, Amenity Preferences, and Urban Planning

2.1. House Prices Reflect Amenity Preferences

Theories in urban economics, in particular *urban amenities theory*, provide a basis for understanding the drivers of housing demand. Urban amenities theory posits that land values predominantly explain house prices, and proximity to amenities, in turn, influences land values. Amenity types shown to influence local house prices have included schools [5], public parks and open spaces [6], transit access [7], scenic view [8], air quality [9], and crime rates [10], among others. The urban economics literature also states that distance from the city center influences house prices because city centers naturally offer a generous bundle of economic opportunities, urban amenities, and other attractions [11,12].

The hedonic price method (HPM) is the central analytical tool typically employed in empirical tests of the above theories. As an urban economics approach to evaluating amenity values in cities, it recognizes that the value of a complex commodity (housing in this context) is the sum of implicit values of its utility-bearing components, and the value (or price) reveals the utility generated by the consumption of a house and the associated bundle of amenities. The term 'implicit values' signifies the prices indirectly revealed through the estimates of consumers' willingness to pay for specific non-market characteristics of houses. The HPM thus predicts that the overall value of a house, given its structural, locational, and neighborhood characteristics, is determined by the consumer's willingness to pay.

Rosen [13] illustrated the theoretical underpinnings of the HPM by mapping out the way hedonic prices represent the joint envelope of bids (demand) and offers (supply). Housing is a multidimensional commodity, a 'differentiated product' in Rosen's terminology, and each household's bid function for a particular housing characteristic is determined by the utility generated through the consumption of some quantity of the characteristic, and other variables that influence tastes and preferences, given a budget constraint (utility optimization). On the other hand, each producer's offer function represents the quantities produced of each housing characteristic, subject to the costs associated with the production process (profit maximization). The equilibrium price and quantity are determined based on the bid price, quantity demanded, offer price, and quantity supplied.

The theoretical model Rosen specified has two stages. The first stage involves estimating the hedonic equation to generate implicit prices by regressing the price of a house according to its characteristics. The second stage evaluates the empirically derived implicit price of each characteristic against a particular bundle of characteristics. Because the hedonic price function is jointly derived from both the supply and the demand sides, Rosen argued that the entire set of implied prices guides the location decisions of both consumers and producers in a 'characteristics space.' Both the implicit prices and the demand and supply parameters trace an inverse demand curve, that is, the marginal willingness to pay function. In equilibrium, the household bid function is tangent to the hedonic price function [14]. Therefore, the vast majority of hedonic applications employ the standalone 'first stage,' or 'reduced form,' hedonic equation [15]. These theoretical foundations are further discussed in Rosen [13] and summarized in Sheppard [16] and Follain and Jimenez [17].

The hedonic model represents price as a function of housing characteristics thus:

$$P_i = \beta_0 + \sum_{k=1}^{x} \beta_k z_{ik} + \varepsilon_i \quad i = 1, \ldots, n \tag{1}$$

where

P_i is the price of the ith house;
β_0 is the intercept term parameter;
β_k is the parameter of the kth variable;
x is the number of unknown parameters to be estimated;
z_{ik} is the kth housing characteristic associated with β_k;
ε_i is an independently and identically distributed random error term; and
n is the number of observations.

The term z_{ik} captures a broad set of housing characteristics:

1. Structural characteristics, such as the number (and type) of rooms, floor area, dwelling type, age, heating and cooling provision, and lift access in multistorey buildings;
2. Locational characteristics such as access to educational and health services, recreational facilities and open spaces, and distance from the city center (as a proxy for the amenity bundle in the CBD); and
3. Neighborhood characteristics, such as socioeconomic status, crime rate, and air and noise pollution in the local neighborhood, i.e., suburb, postcode, or based on distance from the house.

Hedonic studies often include a range of locational attributes to assess the effects of these amenities on house prices. Policymakers are aware of locations and the distribution of amenities in a city; however, what essentially matters in terms of resident wellbeing is whether such provision is beneficial to them or 'generates utility' in an economic sense. Some amenities may well be located in specific city locations with limited use by residents. The coefficient estimates of the hedonic model reflect these positive and negative effects. In fact, a long tradition of analyzing locational amenity values using the HPM has produced a rich knowledge base. Table 1 summarizes a sample of recent hedonic analyses focusing on specific amenities, the provision of which falls within the remit of the government.

Table 1. A Sample of Recent Hedonic Analyses on Transport Access, Public Parks, and School Quality.

Amenity	Study	City	Property Type	Period Covered	Number of Observations	Summary Findings
Transport access	Higgins and Kanaroglou [7]	Toronto, Canada	Single-detached homes	2001–2003, 2010–2014	2000 (sales)	Transit-oriented development (TOD) was capitalized into land values, though the maximum amount and spatial impact area of this capitalization differed by the TOD context.
	Mulley [18]	Sydney, Australia	Residential housing	2006	1598 (sales)	Accessibility by car and accessibility to employment along the transitway partly determined property prices. Accessibility effects varied significantly over the geographical space.
	Wang et al. [19]	Cardiff, Wales	Residential housing	2000–2009	12,887 (sales)	The number of bus stops within walking distance (300–1500 m) of property was positively associated with the property's observed sale price. The properties with higher market prices tended to benefit more from spatial proximity to bus stop locations compared with their low-priced counterparts.
	Wang et al. [20]	Shanghai, China	Two-bedroom/one-bathroom apartments	2012–2013	2575 (average asking rents)	A residential complex community's adjacency to the nearest Shanghai Metro station tended to correlate positively with its average asking rent.
Public parks and green areas	McCord et al. [6]	Belfast, Ireland	Residential housing	2011	3854 (sales)	Urban green spaces had a significant positive impact on nearby residential properties' sale prices for the terrace and apartment sectors. Terrace and apartment properties located closer to public green spaces achieved increases in their sale price of up to 49 percent compared to otherwise similar properties. Adjacency to green open space produced significant property value premiums in only two of the four housing types analyzed, with limited statistically significant proximate effects evident for the detached and semi-detached sectors.
	Herath et al. [21]	Vienna, Austria	Apartments	2009–2010	1651 (sales)	Base Model: the price of constant-quality apartments declined by approximately 0.13–0.26% with every 1% increase in distance from the greenbelt. Spatial Model (SEM): The corresponding decrease in price was 0.13%.
	Franco and Macdonald [22]	Lisbon, Portugal	Two-bedroom apartments	2007	11,617 (listing prices)	Residential apartment prices reflected the effect of proximity to both large urban forests and smaller neighborhood parks. There was a positive value associated with tree canopy coverage, in that a 1 km^2 increase in the relative size of tree canopy was valued at 0.20% of dwelling price, or approximately €400 per dwelling.
	Belcher and Chisholm [23]	Singapore	Public housing apartments	2013–2014	15,962 (sales)	On average, managed, spontaneous, and high conservation value vegetation had positive effects on property selling price, accounting for 3% of the average property's value, or a total of $179 million Singapore dollars for all public housing apartments sold over 13 months. These effects were almost entirely driven by managed vegetation, which had positive marginal effects on the price for 98.1% of properties.

Table 1. *Cont.*

Amenity	Study	City	Property Type	Period Covered	Number of Observations	Summary Findings
School quality	Turnbull et al. [5]	Orange County, Florida	Single-family houses	2001–2012	127,120 (sales)	House prices reflected the effect of school quality. Increasing the variance in test scores by one standard deviation reduced house prices by 0.7%. The capitalization effects tended to be stronger in higher-income neighborhoods.
	Chung [24]	Seoul, South Korea	Apartments	2004–2012	28,182 (sales), 10,259 (rents)	School choice reform (allowing students to choose high schools outside as well as within their school districts) reduced housing prices in a high-performing school district relative to the housing prices in a low-performing school district by approximately 10–27%.
	Sah et al. [25]	San Diego County, California	Residential housing	2010–2011	20,000 (sales)	There was a school proximity penalty for public elementary schools. Spatially dividing the sample area into a coastal and an inland region led to a delayed proximity premium for the inland region and a very strong proximity penalty effect for the coastal region.
	Wen et al. [26]	Hangzhou, China	Multi-layer and high-rise housing	2011–2013	660 (average prices)	Surrounding housing prices reflected the quality of basic education in the neighborhood. Primary and secondary schools had significant school district effects.

2.2. Amenity Preferences of Residents Inform Urban Planning Policy

Residents' preferences for amenities are often unrecognized by planners and hence ignored in prioritizing needs. However, scholars argue that prudently implemented evidence-based practice could enrich the field of planning by linking research to practice [27,28]. As discussed in Section 2.1, house price is a summary measure of the desirability of a location based on a bundle of nearby amenities. By addressing the deficits of current methods for assessing urban amenity values (discussed next in Section 3), such models can generate new fine-grained information about the utility of amenities at the local level.

Research on amenity preferences suggests two areas of pertinence to urban planning: enabling evidence-informed planning ('instrumental view') [29] and enriching the knowledge environment ('enlightenment view') [30].

Evidence-informed planning gains direct knowledge from research and analysis of where different interventions are likely to be socially and economically effective. In a planning context, the values ascribed to amenities change over time; hence, monitoring the benefits becomes paramount in preserving the locational value. Three primary outcomes using the improved tools derived from evidence-informed planning include precise identification of local areas well served by valuable amenities, identification of areas with poor amenities, and assessment of the values of specific infrastructure projects. A promising application to infrastructure is in the field of infrastructure funding and value capture. The appropriate use of evidence-informed policy at the research and planning interface can have far-reaching positive wellbeing outcomes for urban residents as cities develop within a system of ongoing development processes. The experiential tools presented here are particularly useful in a contemporary policy environment that favors evidence-based planning [30] and quantitative evaluation methods [31].

The enlightenment view specifies that the development of accurate new data enriches the knowledge environment within which planners make decisions. If the planning environment is such that planners are not often influenced by new data but mainly by political and financial considerations [32], the proposed framework could still contribute meaningfully by enlightening the planning environment. This viewpoint emphasizes the need for developing a knowledge base of planning case studies and examples that provides a deeper understanding of the conditions within which different interventions might be useful. The positioning of research in this way improves the capacity to make use of the available evidence. Such a knowledge base also enables a greater awareness amongst the various stakeholders, including bureaucrats, planners, and the public. It informs broader public debate, promoting an 'enlightened society' as a way of engaging the public in the planning process. Planners can exploit this knowledge to prioritize and communicate planning needs in localities and to negotiate consensus around their planning goals.

As a revealed preference method, the HPM has some traits that make it appropriate in the quest for understanding amenity values based on both the 'evidence-based planning' and 'enlightenment' viewpoints [33]. Originally, the revealed preference theory was motivated by the unobservability of preferences [34]. For instance, in the alternative 'stated preference' methods, the respondents may state their intentions about the worth of some amenity. However, these are not the behaviors that are observed (or revealed) in actual markets. This perceived divergence from market prices, known as 'hypothetical bias,' is a result of the imaginary nature of stated preference experiments that do not entail the respondents actually making the choices or executing the behaviors they state [35]. Often the respondents are unable to predict their own actual market behavior accurately in a hypothetical environment.

In contrast, the HPM generates more realistic resident preferences. The HPM estimates the values of housing attributes, including location, using actual house prices in different locations. In an urban planning context, the price premiums paid by residents of an area themselves reveal the attractiveness of localities. One could thus view the HPM as an innovative planning technology in the context of urban policymaking where stakeholder engagement is indirectly incorporated. The application of a relatively objective empirical method also makes the policymaking process more transparent.

Though standalone hedonic studies (see examples in Table 1) may be robust in their own right, they do not provide a reliable and consistent basis for understanding the values of locational amenities more broadly in an applied planning context. These inconsistencies are primarily due to three main limitations in hedonic modeling: measurement, specification, and scale issues.

3. Developing a Coherent Analytical Framework

Despite the potential contribution of the hedonic theory in a planning context, existing modeling instruments investigating amenity values have three main limitations. This section discusses the specifics of these limitations and develops a coherent analytical framework incorporating several essential improvements.

3.1. Addressing Measurement Issues

The *first limitation* refers to the measurement issues in identifying and accurately calculating locational variables. We highlight two specific respects in which the conventional assumptions of the HPM are arguably flawed.

The first assumption is that a distance threshold of 400 m (known as 'Ped-Shed') is appropriate for measuring the effects of proximity to transport nodes on house prices [36–38]. This assumption rests on the assertion that 400 m 'represent a comfortable walk for most people under normal conditions' [39] (p. 322). A notable concern here is that it takes no account of local geography. Thus, in areas lacking pavements or with steep gradients, people are unlikely to walk as much as where paths are more 'walkable.' This practicality means that conventional HPM applications are likely to produce misleading findings. For instance, Herath [40] compared 100 m, 200 m, and 400 m distance thresholds from bus stops in a study of Sydney with contrasting results. The proposed framework systematically tests this particularly problematic assumption of the 400 m distance threshold commonly used to measure the effect of proximity to transport nodes.

The second questionable assumption is that a variable measuring 'straight-line' distance can capture the house price effects of distance from an amenity. For instance, house price models employing variables measuring the distance from the CBD generally assume straight-line distance. However, scholarly research conceptualizes distances between different points in space in three different ways [41–44]. First, *straight-line distance* computes distance 'as the crow flies,' and disregards the actual road network in a city. Second, *travel distance*, also termed 'road-network distance,' measures the distance between two locations along the actual road network, generally considering the shortest path among available routes. Third, *overland distance* reflects travel times or travel costs, yet again allowing for the fastest or lowest-cost path between two points. Unless the model incorporates the most representative and context-specific distance variable amongst these alternatives, this variable may be statistically insignificant in the model, or the model may lack explanatory power.

3.2. Addressing Specification Issues

The *second limitation* is the assumption that a variable measuring linear distance from the CBD can explain the structure of a city. Hedonic analyses regularly assume that linear distance adequately captures the CBD's influence on house prices [45], although an alternative functional form suggests a quadratic function, which translates as an inverted U-shaped curve [46]. The latter reflects an understanding that residents may want to reside neither too close to the city center, because of negative externalities such as air pollution and crime, nor too distant from it, as jobs are concentrated in the CBD. Therefore, empirical assessments need to consider variants of the polynomial distance function, namely, linear, quadratic, and cubic functions.

3.3. Applying to the Local Scale

The *third key limitation* discussed in this study is the recent proposition that modeling house prices using 'local' regression methods is beneficial [47]. The 'global' model [48] that focuses on modeling *spatial dependence*[1] in the data is dominant within the empirical literature[2]. Yet, the local model [49] that incorporates *spatial heterogeneit*[3] is attracting increasing interest [47,50]. The former derives inferences based on the entire data sample (i.e., aggregate study area), although the latter focuses on local data subsamples, uncovering spatial relationships usually hidden in global models. If the global model produces a significant effect for a characteristic, does that hold across all the local areas? Not necessarily. Some local areas may exhibit a positive and significant effect, some a negative and significant effect, and others an insignificant effect. Conversely, an insignificant effect for a characteristic from a global model might also not hold across all local areas—some areas might have a positive and significant effect, some might have a negative and significant effect, while others might have an insignificant effect. In other words, the common use of *global* regression models is problematic because they omit hidden yet important local dynamics; local models provide a more pertinent modeling strategy capable of producing useful information about the values of local amenities.

The geographically weighted regression (GWR) is a common technique for estimating the local model. The GWR expands the above hedonic model (1) to a form that allows for local variations in the parameter values, taking into account the coordinates of individual regression points. If the dependent variable has the spatial coordinates (u_i, v_i), we can rewrite the model (1) as the following GWR local model:

$$P_i\,(u_i, v_i) = \beta_0(u_i, v_i) + \sum_{k=1}^{x} \beta_k(u_i,\, v_i) z_{ik} + \varepsilon_i \quad i\ = 1,\ \ldots,\ n \tag{2}$$

where

$P_i(u_i, v_i)$ is the price of the ith house;
$\beta_0(u_i, v_i)$ is the location-specific intercept term parameter;
$\beta_k(u_i, v_i)$ is the location-specific parameter of the kth variable;
x is the number of unknown local parameters to be estimated;
z_{ik} is the kth housing characteristic associated with β_k;

ε_i is an independently and identically distributed random error term; and
n is the number of observations.

To analyze spatially varying relationships, the GWR model embodies spatial heterogeneity, presumed to be discontinuous or continuous. Parametric methods often achieve a discontinuous demarcation of local areas with fixed bandwidth, a measure of local scale for each conditional relationship. The use of a fixed bandwidth results in the simpler moving-window regression model: the GWR moves a search window from one observation to the next, identifying all the observations within the window. A cross-validation score, a method of reducing variability within sub-samples of data, determines the window size. For local areas with uneven spacing, an adaptive bandwidth is appropriate [49]. In this case, the bandwidth reflects a fixed local sample density instead of a fixed distance across the entire sample, i.e., 'an optimal bandwidth in an adaptive form' [51]. Multiscale GWR [52] relaxes GWR's assumption of a single bandwidth constant across all conditional relationships in the model. This is a further step towards local models that incorporate local scales and dynamics. Multiscale GWR is demonstrated to be superior to GWR in terms of producing more accurate estimates.

As GWR is a local model, relationships are estimated locally and mapped across space. Before proceeding with mapping, a random distribution of residuals validates the statistical robustness of the analysis. It is only worthwhile mapping the statistically significant variables based on the typical t-value for hypothesis-testing ($t > 2$). A comparison of the global and local models indicates statistical performance using R^2 and Akaike's information criterion (AIC) values[4].

The flaws of the house price models mentioned above—measurement, specification, and scale issues—problematically skew our understanding of the urban structure and spatial distribution of amenities [53]. Therefore, these analyses may result in misinformed policy interventions and poor amenity planning decisions [31].

3.4. A Coherent Analytical Framework

The improved conceptual design presented in Figure 1 sets out a coherent analytical framework. The first step is to identify the potential variables, i.e., housing values (proxied by prices or rents) and housing characteristics (see Stage 1). The incorporation of structural variables such as the number of bedrooms and dwelling type into the model specification is relatively straightforward. Thereupon, based on the addresses of houses, their exact locations in longitude and latitude (x, y) coordinates are generated using geographic information software (GIS) (e.g., ArcGIS). These enable the calculation of location variables in relation to various amenities in the city.

Notably, the proposed framework facilitates the identification of context-relevant location variables, computed in the right measurement scales (Stage 2, a). It lays out the need to simultaneously test and compare different model specifications to represent the relationships accurately (Stage 2, b). Model diagnostics such as the statistical significance of the variables and goodness of fit statistics (i.e., the overall performance of the models) assist in choosing the most relevant variables. For the selection problem of locational variables related to distance measurement (e.g., proximity to transport nodes, distance to amenity), the simultaneous consideration of different candidate variables is required. If different ordered tests result in divergent model specifications, the attributes of local geography based on ground-truthing provide clues on context-relevant variables. The proposed procedures improve the reliability and performance of the house price model in a preliminary step by comparing the implications of different location variables on findings. As such, this analytical framework overcomes the shortcomings discussed in Sections 3.1 and 3.2 above by empirically testing and appropriately adjusting the traditional assumptions related to the distance measurement of the locational variables.

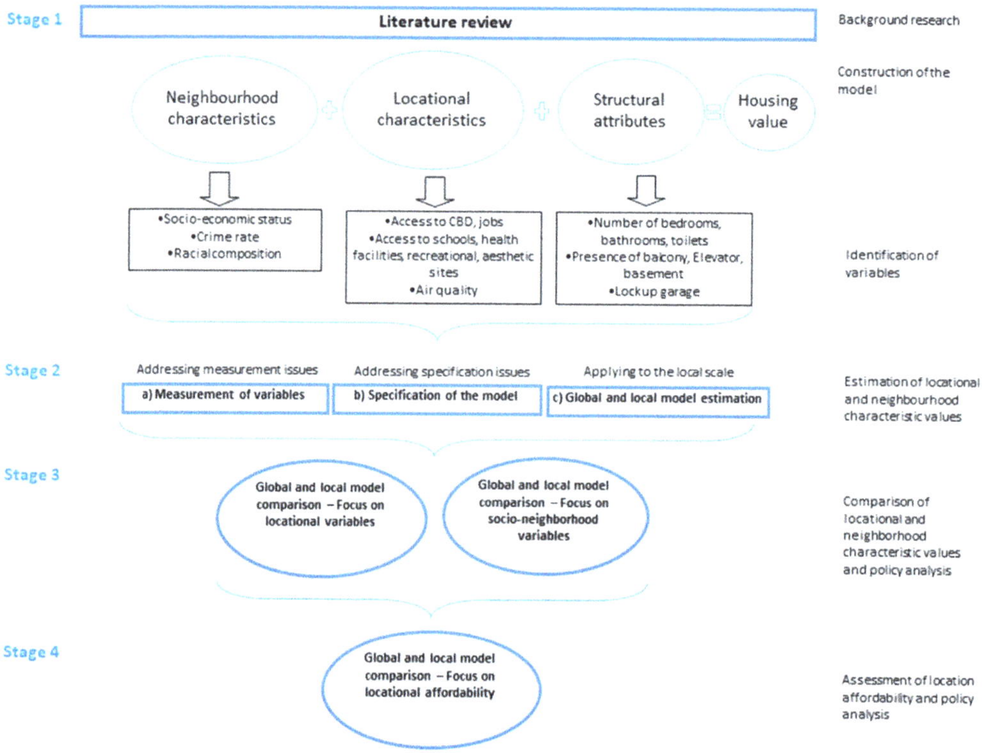

Figure 1. A Coherent Analytical Framework.

As discussed in Section 3.3, the framework also integrates house prices in a parallel *local* regression model, i.e., estimated for local subsamples of data, as a more reliable method to capture specific values of local amenities (Stage 2, c). This improvement is critical given the models need to be consistent and appropriate for analyzing the local context from a planning perspective. In addition, the two types of models implemented (i.e., global and local) provide a sound framework for scrutinizing amenity values at both the aggregate city level (city-wide patterns) and local level (local area trends). By comparing global model and local model estimates side-by-side, interesting patterns at the local scale and characteristics more appropriate at the broader study level would be revealed.

The framework thus helps generate information on a range of planning matters, including investigating the values of particular locational amenities and socio-neighborhood characteristics (Stage 3). Since these models incorporate spatial fixed effect variables, such as districts, suburbs, or postcodes, they also indicate the relative values of localities determined by specific amenity values. This new knowledge provides the possibility to detect desirable residential locations and low-value localities in different parts of a city (Stage 4). Based on the findings, governments may direct residential developments into locations with valuables bundles of amenities, or, if city density is becoming unmanageable, preserve these areas from over-development through zoning, council rates, or other means. Overall, this framework represents a comprehensive and innovative approach that challenges the traditional assumptions of house price models, applies variants of the model to understand city-wide and local housing market dynamics, and has the potential to empirically generate policy-significant information regarding the values of locational and neighborhood amenities in cities.

4. Conclusions and Policy Implications

Scholarly work over the last few decades has scrutinized the non-market monetary values of urban amenities using the hedonic price method (HPM), including research with significant planning implications. However, this research falls short of establishing the theoretical interconnections between house prices, the amenity preferences of residents, and urban planning. In this article, we put forward a framework termed a *consumer preference-based approach* to urban amenity provision to map out how hedonic prices reveal the *values* that residents ascribe to nearby amenities and how planners could incorporate this information into their decision-making. This new framework is based on a house price model typically used to explain the demand for, and the prices of, housing, and hence generates incremental and transformative gains in the urban economics and urban planning disciplines. The improvements in potential reliability of the model lead to incremental gains within urban economics, and the application of a new and consistent hedonic approach for local amenity planning results in transformative gains in the urban planning discipline.

As a revealed preference method, the HPM at first glance appears to be an appropriate methodology to assess amenity values. However, several concerns need resolving to utilize these applications appropriately in a planning context: (1) measurement issues associated with identifying and accurately calculating locational variables, (2) model misspecification issues, and (3) the problematic common use of global regression models. A robust, integrated, transparent, and systematic analytical framework is needed to empirically test a range of model assumptions and eventually improve the precision of the model.

This article aimed to advance existing approaches to understanding the value of urban locations by focusing attention on amenities, particularly as they affect residential land values at the local level. The proposed conceptual framework integrates hedonic price theory and cutting-edge econometric tools but applies these as planning tools, demonstrating improved analytical models to reliably identify the values of high-demand amenities in cities. Despite the long-standing practice of explaining local spatial disparities in the housing markets based on *global* hedonic models, the framework further evaluates the usefulness of *local* regression methods that are intuitively preferred when location-specific information is required for housing and urban planning purposes. These improvements are innovative in their ability to address deficits in current methods of assessing urban amenity values and generate new local fine-grained information about the utility of such amenities.

This is the first paper to explicitly demonstrate the contribution of the separate theoretical field of urban economics to the planning literature and to lay out the conceptual alignment between house prices, amenity preferences, and amenity provision (Section 2). It also recognizes the important deficits that exist in the conventional hedonic applications and specific improvements within an integrated framework (Section 3). Future research will focus on an extended piece of work testing this framework using a case study, including tests of different measurements and specifications, location and neighborhood characteristics, and local and global model estimates.

In policy terms, the framework develops and tests an objective and data-driven mechanism that assesses spatial equality in terms of urban amenities in smaller localities such as suburbs. The creation of more accurate and dynamic data-driven prototypes could inform a broad range of spatial planning and policy interventions to help governments address the following key questions:

- In which areas should housing be provided (or zoned)? The proposed suite of models can benchmark local area amenity availability against city-wide amenities. The improved models could, therefore, facilitate precise identification of localities well served by valuable amenities, with capacity and potential for increased housing. These more reliable measurements of valuable amenities may indicate possible directions of urban growth processes, enable more efficient use of existing urban amenities, encourage development within existing urbanized areas, and minimize the environmental impact of residential development.

- In which areas should urban amenities be provided? Again, this proposed framework could help identify low-value localities with poor amenities. The operation of the housing market in major cities embeds inequality, and there seems little interest in addressing these disparities. While some inequality is inherent in amenity provision, the integrated tools proposed here could reliably identify where investment in amenities would pay off best, an essential element when the value for money is a paramount concern. This new knowledge may facilitate the distribution of public funds and target investment to improve amenities in areas identified as optimal from an amenities point of view, improving residents' wellbeing and enhancing spatial connectivity. Identifying localities that lack amenities may also provide a substantial incentive to the private sector to invest in transit-oriented development. As such, the proposed tools offer another source of value, providing at least the potential of an additional means of reducing urban inequality. It is also important to note that improving the most vulnerable areas may have a negligible effect on house prices at the aggregate city level.
- What is a community's willingness to pay for specific infrastructure assets? The improved models may assist in evaluating these assets and developing value capture mechanisms through rates, levies, charges, or other means. They can explain the effects of specific infrastructure on nearby property values and help assess associated urban productivity outcomes related to amenity projects and value creation opportunities.
- What are the effects of adverse environmental characteristics? The suite of proposed models could detect low-value undesirable localities with dis-amenities requiring intervention.

From a planning practice perspective, elements within this analytical framework could be translated into a useful toolkit for urban planners to assist and inform urban planning decision-making. An important area of application is in the provision of urban parks. Production costs of urban parks are often known and easily measurable (costs of planting trees, hedge cuts, etc.), but benefits are more difficult to assess. These models evaluate the benefits of protecting existing urban green areas and providing new urban parks within the planning processes to facilitate long-term urban development strategies. For instance, a local council may assess how the benefits could change with different configurations of park sizes. Mahmoudi et al. [54] estimated that an expansion of a pocket park from 0.4 ha to 1 ha resulted in $0.9 M in private benefits being capitalized in property values. These values are useful in cost-benefit analyses to determine the types of urban parks that are likely to yield benefits. City planners could preserve or enhance housing values by considering the spatial configuration and species composition of open space in residential neighborhoods [55]. By identifying forest land values of planned areas with consistently high visual amenity values, planners could generate the highest potential return on investment for preservation and reforestation [56]. In addition, the capitalized property values associated with key environmental amenities could measure the contribution of local amenities to land tax [57].

The improved models would address a critical gap in the planning practice and policy domains, introducing demand-side analytical tools to uncover local area demand determinants. The information and maps generated could indicate the values of amenities in localities, a piece of information vital for facilitating a more efficient distribution of housing and amenities in cities. Hence, the proposed analytical framework could have a significant and immediate policy and practice implications for creating and preserving a city that is both equitable and sustainable. It is worth noting, however, that social wellbeing motives alone do not determine amenity provision, and economic (e.g., provision of amenities to attract businesses and promote tourism) and political reasons may influence amenity provision to some extent.

Finally, this article contributes to an important area of research in the evaluation of amenities and services in urban localities. While addressing a significant theoretical and empirical gap in the planning literature, the proposed framework provides a potentially robust methodological integration to deal with the shortcomings of present evaluation methods. The framework could inform planning policy by improving the capacity to identify localities well served with amenities and vice versa, thus providing a basis for policy action to enhance spatial equality in cities. The empirical tools presented are particularly suited to a policy environment that aligns with the new trend towards evidence-based planning and quantitative evaluation methods. Due to its data-driven nature, the proposed framework supports an objective assessment that can enhance economic efficiency, productivity, and spatial equity within cities. To that end, the proposed framework seeks greater policy relevance by positioning the measurements in local planning realities.

Funding: This research received no external funding.

Institutional Review Board Statement: Not applicable.

Informed Consent Statement: Not applicable.

Data Availability Statement: Not applicable.

Conflicts of Interest: The author declares no conflict of interest.

Notes

1 *Spatial dependence* is the interdependence of variable values—either prices, housing features or both—among nearby locations.

2 The Equation (1) above estimated for the entire study sample is a 'global model'.

3 This refers to an uneven distribution of variable values within an area that originates from characteristics of demand, supply or institutional barriers.

4 Akaike's information criterion (AIC) is a measure of model fit (a lower value indicating a better model fit).

References

1. United Nations. *World Urbanization Prospects: The 2018 Revision (St/Esa/Ser. A/420)*; United Nations: New York, NY, USA, 2019.
2. Rigolon, A. A Complex Landscape of Inequity in Access to Urban Parks: A Literature Review. *Landsc. Urban Plan.* **2016**, *58*, 160–169. [CrossRef]
3. Payton, S.B.; Ottensmann, J.R. The Implicit Price of Urban Public Parks and Greenways: A Spatial-Contextual Approach. *J. Environ. Plan. Manag.* **2015**, *58*, 495–512. [CrossRef]
4. Myers, D. Building Knowledge About Quality of Life for Urban Planning. *J. Am. Plan. Assoc.* **1988**, *54*, 347–358. [CrossRef]
5. Turnbull, G.K.; Zahirovic-Herbert, V.; Zheng, M. Uncertain School Quality and House Prices: Theory and Empirical Evidence. *J. Real Estate Financ. Econ.* **2017**, *57*, 167–191. [CrossRef]
6. Mccord, J.; Mccord, M.; McCluskey, W.; Davis, P.; McIlhatton, D.; Haran, M. Effect of public green space on residential property values in Belfast metropolitan area. *J. Financ. Manag. Prop. Constr.* **2014**, *19*, 117–137. [CrossRef]
7. Higgins, C.; Kanaroglou, P. Rapid transit, transit-oriented development, and the contextual sensitivity of land value uplift in Toronto. *Urban Stud.* **2017**, *55*, 2197–2225. [CrossRef]
8. Baranzini, A.; Schaerer, C. A Sight for Sore Eyes: Assessing the Value of View and Land Use in the Housing Market. *J. Hous. Econ.* **2011**, *20*, 191–199. [CrossRef]
9. Zheng, S.; Cao, J.; Kahn, M.E.; Sun, C. Real Estate Valuation and Cross-Boundary Air Pollution Externalities: Evidence from Chinese Cities. *J. Real Estate Financ. Econ.* **2013**, *48*, 398–414. [CrossRef]
10. Buonanno, P.; Montolio, D.; Raya-Vílchez, J.M. Housing prices and crime perception. *Empir. Econ.* **2012**, *45*, 305–321. [CrossRef]
11. Herath, S.; Gunther, M. Local Particularities or Distance Gradient. *J. Eur. Real Estate Res.* **2013**, *6*, 163–185. [CrossRef]
12. Mayor, K.; Lyons, S.; Duffy, D.; Tol, R.S. A Hedonic Analysis of the Value of Rail Transport in the Greater Dublin Area. *J. Transp. Econ. Policy* **2012**, *46*, 239–261.
13. Rosen, S. Hedonic Prices and Implicit Markets: Product Differentiation in Pure Competition. *J. Political Econ.* **1974**, *82*, 34–55. [CrossRef]
14. Gross, D.J. Estimating willingness to pay for housing characteristics: An application of the Ellickson bid-rent model. *J. Urban Econ.* **1988**, *24*, 95–112. [CrossRef]
15. Malpezzi, S. Hedonic Pricing Models: A Selective and Applied Review. In *Housing Economics and Public Policy*; O'Sullivan, T., Gibb, K., Eds.; Blackwell Science: Malden, MA, USA, 2003; pp. 67–89.
16. Sheppard, S. Hedonic Analysis of Housing Markets. In *Handbook of Regional and Urban Economics*; Chesire, P.C., Mills, E.S., Eds.; Elsevier: Amsterdam, The Netherlands, 1999.
17. Follain, J.R.; Jimenez, E. Estimating the demand for housing characteristics: A survey and critique. *Reg. Sci. Urban Econ.* **1985**, *15*, 77–107. [CrossRef]
18. Mulley, C. Accessibility and Residential Land Value Uplift: Identifying Spatial Variations in the Accessibility Impacts of a Bus Transitway. *Urban Stud.* **2013**, *51*, 1707–1724. [CrossRef]

19. Wang, Y.; Potoglou, D.; Orford, S.; Gong, Y. Bus Stop, Property Price and Land Value Tax: A Multilevel Hedonic Analysis with Quantile Calibration. *Land Use Policy* **2015**, *42*, 381–391. [CrossRef]
20. Wang, Y.; Feng, S.; Deng, Z.; Cheng, S.; Feng, S. Transit premium and rent segmentation: A spatial quantile hedonic analysis of Shanghai Metro. *Transp. Policy* **2016**, *51*, 61–69. [CrossRef]
21. Herath, S.; Choumert, J.; Maier, G. The value of the greenbelt in Vienna: A spatial hedonic analysis. *Ann. Reg. Sci.* **2015**, *54*, 349–374. [CrossRef]
22. Franco, S.F.; Macdonald, J.L. Measurement and Valuation of Urban Greenness: Remote Sensing and Hedonic Applications to Lisbon, Portugal. *Reg. Sci. Urban Econ.* **2018**, *72*, 156–180. [CrossRef]
23. Belcher, R.N.; Chisholm, R.A. Tropical Vegetation and Residential Property Value: A Hedonic Pricing Analysis in Singapore. *Ecol. Econ.* **2018**, *149*, 149–159. [CrossRef]
24. Chung, I.H. School Choice, Housing Prices, and Residential Sorting: Empirical Evidence from Inter-and Intra-District Choice. *Reg. Sci. Urban Econ.* **2015**, *52*, 39–49. [CrossRef]
25. Sah, V.; Conroy, S.J.; Narwold, A. Estimating School Proximity Effects on Housing Prices: The Importance of Robust Spatial Controls in Hedonic Estimations. *J. Real Estate Financ. Econ.* **2016**, *53*, 50–76. [CrossRef]
26. Wen, H.; Xiao, Y.; Zhang, L. School district, education quality, and housing price: Evidence from a natural experiment in Hangzhou, China. *Cities* **2017**, *66*, 72–80. [CrossRef]
27. Krizek, K.; Forysth, A.; Slotterback, C.S. Is There a Role for Evidence-Based Practice in Urban Planning and Policy? *Plan. Theory Pract.* **2009**, *10*, 459–478. [CrossRef]
28. Faludi, A.; Waterhout, B. Introducing Evidence-Based Planning. *DisP-Plan. Rev.* **2006**, *42*, 4–13. [CrossRef]
29. Higgins, K.; Canavan, J.; Coen, L. Mapping the field in evidence-informed policy and practice: International perspectives. *EÉvid. Policy J. Res. Debate Pract.* **2014**, *10*, 489–495. [CrossRef]
30. Davoudi, S. Evidence-Based Planning. *DisP-Plan. Rev.* **2006**, *42*, 14–24. [CrossRef]
31. Herath, S. Technology in Urban Planning: Spatial Hedonic House Price Model as an Information Tool. In *Instruments of Planning: Tensions and Challenges for More Equitable and Sustainable Cities*; Leshinsky, R., Legacy, C., Eds.; Routledge: London, UK, 2016; pp. 228–240.
32. Adjei-Poku, B. Rationality and Power in Land Use Planning: A Conceptual View of the Relationship. *J. Plan. Lit.* **2018**, *33*, 45–60. [CrossRef]
33. Caplin, A.; Dean, M. Revealed Preference, Rational Inattention, and Costly Information Acquisition. *Am. Econ. Rev.* **2015**, *105*, 2183–2203. [CrossRef]
34. Samuelson, P.A. A Note on the Pure Theory of Consumer's Behaviour. *Economica* **1938**, *5*, 61–71. [CrossRef]
35. Beck, M.J.; Simon, F.; John, M.R. Can You Ever Be Certain? Reducing Hypothetical Bias in Stated Choice Experiments Via Respondent Reported Choice Certainty. *Transp. Res. Part B Methodol.* **2016**, *89*, 149–167. [CrossRef]
36. Ibeas, Á.; Cordera, R.; dell'Olio, L.; Coppola, P.; Dominguez, A. Modelling Transport and Real-Estate Values Interactions in Urban Systems. *J. Transp. Geogr.* **2012**, *24*, 370–382. [CrossRef]
37. Munoz-Raskin, R. Walking Accessibility to Bus Rapid Transit: Does It Affect Property Values? The Case of Bogotá, Colombia. *Transp. Policy* **2010**, *17*, 72–84. [CrossRef]
38. Alan, T.M. Strategic Analysis of Public Transport Coverage. *Socio-Econ. Plan. Sci.* **2001**, *35*, 175–188.
39. Murray, A.T.; Davis, R.; Stimson, R.J.; Ferreira, L. Public Transportation Access. *Transp. Res. Part D Transp. Environ.* **1998**, *3*, 319–328. [CrossRef]
40. Herath, S. How Far Is Too Far from Public Transport? An Empirical Assessment. In Proceedings of the RICS COBRA AUBEA 2015, Sydney, Australia, 8–10 July 2015.
41. Nilsson, P. The Influence of Urban and Natural Amenities on Second Home Prices. *J. Hous. Built Environ.* **2014**, *30*, 427–450. [CrossRef]
42. Karlsson, V. The Relationship of Housing Prices and Transportation Improvements: Location and Marginal Impact. *Spat. Econ. Anal.* **2011**, *6*, 223–241. [CrossRef]
43. A Sander, H.; Ghosh, D.; van Riper, D.; Manson, S.M. How Do You Measure Distance in Spatial Models? An Example Using Open-Space Valuation. *Environ. Plan. B Plan. Des.* **2010**, *37*, 874–894. [CrossRef]
44. Herath, S.; Jayasekare, A.S. City proximity, travel modes and house prices: The three cities in Sydney. *Neth. J. Hous. Environ. Res.* **2021**, *36*, 407–431. [CrossRef]
45. Osland, L.; Pryce, G. Housing Prices and Multiple Employment Nodes: Is the Relationship Nonmonotonic? *Hous. Stud.* **2012**, *27*, 1182–1208. [CrossRef]
46. Lu, B.; Charlton, M.; Harris, P.; Fotheringham, A.S. Geographically weighted regression with a non-Euclidean distance metric: A case study using hedonic house price data. *Int. J. Geogr. Inf. Sci.* **2013**, *28*, 660–681. [CrossRef]
47. Crespo, R.; Grêt-Regamey, A. Local Hedonic House-Price Modelling for Urban Planners: Advantages of Using Local Regression Techniques. *Environ. Plan. B Plan. Des.* **2013**, *40*, 664–682. [CrossRef]
48. Anselin, L. *Spatial Econometrics: Methods and Models*; Kluwer: Dordrecht, The Netherlands, 1988.
49. Fotheringham, A.S.; Brunsdon, C.; Charlton, M. *Geographically Weighted Regression: The Analysis of Spatiality Varying Relationships*; John Wiley: Chichester, UK, 2002.

50. Tsai, C.-H.; Mulley, C.; Burke, M.; Yen, B. Exploring property value effects of ferry terminals: Evidence from Brisbane, Australia. *J. Transp. Land Use* **2015**, *10*, 119–137. [CrossRef]
51. Harris, P.; Fotheringham, A.S.; Crespo, R.; Charlton, M. The Use of Geographically Weighted Regression for Spatial Prediction: An Evaluation of Models Using Simulated Data Sets. *Math. Geol.* **2010**, *42*, 657–680. [CrossRef]
52. Fotheringham, A.S.; Yang, W.; Kang, W. Multiscale Geographically Weighted Regression (Mgwr). *Ann. Am. Assoc. Geogr.* **2017**, *107*, 1247–1265. [CrossRef]
53. Dziauddin, M.F.; Powe, N.; Alvanides, S. Estimating the Effects of Light Rail Transit (LRT) System on Residential Property Values Using Geographically Weighted Regression (GWR). *Appl. Spat. Anal. Policy* **2014**, *8*, 1–25. [CrossRef]
54. Mahmoudi, P.; MacDonald, D.H.; Crossman, N.D.; Summers, D.M.; van der Hoek, J. Space matters: The importance of amenity in planning metropolitan growth. *Aust. J. Agric. Resour. Econ.* **2012**, *57*, 38–59. [CrossRef]
55. Cho, S.H.; Poudyal, N.C.; Roberts, R.K. Spatial Analysis of the Amenity Value of Green Open Space. *Ecol. Econ.* **2008**, *66*, 403–416. [CrossRef]
56. Chadourne, M.; Cho, S.H.; Roberts, R.K. Identifying priority target areas for the Knoxville-Knox County hillside and ridgetop protection plan: Using the value of visual amenity during the real estate boom of 2002-2007 and the recession of 2008. *Ann. Reg. Sci.* **2013**, *50*, 911–934. [CrossRef]
57. Gibbons, S.; Mourato, S.; Resende, G.M. The Amenity Value of English Nature: A Hedonic Price Approach. *Environ. Resour. Econ.* **2013**, *57*, 175–196. [CrossRef]

land

MDPI

Article

Factors Influencing Large Real Estate Companies' Competitiveness: A Sustainable Development Perspective

Bo Li [1], Rita Yi Man Li [2,*] and Thitinant Wareewanich [1]

[1] Chakrabongse Bhuvanarth International Institute for Interdisciplinary Studies, Rajamangala University of Technology Tawan-Ok, Bangkok 10400, Thailand; bo.li@rmutto.ac.th (B.L.); thitinant_wa@rmutto.ac.th (T.W.)

[2] Sustainable Real Estate Research Center, Hong Kong Shue Yan University, Hong Kong, China

* Correspondence: ymli@hksyu.edu

Abstract: Strict policy control and real estate market downturn affects large-scale real estate enterprises performance. We surveyed large Chinese real estate enterprises and the internal factors that affect their competitiveness. Verified by the hierarchical regression and structural equation modelling approach, the results mainly show that profitability, capital ability, management and operation ability, human resource ability, brand name, and innovation ability play positive roles in the competitiveness of large real estate enterprises. Management and operation ability plays an intermediary role between human resources and the improvement of competitiveness. Real estate enterprises' capital sources play an intermediary role between brand names and the improvement of competitiveness. Moreover, landbank area and quality and sales are three major factors that impact the competitiveness improvement of real estate enterprises, while the ability for marketing innovation and the payment collection of enterprises has a relatively small impact. All in all, this paper provides practical implications concerning factors that affect the competitiveness of large real estate enterprises. The findings are helpful to improve the sustainable development of real estate enterprises in the future. As research on factors that affect large-scale real estate enterprises is scarce, this study aims to fill this gap.

Keywords: large real estate enterprise; comprehensive competitiveness; sustainable development; structural equation model

Citation: Li, B.; Li, R.Y.M.; Wareewanich, T. Factors Influencing Large Real Estate Companies' Competitiveness: A Sustainable Development Perspective. *Land* **2021**, *10*, 1239. https://doi.org/10.3390/land10111239

Academic Editors: Elizelle Juanee Cilliers and Sarel Cilliers

Received: 6 October 2021
Accepted: 8 November 2021
Published: 12 November 2021

1. Introduction

The real estate industry is an economic pillar industry in China [1]. In recent years, the People's Republic of China government has issued a series of real estate-related policies to stabilize the real estate market. The government requires enterprises to adhere to "houses are for accommodation, not for speculation" guidelines, and real estate should not be used as a short-term means to stimulate the economy [2]. At the beginning of 2020, due to the COVID-19 outbreak [3], the Chinese economy and real estate industry suffered, and real estate sales showed a cliff-like decline [4]. At present, with the effective control of the epidemic in China, the economy is recovering gradually. Still, the main policy direction of "real estate is not for speculation" remains unchanged, and real estate companies' finances remain tight [2]. All these changes highlight the importance of enhancing enterprises' competitiveness as a critical path for the sustainable development of the real estate industry.

The research objective of this paper is to study the factors that affect large-scale real estate enterprises' competitiveness. As shown in Table 1, the National Bureau of Statistics of China defined large real estate enterprises in 2018 [5]. We define large real estate enterprises according to the two indicators of operating income and total assets in the table.

Table 1. Standards for the scale of real estate enterprises.

Indicator Name	Unit (CNY)	Large	Medium	Small	Miniature
Operating income (Y)	Ten thousand	$Y \geq 200{,}000$	$1000 \leq Y < 200{,}000$	$100 \leq Y < 10{,}000$	$Y < 100$
Total assets (Z)	Ten thousand	$Z \geq 10{,}000$	$5000 \leq Z < 10{,}000$	$2000 \leq Z < 5000$	$Z < 2000$

Research on the competitive advantage and competitiveness of real estate enterprises is relatively scarce in academia. Nevertheless, it is of high importance among real estate business operators and entrepreneurs, government departments and industry associations within the real estate industry [6–11]. In the report "Research on the Comprehensive Competitiveness of Real Estate Enterprises" issued by the China Real Estate Association [12], the comprehensive competitiveness of real estate enterprises is defined as market, quality and brand; the core idea sheds light on an enterprise's resource integration and innovation ability. The connotations that correspond to the market include comprehensive market research ability, ability to grasp home purchasers' needs, rapid response to market needs and resource integration. The connotations corresponding to quality include project management, planning and design, engineering, product culture and art quality assurance. The brand name concerns the corporate, project product, corporate personnel images, marketing and integrated communication ability and serviceability. The competitiveness of the large real estate enterprises studied in this paper refers to the fact that a large real estate enterprise has stronger survival and development ability and stronger market influence than its competitors. The comprehensive competitiveness is a dialectical relationship between comprehensive competitiveness and its basic elements, both independent and interrelated [13].

Regarding the basic factors that affect the comprehensive competitiveness of large real estate enterprises, they can be divided into external and internal factors [14]. External factors mainly refer to two aspects of market and policy. The market factors include demand capacity, supply capability and market structure, and the policy factors include land, financial and tax policies. Internal factors mainly refer to the enterprise's factors, which reflect the enterprise's ability to face challenges, etc., and are the fundamental factors that affect the likelihood of survival and development prospects. They are also the internal driving force of an enterprise's development [15]. Paying attention to the internal factors allows internal adjustment easily, even if the external environment is uncontrollable. In China, market and policy factors outside enterprises have a huge impact on real estate enterprises [16,17]. While the influence of the external factors is uncontrollable, internal factors that affect enterprises' competitiveness can be under control by these enterprises. Therefore, in this paper, we focus on the enterprise itself and find the internal factors that affect the comprehensive competitiveness of large-scale real estate enterprises. We try to find out the different effects of different internal factors that impact comprehensive competitiveness and the importance of different factors, summarize them, and suggest large-scale real estate enterprises in China.

In summary, the research question of this paper is to analyze the factors that affect the comprehensive competitiveness of large real estate companies in China, especially from the perspective of the future sustainable development of large-scale real estate enterprises. We use the structural equation model method to study the internal factors affecting the improvement of the comprehensive competitiveness of large real estate enterprises. In Section 2, through literature research and analysis, we put forward research hypotheses. In Sections 3 and 4, we use empirical research to verify the hypotheses. Finally, based on research results, some suggestions are made for large-scale real estate enterprises to improve their comprehensive competitiveness in the future.

2. Theoretical Basis and Research Hypothesis

2.1. Direct Effect Hypothesis

The scale of large real estate enterprises is constantly expanding, and the real estate industry faces a fiercer market environment and increased competition. The comprehensive competitiveness of an enterprise can be reflected by its profitability. Both the enterprise's leaders and external professionals affect the development potential of an enterprise [18]. The high profitability of the real estate industry is an important reason for the continuous increase in real estate businesses, and it is essential to real estate enterprises. It determines whether investors want to invest in the enterprise and the project. Ample funds are the prerequisite for the operation of an enterprise [19]. The stronger the profitability of an enterprise, the stronger the comprehensive competitiveness [20,21]. Many factors can reflect the profitability of a real estate enterprise.

Regarding enterprises' factors that affect competitiveness, they mainly concern the enterprise's internal management system, long-term strategic development goals, the profit model and the existing equity structure. This article mainly evaluates the profitability of large real estate enterprises from the two most direct and important indicators, namely, an enterprise's annual sales and profit margin. Therefore, this research proposes the following hypothesis:

Hypotheses 1 (H1). *It is assumed that profitability plays a positive role in improving the comprehensive competitiveness of large real estate enterprises.*

In China, land price is the main cost of the real estate business. Obtaining high-quality and low-priced land has become the most critical issue for real estate enterprises [22]. As a fund-intensive industry, the real estate industry is highly dependent on sources of funds. At present, one major finance source is bank loans, which has aroused great concern among economists and bankers. Some government requirements are set in different developmental stages of the real estate market [23]. It is increasingly difficult for real estate enterprises to obtain funds from various channels. The financial strength of enterprises determines the entry threshold and is also a key factor in determining whether the enterprises can remain competitive in the industry. Therefore, the land bank and financial sources are very important to the comprehensive competitiveness of real estate enterprises [22]. In this paper, we adopt the total assets, financing ability, land bank quality and area as the proxy for capital ability of large real estate enterprises and propose the following hypothesis:

Hypotheses 2 (H2). *It is assumed that capital ability plays a positive role in improving the comprehensive competitiveness of large real estate enterprises.*

In recent years, with an increase in tightening the regulation of the real estate industry, real estate enterprises are facing greater operating pressure. Enterprises pay more and more attention to project development plans and costs and begin to focus more on the internal management of project development plans and cost control [24]. In the context of regulation, real estate enterprises are forced to slow down and spend more effort on checking the gains and losses in operation and management. In addition, as the profit margin of real estate enterprises has decreased, the trend of increasing costs has become more and more prominent, and many real estate enterprises are more concerned with internal management. Real estate enterprises' management and operation ability is related to whether the enterprises can have stronger internal sustainable development ability given the fierce market competition and strict policy control. Thus, real estate enterprises must pay more attention to their management and operation ability, and continuously enhance effective operational management through organizational adjustment, process optimization and information technology improvement [25]. The most critical operational indicators for real estate enterprises are project payment cycle, delivery quality, and cost control level. Large-scale real estate enterprises must reform and conform to their management and internal control procedures. Therefore, we put forward the following hypothesis:

Hypotheses 3 (H3). *Management and operation ability play a positive role in improving the comprehensive competitiveness of large real estate enterprises.*

The real estate industry is characterized by strong requirements of professionalism, a large amount of investment, a long investment return period and high investment risk. Employees with solid professional knowledge, strong psychological quality and comprehensive practical ability are important means for real estate enterprises to gain competitive market advantages [26]. At present, many enterprises have difficulties in retaining people. Although enterprises are constantly recruiting employees, many employees resign. This situation reflects poor human resource management. The mobility of talents is an obstacle to the development of enterprises. Therefore, smart human resource managers should pay more attention to retaining and educating employees than recruiting employees [27]. This paper takes the employee turnover rate and the average education level of employees as two evaluation indicators of human resource ability. The lower the employee turnover rate, the higher the average educational level of employees and the higher the operation level of enterprises. When the human resources of real estate enterprise is enhanced, the operational level will be improved accordingly. We thus propose the following hypotheses:

Hypotheses 4 (H4). *Human resource ability plays a positive role in improving large real estate enterprises' comprehensive competitiveness and management and operation ability.*

Hypotheses 5 (H5). *Human resource ability plays a positive role in improving large real estate enterprises' management and operation ability.*

Peter F. Drucker [28] once said: "organizations in the 21st century can only rely on brand competition because they have nothing else". The competition in the real estate industry is consistent with the ultimate fate of other industries, that is brand. Whoever can seize the brand advantage will grasp the present and future of the Chinese real estate industry [29]. Although consumers ultimately accept a brand name, a corporate brand name often requires enterprises to comprehensively evaluate the comprehensive competitiveness of the operators who provide products [30]. Given the increasingly fierce market competition, branding has become a sharp weapon for enterprises and products to gain a long-lasting competitive advantage. In recent years, brand building, management, marketing, and strategy have become the most widely spread concepts in the real estate industry. In brand competition, real estate enterprises are not only selling houses but also emotion and culture. The enterprises can obtain high added value and maximize profits through brand dissemination and expansion [31]. The real estate enterprises without brands will pay a heavy price for that and face pressure from stronger brands. The social image, customer reputation and the number of benchmark projects of real estate enterprises can better reflect enterprises' brands. The stronger the brand name, the stronger the enterprises in the capital market, and the difficulty of obtaining finance sources will be reduced. At the same time, a brand name gives enterprises more opportunities to obtain more and higher-quality land resources through the government and partners. Therefore, we hypothesize that:

Hypotheses 6 (H6). *Brand name plays a positive role in improving the comprehensive competitiveness of large real estate enterprises.*

Hypotheses 7 (H7). *Brand name plays a positive role in improving the financial ability of large real estate enterprises.*

After 40 years of development, the Chinese real estate industry has transformed from quantity to quality. High-quality housing is a general trend gradually accepted by the market. Large real estate enterprises have made strategic adjustments and changes [32]. With the support of government policies, real estate technology innovation is developing rapidly, and technology is particularly closely integrated with the real estate industry. From the software data service at the earliest complimentary stage, to the technology empowerment service, and then to the Wework office and the app for the intermediary services, they subvert the traditional service model. Technological innovation is the key driving force for real estate value growth, which is one of the core views in the 2020 China Real Estate Innovation Conference [33]. In addition to technological innovation, product innovation and marketing innovation of real estate enterprises are also very important. Only by continuously catering to the needs of consumers to develop new products and continuously enhancing the ability of marketing innovation can the comprehensive competitiveness of real estate enterprises be improved. Technological innovation, product innovation and marketing innovation have become the most important manifestations of the innovation ability of real estate enterprises [34]. As an enterprise's product innovation ability increases, its products will be recognized by more consumers. If the marketing innovation ability is constantly enhanced, the enterprise's sales will be improved. With the continuous enhancement of technological innovation ability, the project construction cost will be significantly optimized, and the profit margin of the enterprise will also be improved. Therefore, we propose the following hypotheses:

Hypotheses 8 (H8). *Innovation ability plays a positive role in improving the comprehensive competitiveness of large real estate enterprises.*

Hypotheses 9 (H9). *Innovation ability plays a positive role in improving the profitability of large real estate enterprises.*

2.2. Intermediary Effect Hypothesis

According to H3 and H5, the stronger the management and operation ability of large real estate enterprises, the stronger the comprehensive competitiveness of the enterprises. The stronger the human resource ability of enterprises, the stronger the management and operation ability of the enterprises. Therefore, we put forward the following hypothesis given the intermediary role of an enterprise's management and operation ability:

Hypotheses 10 (H10). *Management and operation ability play an intermediary role in human resource ability and improving the comprehensive competitiveness of large real estate enterprises.*

According to H1 and H9, the stronger the profitability of large real estate enterprises, the stronger the comprehensive competitiveness of the enterprises. The stronger the innovation ability of enterprises, the stronger the profitability of the enterprises. Therefore, we put forward the following hypothesis given the intermediary role of an enterprise's profitability:

Hypotheses 11 (H11). *Profitability plays an intermediary role in innovation ability and the improvement of the comprehensive competitiveness of large real estate enterprises.*

According to H2 and H7, the stronger the capital ability of large real estate enterprises, the stronger the comprehensive competitiveness of the enterprises. The stronger the brand-ability of enterprises, the stronger the capital ability of the enterprises. Therefore, we put forward the following hypothesis given the intermediary role of an enterprise's capital ability:

Hypotheses 12 (H12). *Capital ability plays an intermediary role in brand-ability and improving the comprehensive competitiveness of large real estate enterprises.*

2.3. Summary of Research Hypothesis

Based on the above analysis, there are nine direct effect and three intermediary effect hypotheses, as summarized in Table 2.

Table 2. The summary of research hypotheses.

Effect	Number	The Content of Hypotheses
Direct effect	H1	Profitability plays a positive role in improving the comprehensive competitiveness of large real estate enterprises.
	H2	The capital ability plays a positive role in improving the comprehensive competitiveness of large real estate enterprises.
	H3	The management and operation ability plays a positive role in improving the comprehensive competitiveness of large real estate enterprises.
	H4	The human resource ability plays a positive role in improving the comprehensive competitiveness of large real estate enterprises.
	H5	Human resource ability plays a positive role in improving large real estate enterprises' management and operation ability.
	H6	Brand-ability plays a positive role in improving the comprehensive competitiveness of large real estate enterprises.
	H7	Brand-ability plays a positive role in improving the capital ability of large real estate enterprises.
	H8	The innovation ability plays a positive role in improving the comprehensive competitiveness of large real estate enterprises.
	H9	The innovation ability plays a positive role in improving the profitability of large real estate enterprises.
Intermediary effect	H10	The management and operation ability plays an intermediary role in human resource ability and improving the comprehensive competitiveness of large real estate enterprises.
	H11	Profitability plays an intermediary role in innovation ability and the improvement of the comprehensive competitiveness of large real estate enterprises.
	H12	The capital ability plays an intermediary role in brand-ability and improves the comprehensive competitiveness of large real estate enterprises.

2.4. Conceptual Model

As shown in Figure 1, a conceptual model of the internal factors influencing the comprehensive competitiveness of large real estate enterprises is established, which is in line with the above research hypotheses.

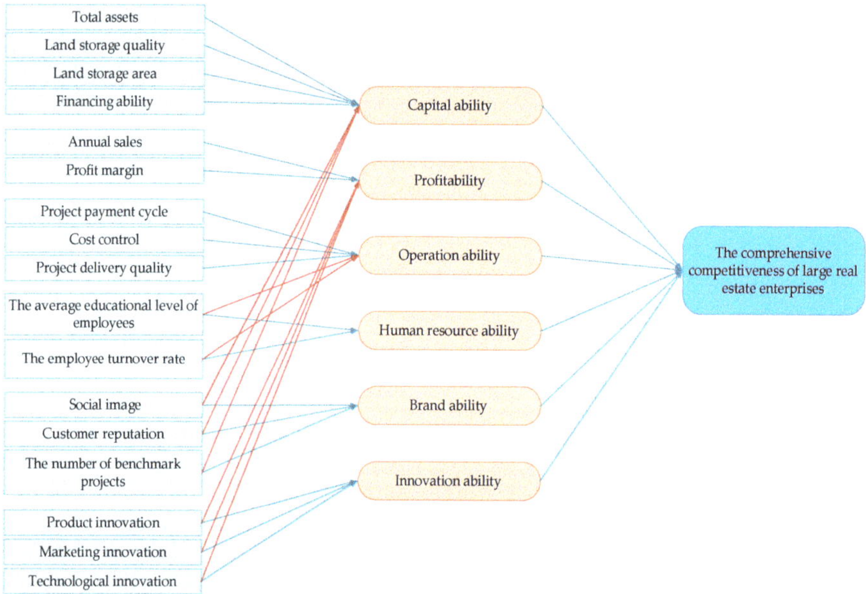

Figure 1. The conceptual model of the internal factors influencing the comprehensive competitiveness of large real estate enterprises.

3. Research Design

3.1. Research Sample and Variable Measurement

In this research, we collected data through questionnaires to study the impact of the 17 observed variables (Figure 1) on improving the comprehensive competitiveness of large real estate enterprises. Because personnel in different professions or functional departments have different perceptions of improvement of the comprehensive competitiveness of large real estate companies in the future. To maintain the objectivity of the questionnaire, the respondents of the questionnaire are managers of large real estate enterprises in China, with nine from each enterprise. They are the main responsible personnel of the enterprise's investment, finance, operations, human resources, cost and other departments. The questionnaire results were collected and summarized through online and on-site filling. In total, 260 questionnaires were collected, and 236 valid questionnaires were used for further analysis after excluding some incomplete and invalid questionnaires. The basic information statistics of the respondents are shown in Table 3.

To make the questionnaire meet the basic requirements of scientific research, according to the basic procedures of the questionnaire survey and previous studies, the top ten real estate enterprises in China in 2020 were selected for pre-survey before the formal questionnaire survey [29]. Based on the pre-survey results, professionals and scholars in this field are invited to formulate a formal questionnaire for this research. The questionnaire of this paper is mainly composed of the following parts: the first part is the survey introduction, which introduces the purpose and significance of this survey, as well as the promise to guarantee the confidentiality of the personal information and questionnaire information of the respondents; the second part is the instructions and guidelines for filling in the questionnaire; the third part is the central part of the questionnaire, including the basic information of the respondents and measurement of internal factors affecting the improvement of the comprehensive competitiveness of large real estate enterprises.

Table 3. The basic information of the respondents.

Sample Information	Options	Proportion (%)	Sample Information	Options	Proportion (%)
Age	25–30	28.39	Working years	1–5	21.41
	31–35	30.93		6–10	33.9
	36–40	20.76		11–15	24.58
	41–45	11.44		15–20	11.02
	46–50	8.47		More than 21	9.1
Gender	Male	72.46	Job category	Finance related	9.75
	Female	27.54		Cost related	12.71
Educational background	Junior college	5.08		Investment-related	25.42
	Bachelor	68.64		Operation related	27.97
	Master	25		Marketing related	27.97
	Ph.D.	1.27		Engineering related	22.88
				Human resources and administration related	5.08

The questionnaire questions include various indicators such as the enterprise's total assets, land bank area, financing level, payment cycle and customer reputation. The Likert scale [35,36] is used in the questionnaire, reflecting the degree of agreement among respondents to the questions set in the questionnaire. Each question has a degree value of 1–5: 1 means "completely disagree", 2 means "relatively disagree", 3 means "neutral", 4 means "relatively agree" and 5 means "completely agree". The larger the value, the greater the degree of agreement. On the other hand, the respondents need to rank the importance of each factor in the mentioned questions [37]. Each question has A–E degree value: A means "completely unimportant", B means "relatively unimportant", C means "generally important", D means "relatively important" and E means "particularly important"—that is, the larger the value, the more important.

3.2. Reliability and Validity Test

The SPSS and AMOS statistical software was used for data analysis.

3.2.1. Reliability Test

We used the reliability test to evaluate the reliability of the questionnaire and the stability and consistency of the questionnaire results [38]. The measurement index of the reliability test is usually the reliability coefficient, Cronbach's α coefficient. In the testing process, the reliability coefficient first should be analyzed: if the reliability is higher than 0.8, the reliability is good; if it is between 0.6 and 0.8, the reliability is acceptable; if it is less than 0.6, the reliability is poor. The second is to analyze the corrected item-total correlation (CITC). If the CITC value is lower than 0.3, the item can be deleted. At last, if the Cronbach's α coefficient of the item deleted is significantly higher, the item can be re-analyzed [39].

The reliability test of the questionnaire on internal factors affecting the improvement of the comprehensive competitiveness of large real estate enterprises is shown in Table 4. The reliability coefficient is 0.904, which is greater than 0.9. The CITC values are greater than 0.3. For the Cronbach's α coefficient, if an item is deleted, the reliability coefficient will not increase much after any question items are deleted, so no item needs to be deleted. In conclusion, the reliability of the questionnaire data is excellent and can be used for further analysis.

Table 4. Reliability test of the questionnaire.

Name	CITC.	The Cronbach's α Coefficient if Item Deleted	Cronbach's α Coefficient
Total assets	0.467	0.905	0.904
Land storage area	0.555	0.899	
Land storage quality	0.435	0.902	
Financing ability	0.382	0.906	
Annual sales	0.491	0.901	
Profit margin	0.537	0.900	
Payment cycle	0.350	0.904	
Cost control	0.437	0.903	
Delivery quality	0.493	0.901	
The average educational level of employees	0.371	0.903	
The employee turnover rate	0.451	0.902	
Social image	0.603	0.898	
Customer reputation	0.483	0.901	
The number of benchmark projects	0.696	0.896	
Product innovation	0.723	0.896	
Marketing innovation	0.609	0.898	
Technological innovation	0.546	0.900	

3.2.2. Validity Test

Validity analysis is used to study the design rationality of quantitative data, especially attitude scale questions. Generally, the KMO, communality (common factor variance), variance interpretation rate and factor loading coefficient are introduced to study whether the item is reasonable and meaningful [40]. In the process of testing, the KMO value is analyzed first to test the validity: if the value is higher than 0.8, it means that the validity is high; if the value is between 0.6 and 0.8, the validity is acceptable; if the value is less than 0.6, the validity is poor. The second is to analyze the communality value used to exclude unreasonable research items and the variance interpretation rate value used to indicate the level of information extraction. If the communality value corresponding to an analysis item is lower than 0.4, the item can be deleted. If the cumulative variance interpretation rate after rotation is greater than 50%, the amount of information on the question items can be effectively extracted. Finally, some attention should be paid to the corresponding relationship between the question items and the factor loading coefficient. If the corresponding relationship is consistent with the psychological research expectation, the validity is high [41].

From Table 5, we can see the validity test of the questionnaire on internal factors affecting the improvement of the comprehensive competitiveness of large real estate enterprises. The KMO value is 0.936, greater than 0.6, indicating the validity of the data; the communality values of all research items are higher than 0.4, indicating that the research item information can be effectively extracted. The variance interpretation rate values of the three factors are 27.927%, 21.289% and 12.403%, respectively. After rotation, the cumulative variance interpretation rate is 61.619%, which is greater than 50%, indicating that the amount of information in the research item is valid. Concurrently, all research items have factor loading coefficients with an absolute value greater than 0.4, which means that the items and factors have a corresponding relationship. The relationship is positive, which is in line with expectations. In summary, the research data are highly valid and can be used for further analysis.

Table 5. Validity test of the questionnaire.

Name	Factor Loading Coefficient			Communality
	Factor 1	Factor 2	Factor 3	
Total assets	0.163	0.640	0.326	0.542
Land storage area	0.219	0.734	0.313	0.685
Land storage quality	0.272	0.782	0.071	0.690
Financing ability	0.263	0.684	0.109	0.549
Annual sales	0.240	0.615	0.345	0.555
Profit margin	0.513	0.503	0.002	0.517
Payment cycle	0.552	0.560	0.140	0.638
Cost control	0.579	0.372	0.164	0.501
Delivery quality	0.676	0.248	0.342	0.635
The average educational level of employees	0.376	0.280	0.595	0.574
The employee turnover rate	0.154	0.125	0.805	0.688
Social image	0.593	0.244	0.546	0.710
Customer reputation	0.685	0.253	0.297	0.622
The number of benchmark projects	0.655	0.287	0.359	0.641
Product innovation	0.802	0.194	0.135	0.700
Marketing innovation	0.672	0.237	0.206	0.550
Technological innovation	0.774	0.232	0.164	0.680
Eigen value (Unrotated)	8.161	1.270	1.043	
Variance interpretation rate % (Unrotated)	48.009%	7.473%	6.138%	
Cumulative variance interpretation rate % (Unrotated)	48.009%	55.482%	61.619%	
Eigen value (Rotated)	4.748	3.619	2.109	
Variance interpretation rate % (Rotated)	27.927%	21.289%	12.403%	
Cumulative variance interpretation rate % (Rotated)	27.927%	49.216%	61.619%	
KMO		0.936		

4. Research Results

4.1. *Descriptive Statistics and Correlation Analysis of Variables*

The structural equation model constructs the relationship between each potential variable and its observed variable [42–45]. It shows the influence degree of each observed variable on other variables. The specific and intuitive influence coefficients can be obtained after modelling analysis and data processing through Amos, as shown in Table 6.

First, the observed variables of capital ability are total assets, land storage area, land storage quality and financing ability of enterprises. It can be seen from Table 6 that the area of the land bank has the greatest impact on the improvement of the comprehensive competitiveness of large real estate enterprises (0.994), followed by the landbank quality (0.939); the level of financing ability and total assets are ranked last. Overall, the capital ability has a noticeable influence on improving the comprehensive competitiveness of an enterprise, which belongs to the first gradient level among several potential variables. Therefore, large real estate enterprises must pay attention to landbank area and quality indicators, and then financing ability and total assets. These affect real estate enterprises' comprehensive competitiveness and will affect the long-term development of enterprises.

Table 6. The influence coefficients between the potential variable and the observed variable.

Potential Variable	Observed Variable	Standardized Coefficient
Capital ability	Total assets	0.845
	Land storage area	0.994
	Land storage quality	0.939
	Financing ability	0.860
Profitability	Annual sales	0.981
	Profit margin	0.845
Management and operation ability	Payment cycle	0.723
	Cost control	0.854
	Delivery quality	0.828
Human resource ability	The average educational level of employees	0.685
	The employee turnover rate	0.881
Brand-ability	Social image	0.736
	Customer reputation	0.841
	The number of benchmark projects	0.809
Innovation ability	Product innovation	0.818
	Marketing innovation	0.648
	Technological innovation	0.788

Second, the observed variables of profitability are corporate annual sales and profit margin. Compared with a profit margin (0.845), annual sales have a greater impact on the comprehensive competitiveness of large real estate enterprises (0.981), which also reflects why real estate enterprises want to engage in large-scale business today—after the increase in sales, the comprehensive competitiveness of enterprises will be improved. Many real estate enterprises are struggling to reach CNY 100 billion and CNY 200 billion in sales. At the same time, some real estate enterprises prefer to sacrifice profits and obtain larger sales by acquiring more land.

Third, the observable variables of operational ability are the payment cycle, delivery quality and cost control. Among the three variables, the payment cycle has the most negligible impact on the comprehensive competitiveness of large real estate enterprises (0.723), and the impact of delivery quality (0.828) and cost control (0.854) is comparable. At present, many scholars have pointed out that the future development of the real estate industry will test enterprises' operational ability. The management and operation ability also play a significant role in enhancing the comprehensive competitiveness of enterprises. Take China Overseas Land and Investment Ltd. as an example—although it is a state-owned enterprise, it has a rigorous cost control level which is well-known in the industry, and this lays the foundation for its comprehensive competitiveness in the top ten Chinese real estate enterprises.

Fourth, human resource ability refers to the average educational level of employees and employee turnover rate. Compared with the average educational level of employees (0.685), the employee turnover rate has a greater influence on the comprehensive competitiveness of large real estate enterprises (0.881). This reflects that employees' experiences are more important to the enterprise than their educational background. The importance of employee turnover rate to enterprises is self-evident, and enterprises with too high an employee turnover rate will certainly not be too competitive. Among the top 30 real estate enterprises in China, the employee turnover rates of Poly Developments and Holdings Co., Ltd. and Longfor Group Holding Co., Ltd. are relatively low, and they have strong comprehensive competitiveness in the industry.

Fifth, the observed variables of brand-ability are corporate social image, customer reputation and the number of benchmark projects. Customer reputation has the most apparent impact on improving the comprehensive competitiveness of large real estate enterprises (0.841), followed by the number of benchmark projects (0.809). This reflects the importance of customer reputation to the comprehensive competitiveness of enterprises. A house is essentially a commodity. Buyers have an intuitive feeling about the quality of the commodity after moving in, and this feeling will affect the real estate enterprises' reputation in the end. Only by continuous accumulation in good customer reputation can the enterprise's brand-ability be established, and then the enterprise's comprehensive competitiveness will be enhanced. This also explains well that the first factor for many buyers in choosing a house is to look at the enterprise's reputation because people believe that the quality of products made by a company with a good reputation can be guaranteed. Meanwhile, the number of benchmark projects is also very important because brand-ability is promoted by creating multiple benchmarking projects in many cases. In the Chinese real estate industry, enterprises such as Vanke Co., Ltd. and Greentown China Holdings Ltd. have very strong brand-ability. People recognize the products of these companies, so their comprehensive competitiveness in the industry is very strong.

Sixth, the observed variables of innovation ability are product innovation, marketing innovation and technological innovation. In contrast to product innovation and technological innovation, marketing innovation has the least impact on the comprehensive competitiveness of large real estate enterprises (0.648), and marketing innovation is also the variable with the least impact among the 17 observed variables. On the one hand, the marketing model of the real estate industry has been very mature, and it is relatively challenging to innovate. On the other hand, no matter how innovative the marketing method is, it is impossible to ultimately decide whether the buyers choose to buy the enterprise's products. However, there are some differences between product innovation and technological innovation. Real estate enterprises must constantly polish their development technology and product design, and constantly strengthen product innovation and technological innovation, to better cater to the demands of buyers in the market. In addition, the requirements of the government for energy conservation, environmental protection, green buildings, sustainable development and other concepts should also comply.

4.2. Structural Model

The influence of the 17 observed variables shown in Figure 1 on the improvement of the comprehensive competitiveness of large real estate companies was obtained through the questionnaire. Then, the reliability of the questionnaire and the stability and consistency of the questionnaire results were verified (Section 3.2). Finally, the structural equation model was used to construct the calculation process, and the influence coefficients shown in Table 6 were obtained after modelling analysis and data processing through the Amos software (Section 4.1). Next, Figure 2 summarizes the structural equation model results for the internal factors influencing the improvement of the comprehensive competitiveness of large real estate companies shown in Table 6.

The validity of the 12 hypotheses mentioned above can be tested through the model. Among them, capital ability, profitability, management and operation ability, human resource ability, brand-ability and innovation ability have positive effects on the comprehensive competitiveness of large real estate enterprises. Meanwhile, management and operation ability play an intermediary role between human resource ability and the comprehensive competitiveness of large real estate enterprises. Profitability plays an intermediary role between innovation ability and the improvement of the comprehensive competitiveness of large real estate enterprises. Capital capability plays an intermediary role between brand-ability and the comprehensive competitiveness of large real estate enterprises.

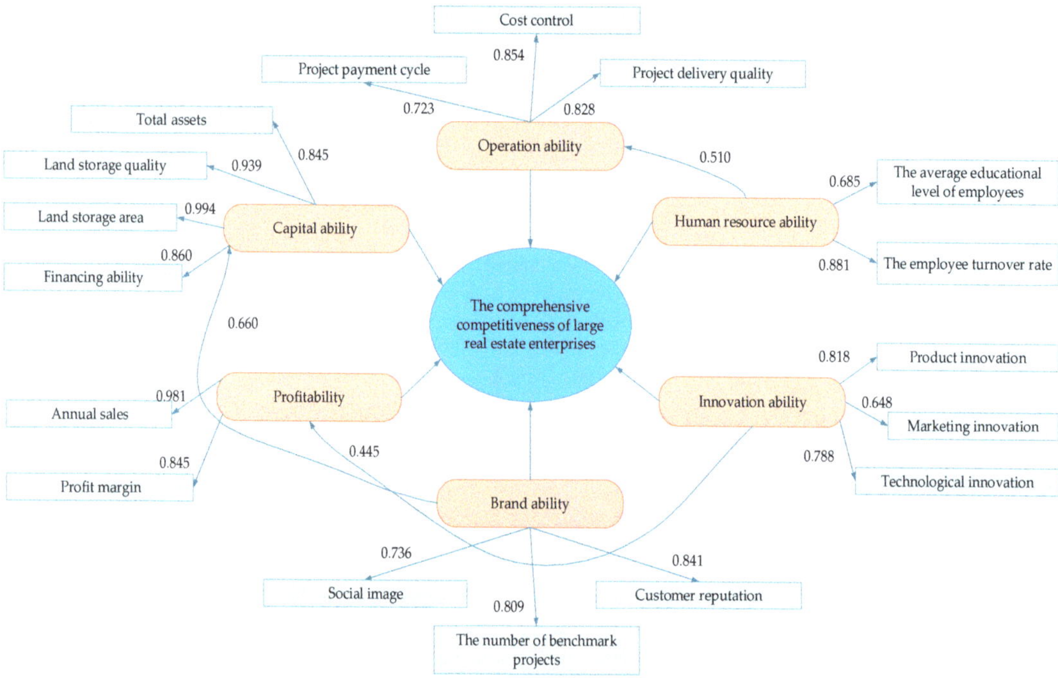

Figure 2. Results of structural equation model for internal factors affecting the comprehensive competitiveness improvement of large real estate enterprises.

4.3. Hypothesis Test

4.3.1. Analysis of the Direct Effect

Based on the results in Figure 2, we can draw the following conclusion:

The standardized coefficients of annual sales and profit margin on improving the comprehensive competitiveness of large real estate enterprises are 0.981 and 0.845, respectively, indicating that corporate profitability has a significant positive effect on improving the comprehensive competitiveness of large real estate enterprises. Thus, H1 is confirmed.

The standardized coefficients of the total assets, land storage area, land storage quality and financing ability on the improvement of the comprehensive competitiveness of large real estate enterprises are 0.845, 0.994, 0.939 and 0.860, respectively, indicating that corporate capital ability has a significant positive effect on the improvement of the comprehensive competitiveness of large real estate enterprises. Thus, H2 is also confirmed.

The project delivery quality, cost control and payment cycle have standardized coefficients of 0.828, 0.854 and 0.723 for the improvement of the comprehensive competitiveness of large real estate enterprises, respectively, indicating that corporate management and operation ability has a significant positive effect on the improvement of the comprehensive competitiveness of large real estate enterprises. Therefore, H3 has been confirmed.

The standardized coefficients of the average educational level of employees and the employee turnover rate for the improvement of the comprehensive competitiveness of large real estate enterprises are 0.685 and 0.881, respectively, indicating that corporate human resource ability has a significant positive effect on the improvement of the comprehensive competitiveness of large real estate enterprises. As such, H4 is confirmed.

The number of benchmark projects, social image and customer reputation have standardized coefficients of 0.809, 0.736 and 0.841 for the improvement of the comprehensive competitiveness of large real estate enterprises, respectively, indicating that corporate brand-ability have a significant positive effect on the improvement of the comprehensive competitiveness of large real estate enterprises. Therefore, H5 has been verified.

The standardized coefficients of product innovation, marketing innovation and technological innovation for the improvement of the comprehensive competitiveness of large real estate enterprises are 0.818, 0.648 and 0.788, respectively, indicating that corporate innovation ability has a significant positive effect on the improvement of the comprehensive competitiveness of large real estate enterprises. Consequently, H6 has been verified.

The standardized coefficient of brand-ability for capital ability is 0.760, indicating that brand-ability has a significant positive effect on the improvement of capital ability. H7 has been confirmed.

The standardized coefficient of innovation ability for profitability is 0.736, indicating that innovation ability has a significant positive effect on the improvement of profitability. H8 has been confirmed.

The standardization coefficient of human resource ability for management and operation ability is 0.810, indicating that human resource ability has a significant positive effect on the improvement of management and operation ability. H9 has been confirmed.

4.3.2. Analysis of the Intermediary Effect

The intermediary variable is the third variable compared to the independent variable and dependent variable. When considering the influence of independent variable X on dependent variable Y, if variable X affects Y by influencing variable M, then the variable M is the intermediary variable [46]. This paper uses the three-part intermediary test method proposed by Wen et al. [47] to test the intermediary relationship. The specific steps are as follows. Firstly, the correlation between the independent variable X and the intermediary variable M is obtained through regression analysis to verify the significant correlation between the two variables. Secondly, the correlation between the independent variable X and the dependent variable Y is obtained by regression analysis to verify the significant correlation between the two variables. Finally, the intermediate variable M is added to the second step of the regression analysis, and the verification result is obtained. Suppose the standardized regression coefficient of variable X to variable Y in the third step is smaller than the coefficient of variable X to variable Y in the second step, and variable M is significantly correlated with variable Y. In that case, there is an intermediary relationship among the three variables. In the third step of the test, if variable X has no significant effect on variable Y, then variable M plays a completely intermediary role; if variable X has a significant effect on variable Y, variable M plays a partially intermediary role.

The enterprise's management and operation ability are intermediary variables, the employee turnover rate and average educational level of employees are the independent variables, and the improvement of the comprehensive competitiveness of large real estate enterprises is the dependent variable. According to the previous analysis, there is a significant positive correlation between the variables, which meets the first and second steps of the three-step test. After adding the intermediary variable, the intermediary effect test is shown in Table 7.

After adding management and operation ability as the intermediary variable, we obtain the following conclusions: the standardized regression coefficient of the human resource ability on the improvement of the comprehensive competitiveness of large real estate enterprises has decreased. However, it is still significant ($0.701 < 0.881$, $p < 0.001$). Moreover, as an intermediary variable, management and operation ability still has a significant effect on improving the comprehensive competitiveness of large real estate enterprises. This shows that management and operation ability play a partially intermediary role between human resource ability and the comprehensive competitiveness of large real estate enterprises. Accordingly, H10 is verified.

Table 7. Test results of the intermediary effect of management and operation ability.

Intermediary Variable	**Management and Operation Ability**	
Dependent variable	The improvement of the comprehensive competitiveness of large real estate enterprises	
Independent variable	Employee turnover rate	The average educational level of employees
	Step 1: Direct influence effect	
Independent variable	0.881	0.685
Adjusted R^2	0.747	0.653
	Step 2: Intermediary influence effect	
Independent variable	0.701	0.657
Intermediary variable	0.582	0.433
Total R^2	0.658	0.613
ΔR^2	0.089	0.040

Corporate profitability is used as an intermediary variable; product innovation, marketing innovation and technological innovation are the independent variables; and the improvement of the comprehensive competitiveness of large real estate enterprises is the dependent variable. According to the previous analysis, there is a significant positive correlation between the variables, which meets the first and second steps of the three-step test. After adding the intermediary variable, the intermediary effect test is shown in Table 8.

Table 8. Test results of the intermediary effect of profitability.

Intermediary Variable	**Profitability**		
Dependent variable	The improvement of the comprehensive competitiveness of large real estate enterprises		
Independent variable	Product innovation	Marketing innovation	Technological innovation
	Step 1: Direct influence effect		
Independent variable	0.818	0.648	0.788
Adjusted R^2	0.723	0.611	0.701
	Step 2: Intermediary influence effect		
Independent variable	0.823	0.657	0.801
Intermediary variable	0.682	0.565	0.726
Total R^2	0.645	0.509	0.694
ΔR^2	0.078	0.102	0.007

After adding profitability as the intermediary variable, the following conclusions can be drawn: the standardized regression coefficient of innovation ability on improving the comprehensive competitiveness of large real estate enterprises has risen; it is no longer significant (0.823 > 0.818, $p < 0.001$). This shows that profitability cannot play an intermediary role between the innovation ability and the improvement of comprehensive competitiveness of large real estate enterprises. Therefore, H11 is not tenable.

According to the same method, it can be verified that H12 is established, that is, capital ability plays a partial intermediary role between brand-ability and the improvement of comprehensive competitiveness of large real estate enterprises.

4.3.3. Summary of Hypothesis Test Results

Through empirical research and analysis, we found that the improvement of comprehensive competitiveness of large real estate enterprises is affected by profitability, capital ability, management and operation ability, human resource ability, brand-ability and innovation ability. This study's main purpose is to clarify further the logical relationship and influence mechanism between these variables. Based on the 236 valid samples obtained from the formal investigation, all the hypotheses of this study were tested by statistical analysis methods commonly used in empirical research. The test results are shown in Table 9.

Table 9. The summary of hypothesis test results.

Effect	Number	The Content of Hypotheses	Test Results
Direct effect	H1	Profitability plays a positive role in improving the comprehensive competitiveness of large real estate enterprises.	Confirmed
	H2	The capital ability plays a positive role in improving the comprehensive competitiveness of large real estate enterprises.	Confirmed
	H3	The management and operation ability plays a positive role in improving the comprehensive competitiveness of large real estate enterprises.	Confirmed
	H4	The human resource ability plays a positive role in improving the comprehensive competitiveness of large real estate enterprises.	Confirmed
	H5	Human resource ability plays a positive role in improving large real estate enterprises' management and operation ability.	Confirmed
	H6	Brand-ability plays a positive role in improving the comprehensive competitiveness of large real estate enterprises.	Confirmed
	H7	Brand-ability plays a positive role in improving the capital ability of large real estate enterprises.	Confirmed
	H8	The innovation ability plays a positive role in improving the comprehensive competitiveness of large real estate enterprises.	Confirmed
	H9	The innovation ability plays a positive role in improving the profitability of large real estate enterprises.	Confirmed
Intermediary effect	H10	The management and operation ability plays an intermediary role in human resource ability and the improvement of the comprehensive competitiveness of large real estate enterprises.	Partial Confirmed
	H11	Profitability plays an intermediary role in innovation ability and the improvement of the comprehensive competitiveness of large real estate enterprises.	Not support
	H12	The capital ability plays an intermediary role in brand-ability and the improvement of the comprehensive competitiveness of large real estate enterprises.	PartialConfirmed

5. Conclusions

With this paper, we attempted to study the internal factors influencing the comprehensive competitiveness of China's large real estate companies in the future. We conducted a questionnaire survey on Chinese large real estate enterprises. The Amos structural equation method was used for modelling analysis and data processing. Then, the hierarchical regression analysis method was introduced to verify 12 research hypotheses, of which nine hypotheses were established, two were partially established, and one was invalid. Based on our study, the following suggestions are proposed for large real estate enterprises to enhance their comprehensive competitiveness in the future.

First, land resources integration should be improved, which means the increase in the land storage area and the improvement of land storage quality. Land resources are an essential core competitive resource for real estate enterprises, and whether land resources are adequate or not will directly affect the development of enterprises. Reserving a large amount of high-quality land can effectively increase the scale of development projects, achieve economies of scale and reduce development costs. For a real estate enterprise, only by carrying out proper land resource reserves can it lay the foundation for future development. Therefore, real estate enterprises must continuously improve their ability to integrate land resources.

Second, some importance should be attached to human resource management. Land resources and capital resources are important to core competition resources for real estate enterprises, but they are only necessary conditions for cultivating the core competitiveness of enterprises. Combined with the impact of human resources, the development of enterprises will be improved. Real estate enterprises belong to an industry with a wide range of business categories, which involves planning, design, building construction, supervision, garden landscape, enterprise management, cost accounting, marketing planning, evaluation, etc. At present, what real estate enterprises urgently need are compound talents with "one specialization and multiple abilities". In the new economic era, the competition between enterprises is the competition of talents. If real estate enterprises want to win in the fierce market competition, they must invest in the development of human resources and improve the ability of human resources integration. Moreover, they must maintain the team's stability and lower the employee turnover rate, thereby improving the overall competitiveness of the enterprises.

Third, brand-building should be paid attention to. The competition in the real estate industry has changed from the competition of scale advantages, the competition of specialization, and technology to the competition of corporate brands. Branding is an intangible force, playing a decisive role in improving the comprehensive competitiveness of real estate enterprises, and it is an important manifestation of whether real estate enterprises form core competitiveness. In the fierce competition of the real estate market, the brand has an increasing influence on consumers' house purchase decisions, and it will take a more critical role in the competition of real estate enterprises.

Fourth, corporate financing ability and management and operation ability should be enhanced. Enhancing financing capacity is an essential guarantee for improving the comprehensive competitiveness of enterprises. In contrast to enterprises in other industries, real estate enterprises are highly capital-intensive operators. Since most of the completed real estate projects are mainly held, cash can be recovered through rents, mortgages and other income, which leads to a more extended recovery period of capital and higher requirements for the operation of the capital chain. Therefore, a strong financing ability is an important guarantee for enhancing the competitiveness of real estate companies. On the other hand, improving the management and operation ability of an enterprise means that all enterprise links are required to be in a state of efficient operation, which is an expression of the comprehensive competitiveness of the enterprise.

In terms of the management field, the improvement of comprehensive competitiveness of large real estate enterprises is affected by the profitability, capital, operation, human resource, brand and innovation ability. The hypothetical research results proposed in this paper clarify the logical relationship and influence mechanism between these variables. The research results of this paper provide practical enlightenment for enhancing the competitiveness of large-scale real estate enterprises. It is helpful to improve the sustainable development of the real estate industry in the future to a certain extent.

In terms of academic contribution, as research on factors that affect large-scale real estate enterprises is scarce, this work aims to fill this gap. Compared with previous studies on the improvement of the comprehensive competitiveness of large real estate enterprises, this paper attempts to quantify the internal factors affecting the improvement of comprehensive competitiveness by using the structural equation model for the first time. It provides theoretical support for the sustainable development research of large real estate enterprises.

There are still some works to be completed in the future, considering some limitations of this paper. Although we focus on the internal influencing factors of comprehensive competitiveness, large real estate enterprises in China are faced with complex external factors such as market and policy. Improving the comprehensive competitiveness of real estate enterprises in the complex external environment still requires more research in the future. In addition, this paper only discusses the one-way relationship between the profitability, capital ability, management and operation ability, human resource ability, brand-ability, innovation ability of the enterprises and the improvement of the comprehensive competitiveness of large real estate enterprises. However, there actually may be a complex two-way relationship between them. Exploring the reverse impact of the improvement of comprehensive competitiveness on these abilities may increase the objectivity of the research results and provide another avenue for future research.

Author Contributions: B.L. and R.Y.M.L. conceived and designed the study; B.L. completed the paper in English and R.Y.M.L. revised the important intellectual content; R.Y.M.L. provided a comprehensive English revision. T.W. gave valuable research advice and revised the manuscript. All authors have read and agreed to the published version of the manuscript.

Funding: This research received no external funding.

Data Availability Statement: Not applicable.

Acknowledgments: The authors wish to thank the anonymous reviewers, whose insightful comments and helpful suggestions significantly contributed to improving this paper.

Conflicts of Interest: The authors declare no conflict of interest.

References

1. Cai, Z.; Liu, Q.; Cao, S. Real estate supports rapid development of China's urbanization. *Land Use Policy* **2020**, *95*, 104582. [CrossRef]
2. Cai, Z. The impact of COVID-19 on the housing market and the trend of differentiation. *China Policy Rev.* **2020**, *119*, 116–131. (In Chinese)
3. de la Fuente-Mella, H.; Rubilar, R.; Chahuán-Jiménez, K.; Leiva, V. Modeling COVID-19 Cases Statistically and Evaluating Their Effect on the Economy of Countries. *Mathematics* **2021**, *9*, 1558. [CrossRef]
4. Chen, S.; Li, G. Evaluation of Impacts on China's Economy and Fiscal Countermeasures of COVID-19. *Res. Econ. Manag.* **2020**, *41*, 11–20. (In Chinese) [CrossRef]
5. National Bureau of Statistics of China. Notice on Printing and Distributing "the Measures for the Division of Large, Medium, Small and Micro Enterprises in Statistics". 2018. Available online: http://www.stats.gov.cn/tjgz/tzgb/201801/t20180103_1569254.html (accessed on 10 December 2020).
6. Sun, H.; Fan, Z.; Zhou, Y.; Shi, Y. Empirical research on competitiveness factors: Analysis of real estate industry of Beijing and Tianjin. *Eng. Constr. Archit. Manag.* **2010**, *17*, 240–251. [CrossRef]
7. Zhang, X.; Shen, L.; Wu, Y.; Fan, L.C. Competitiveness assessment for real estate enterprises in china: A model-procedure. *Int. J. Strateg. Prop. Manag.* **2009**, *13*, 229–245. [CrossRef]
8. Wang, Y.; Wang, E.T. Research on Core Competitiveness of Real Estate Enterprise Based on Factor Analysis and Clustering Analysis. *Appl. Mech. Mater.* **2014**, *457*, 1323–1329. [CrossRef]
9. Nuan, C.; Ji, C.; Chen, Z. China city real estate industry competitiveness evaluation basing on Mutation series method—Demonstration from 35 large and medium city plate data. *Spec. Zone Econ.* **2011**, *8*, 254–256. (In Chinese)
10. Zheng, L.; Zhang, J. Analysis of the Real Estate Enterprise's Core Competitiveness Based on Sustainable Development. *Manag. Sci. Eng.* **2011**, *5*, 132–134.
11. Luo, S. A Study of Listed Real Estate Enterprise's Comprehensive Competitiveness Evaluation Based on Rough Sets and Entropy Weight. Master's Thesis, Chongqing University, Chongqing, China, 2011.
12. Mo, J. Research Report on the Competitiveness of Chinese Real Estate Enterprises. 2007. Available online: http://www.gov.cn/govweb/fwxx/wy/2007-09/19/content_754091.htm (accessed on 20 December 2020).
13. Chen, B. Study on Real Estate Comprehensive Competition Analysis and Strategies. Ph.D. Thesis, Tianjin University, Tianjin, China, 2009.
14. Li, V. A Methodology to Assess the Competitiveness of Real Estate Developers in China. Ph.D. Thesis, Queensland University of Technology, Brisbane, Australia, 2011.
15. Jiang, J. A Study on the Regulation Policies on China's Real Estate Market. Ph.D. Thesis, Huazhong University of Science and Technology, Wuhan, China, 2012.

16. Wu, F. China's recent urban development in the process of land and housing marketization and economic globalization. *Habitat Int.* **2001**, *25*, 273–289. [CrossRef]
17. Yu, H. China's house price: Affected by economic fundamentals or real estate policy? *Front. Econ. China* **2010**, *5*, 25–51. (In Chinese) [CrossRef]
18. Wei, X.; Tamura, T.; Oidaira, T. Financial Analysis Concerning Medium-Scale Enterprises in Chinese Real Estate Industry. *J. Jpn. Manag. Diagn. Assoc.* **2017**, *17*, 108–114. [CrossRef]
19. Oubala, K. Is Morocco an Attractive Destination for Foreign Investors Looking to Invest in the Residential Real Estate Segment? Master's Thesis, Massachusetts Institute of Technology, Cambridge, MA, USA, 2011. Available online: http://hdl.handle.net/1721.1/68177 (accessed on 2 November 2021).
20. Ye, X. Risk analysis in the process of real estate enterprise project investment. In Proceedings of the 2011 Fourth International Joint Conference on Computational Sciences and Optimization, Kunming and Lijiang City, China, 15–19 April 2011; pp. 731–736. [CrossRef]
21. Gibler, K.M.; Lindholm, A.-L. A test of corporate real estate strategies and operating decisions in support of core business strategies. *J. Prop. Res.* **2012**, *29*, 25–48. [CrossRef]
22. Yan, J.; Lei, F.; Mo, D.; Xue-qin, T. Housing market cycle in China: An empirical study based on stock-flow model. In Proceedings of the 2010 International Conference on Management Science & Engineering 17th Annual Conference Proceedings, Melbourne, VIC, Australia, 24–26 November 2010; pp. 1671–1676. [CrossRef]
23. Liu, Y. Motivation, Dilemma and Countermeasures of Foreign Bond Financing of Real Estate Enterprises? Based on Perspective of Financial Internationalization. *Proc. Bus. Econ. Stud.* **2020**, *3*. [CrossRef]
24. Wen, C.; Wang, X. Analysis of Influencing Factors of the Value of Real Estate Companies Based on the Value Chain. In Proceedings of the 2014 International Conference on Construction and Real Estate Management, Kunming, China, 27–28 September 2014; pp. 627–636. [CrossRef]
25. Meng, F.; Wang, X.; Chen, H.; Zhang, J.; Yang, W.; Wang, J.; Zheng, Q. The influence of organizational culture on talent management: A case study of a real estate company. *J. Chin. Hum. Resour. Manag.* **2016**, *7*, 129–146. [CrossRef]
26. Gibb, S.; Zhang, S. Guanxi influence and talent management in Chinese organizations; evidence from the real estate sector. *Hum. Resour. Dev. Int.* **2017**, *20*, 79–98. [CrossRef]
27. Shi, M.Y. *Sectoral Booms and Misallocation of Managerial Talent: Evidence from the Chinese Real Estate Boom*; IMF Working Paper; International Monetary Fund: Washington, DC, USA, 2018; p. 18.
28. Drucker, P.F. *The Essential Drucker*; Routledge: London, UK, 2020.
29. China Real Estate Think Tank. Release of research results on brand value of Chinese real estate enterprises. *China Real Estate Bus.* **2021**, *9*, 1–6. [CrossRef]
30. Jiang, H.; Zhao, Y. China Aoyuan Dai Yuancheng: Brand is the "moat" of comprehensive strength. *China Bus.* **2020**, *T12*, 1–16. [CrossRef]
31. Liu, Z. Design and Application of Brand Value Evaluation Method of Real Estate Enterprises Based on Brand Value Chain. Master's Thesis, Zhejiang University, Hangzhou, China, 2008.
32. Zhou, G.; Liu, H. Mechanism and effect evaluation of the impact of sales restriction policy on China's real estate market. *China Real Estate* **2021**, *15*, 8–15. [CrossRef]
33. Liu, H. Technological Innovation is the Key Driving Force for the Growth of Real Estate Value. 2020. Available online: https://www.sohu.com/a/421900984_120845342 (accessed on 30 December 2020).
34. Shi, J.; Wang, Y.; Wu, W. The crowding-out effect of real estate markets on corporate innovation: Evidence from china. In Proceedings of the Asian Finance Association (AsianFA) Conference, Bangkok, Thailand, 30 August 2016. [CrossRef]
35. Albaum, G. The Likert Scale Revisited. *Mark. Res. Society. J.* **1997**, *39*, 1–21. [CrossRef]
36. Taherdoost, H. What is the best response scale for survey and questionnaire design; review of different lengths of rating scale/attitude scale/Likert scale. *Int. J. Acad. Res. Manag.* **2019**, *8*, 1–10. Available online: https://ssrn.com/abstract=3588604 (accessed on 2 November 2021).
37. Duan, K.; Zhang, C.; Li, J.; Zhang, R.; Zhang, Y. Boundary-Spanning Search for Knowledge, Knowledge Reconstruction and the Sustainable Innovation Ability of Agricultural Enterprises: A Chinese Perspective. *Agriculture* **2020**, *10*, 39. [CrossRef]
38. Taherdoost, H. Validity and reliability of the research instrument; how to test the validation of a questionnaire/survey in a research. *Int. J. Acad. Res. Manag.* **2016**, *5*, 28–36. [CrossRef]
39. Gim Chung, R.H.; Kim, B.S.; Abreu, J.M. Asian American multidimensional acculturation scale: Development, factor analysis, reliability, and validity. *Cult. Divers. Ethn. Minority Psychol.* **2004**, *10*, 66–80. [CrossRef] [PubMed]
40. Mooi, E.; Sarstedt, M.; Mooi-Reci, I. Principal component and factor analysis. In *Market Research*; Springer: Berlin/Heidelberg, Germany, 2018; pp. 265–311.
41. Feng, B.; Chen, M. The Impact of Entrepreneurial Passion on Psychology and Behavior of Entrepreneurs. *Front. Psychol.* **2020**, *11*, 1733. [CrossRef]
42. De Carvalho, J.; Chima, F.O. Applications of structural equation modeling in social sciences research. *Am. Int. J. Contemp. Res.* **2014**, *4*, 6–11. Available online: http://www.aijcrnet.com/journals/Vol_4_No_1_January_2014/2.pdf (accessed on 2 November 2021).

43. Wang, J.; Li, B. Impact of Repatriate's Knowledge Transfer on Enterprise Performance: The Mediating Effect of Ambidexterity Innovation. *J. Syst. Sci. Inf.* **2016**, *4*, 56–67. [CrossRef]
44. González-Valero, G.; Vidal-Conti, J.; Zurita-Ortega, F.; Palou-Sampol, P. Active Time in Cooperative Activities, Quality of Life and Body Mass Index in Individuals with Intellectual Disabilities. A Model of Structural Equations. *Sustainability* **2021**, *13*, 2341. [CrossRef]
45. García Alcaraz, J.L.; Martínez Hernández, F.A.; Olguín Tiznado, J.E.; Realyvásquez Vargas, A.; Jiménez Macías, E.; Javierre Lardies, C. Effect of Quality Lean Manufacturing Tools on Commercial Benefits Gained by Mexican Maquiladoras. *Mathematics* **2021**, *9*, 971. [CrossRef]
46. García-Martínez, I.; Landa, J.M.A.; León, S.P. The Mediating Role of Engagement on the Achievement and Quality of Life of University Students. *Int. J. Environ. Res. Public Health* **2021**, *18*, 6586. [CrossRef]
47. Wen, Z.; Hou, J.; Lei, Z. A Comparison of moderator and mediator and their applications. *Acta Psychol. Sin.* **2005**, *37*, 268–274. (In Chinese)

land

MDPI

Article

Integrated Planning: Towards a Mutually Inclusive Approach to Infrastructure Planning and Design

Dario Hernan Schoulund [1,*], Carlos Alberto Amura [2] and Karina Landman [3]

1 Department of Architecture, Faculty of Engineering, Built Environment and Information Technology, Hatfield Campus, University of Pretoria, Pretoria 0002, South Africa
2 Department of Constructions and Structures, School of Civil Engineering, Engineering Faculty, Las Heras Campus, University of Buenos Aires, Buenos Aires 1126, Argentina; ingamura@fi.uba.ar
3 Department of Town and Regional Planning, Faculty of Engineering, Built Environment and Information Technology, Hatfield Campus, University of Pretoria, Pretoria 0002, South Africa; karina.landman@up.ac.za
* Correspondence: Dario.schoulund@up.ac.za; Tel.: +27-82299-8947

Abstract: Increasingly independent fields of specialization, civil engineering, and urban design find themselves practicing in isolation on the same urban issues. The result surfaces on the relative qualities of public spaces: projects that are functionally successful but spatially poor, and vice versa This is critical in the global south, where infrastructure is prioritized, and politicized, as the key driver of change but often heedless of spatial consequences. The present study explores the dynamics of integration between logics arising from technical and spatial fields, and the planning processes under which such integration is feasible. An urban design/infrastructural project in Argentina, stalled for more than two decades under regulatory policies, was selected as a case study. An overview and background of the adopted planning/design methodologies are followed by a structural/spatial analysis, focusing on type, logistics, and construction on the one hand, and on indicators of successful public spaces on the other: access, uses, comfort and image. Aspects that a priori appeared as inevitable compromises found a common, but the critically logical ground in which urban and structural thinking complemented each other. More than a functional asset, infrastructure presents an opportunity to re-think the future of the built environment as a typology that could be conceived, designed and evaluated, on the same terms as successful public spaces.

Keywords: integrated spatial planning; urban infrastructure; structural design; urban design

Citation: Schoulund, D.H.; Amura, C.A.; Landman, K. Integrated Planning: Towards a Mutually Inclusive Approach to Infrastructure Planning and Design. *Land* **2021**, *10*, 1282. https://doi.org/10.3390/land10121282

Academic Editors: Sarel Cilliers and Elizelle Juanee Cilliers

Received: 8 October 2021
Accepted: 19 November 2021
Published: 23 November 2021

1. Introduction

There is a tendency to conceive of projects in terms of very distinct and often separated phases. This is especially true of urban infrastructure projects. While this may assist to demarcate the role of various professionals involved, it also tends to create a silo approach [1] and an unhealthy separation of issues related to spatial design on the one hand and technical considerations on the other. Such a separation has often adversely influenced the quality of the built environment and the potential for quality of life in cities. This article focuses on the role of integrated spatial planning and design to align the contributions from different professional fields and move towards more integrated urbanism with a stronger sense of place. This is illustrated through an urban design/infrastructural project in Argentina developed by the Municipality of Villa Carlos Paz. The objective was to bridge the development divide between the two central precincts (east and west) separated by the San Antonio River. The project presents an attempt to integrate urban and structural design as part of a simultaneous process and not distinct stages of development. This process is discussed through an introduction of the project context and planning/design process, followed by a detailed discussion of the choice and nature of the structure and the spatial solutions. The paper argues for a trans-disciplinary perspective towards infrastructure and advocates that structural and spatial considerations should not be mutually exclusive but should be aligned to enhance both the efficiency and the experience of urban infrastructure projects.

1.1. A Spatial or Functional Approach: Mutually Exclusive?

The current 'infrastructure turn', where emerging global infrastructure practices are supported by a new set of discursive, political, and technical arrangements, tends to displace and override spatial planning practice [2]. In addition, fragmented knowledge of infrastructure across different disciplines compromises the development of robust planning strategies [3]. Even when the bond between infrastructures and cities is tight as ever, the relationship between planning/urban design and infrastructure remains noncomprehensive and nonstrategic [4]. Part of the problem lies in that infrastructure is often framed as an abstraction [5], prioritized—and politicized—as an essential driver of change, but heedless of spatial implications.

The institutional and epistemological gaps between spatial and technical disciplines often reflect on regulatory planning approaches, and consequently in the quality of the urban experience itself. On the one hand, the planning and urban design professions' interest in infrastructure has declined in deference to specialists [4], leaving key roles to professions mostly concerned with efficiencies, such as engineers, financers, managers, and public work officials. On the other hand, infrastructures' technical and organizational aspects are unappealing to the architect, as they require a "different habit of mind about design" [6] (p. 264). Thus, architects and urban designers have focused overwhelmingly on the design of spaces within envelopes rather than the networked infrastructures that bind and configure them [7] (p. 18), giving rise to infrastructural urbanism [8], where reductionist spatial engineering asserts itself.

While planning tends to deal more with the two-dimensional spatial organization and regulation of the city, urban design focuses on the three-dimensional quality and organization of the built environment. Integrated planning, or what we refer to as inclusive planning, presents an alternative defined by horizontal and vertical integration of governance levels. As opposed to regulatory planning, integrated planning relies on appropriate methodologies, management, public involvement, and coordination that respond and adapt to local conditions [9]. This necessitates planners to move beyond their traditional functions to a more communicating and mediating role [10]; or to what has been called participatory design [11] or co-design [12]. These approaches emphasize the role of the planner/urban designer in working with relevant stakeholders to encourage a more positive and sustainable outcome. Although participatory planning has shown to be beneficial, developers and planners often neglect this and fail to involve all the stakeholders [13]. At the same time, it reflects the increasing focus in urban design theory and practice to acknowledge urban design as both product and process [14–16], where it is recognized that the process plays a very important part to create quality places. Four process dimensions, namely design, development, space in use, and management play an important role to shape place over time [15].

Due to logistical challenges, however, these processes are not exempt from risks: Holden [17] warns that preconditions for integration must be present across normative dimensions. Similarly, the 'tree planning approach' [18] proves to be a barrier to institutional integration. In addition, there is often not enough time spent in the planning or design phase, as projects need to be delivered very rapidly [19]. Another concern relates to the measurement of design outcomes or performance measurement [19,20]. Some of these challenges are evident in developing countries such as South Africa, where there is a lack of integration between various departments and levels of governance [21].

South America also struggles with these challenges, resulting primarily from disjointed management and less so from professional capacity [22]. Urban governance is a major challenge in itself [23], characterized by an abundance of overlapping authorities, competing departments, and uncoordinated efforts and policies. The absence of proper reviewing mechanisms, like design review in the UK [24], further contributes to the 'end product oriented' approach. Additionally, urban design, and its potential coordinating role [25], remain a weak discipline in Argentina, often associated with exclusive developments [26], or in cases seen as a luxury [27].

In the selected case, an approach that focused on participation among departments and stakeholders was used as an alternative to hitherto unfruitful proposals under regulatory approaches. Admittedly, "negotiation, as much as collaboration, was expected" [28]. This analysis illustrates the dynamics of a process where urban design and structural engineering found a common ground, and the benefits of a broad base professional/public forum that ultimately defined a feasible proposal.

1.2. Project Background

The project is located in the town of Villa Carlos Paz (population 90,000), Córdoba province, Argentina (Figure 1). This town is the second most touristic place in the country, with occupancies often reaching as many as 1 million visitors during the high season (January and February). The layout of the town can be described as two similar-sized urban areas divided by the San Roque Dam and the San Antonio River. The river delineates the two halves of the downtown area, connected by the 1889 bridge ('old bride') (Figures 2 and 3). The downtown area was originally established on the west bank, but over the years expanded to the east bank, where it thrived, overshadowing the former west-central precinct.

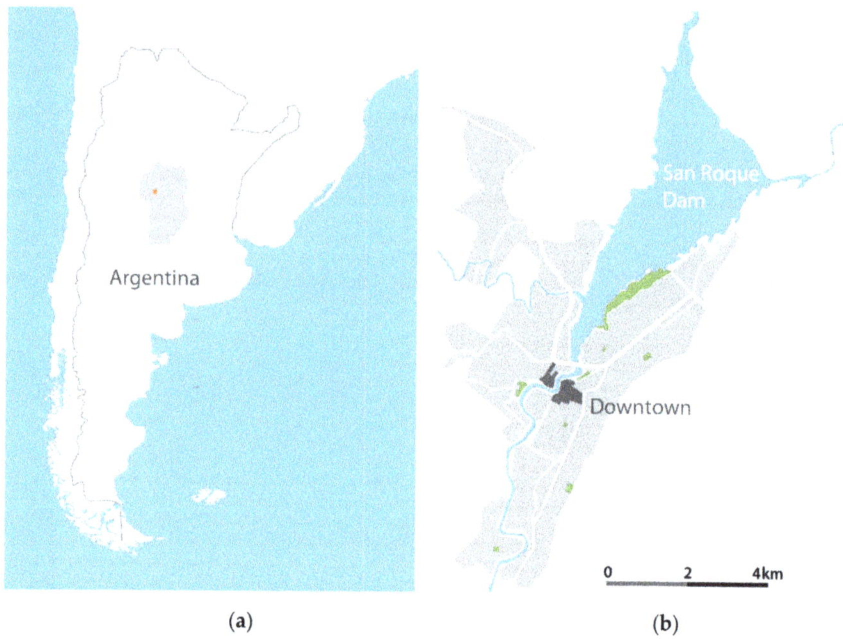

Figure 1. Location maps. (**a**) Argentina and Cordoba Province. (**b**) Villa Carlos Paz, the San Roque Dam, and the Downtown area on both sides of the San Antonio River.

Figure 2. The existing 1889 bridge, its supporting piles account for 45% of the drainage section. (Credit: Reproduced with permission from the author, Eng. Gerónimo Cáffaro).

Figure 3. Aerial view taken from the east precinct, showing the San Antonio River and the 1889 bridge. Mixed-use, mostly retail and residential, occurs in low to medium densities.

The general objective of this project was to bridge the increasingly marked development divide between the two central precincts. The need was identified during the 90s when the town grew rapidly in popularity. Although several proposals were developed in the subsequent years, they did not meet enough political/financial traction, or public acceptance [28]. Additionally, these proposals focused on replacing the existing old bridge with wider and modern structures. Even when the old bridge has proven structurally unsound and hydraulically problematic, its replacement remains a sensitive heritage issue. These aspects compounded, resulting in the project stalling for more than two decades.

1.3. Planning Methodology

From 2014, the municipal planning department managed an integrated process. The methodology aligns to some degree with that proposed by Yigitcanlar and Teriman [10], consisting of a series of re-evaluative steps that belong to two distinct phases: definition and confirmation (Figure 4). In the definition phase, the integration places particular emphasis on 'participatory design' as an extension of participatory mapping techniques. A neutral-role format was employed for a series of design workshops that culminated in the urban vision and later on the design parameters of the two proposed bridges, promenades, and public spaces. The rationale behind focused on removing the boundaries among the three main stages of regulatory planning (Figure 5), and the boundaries of professional fields. In this manner, the identification of the issue, the 'what-to-do part' of the process, benefited from a broader perspective in which the wealth of local knowledge formed a key contribution.

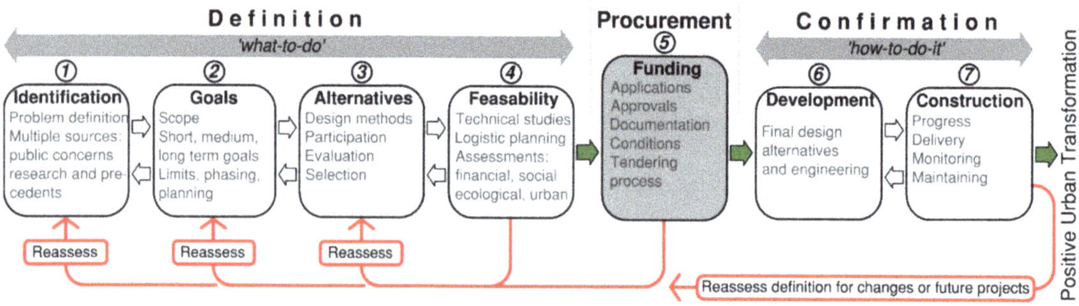

Figure 4. The integrated model was employed, placing particular emphasis on the definition.

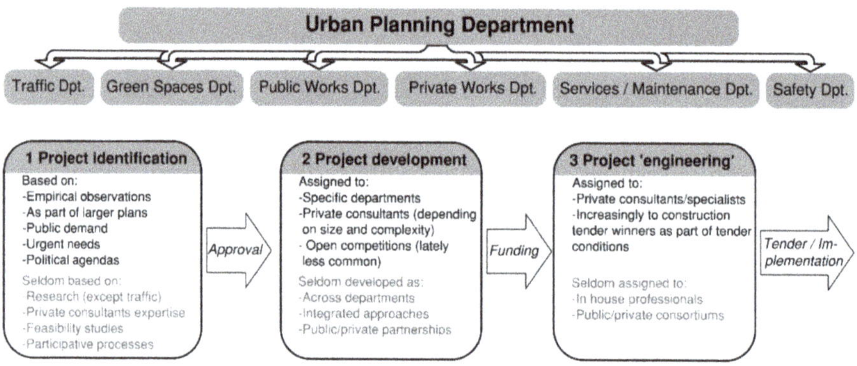

Figure 5. Three main stages of regulatory planning are typically assigned to specific departments. The many variables make it difficult, if not impossible, to expect a resulting integrated proposal.

The first three planning steps of definition, comprised of a series of information and design workshops, where public officials, private consultants, and the public, in general, took part. The focus was on reframing the project from an engineering issue (conceived and developed by specialists) to an urban design issue (conceived by many). Stakeholders also included urbanists, project managers, architects, landscape architects, engineers (traffic, civil, and hydraulic), politicians, and historians. Subsequently, during the confirmation phase, the focus was specifically on urban and structural design, the 'how-to-do-it' part of the process. Facilitators encouraged participation in all aspects of design solutions: engineers were encouraged to design, and share ideas with the 'designers' (urban designers/architects), in turn, designers were prompted to conceive and participate in technical solutions. The exchange of ideas was at the core of the process, and an experimental element, which in regulatory terms is considered a risk, was embraced with motivation.

2. Methods

A case study research approach was employed, together with a documentary data collection method. An overview of the project and a background of the adopted planning/design methodologies places the process within integrated planning boundaries. To establish the relationship between technical efficiency and spatial considerations, the main structural features (type, geometry, and implemented solutions) were analyzed with criteria of successful public spaces and a range of spatial methodologies for understanding public space and the performance of the built environment. There has been extensive research on what constitutes successful public spaces or precincts. For example, Montgomery notes that the components contributing to a sense of place relate to activity (including access, diversity, vitality, use, etc.), form (including permeability, landscaping, and scale), and image (including symbolism and memory) [29]. The Project for Public Spaces' Place Diagram highlights five key qualities, namely sociability, uses and activities, access and linkages, comfort, and image [30]. Carmona identified 10 aspects of successful spaces: evolving, diverse, free, delineated, engaging, meaningful, social, balanced, comfortable, and robust [15]. These were reduced to four main indicators for the consideration of spatial quality, namely accessibility, uses, comfort, and image.

The evaluation of proposals/alternatives centered on structural and spatial qualities, followed in order of importance by management/logistical aspects, and lastly by financial considerations. Comparative charts were used as an indicative tool to evaluate and compare these aspects (Figure 6). The intention was to identify and combine the benefits of alternatives. In this simple tool, the criteria for structural design and the desired spatial qualities were outlined. Budget played no significant role, which resulted in design flexibility, variety, and choice. This approach was possible since budget allocations to (small) municipalities are irrespective of specific cost estimates but dependent on the project's relevance.

	Structural			Spatial						
Alternative	Logistics	Construction	Type	Access	Uses	Comfort	Impact	Management Logistics	Financials	Score
	local sourcing import items timeframes flood seasons	hidraulic feasability geological feasability profesional and work-force capacity	clarity definition image	connected welcoming safe convenient	diversity 24 hours democratic celebratory	walkable clean universal access inviting	real state commerce historic natural touristic	Implementation timeframes / political cycles funding opportunities Affected areas and upgrades	% ratio import/local actual price ancillary works downtime impact costs	
#										
#										

Figure 6. Comparative charts were used to evaluate alternatives, as an indicative tool only.

Accessibility was measured through pedestrian counting recorded on January 2018, 2019, and 2020, during the most popular weekend of the high season. Comfort and image were recorded through a documentary data collection method in which publications concerning the project (broadcast, online publications, paper publications) were analyzed for the same period. Finally, direct observations attempted to connect the various uses to key structural/spatial features. This reflects the importance of trying to measure the performance of the built environment in various ways [20,31], despite the difficulties associated with such performance measurements [32].

3. Results

3.1. Urban Vision

The first collective outcome defined the urban vision: a staged construction plan in which the old bridge was not affected in principle. The sequence entitled: (1) building of a pedestrian bridge (completed), (2) building of a mixed bridge of four traffic lanes, and (3) removing of the old bridge. The advantages of this plan are numerous: first, the three phases are independent, secondly, the decision to remove the old bridge—a sensitive point—could be delayed until consensus is reached or be left untouched, and lastly, the funding procurement model resulted in better chances of adjudication. The vision for the area involves significant public space reconfiguration (roads, green spaces, promenades, and sidewalks), leading to a more pedestrian-friendly environment (Figure 7).

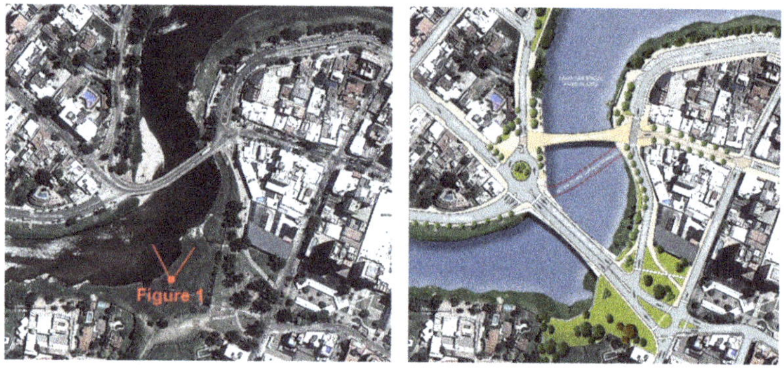

(a) (b)

Figure 7. (**a**) The downtown area connected by the existing bridge of 1889. (**b**) The urban vision shows substantial public space reconfiguration, a proposed mixed bridge (bottom), and the implemented pedestrian bridge (top). (Credit: Provided by the planning department of Villa Carlos Paz (public domain)).

3.2. Pedestrian Bridge

The first project set for implementation, the pedestrian bridge, was envisioned as a public space across the San Antonio River. Yet the site constraints, seismic and flooding conditions, and the restrictions imposed on the geometric envelope presented a considerable engineering challenge that risked overriding key spatial considerations. The design premises were defined as a set of conditions:

- A comfortable pedestrian deck of 6 m of minimum width;
- A maximum height of 12.5 m above the 100-year flood line;
- A maximum of two intermediate supports (3 spans);
- An attractive, contextual, and memorable architectural language;
- A lookout point to the surroundings.

3.2.1. Bridge Typology

The San Antonio River is prone to flooding for 6 months of the year. Its drainage section on the location of the bridge, measured at 100-year flood recurrence (2100 m^3/s), is 72 m. The metamorphic bedrock sits at ± 6 m below shifting layers of alluvial boulders, gravels, and sands. These two aspects challenged the practicalities of building intermediate supports. Thus, two options presented themselves: to build two intermediate supports of efficient hydrodynamic profiles (for a 3-span deck), or a single 72 m span. The construction of intermediate supports offset any apparent savings derived from the simplicity of the superstructure while adding a series of constructive and hydraulic constraints. A single 72 m span avoids interference with water flow while establishing the lowest possible level for the deck above the 100-year flood line (Figure 8). In terms of accessibility, it provides a smoother transition, allowing the use of frontal ramps instead of stairs due to the reduced depth of the deck.

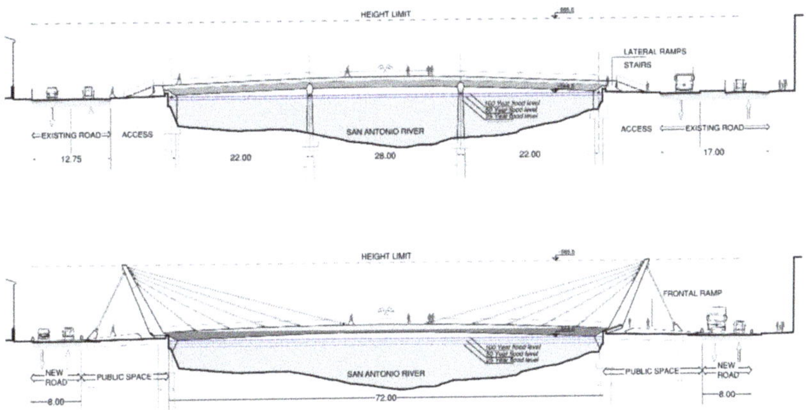

Figure 8. Top: early 3-span proposal: accesses were uninviting as ramps could only occur laterally. **Bottom**: selected single-span cable-stayed typology, together with new public spaces and road reconfigurations for friendlier accesses.

3.2.2. Structure

The bridge's substructure consists of concrete pile foundations and abutment walls, integral concrete masts, and deep active rock anchorages for backstays. The superstructure consists of a metallic web of fixed longitudinal and transversal beams coinciding with the stays' attachments (Figure 9). All stay cables are composed of parallel tendons of 7, 12 (deck), and 19 strands (backstays). Due to the very low angle of the central cables (15°) and to seismic deformations, fork anchorages at both ends were preferred to minimize fatigue at the terminal. It was therefore decided, for consistency, to utilize these attachments in all cables in what was dubbed 'honest engineering'. The bridge was modeled using the FEA system SAP2000 (V.16), considering the AASHTO LRFD Bridge Design Specification norm [33].

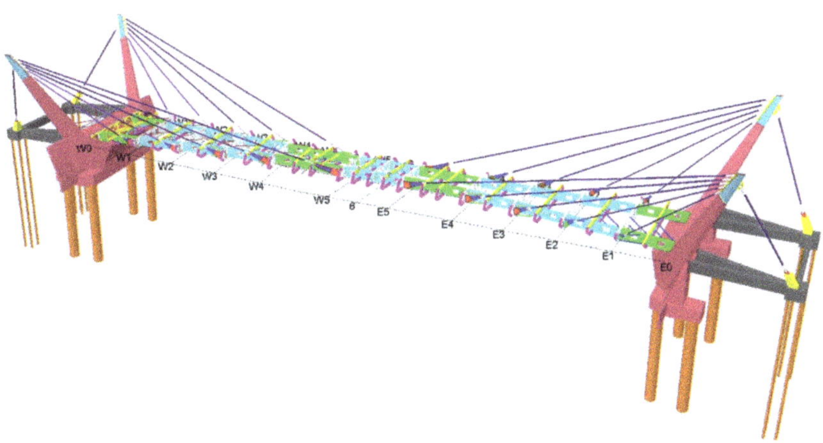

Figure 9. 3D BIM model of the structure.

3.3. *Spatial Experience*

The deck was to create a lookout point free of structural interference. To accentuate this premise, the central structural modules stretched from 6 m to 8 m. This small change enlarges the lookout zone from 18 m to 23 m (measured at eye level), which on a 72 m crossing represents just under a third of the total (Figure 10). Structurally, the larger modules coincide with the lighter zone of the deck; the modules are longer but narrower. The area is further celebrated by the 1m longitudinal bulge that confers a sense of mastery over the environment while preventing debris impacts during floods.

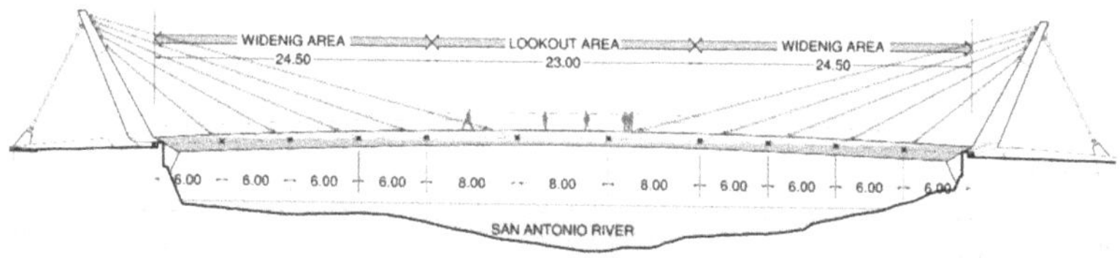

Figure 10. The central lookout area, higher and clear of visual interference for almost a third of the crossing.

The deck's width, first defined at 6mts, offered a generous crossing but created a blunt transition to the access public spaces. To address this point, the entry areas widened to 11.4 m, in this way, the deck 'opens' to the city together with the outward leaning masts. This geometry creates zones within the deck in which users can linger without interfering with the central flow. A variety of activities is then encouraged in line with diverse public spaces. Structurally, the extra mass of the enlarged deck coincides with the more efficient stays of greater vertical components; the logic of the load diagram matches the geometry of the deck (Figures 11 and 12).

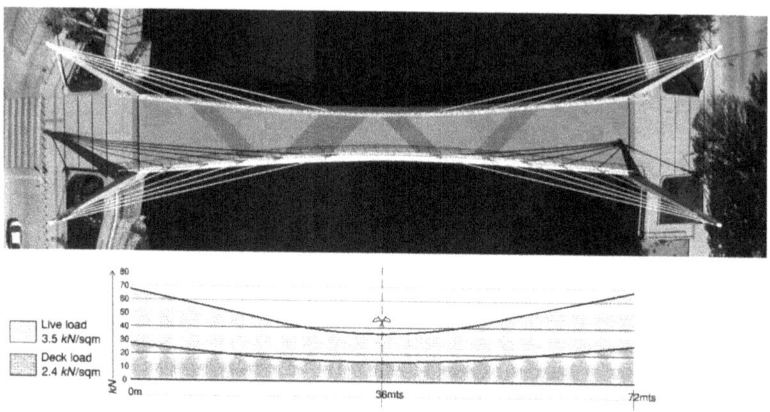

Figure 11. Top: the deck geometry in the plan widens from 6mts at the center to 11.4 m at the entry points. **Bottom**: the load diagram places most of the mass closer to the abutments and onto the more efficient stays.

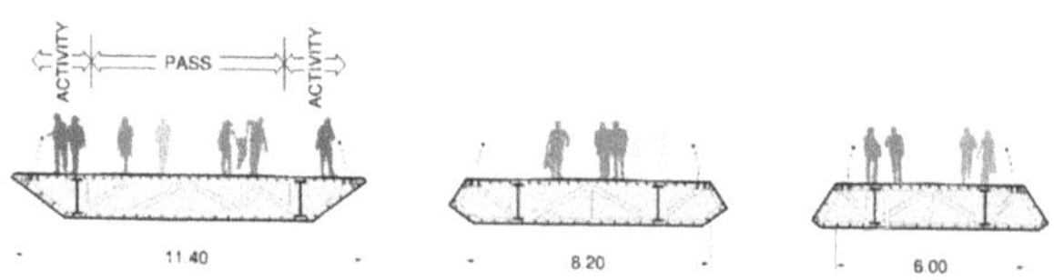

Figure 12. Deck's cross-sections at different instances: from left to right, entry points, quarter of span, and at the center.

The concrete masts lean backward (29°) and outwards (18°) in permanent flexion. They are integral to the abutments and blend in geometrically with the adjacent stone retaining walls. The failure of one backstay/anchorage was modelled as a worst-case scenario; in such an event, the leaning concrete mass of each mast (42.5 metric tons) plays an important balancing role. These aspects align to open the entry areas: the masts, with their upward and outward gesture resemble 'open arms' (Figures 13 and 14). This also highlights the importance of spatial experience and symbolic interpretation of the structure.

(a) (b)

Figure 13. (**a**) Exposed stay attachments as 'honest' engineering themes. (**b**) Various activities were observed in the wider areas of the deck, without interfering with the central pedestrian flow.

(a) (b)

Figure 14. (**a**) Masts raising upwards and outwards resemble welcoming 'open arms'. (**b**) Masts, abutment walls, and foundations form an integral element.

3.3.1. Usage, Access, Comfort, and Image

Usage was registered through a comparative pedestrian counting (full crossings) during the most popular weekend of the high season; first in January 2018, then in January 2019, and finally in January 2020 (January 2021 was omitted due to lockdown). The high season was selected since a requirement of the project was to become an 'attraction'. We employed two assistants per abutment: 4 in 2018 and 2019 (old bridge), and 8 in 2020 (both bridges), discreetly stationed on each sidewalk/side, counting only exiting pedestrians. The counting took place from 7:00 a.m. until 12:00 a.m. for three consecutive days using a mobile counting application. All nine days' results are shown in Figure 15.

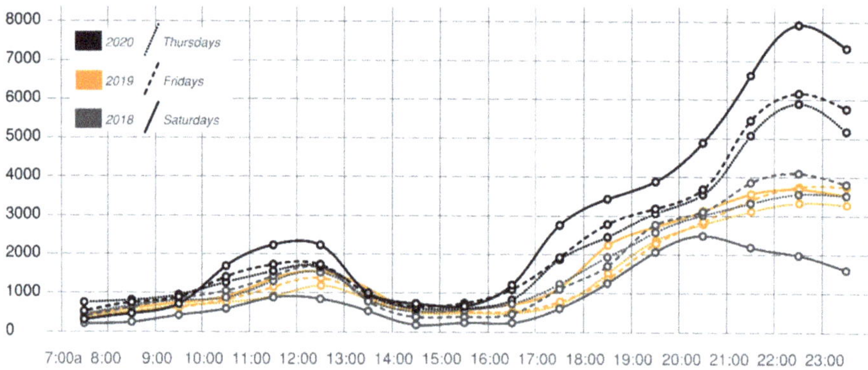

Year	Total crossings (3 days)	To old centre	To new centre	Using old bridge	Using new bridge	% + - (from previous year)
2018	72,510	34,665 (48%)	37,845 (52%)			
2019	78,122	38,068 (49%)	40,054 (51%)			+ 7.7%
2020	124,619	55,630 (45%)	68,989 (55%)	44,356 (36%)	80,263 (64%)	+ 59.5 %

Figure 15. Periods of activity relate to summer temperatures, decreasing during the hot afternoons and increasing during the nights, and well into the next day.

Pedestrian activity increased for 2020, peaking consistently on Fridays and Saturdays, except for the year 2018 (Saturday) due to unstable weather (drizzling but pleasant temperature). Interestingly, not all users crossed the bridge; many would enter, take photographs, observe, linger and return. We then employed one assistant per abutment to discount "visitors" from "crossers"; therefore, more people than counted visited the immediacy of the project.

3.3.2. Documentary Data

Reporting of the project in public media was classified as negative, neutral, and positive. On the negative aspect, the project was criticized for not being an essential priority in a town that lacks basic services in some areas [18], and for traffic and services disruptions spanning two years [34,35] (Figure 16). The negative phase moved into the expectation phase before completion, where reporting shifted to the nature of the project, its scale, and its meaning. Imminent inauguration, coinciding with the high season, created some positive coverage of the political dimension [36,37] (Figure 17). Finally, once inaugurated, an element of novelty took place. Neutral reports that indirectly included the project as a referential place started to appear, and several events and small rites were observed: engagements, wedding photography, lovelocks, T.V. interviews, promotions, and a public protest (Figure 18).

Figure 16. (left to right) Road disruption during construction: the main source of negative reporting, affecting pedestrian and bicycle traffic as well as key trunk services for almost two years. (Credit: Left: From public television broadcast (Cordoba Governance); Centre and Right: reproduced with permission from the author Mr. Mario Rojas (Centedario digital publication)).

Figure 17. (**left**) Political promotion during the inauguration ceremony. (**center** and **right**) The bridge is used more as a public space. (Credit: Left: From public television broadcast (Cordoba Governance)).

Figure 18. The project as a democratic venue: (**left**) Hairdressers' marathon (charity event). (**center**) Public protest on water management. (**right**) T.V. interviews using the bridge as a referential place. (Credit: Left and Center: reproduced with permission from the author, Mr. Mario Rojas; Right: Still from public television broadcast (Cordoba Governance)).

3.4. Fusing Structural and Spatial Considerations

Almost in all cases, the preferred structural solutions were also the higher scoring ones in terms of spatial considerations. Good structural solutions tended to be financially reasonable: a minor extra cost represented a significant increase in spatial value. This point is particularly clear on the new public spaces: they play key roles in calming traffic, in defining a safe pedestrian area, and in protecting the backstays against potential impacts. These spaces cannot be separated from the bridge: they are part of the structure (Figure 19). The same logic extends to material selection: the green islands' edges are designed as continuous benches, but they are constructed as reinforced concrete barriers.

Figure 19. Pedestrian view looking towards the east precinct from $\frac{1}{4}$ of the span.

This common ground between urban and structural design presented itself on each composing aspect, from major features to the detail of smaller parts. Similarly, the larger area of the project, which involved the design of promenades, parking bays, a bus stop, road/traffic reconfigurations, light fittings, and green spaces, responded—and was guided—by the same spatial/functional premise (Figures 20 and 21). It, therefore, represents a fusion of spatial and structural considerations to become more than just a bridge but a place of encounter, crossing, and reflection.

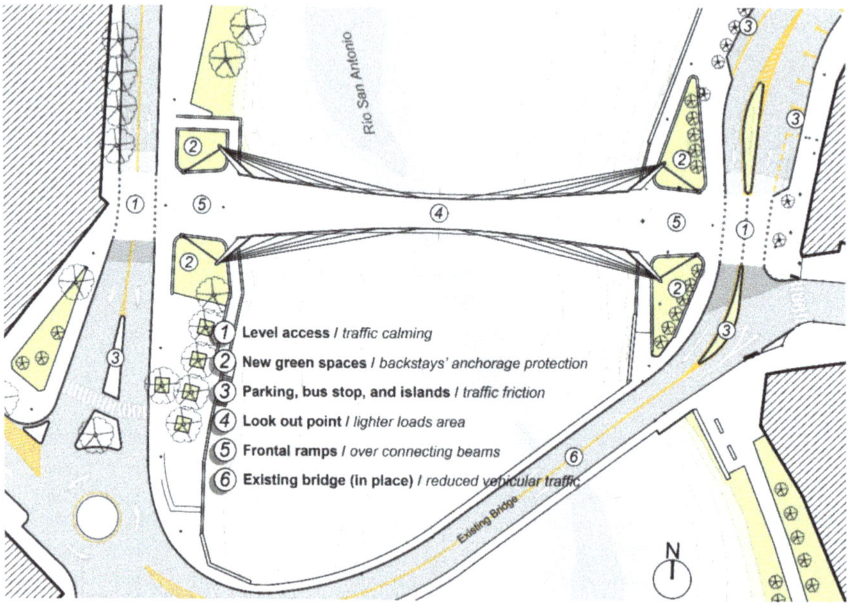

Figure 20. Ancillary works included traffic reconfiguration, parking, pedestrian crossings, public spaces, and promenades.

Figure 21. The completed project was inaugurated in 2020, with the 'old bridge' partially visible behind (upstream).

4. Discussion

4.1. Towards a Mutually Inclusive Approach to Infrastructure Planning and Design

Despite its short normal operational lifespan, essentially less than 3 months before the COVID-19 lockdown, several interesting events occurred in and around the project. At the same time, increased pedestrian traffic aligns with the main intention of equalizing both halves of the central precinct: to bring life to the west bank. Two logics needed to be fulfilled: the logic of urban design as manifested in successful public spaces and the logic of efficient structural engineering. In this project, these arguments have proven to be the same, suggesting that at no point does any design aspect reflect a trade-off or a compromise. Sound structural solutions tended to be superior spatial solutions. This approach in which infrastructure is public space requires a shift in perceptions: it implies that public space, instead of infrastructure [38], is the medium that binds the city together, defined by the constant combination of multiple components, including infrastructure. Therefore, the importance of aligning and fusing structural efficiency and spatial considerations should benefit under the more detailed understanding of urban design.

In the present case, the importance of coordination at the Municipal level, where professional, scientific, and educational institutions participate, was central [9]. Secondly, participatory design [10] became the vehicle through which design professionals sourced contextual creative material. This also highlights the value of participatory design; both as a means to give voice to the users of space but also as a reflective practice [11]. It also echoes the importance of working with the place to find the value-adding role of each element and member of the system towards evolutionary sustainability [39]. Lastly, due to the self-evaluating possibilities embedded within the process, changes and reconsiderations were possible, even during construction, which emphasizes the important role of adaptive planning and design to enhance the quality of the built environment.

4.2. Aligning Technical and Spatial Thinking

For decades, authors have noted that creativity in the instruction of engineering is not encouraged [40–43]. Leonhardt [44] suggested that the profession only attracted "persons who have a talent for rationalism and logical thinking, but no sensitive feeling". However, the experience of this process tells otherwise: engineers are highly creative in technical terms and resourceful when discussing various alternatives, which is often during the early stages of project development and precisely when their participation is typically absent. Creativity is thus not only limited during training as part of a critical foundation but also systemically avoided during regulatory practice. As Christian Menn [45] points out, "The most challenging aspect of a bridge engineer's work is conceptual design". If engineers are not exposed to an ambit conducive to creative challenges, their work is relegated to the 'engineering' of a given project.

On the other hand, urban designers considered themselves designers, mostly due to their architectural backgrounds, but admittedly not trained—or interested—n technical aspects of functional typologies. As opposed to engineers, their intuition for form and balance is only superficially related to technical considerations, as it is not conceptualized from such sources. Instead, designers are particularly good at identifying avenues of creative thinking within the multiple issues touching urban projects, even political or financial for instance. In this case, designers asked the 'right' questions, which triggered engineers' creativity with multiple answers/solutions. Thus, the combination of designer and engineer is not merely a partnership set up to build; it is a method for interpreting public and political interests to their ultimate physical solution. Arguably, this project moved from being an engineering issue to an urban design one. It found no clear direction as long as it was framed within technical boundaries. Only when considered from a wider urban design perspective the appropriate answers start to present themselves; an approach to design that is "not always about meeting the exact standards and having the right answer" [46] (p. 265), but working with the story of the place and conscious interventions in the right place to create system-wide effects [39].

This reiterates the importance of not only focussing on the product of urban design but also the process. Focussing on the process has the potential to add significant value in the Global South to acknowledge the specific context through the various dimensions of placemaking, namely the spatial, procedural, and psychological dimensions [47]. The spatial dimension was reiterated through the strong visual qualities of the structure and the places created around the bridge. The procedural dimension was addressed through the participatory design process. Finally, the psychological dimension has been acknowledged through the positive responses by the users of the space. This allowed for the nuances of language to allow meaning to be relayed in particular ways unique to the context, a way to address the specific context of the Global South [48].

4.3. Key Considerations for Infrastructure

Whether the infrastructure is different from other types of urbanization has been questioned [2]. The relation between its engineering nature and the spatial implications on urban form, regions, or even countries, defines the concept of 'infrastructure space' [46]. This connection cannot be isolated because infrastructure is not an end in itself, nor is a spatial vision possible without its supporting networks. Yet in South America, infrastructure represents primarily provision, prioritized, and politicized at all costs above other components of the urban and natural environment. The implications, when understood, are regrettably accepted as a fact-of-life, in which "one person's infrastructure is another's person difficulty" [49]. Public space becomes an increasingly contested arena, where components deriving from different disciplines find little coordination and coherence, or no place at all.

It is, therefore, necessary to review infrastructural projects under the urban design concern. This premise underlies the present case: the structural type promotes the diversity of spatial features, but the type responds to a greater urban strategy. Such alignment between structure and space suggests that professional specialization and regulatory planning retain the underlying logic between these fields. After all, a bridge is a particular case whose landmark significances, both figurative and literal, have not changed over the years. Other typologies that had a structuring presence in the past, like water provision and distribution, are now largely backgrounded. Conversely, underground stormwater networks are surfacing with a strong image and structuring capacity. These types of interventions highlight the potential of using infrastructure to increase a sense of place and sustainability. A recent report by Cambridge University [50] (p. 13) emphasizes the role that infrastructure can play towards place-making and greater sustainability: an example from Cape Town is presented in which hard infrastructure (Bus Rapid Transit) is combined with the creation of public spaces and facilities. At another level, a greater focus on green infrastructure in cities can also support both the goals of sustainability and a sense of place [51]. If a balanced approach to public space design prevails, one in which all its intrinsic components are encouraged to be in dialogue, we can expect to be on the right track to a more meaningful urban environment.

5. Conclusions

Infrastructure planning tends to focus too much on function at the expense of spatial quality. On the other hand, the planning professions tend to focus too much on regulation, while urban designers are often only fixated on the product and its aesthetic qualities. This article examined a specific urban design/infrastructure project in Argentina developed by the Municipality of Villa Carlos Paz. The project involved the building of a pedestrian bridge to address the divide between two central precincts in the city. The discussion indicated that the project did not only connect the two parts of the city but also bridged the gap between urban and structural design through the planning process, design, and implementation of a project that was, under regulatory policies, unprovable. It also illustrated the possibilities of achieving good results when the barriers of communication and coordination—a key challenge of the Global South-are displaced to create space for dialogue and collaboration: through a process-oriented approach based on inclusive planning. This involves inclusivity at three levels: (1) including a focus on both the product and process of urban design, (2) involving all the relevant professionals of the built environment in all phases of the project to allow co-design, and (3) working with municipalities, communities and the relevant stakeholders to enable a participatory design that would address the context-specific issues towards appropriate placemaking in the Global South. Restricted funding and fewer projects make it even more important than the money spent on placemaking projects in the Global South should be able to create accessible, useful, comfortable, and meaningful places.

The main role of inclusive planning centers on facilitating the alignment of design logic arising from different professional fields and on guaranteeing a positive degree of contextualization. At the same time, this approach to planning promoted the articulation of the many voices' concerns and ideas into genuine sources of creative thinking. The fusion of urban and structural design shows the potential to increase the sense of place and contribute to greater urban sustainability. What is promising, therefore, is the possibility of integrated typologies that challenge the idea of infrastructure and public space as separate components of the urban experience. Critical for the future of the built environment, such an approach could also revitalize the declining relevance of the traditional public space by attaching its design principles to the infrastructures of daily life. Given the increased importance and multiple demands placed upon urban infrastructure, integrated spatial planning represents an alternative for conceiving, developing, and enhancing projects both functionally and spatially.

Author Contributions: All three authors participated actively in this study. D.H.S. provided the urban design and architectural perspective from which the project emerged. C.A.A. provided the engineering counterpart, and K.L. a combination of planning framework, methodology design, and research structure. Conceptualization, D.H.S. and C.A.A.; methodology, D.H.S., C.A.A., and K.L.; software, C.A.A.; validation, D.H.S., C.A.A., and K.L.; formal analysis, D.H.S., C.A.A., and K.L.; investigation, D.H.S.; resources, D.H.S. and C.A.A.; data curation, D.H.S., C.A.A., and K.L.; writing—original draft preparation, D.H.S.; writing—review and editing, D.H.S. and K.L.; visualization, D.H.S. and C.A.A.; supervision, K.L.; project administration, D.H.S.; funding acquisition, n/a. All authors have read and agreed to the published version of the manuscript.

Funding: This research received no external funding.

Institutional Review Board Statement: Not applicable.

Informed Consent Statement: Not applicable.

Data Availability Statement: Not applicable.

Acknowledgments: We would like to thank the assistance of the Director of the Urban Planning Department of Villa Carlos Paz, Liliana Bina, for her time, feedback, references, and information provided.

Conflicts of Interest: The authors declare no conflict of interest.

References

1. Olagunju, A.; Gunn, J.A.E. Challenges to integrating planning and policy-making with environmental assessment on a regional scale—A multi-institutional perspective. *Impact Assess. Proj. Apprais.* **2016**, *34*, 1–18. [CrossRef]
2. Dodson, J. The Global Infrastructure Turn and Urban Practice. *Urban Policy Res.* **2017**, *35*, 87–92. [CrossRef]
3. McArthur, J. Comparative infrastructural modalities: Examining spatial strategies for Melbourne, Auckland and Vancouver. *Environ. Plan. C Politi-Space* **2018**, *36*, 816–836. [CrossRef]
4. Neuman, M.; Smith, S. City Planning and Infrastructure: Once and Future Partners. *J. Plan. Hist.* **2010**, *9*, 21–42. [CrossRef]
5. Addie, J.-P. Theorizing suburban infrastructure: A framework for critical and comparative analysis. *Trans. Inst. Br. Geogr.* **2016**, *41*, 273–285. [CrossRef]
6. Easterling, K. Split screen. In *Infrastructure Space*; Ruby, I., Ruby, A., Eds.; Ruby Press: Berlin, Germany, 2017; pp. 26–272.
7. Graham, S.; Marvin, S. *Splintering Urbanism: Networked Infrastructures, Technological Mobilities and the Urban Condition*; Routledge: London, UK; New York, NY, USA, 2001; pp. 1–28.
8. Dodson, J. The 'Infrastructure Turn' in Australian Metropolitan Spatial Planning. *Int. Plan. Stud.* **2009**, *14*, 109–123. [CrossRef]
9. Milojevic, B. Integrated Urban Planning in Theory and Practice. In Proceedings of the XIII Contemporary Theory and Practice in Construction Conference 2018, Banja Luka, Bosnia and Herzegovina, 24–25 May 2018; pp. 323–337. [CrossRef]
10. Yigitcanlar, T.; Teriman, S. Rethinking sustainable urban development: Towards an integrated planning and development process. *Int. J. Environ. Sci. Technol.* **2014**, *12*, 341–352. [CrossRef]
11. Bannon, L.; Pelle, E. Design matters in participatory design. In *Routledge Handbook of Participatory Design*; Routledge: New York, NY, USA, 2013; pp. 37–63.
12. Garth Britton, G. Co-Design and Social Innovation: Connections, Tensions and Opportunities. Routledge Studies in Social Enterprise and Social Innovation; Routledge: London, UK; New York, NY, USA, 2017.
13. Cilliers, E.J.; Timmermans, W.; Goorbergh, F.V.D.; Slijkhuis, J.S.A. The Story Behind the Place: Creating Urban Spaces That Enhance Quality of Life. *Appl. Res. Qual. Life* **2014**, *10*, 589–598. [CrossRef]
14. Madanipour, A. *Design of Urban Space: An Inquiry into a Socio-Spatial Process*; John Wiley & Sons: New York, NY, USA, 1996.
15. Carmona, M. The Place-shaping Continuum: A Theory of Urban Design Process. *J. Urban Des.* **2013**, *19*, 2–36. [CrossRef]
16. Lang, J. *Urban Design: A Typology of Procedures and Products*; Routledge: London, UK, 2017.
17. Holden, M. Is Integrated Planning Any More Than the Sum of Its Parts? Considerations for Planning Sustainable Cities. *J. Plan. Educ. Res.* **2012**, *32*, 305–318. [CrossRef]
18. Derrible, S. Urban infrastructure is not a tree: Integrating and decentralizing urban infrastructure systems. *Environ. Plan. B Urban Anal. City Sci.* **2016**, *44*, 553–569. [CrossRef]
19. Carmona, M.; Sieh, L. *Measuring Quality in Planning: Managing the Performance Process*, 1st ed.; Routledge: London, UK; New York, NY, USA, 2004. [CrossRef]
20. Shach-Pinsly, D.; Capeluto, I. From Form-Based to Performance-Based Codes. *Sustainability* **2020**, *12*, 5657. [CrossRef]
21. Merrifield, A.; Oranje, M.; Fourie, C. Dialogue or Control? The Future of Planning in South Africa 2008. Prepared for the Presidency and GTZ. Available online: https://www.researchgate.net/publication/335137594_Dialogue_or_Control_The_future_of_planning_in_South_Africa (accessed on 20 September 2021).
22. Cordoba-Aguilar, H. Urban Planning in Latin America: Myth or Reality. *Int. J. Urban Des.* **2019**, *2*, 53–63. [CrossRef]
23. Storper, M. Governing the Large Metropolis. *Territ. Politi-Gov.* **2014**, *2*, 115–134. [CrossRef]
24. Carmona, M. Marketizing Design Review, Urban Maestro Program. United Nations Human Settlements Programme (UN-Habitat). 2020. Available online: www.urbanmaestro.org (accessed on 10 November 2021).
25. Madanipour, A. Roles and Challenges of Urban Design. *J. Urban Des.* **2006**, *11*, 173–193. [CrossRef]
26. Centner, R. Conflictive sustainability landscapes: The neoliberal quagmire of urban environmental planning in Buenos Aires. *Local Environ.* **2009**, *14*, 173–192. [CrossRef]
27. Klaufus, C.; Van Lindert, P.; Van Noorloos, F.; Steel, G. All-Inclusiveness versus Exclusion: Urban Project Development in Latin America and Africa. *Sustainability* **2017**, *9*, 2038. [CrossRef]
28. Bina, L.; (Municipality of Villa Carlos Paz, Córdoba, Argentina). Personal communication, 2021.
29. Montgomery, J. Making a city: Urbanity, vitality and urban design. *J. Urban Des.* **1998**, *3*, 93–116. [CrossRef]
30. Project for Public Spaces. *How to Turn a Place Around*; Project for Public Spaces: New York, NY, USA, 2000.
31. Gehl, J.; Svarre, B. *How to Study Public Life*; Island Press: Washington, DC, USA, 2013.
32. Baker, D.C.; Sipe, N.G.; Gleeson, B.J. Performance-based planning: Perspectives from the United States, Australia, and New Zealand. *J. Plan. Educ. Res.* **2006**, *25*, 396–409. [CrossRef]
33. American Association of State Highway and Transportation Officials. *AASHTO LRFD Bridge Design Specifications*; American Association of State Highway and Transportation Officials: Washington, DC, USA, 2007.
34. LM Interior. Available online: https://interiorlmdiario.com.ar/contenido/1529/villa-carlos-paz-restringen-el-transito-por-la-obra-del-puente-peatonal (accessed on 4 June 2021).

35. La Jornada Web. Available online: https://lajornadaweb.com.ar/2019/10/29/por-las-obras-del-nuevo-puente-peatonal-cortan-el-transito-en-el-carena-hacia-el-centro-viejo (accessed on 4 June 2021).
36. El Diario de Carlos Paz. Available online: https://www.eldiariodecarlospaz.com.ar/sociedad/2019/12/23/se-inauguro-el-nuevo-puente-peatonal-de-carlos-paz-79316.html (accessed on 3 June 2021).
37. Telefe Channel. Available online: https://cordoba.telefe.com/cordoba/schiaretti-inauguro-en-carlos-paz-un-nuevo-puente-peatonal-sobre-el-rio-san-antonio (accessed on 3 June 2021).
38. Foster, N. "Striving for Simplicity". Louisiana Channel, Louisiana Museum of Modern Art. Sir Norman Foster Was Interviewed by Marc-Christoph Wagner in 2015. Available online: https://www.youtube.com/user/TheLouisianaChannelvideos (accessed on 14 February 2021).
39. Mang, P.; Haggart, B.; Regenesis. *Regenerative Development and Design: A Framework for Evolving Sustainability*; John Wiley & Sons: Hoboken, NJ, USA, 2016.
40. Brohn, D.M.; Cowan, J. Teaching towards an improved understanding of structural behaviour. *Struct. Eng.* **1977**, *55*, 9–17.
41. Teng, J.G.; Song, C.Y.; Yuan, X.F. Fostering Creativity in Students in the Teaching of Structural Analysis. *Int. J. Eng.* **2004**, *20*, 96–102.
42. Aparicio, A.C.; Ruiz-Teran, A.M. Tradition and Innovation in Teaching Structural Design in Civil Engineering. *J. Prof. Issues Eng. Educ. Pact.* **2007**, *133*, 340–349. [CrossRef]
43. Gauvreau, P. Sustainable education for bridge engineers. *J. Traffic Transp. Eng.* **2018**, *5*, 510–519. [CrossRef]
44. Leonhardt, F. Aesthetics of Bridge Design. *PCI J.* **1968**, *13*, 14–31. [CrossRef]
45. Menn, C. An approach to bridge design. *Eng. Struct.* **1991**, *13*, 106–112. [CrossRef]
46. Ruby, I.; Ruby, A. *Infrastructure Space*; Ruby Press: Berlin, Germany, 2017.
47. Strydom, W.J.; Puren, K. An Exploration of the Dimensions of Place-Making: A South African Case Study. *Int. J. Urban Civ. Eng.* **2016**, *10*, 1324–1332.
48. Beza, B.B.; Freeman, C.G.; Fullaondo, D.; Mejía, G. Place? Lugar? Sitio? Framing place and placemaking through Latin American contexts. In *Urban Space: Experiences and Reflections from the Global South*; Pontificia Universidad Javeriana: Santiago de Cali, Colombia, 2018; pp. 19–44. [CrossRef]
49. Star, S.L. The ethnography of infrastructure. *Am. Behav. Sci.* **1999**, *43*, 377–391. [CrossRef]
50. Maidment-Blundell, L. *Infrastructure, Place-Making and Sustainability. A Report for Places for People*; University of Cambridge: Cambridge, UK, 2020; Available online: https://www.cchpr.landecon.cam.ac.uk/files/documents/DownloadTemplate_49.pdf (accessed on 30 October 2021).
51. Cilliers, J.; Cilliers, S. Planning for Green Infrastructure: Options for South African Cities 2018; SACN Report. Available online: https://www.researchgate.net/publication/317303645_Planning_for_Green_Infrastructure_Options_for_South_African_Cities (accessed on 4 October 2021).

land

Article

How Does the Concept of Resilient City Work in Practice? Planning and Achievements

Chuan Wang *, Xinhua Li and Siheng Li

School of Architecture, Southeast University, 2 Sipailou, Nanjing 210096, China; xinhua_li@seu.edu.cn (X.L.); sihengli@seu.edu.cn (S.L.)
* Correspondence: chuanwang@seu.edu.cn

Abstract: In the past decade, resilient cities (RCs) have gained extensive attention in academic and political debates as a vision of urban futures. In particular, with the support of the Rockefeller Foundation's Resilient City 100 Program (RC100), a number of cities worldwide have pushed this concept forward from theory to practice through their RC plans/strategies. However, there is widespread doubt regarding how much this holistic idea of the future built environment contributes to urban practice. After developing a scoring evaluation matrix based on the synthesis of existing RC assessment frameworks, this review scrutinizes the plans, reports, city leaders' speeches, official websites and academic reviews of five representative resilient cities and investigates their motivations, planning and achievements. The results demonstrate a huge theoretical and practical gap in RC: while RC plans attempt to expand as comprehensively as possible from cities' initially narrow motivations, their achievements in implementation are limited. Although RC provides more holistic solutions to the cities, the limited resources mean that cities have to prioritize their urgent issues in their everyday practice. This paper calls for designating more feasible and specific features in RC visions and maintaining regular alignments from planning to actions in future RC practice.

Keywords: resilient city; theory and practice; planning practice; future built environment; scoring evaluation matrix; assessment framework; urban resilience; RC100

Citation: Wang, C.; Li, X.; Li, S. How Does the Concept of Resilient City Work in Practice? Planning and Achievements. *Land* **2021**, *10*, 1319. https://doi.org/10.3390/land10121319

Academic Editors: Elizelle Juanee Cilliers and Sarel Cilliers

Received: 5 November 2021
Accepted: 27 November 2021
Published: 1 December 2021

1. Introduction

With the increasing proportion of cities in human settlements, the giant and complex system of cities has encountered increasing uncertainties and risks [1,2]. In the past two decades, terrorist attacks, extreme weather, financial crises, global warming and other crises have significantly impacted our vision of the future built environment. In the face of various natural and human disasters, the vulnerability of cities has become a major problem impeding sustainable urban development [3]. Resilient city (RC), a fashionable concept in urban planning, emerged as an attempt to resolve this problem. Scholars aim to use this concept to improve the resistance, recovery and adaptability of complex urban systems and to enhance the predictability and guidance for urban planning [3–6]. Furthermore, after years of advocacy and debates, cities have gradually adopted and translated RC into their master plans, strategic plans and community plans [7–9]. In particular, with the support of the Rockefeller Foundation's Resilient City 100 Program (RC100) since 2013, a large number of cities worldwide have pushed this concept forward from theory to practice through their RC plans/strategies [10–13].

However, while many city authorities increasingly use the RC concept in envisioning the future built environment, some scholars have questioned how much content of RC can be implemented in reality and are concerned about the gap between its theory and practice [14–17]. In contrast to the heated theoretical discussions, RC practice still has not demonstrated convincing models. Some scholars have warned that RC may become another fuzzy concept such as "sustainable development" in the hands of policymakers who change RC policy standards flexibly according to different situations [18–20]. Empirical data also demonstrate that the same plan or urban policies can be interpreted selectively in the implementation stage by different government agencies [9,20]. In essence, RC, as a systematic concept for future cities, can only have a positive impact on the built environment if its multiple and integrated dimensions are considered. The limited achievements mean that the initial comprehensive RC plans may not perform as well in reality as expected in the planning.

Therefore, this research aims to explore the gap between the planning and implementation of RC through an analysis of the empirical data of five representative RC cities. Although some individual RC cases have been critically reviewed and assessed [9,12,21–25], the uniqueness of each city means that a systematic and comparative review of RC planning and implementation is still needed. To fill this gap, this research compares the RC planning and implementation of five cities using a newly synthesized scoring matrix evaluation framework and detailed narratives of each city's RC trajectory. This research attempts to answer how the RC concept lands in urban practice on a global scale.

In the next sections, this paper first reviews the development of resilience in the academic circles, particularly on its trajectory associated with cities. This is the theoretical foundation for understanding the changes and mismatches of RC in theory and practice. Second, five representative RC cities are selected as the resources of the empirical data. They are analyzed through a comprehensive scoring evaluation matrix of 26 indicators in seven clusters developed based on the grounded analysis of existing assessment frameworks. The scored results are accompanied by the narratives of each case to demonstrate the differences among the motivations, planning and achievements of representative RC cities. This reveals that, although authorities usually make their RC plans inclusive and comprehensive, the implementation is limited to easy or urgent parts. Through an in-depth analysis of the relationship among motivations, planning and achievement, this study demonstrates the contribution and constraints of RC for urban practice and suggests future development directions.

2. Expanding Dimensions of the Resilient City

The concept of resilience first appeared in ecological scholarship in 1973, when C. S. Holling used the term "resilience" to describe the ability of ecological systems to resist and/or adapt to a particular disturbance and recover its normal functioning or state of balance, which may set the initial baseline or a new situation [26]. This concept challenged the then-dominant stable equilibrium theory in ecology, with an emphasis on non-linear dynamics and uncertainty. It later extended to social systems to describe the interaction between disturbance and reorganization in social-ecological synergies [27,28].

The concept of resilience was then introduced to the built environment, focusing on the relationship between ecological systems in the city [29] and cities' responses to natural disasters [3,6]. The development of social resilience has motivated urban scholars to regard cities as giant social-ecological systems, emphasizing integrated feedback of urban systems and cross-scale dynamic interaction of urban elements [30]. RC research began to focus on how the urban system achieves a new balance through the utilization, preservation, release and reorganization of social cycles [31]. Along with public awareness shifting to financial austerity, class conflicts and populist movements after the 2008 financial crisis, more scholars have begun associating the RC concept with sociology, political science and economics to seek new urban solutions related to social equality, political empowerment and economic optimization [13,32,33]. Resilience's connotation with dynamics, co-evolution and elasticity makes this concept an ideal concept for dealing with unpredictable changes in the future built environment. Considering the increasing corpulent content of RC, some scholars have attempted to organize them into coherent frameworks [4,5,34]. For instance, Desouza and Flanery proposed a resilient urban evolution mechanism based on a complex adaptive system by simplifying the model of the interactive relationship between the urban system and the agent as the two-element attributes of "materiality" and "sociality" [4]. However, these attempts have not prevented the continuous expansion of the RC theory, particularly after the involvement of the wheeling-out power of NGOs, including the United Nations Office for Disaster Risk Reduction, the World Bank and the Rockefeller's Foundation [35,36]. A recent review shows that the scope of RC is still continuously shifting and expanding in academic discourse [37], tending to cover increasingly important aspects of the economy, infrastructure, society and ecology for the future built environment.

3. Materials and Methods

3.1. Case Selection and Data Resources

This research selects five cities as empirical data to analyze the gap between the planning and implementation of RC, according to cities' planning activities, data accessibility and regional representativeness. First, each selected city has either specific RC plans or at least a substantial part influenced by the RC concept in the city's master or strategic plans. As the concept of "resilience" is becoming increasingly popular in urban planning and management, many city plans contain "resilient" content. For this research, at least a designated section is needed to be regarded as a planning action for RC. Although almost all RC100 cities contain a comprehensive RC plan as required in their RC100 participation, this selection attempts to extend beyond the RC100 inventory to represent a more diverse scope. Therefore, every selected city has at least one RC plan other than its RC100 plan.

Second, sufficient data can be retrieved to assess representative cities' RC implementation. This part is more critical for this selection procedure, as implementation data are usually more fragmented than planning documents. Four types of resources were retrieved and synthesized for this assessment: cities' official websites on their RC actions and achievements, recent academic reviews on specific cities' RC practice, city leaders' recent speeches or interviews on RC and cities' own review reports. News reports are generally excluded in this assessment, as our preliminary study discovered that their links with resilience heavily depend on journalists' own interpretations. Admittedly, languages are a constraint in this selection process, but the authors have tried their best to obtain access to all available data with the help of Google Translate's website translation service. The assessed documents span from 2010 to September 2021, with a focus on the past three years.

As a result, the selected cities are London, Addis Ababa, Rotterdam, New York and Surat (Table 1). This selection covers cities in different continents (Europe, North America, Asia and Africa) and different developing stages (developed and developing countries). It should be noted that only (at least regionally) large cities were selected because of the scarcity of open data of smaller cities for this analysis.

Table 1. Representative RC Cities and Their Practice Data Resources.

City	Planning Documents	Implementation Data [1]
London	London Resilience Partnership Strategy 2020 London City Resilience Strategy 2020 (RC 100)	City Official Website: https://www.london.gov.uk/ Academic Reviews: [8,38] City Leaders' Speech: Deputy Mayor for Environment & Energy Shirley Rodrigues's Speech to Resilience First, 13 January 2021 [39]
Addis Ababa	Addis Ababa, Enhancing Urban Resilience 2015 Addis Ababa Resilience Strategy 2020 (RC 100)	City Official Website: https://resilientaddis.org/ Academic Reviews: [40–47] City Leaders' Interview: An interview with Chief Resilience Officer [48]
Rotterdam	Rotterdam Climate Change Adaptation 2013 Rotterdam Resilience Strategy 2016 (RC 100)	City Official Website: https://www.resilientrotterdam.nl; https://www.rotterdam.nl Academic Reviews: [9,12,24,49]

Table 1. *Cont.*

City	Planning Documents	Implementation Data [1]
New York	One NYC—A Stronger, more resilient New York 2013 One NYC—The Plan for a Strong and Just City 2015 (RC 100)	City Official Website: https://www1.nyc.gov; https://nyc-oem.maps.arcgis.com/ Academic Reviews: [25,50]
Surat	Surat City Resilience Strategy 2011 Surat Resilience Strategy 2017 (RC 100)	City Official Website: https://www.suratmunicipal.gov.in/ Academic Reviews: [51–53] Governmental Reviews: [54]

[1] The accessing details of city official websites are available in Supplement S1.

3.2. Assessment Framework

To analyze the gap between the planning and implementation of RC, this research establishes an assessment framework based on the comparison and synthesis of existing assessment methods. Although relevant to and inspiring for the RC concept, current assessment frameworks on similar concepts, such as "Goal 11 (sustainable cities and communities) of the Sustainable Development Goals" [55] and "Disaster Resilience Scorecard for Cities" [56], have different priorities and coverage for cities. Therefore, this research principally focuses on the current assessment methods directly named after "resilient city" or "city/urban resilience" to avoid further expansion of this research target.

Existing RC assessment frameworks primarily focus on two interactive directions: process and content. The process framework focuses on the different stages of urban resilience and usually uses flow maps to demonstrate the relationships between different components [4,5,29,34]. This direction typically closely represents the nature of the RC's complexity, diversification, dynamic decision-making and uncertainty. However, its complexity and ambiguity make it difficult to measure the empirical data. Second, content-based frameworks include different ideas that can be associated with the enhancement of urban resilience. This direction is usually summarized in an indexed evaluation system that attempts to quantify each indicator that represents an idea. Through the standardization and weighting of the index data, the results of the superimposed index data are used as the standard for evaluating urban resilience [57]. However, it is difficult to reach a consensus on how to determine the weights of indicators within a complex system. In particular, different cities encounter different environmental challenges, which hinders unified standardization and weighting to adapt to different environments. Instead of standardization within predetermined structures, this research uses a grounded approach to build an RC assessment framework based on the synthesis of existing content-based assessment frameworks [58,59]. This enables this research to assemble a comprehensive framework to cover as many aspects of RC-related contents as possible.

The analyzed resources cover the most cited literature on RC assessment in academic databases (Web of Science, Scopus and Google Scholar) as well as the widely applied assessment methods in city plans. It should be noted that two types of assessment literature were excluded from this synthesis but used as supplementary references. The first type includes assessment frameworks that focus only on specific aspects of RC, e.g., community disaster mitigation [60] and the relationship between ecology and planning [29]. The second is the frameworks' scope that is too broad to assess the RC practice. Acknowledging that many indicators (e.g., GDP per capita, employment rate and crime rate) can be linked to RC assessment, this research contends that it is still difficult to determine their exact contribution to RC practice. As a result, this research selects and compares five representative assessment frameworks [3,5,61–63] and lists their contents pertaining to RC in detail (Table A1).

After combining similar ideas, a 26-indicator table is created as the evaluation matrix to assess different aspects of the planning and implementation of RC (Table 2). Furthermore, a grounding method [64] is used to categorize the indicators into seven sections: public participation, social service, robust economics, hazard management, institutional optimization, physical/engineering and natural enhancement. Instead of dividing RC into equally balanced aspects (e.g., social-physical [4] or economic, environmental, security, social and political [63]), this categorization aims to provide groups of ideas pertaining to RC for easier understanding in the later analysis. Therefore, clustering is loosely gathered depending on the closeness and commonality of the listed indicators. The matrix of indicators represents a broad but assessable scope of RC for envisioning the future built environment. This summary lays the foundation for the analysis of selected RC cases.

Table 2. RC Assessment Framework.

Clusters	Indicators
Public Participation (PP)	1. Educate developers and the public about hazard mitigation 2. Encourage community involvement and citizen participation 3. Collective learning process from past urban hazards
Social Service (SS)	4. Enforce security and laws to reduce crimes 5. Set ensured human security for providing basic living needs 6. Assist vulnerable neighborhoods and populations (increase social equity) 7. Enhance robust public health systems and emergency medical care
Robust Economics (RE)	8. Diversify livelihoods that can mitigate business interruption impacts 9. Adopt sustainable economic initiatives 10. Use/facilitate alternative energy (e.g., solar and wind to reduce greenhouse gas emission)
Hazard Management (HM)	11. Identify, monitor and assess hazards and vulnerability 12. Land use and planning management for natural hazard-prone areas 13. Enhance building hazard resistance by appropriate construction regulations 14. Build effective emergency response services/initiatives
Institutional Optimization (IO)	15. Increase capacity of governmental institutions through wide collaboration 16. Adopt a flexible planning system and adaptive design process 17. Employ agile city management for uncertainty and challenges 18. Effective coordination with other government bodies
Physical/Engineering (PE)	19. Set robust protective infrastructure with regular maintenance 20. Build/optimize a distributed or decentralized hazard mitigation system 21. Build/enhance reliable ICT infrastructure (communication network) 22. Use/optimize a sustainable urban form 23. Optimize diverse, effective, affordable, sustainable transport
Natural Enhancement (NE)	24. Conserve, manage and protect ecosystems 25. Optimize urban blue-green ecological networks within compact cities 26. Maintain diversity in biological systems

This assessment framework uses a scoring system to measure each indicator in later analysis: 3 indicates that this point is well elaborated in planning or largely put into city actions; 2 indicates that this idea is partially included in planning proposals or practiced to some extent; 1 indicates that only a limited part of this point is mentioned in the planning stage or very little is realized in terms of implementation; 0 denotes that this point is not mentioned or achieved at all. Compared with its existing counterparts, this assessment framework not only has a wider coverage of most aspects that cities have planned and practiced pertaining to RC but also demonstrates a relatively feasible approach to measure each indicator in the framework.

3.3. Comparison and Narratives of Empirical Data

For the analysis of each city, this research focuses on the comparison of three aspects: motivations, planning and achievements. First, early-stage documents are traced to reveal the initial motivations for adopting the RC concept or participating in the RC100. This part aims to answer why this city started to use the concept of resilience in its city planning and management, particularly before the formal names of the plans included the term resilience. Second, this research uses the assessment framework to determine the content coverage of two selected RC plans for each city (the planning documents column in Table 1). This section aims to discover what is envisioned in cities' plans when the RC concept is introduced as well as their priorities in these strategies compared with earlier motivations. Thirdly, achievements are summarized in the same framework using synthesized information resources. Admittedly, it is difficult to check every detail of the city's achievements. The synthesis of multiple data (the achievement data column in Table 1) can at least identify the key areas that selected cities have achieved under the RC concept. To make the cross-city data comparable, this research set the same standard for assessing achievement data for each city: only the ideas that have been translated into projects, policies, funds or programs with real actions and named after or described as "resilient" are accounted as the achievements. Besides, implementation actions before the RC plans can still be regarded as achievements if they are officially associated with the RC concepts.

It should be noted that this scoring criteria principally focuses on the scope that the RC practice has covered but cannot quantify to what extent one specific idea has been realized. Some large gaps in the degree of implementation are discussed as narratives of each city's RC journey to additionally support conclusions. Furthermore, this research briefly compares the narratives of each city from initial motivations to implementation and highlights some controversial indicators.

4. Results

This research employs the scoring evaluation matrix developed in Section 3.2 to assess every indicator in the most representative RC plan, RC 100 plan and planning achievements of each city. The detailed scoring basis is available in Supplement S1 for each number. This resulting matrix (Table 3) provides comparable and computable data between planning and achievements, as well as among different cities, as the foundation to analyze RC planning and achievements.

Table 3. Scoring results of representative cities' RC planning and achievements.

Clusters	No.	Indicators	London			Addis Ababa			Rotterdam			New York			Surat			Average Score	Score Standard Deviation
			London Resilience Partnership Strategy 2020	*London City Resilience Strategy 2020 (RC100)*	*London RC Achievements*	*Addis Ababa, Enhancing Urban Resilience 2015*	*Addis Ababa Resilience Strategy 2020 (RC100)*	*Addis Ababa RC Achievements*	*Rotterdam Climate Change Adaptation 2013*	*Rotterdam Resilience Strategy 2016 (RC100)*	*Rotterdam RC Achievements*	*One NYC-A Stronger, More Resilient New York 2013*	*One NYC-The Plan for a Strong and Just City 2015 (RC100)*	*New York RC Achievements*	*Surat City Resilience Strategy 2011*	*Surat Resilience Strategy 2017 (RC100)*	*Surat RC Achievements*		
PP	1	Educate developers and the public about hazard mitigation	3	3	0	2	3	1	2	2	3	0	2	3	2	2	2	2.00	0.97
	2	Encourage community involvement and citizen participation	3	3	2	2	3	1	3	3	3	3	3	2	1	2	2	2.40	0.71
	3	Learn collectively from past urban hazards	1	0	2	2	1	2	2	0	0	0	0	0	3	3	1	1.13	1.09
SS	4	Enforce security and laws to reduce crimes	0	1	1	1	0	1	0	3	0	0	0	0	0	1	2	0.67	0.87
	5	Set ensured human security for providing basic living needs	0	1	1	2	3	1	0	0	0	1	2	0	1	2	0	0.93	0.93
	6	Assist vulnerable neighborhoods and populations (increase social equity)	0	1	0	0	1	0	1	0	1	2	2	1	1	2	0	0.80	0.75
	7	Enhance robust public health systems and emergency medical care	0	0	0	1	3	2	1	2	0	3	1	0	2	2	2	1.27	1.06
RE	8	Diversify livelihoods that can mitigate business interruption impacts	0	2	2	0	2	1	0	2	1	1	2	2	0	2	0	1.13	0.88
	9	Adopt sustainable economic initiatives	0	0	3	2	3	2	3	3	0	1	0	0	0	0	2	1.27	1.29
	10	Use/facilitate alternative energy (solar, wind, etc. to reduce greenhouse gas emission)	0	1	1	3	3	2	0	3	3	1	0	0	3	2	0	1.47	1.26

Table 3. Cont.

Clusters	No.	Indicators	London			Addis Ababa			Rotterdam			New York			Surat			Average Score	Score Standard Deviation
			London Resilience Partnership Strategy 2020	*London City Resilience Strategy 2020 (RC100)*	*London RC Achievements*	*Addis Ababa, Enhancing Urban Resilience 2015*	*Addis Ababa Resilience Strategy 2020 (RC100)*	*Addis Ababa RC Achievements*	*Rotterdam Climate Change Adaptation 2013*	*Rotterdam Resilience Strategy 2016 (RC100)*	*Rotterdam RC Achievements*	*One NYC-A Stronger, More Resilient New York 2013*	*One NYC-The Plan for a Strong and Just City 2015 (RC100)*	*New York RC Achievements*	*Surat City Resilience Strategy 2011*	*Surat Resilience Strategy 2017 (RC100)*	*Surat RC Achievements*		
HM	11	Identify, monitor, and assess hazards and vulnerability	3	3	3	3	3	2	3	2	2	2	1	2	3	3	2	2.47	0.62
	12	Build effective emergency response services/initiatives	3	1	3	3	3	2	1	1	0	3	1	2	0	0	2	1.67	1.14
	13	Enhance building hazard resistance by appropriate construction regulations	1	3	2	1	3	0	3	2	2	3	3	3	1	2	0	1.93	1.06
	14	Land use and planning management for natural hazard prone areas	0	0	0	2	1	0	0	0	1	3	1	2	3	0	2	1.00	1.10
IO	15	Increase capacity of governmental institutions by wide collaboration	3	0	2	2	2	2	3	3	3	3	1	2	2	1	0	1.93	1.00
	16	Adopt a flexible planning system and adaptive design process	0	3	1	1	1	0	2	1	0	2	1	0	0	0	0	0.80	0.91
	17	Employ agile city management for uncertainty and challenges	2	3	0	1	1	1	3	2	1	0	0	1	0	0	0	1.00	1.03
	18	Effective coordination with other government bodies	3	3	0	2	2	2	3	3	3	2	1	2	2	1	0	1.93	1.00
PE	19	Set robust protective infrastructure with regular maintenance	0	3	2	2	2	2	3	3	2	3	3	3	3	3	3	2.47	0.81
	20	Build/optimize distributed or decentralized hazard mitigation system	2	3	2	0	0	0	0	0	1	2	2	2	1	2	1	1.20	0.98

Table 3. *Cont.*

Clusters	No.	Indicators	London			Addis Ababa			Rotterdam			New York			Surat			Average Score	Score Standard Deviation
			London Resilience Partnership Strategy 2020	*London City Resilience Strategy 2020 (RC100)*	*London RC Achievements*	*Addis Ababa, Enhancing Urban Resilience 2015*	*Addis Ababa Resilience Strategy 2020 (RC100)*	*Addis Ababa RC Achievements*	*Rotterdam Climate Change Adaptation 2013*	*Rotterdam Resilience Strategy 2016 (RC100)*	*Rotterdam RC Achievements*	*One NYC-A Stronger, More Resilient New York 2013*	*One NYC-The Plan for a Strong and Just City 2015 (RC100)*	*New York RC Achievements*	*Surat City Resilience Strategy 2011*	*Surat Resilience Strategy 2017 (RC100)*	*Surat RC Achievements*		
	21	Enhance reliable ICT infrastructure (communication network)	1	1	1	0	3	2	2	3	3	3	0	0	0	0	0	1.27	1.24
	22	Optimize sustainable urban form	0	1	0	0	2	2	3	2	0	0	0	0	2	1	0	0.87	1.02
	23	Provide diverse, effective, affordable, sustainable transport	0	0	1	3	3	3	0	0	0	3	1	1	3	3	2	1.53	1.31
NE	24	Conserve, manage, and protect ecosystems	0	0	0	1	3	2	0	3	1	2	0	1	0	3	1	1.13	1.15
	25	Optimize urban blue-green ecological networks within compact cities	0	2	2	0	2	2	3	3	0	2	1	2	2	2	1	1.60	0.95
	26	Maintain diversity in biological systems	0	0	1	0	2	0	1	1	2	0	0	0	0	0	0	0.47	0.72

Scoring Criteria: 3 = Well elaborated/practiced; 2 = Partially included/practiced; 1 = Mentioned/practiced to a limited extent; 0 = Not mentioned or practiced at all. For a better visualization, the cell with the higher value is assigned with darker color.

4.1. Different Focuses

The assessment results of five representative RC cities (Table 1) demonstrate that indicators are not equally treated in RC planning and implementation. Furthermore, cities also focus on what is more urgent and necessary for their urban development.

First, different indicators were weighed variously in the planning and implementation procedures. The diverse average scores of each indicator in Table 1 show that the ideas that scholars associate with the RC concept are treated drastically differently in the five cities. The high average scores and low standard deviation of some indicators in Table 3 mean some ideas have reached wide consensus in both RC planning and implementation, such as No. 2 "Encourage community involvement and citizen participation", No. 11 "Identify, monitor, and assess hazards and vulnerability" and No. 19 "Set robust protective infrastructure with regular maintenance". These ideas are widely accepted in almost every RC-related plan and are implemented widely in practice. In contrast, some widely discussed ideas are largely ignored in cities' actions. To illustrate, No. 4 "Enforce security and laws to reduce crimes" is only elaborated in Rotterdam's RC100 plan but is surprisingly not mentioned at all in its RC plan or its implementation. Furthermore, No. 16 "Adopt flexible planning system and adaptive design process" is an important approach to realize London' resilience but is almost entirely ignored elsewhere. In addition, No. 26 "Maintain diversity in biological systems" has been widely discussed in academic discourse and in the policymaking of RC [61,65] but has been ignored in practice. Although some ideas have been eagerly included as part of the RC concept in academic circles, cities do not have much interest in linking them with resilience. For instance, enhancing biodiversity is a critical component in London metropolitan planning and governance policies through London plans and various urban greening guides and funds [66,67], but it is not included in any of London's RC plans at all.

Second, even regarding the same idea, five cities have different approaches toward including them as a part of their RC trajectory. Since the random allocation of the 0–1–2–3 scoring numbers has a standard deviation of 1.12, the high score standard deviation numbers (>1.2 in the column of standard deviation in Table 3) indicate that these indicators are highly polarized in urban practice. Addis Ababa and Surat elaborated No. 23. "Provide diverse, effective, affordable, sustainable transport" in their plans and take real actions in their urban development (relatively high achievement score in Table 3). In contrast, Rotterdam never associates its excellent public transportation system with the RC concept in an official tone [68]. This huge divergence between cities is also observed regarding No. 9 "Adopt sustainable economic initiatives", No 10 "Use/facilitate alternative energy (e.g., solar and wind to reduce greenhouse gas emission)" and No. 21 "Enhance reliable ICT infrastructure (communication network)". Because cities are in different development stages, every city encounters distinctive urgent problems related to the next step of urban development. For instance, globally top cities (London and New York) merely interpret the RC concept in their own ways for urban practice and keep this concept slim in their city plans, while cities in developing countries (Addis Ababa and Surat) made ambitious plans to seize the opportunities for urban development. This means it is an impossible task to build a universal consensus for RC practice.

4.2. Broader Scope in Encounter with the RC100 Program

Comparing the two plans on RC, the RC100 plans generally have a wider scope than the city's own RC plans. Figure 1 graphically illustrates and presents the average scores of each cluster in the planning and implementation of five cities and shows the comparison among different cities and within the same city. This shows that the scope of planning has expanded with the intervention of the RC100 program. The only exception is New York, whose RC plan developed immediately after 2012 Hurricane Sandy has a wider scope than its 2015 plan. As the most prosperous and powerful metropolitan in the world, New York does not care for sponsorship from the RC100 program. Although linked to RC100's official website as a key example, "One New York: The Plan for a Strong and Just City" seems like a perfunctory submission for the Rockefeller's Foundation since it is the only plan without any indication of its RC100 participation.

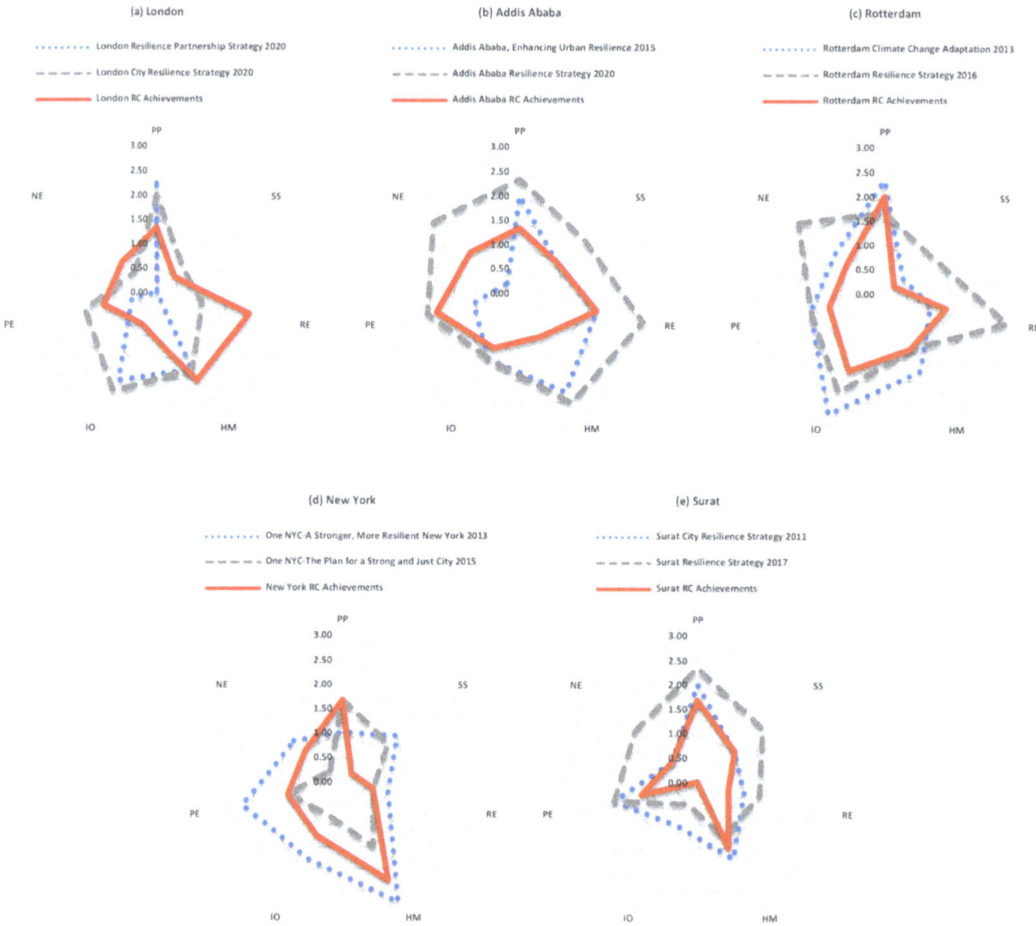

Figure 1. Clusters of focus in five representative cities. (The scores are the average scores for each cluster of each city based on the data in Table 3).

All of the other four RC plans have expanded their scopes, particularly regarding the Robust Economics cluster and the Natural Enhancement cluster (Figure 1). Rotterdam and Addis Ababa included the term resilience in their plans' names years before the RC concept emerged for its water-centered climate change policies and economic re-orientation, respectively [9,69]. However, the RC100 program accelerated its original trajectory to a wider coverage [12,70,71]. While Rotterdam's own RC plan in 2013 focused on the analysis of and potential solutions to five typical hazards, its RC 100 plan in 2016 extended to an all-encompassing plan that attempted to solve wide social, economic and ecological problems. The most comprehensive plan (highest average score of 2.15)—Addis Ababa Resilience Strategy 2020—puts forward three pillars, 13 goals and 48 actions to build a "thriving, safe, intelligent, green and healthy resilient city" [70]. Similarly, when the RC100 program came to London, both the LRF and Mayor's Office for Policing and Crime (MOPAC) proposed different plans [8]. The MOPAC proposal to tackle cyber-crime and emerging digital threats was selected by political authorities in 2014 [72]. However, the lengthy preparation of London's RC plan has changed its initial narrow goal to a much broader scope: traditional disaster mitigation and management have been gradually expanded by a more encompassing vision of climate-change adjustment; MOPAC's cyber-security issues have been expanded to more general social services; and some ideas on flexible planning and design processes and infrastructure maintenance were added. Plans on the encouragement of the RC100 program have become increasingly comprehensive with the ambition to solve more problems.

4.3. Gap between Planning and Implementation

The scope of implementation in reality is usually lower than the content of planning in both city-level (Figure 1) and indicators' comparison (Figure 2). Because this assessment only measures the scopes of RC planning and implementation, the real gap in achievements may be even wider. Figure 2 shows that the scope of most indicators is lower in implementation, particularly for two clusters: Social Service and Institutional Optimization (marked with dotted boxes in Figure 2). Three indicators in Social Service—No. 5 "Set ensured human security for providing basic living needs", No. 6 "Assist vulnerable neighborhoods and populations (increase social equity)" and No. 7 "Enhance robust public health systems and emergency medical care"—represent the ideas that are easy to claim but difficult to implement in reality. Although everyone knows that they are important for the whole society and can significantly boost urban resilience, they require a long-term effort with sufficient financial and labor resources. Similarly, for the three points in Institutional Optimization, it is not easy to change the existing planning, design and governance systems within a short period as RC achievements.

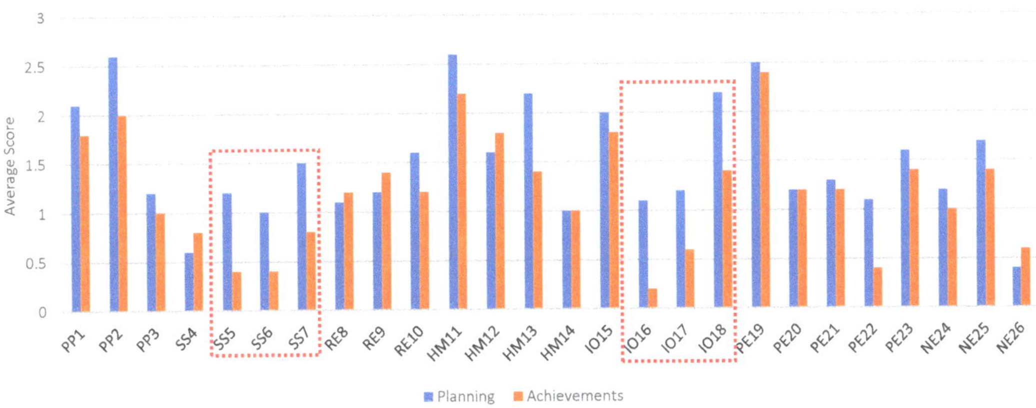

Figure 2. Comparison between the planning and achievements of each indicator. (The average scores are calculated based on the ten RC plans and five cities' achievements for each indicator of the data in Table 3).

Another reason for this gap is that cities prioritize their urgent problems. For instance, both the 2003–2013 structure plan and the 2015 resilient strategy of Addis Ababa allocated a large area of swampy land for flood retention and urban agriculture as an important RC action. However, a so-called "master plan reconciliation" committee of the city administration decided to use the same land for condominium houses. Open areas in the catchment now mainly consist of "residential green space", which implies that they were occupied by informal residents [45]. In the short term, Addis Ababa's housing problem is more urgent than disaster mitigation issues for city authorities. On the one hand, the RC concept helps cities to identify various uncertain risks that cities may encounter by building a complete evaluation structure to prevent blind spots in urban development; on the other hand, municipal authorities must choose the priority of urban development under such a broad plan when resources are limited. Therefore, as shown in Table 3, the implementation content often corresponds to the more urgent parts of cities' needs.

5. Discussion: What RC Brings for Urban Practice?

The assessment results show that RC, as a changing and expanding concept in academic discourse, may not have as considerable of an impact on innovative urban practice as has been claimed. This section critically reflects on the positive features that RC brings to practice and the constraints that prevent good ideas of RC from translating into the improvement of the built environment.

There is no doubt that the RC concept creates a platform for different stakeholders in cities to communicate regarding topics related to resilience [8,73]. In all cases, the journey of making RC plans is a process of building networked governance on strategic mapping of a clear policy and, therefore, setting a clearer agenda with various stakeholders. The forums, workshops and consulting meetings for RC plans are shared platforms to collectively consider the future mode of the built environment in the age of uncertainty. Practitioners benefit from using the RC concept because it empowers the local level to act horizontally against top-down actions. Furthermore, powerful international organizations facilitate intercity sharing for effective and feasible actions on a global scale with financially stimulating initiatives. The RC concept provides a framework for holistically assessing the vulnerability of cities, gaining an opportunity to gather fragmented bureaus and departments to respond to urban challenges together and generating an imaginary knotting point for the future built environment for the general public.

Another noticeable feature of RC is city branding. For many cities, the adoption of RC practice represents a political commitment of city municipalities to post-disaster recovery [50] or is featured in their advertisements for attracting foreign investments and tourists [73,74], regardless of whether the actions have been successfully implemented in reality. The post-Sandy political pressure forced New York to use "resilience" to name its new governance framework, which creatively recalibrated its new and old elements [50]. Ironically, when the most emergency post-storm time passed, New York sought new fashionable labels to substitute the old branding, and RC became a merely specific term for hazard management and flood-related infrastructure (see Supplement S1). Furthermore, for many well-performing cities, RC is a new way to enshrine their advancements among cities and to brand them in a new manner to the general public. For instance, many of Rotterdam's iconic RC achievements (e.g., green roof and water plaza) have been completed or were at least in good progress before the RC concept was introduced to its city plans [24,49]. Similarly, in London, the RC concept has been mainly used for assembling and extolling a number of existing programs instead of bringing new ideas to change the current city planning and development trajectory. Furthermore, RC plans can be an ambitious agenda beyond urban development, such as seeking geopolitical impacts in Addis Ababa. Mayor Diriba Kuma asserted in establishing its RC office [48]: "We want the city to be a model for the region. Addis has already been a leader in promoting climate resilience and green economy and holds a track record of championing resilient initiatives ... " Its Resilient City Office became a propaganda agency to spread the city's achievements that can be linked with RC. Critics argue that the active planning actions in Addis Ababa, including RC plans, represent a propaganda model by the federal government to pursue its geopolitical ambition and influence other countries in the Horn of Africa [75]. Some indicators' wider scopes in implementation compared to planning in Table 1 show that five cities are enthusiastic in advertising some specific resilient-related achievements for city-branding purposes.

However, the shared platform inevitably broadens the scope of the RC concept in practice, as stakeholders tend to input their understanding into RC plans. Furthermore, city branding narratives often ignore the many nuances that RC research initially aims to achieve and merely focus on the positive image for the general public, regardless of its feasibility and effectiveness [73]. Similar to many concepts in planning, the RC concept has become increasingly vague and broad in practice and can be interpreted and implemented differently for stakeholder needs [76]. More precisely, the five cities started their journey of resilience from a relatively small scope of urgent and local problems such as sudden natural disasters or recognized long-term threats (see more details in Supplement S1). New York and Surat responded with their planning actions to the 2012 Hurricane Sandy and the devastating 2006 Surat flood, respectively [50,77]. Rotterdam's location in the low delta area motivates its preparation for climate change, particularly for the sea-level rise [9]. Metropolitan complexity drives London to form the London Resilience Forum (LRF), a network of public and private partners, to prepare for natural hazards and critical infrastructure failure [8]. These motivations have been gradually expanded in the planning stage when more stakeholders were involved and more visions were included, particularly with the intervention of the RC100 program. The meaning and vision of resilience for cities will continue to morph in urban practice, especially as practitioners incorporate their own interpretations and imagination along the way [78]. Although scope expansion itself is not a problem, it amplifies the gap between planning and implementation, as demonstrated in Figure 1.

Therefore, there are concerns that the concept of RC will become an empty signifier in the discourse of the future built environment, as like what happened to "sustainable city", "creative city", and "smart city" [18,79–81]. As Therrien et al. observed in London practice, the "diversity of issues makes it difficult to develop a clear set of interconnected priorities on resilience that reflects a convergence of interests among organizational stakeholders" [8] (p. 7). All claims in the name of RC seem to be an all-winning process, but in many cases, there is an unbalanced consequence that needs to identify winners and losers or those who obtain more privileges in the resilience actions. It would definitely be disappointing if RC is only an umbrella word that merely builds a holistic image of cities and gains wider social support instead of real inspirations for practice. Our assessment results demonstrate that the RC practice actions that are implemented in reality are closely associated with cities' urgent problems rather than how resilient these actions can be for the cities.

6. Conclusions

With the help of the scoring evaluation matrix, this study assessed and analyzed five representative cities' practices from planning to implementation. The results demonstrate that the RC concept is usually interpreted with different focuses depending on the city's own needs when it lands on practice. Although cities' initial motivations for resilience were usually narrow, the shared platform that this concept entails and the city branding trend in practice easily triggers plans to expand as comprehensively as possible, particularly with the intervention of the RC100 Program. The holistic solutions that the RC concept aims to offer to cities always encounter difficulties in the implementation process because the limited resources, particularly in the post-2008 austerity governance, mean that cities can only prioritize their urgent issues in their everyday practice.

Therefore, this paper calls for scholars to pay more attention to researching feasible and specific features of RC that can optimize our future built environment. Instead of adding more content or mapping complex networks of a large collection of components in the name of RC, it is more important to elucidate the contribution of specific features to building our urban futures in the grand narrative of RC. In urban practice, regular alignments from planning to implementation are more important in RC practice than proposing one or another unachievable city-branding plan. If the destination of RC is not clear enough, the deleterious dimension of resilience may secretly do the evil rather than bounce back after disturbance, as expected in reality [15,38].

Supplementary Materials: The following is available online at https://www.mdpi.com/article/10.3390/land10121319/s1, Supplement S1: RC Assessment Raw Data for five representative cities.

Author Contributions: Conceptualization, X.L. and C.W.; methodology, C.W.; software, X.L. and S.L.; validation, C.W.; formal analysis, X.L., S.L. and C.W.; investigation, S.L., X.L. and C.W.; resources, X.L., S.L. and C.W.; data curation, C.W., X.L. and S.L.; writing—original draft preparation, C.W., X.L. and S.L.; writing—review and editing, C.W., X.L. and S.L.; visualization, C.W. and X.L.; supervision,

C.W.; project administration, C.W.; funding acquisition, C.W. All authors have read and agreed to the published version of the manuscript.

Funding: This research was funded by the Jiangsu Innovation and Entrepreneurship Talent Program, the Future Exploration Program of the School of Architecture, Southeast University and the Student Research Training Program, Southeast University.

Conflicts of Interest: The authors declare no conflict of interest. The funders had no role in the study design; in the collection, analyses, or interpretation of data; in the writing of the manuscript; or in the decision to publish the results.

Appendix A

Table A1. Comparison and Summary of Representative RC Assessment Frameworks.

Resource	Urban Hazard Mitigation: Creating Resilient Cities [3]	From Fail-Safe to Safe-to-Fail: Sustainability and Resilience in the New Urban World [61]	Planning the Resilient City: Concepts and Strategies for Coping with Climate Change and Environmental Risk [5]	City Resilience Framework [62]	Defining the Resilient City [63]	Summary
Key Points	Enhancing physical and social elements of cities with seven characters of resilient systems: redundant, diverse, efficient, autonomous, strong, adaptable, collaborative	A non-equilibrium approach; more strategic than normative	An integrative approach with four clusters of concepts: vulnerability analysis matrix, prevention, urban governance, uncertainty-oriented planning	Assessment framework for seven qualities: reflective, flexible, integrated, robust, resourceful, redundant and inclusive	A working definition of resilience for fragile cities and guidance towards specific dimensions and indicators	
Public Participation	Educating developers about mitigation techniques and notifying the public			Education for the public		Educate developers and the public about hazard mitigation
	Community's involvement			Community participation and support; citizen engagement	Strong civil society and community networks	Encourage community involvement and citizen participation
	Facilitate collective learning and self-organization					Learn collectively from past urban hazards
Social Service				Security and laws to reduce crimes		Enforce security and laws to reduce crimes
				Support basic living standards (safer housing, energy, drinking water, sanitation, and food supply)	Ensured human security (incomes, income equality, poverty rates, access to markets and employment, health and nutrition, etc.)	Set ensured human security for providing basic living needs
	Adopt recognized equity standards; assist vulnerable neighborhoods and populations		Increase social equity			Assist vulnerable neighborhoods and populations (increase social equity)
				Robust public health systems and sufficient healthcare and emergency medical care		Enhance robust public health systems and emergency medical care

Table A1. *Cont.*

Resource	Urban Hazard Mitigation: Creating Resilient Cities [3]	From Fail-Safe to Safe-to-Fail: Sustainability and Resilience in the New Urban World [61]	Planning the Resilient City: Concepts and Strategies for Coping with Climate Change and Environmental Risk [5]	City Resilience Framework [62]	Defining the Resilient City [63]	Summary
Robust Economics	Mitigate business interruption impacts	Higher levels of economic diversity		Diverse livelihoods and employment		Diversify livelihoods that can mitigate business interruption impacts
			Boost ecologically friendly economy	Sustainable economics with diversity, integration and competitiveness		Adopt sustainable economic initiatives
			Applying alternative energy Policies and actions to reduce greenhouse gas emission Solar panel buildings			Use/facilitate alternative energy (solar, wind, etc. to reduce greenhouse gas emission)
Hazard Management	Identifying hazards and vulnerability, monitor vulnerability reduction		Vulnerability analysis	Hazard mapping, monitoring, assessment, and warning		Identify, monitor, and assess hazards and vulnerability
				Effective emergency response services	Explicit and cross-sectoral disaster risk reduction initiatives	Build effective emergency response services/initiatives
	Strengthening buildings and public facilities—flood-proofing and wind-proofing existing and new structures through building codes and engineering design			Appropriate building and construction codes, standards and enforcement	Building and/or construction regulations; robust built environments to ensure safety in private domains	Enhance building hazard resistance by appropriate construction regulations
	Avoiding hazard areas—directing new development away from hazardous locations and relocating existing structures and land uses to safer areas		Land use management for natural hazard prone areas	Appropriate land use and zoning with a transparent operation mechanism	Evaluating urban hazards and likely vulnerabilities to inform land use and urban planning	Land use and planning management for natural hazard prone areas

Table A1. *Cont.*

Resource	Urban Hazard Mitigation: Creating Resilient Cities [3]	From Fail-Safe to Safe-to-Fail: Sustainability and Resilience in the New Urban World [61]	Planning the Resilient City: Concepts and Strategies for Coping with Climate Change and Environmental Risk [5]	City Resilience Framework [62]	Defining the Resilient City [63]	Summary
Institutional Optimization	Develop broad hazard mitigation commitment			Wide collaboration with stakeholders and the public	Increase capacity of local government institutions	Increase capacity of governmental institutions by wide collaboration
	Adaptive community practice	Adaptive planning and design model with possible pilot experiments				Adopt a flexible planning system and adaptive design process
			Adaptive governance management for uncertainty			Employ agile city management for uncertainty and challenges
				Effective coordination with other government bodies	Build strong local government linkages	Effective coordination with other government bodies
Physical/Engineering	Using structural approaches such as flood control works, slope stabilization and shoreline hardening			Robust protective infrastructure with regular maintenance	Robust built environments to ensure safety in private and public domains	Set robust protective infrastructure with regular maintenance
	Build distributed hazard mitigation capability	Redundancy by a distributed or decentralized system				Build/optimize distributed or decentralized hazard mitigation system
	Operate networked communications			Reliable communications technology and infrastructure		Enhance reliable ICT infrastructure (communication network)
			Compact urban form with high density			Optimize sustainable urban form
		Good pedestrian transportation	Sustainable transport	Diverse, effective and affordable transport		Provide diverse, effective, affordable, sustainable transport

Table A1. *Cont.*

Resource	Urban Hazard Mitigation: Creating Resilient Cities [3]	From Fail-Safe to Safe-to-Fail: Sustainability and Resilience in the New Urban World [61]	Planning the Resilient City: Concepts and Strategies for Coping with Climate Change and Environmental Risk [5]	City Resilience Framework [62]	Defining the Resilient City [63]	Summary
Natural Enhancement	Conserving natural areas—maintaining and enhancing the functions of wetlands, dunes and forests			Effectively managed and protected ecosystems		Conserve, manage and protect ecosystems
		Provide sustainable ecosystem services within compact cities, blue-green ecological networks	Urban green lands			Optimize urban blue-green ecological networks within compact cities
		Maintain diversity in biological systems				Maintain diversity in biological systems

References

1. Jha, A.K.; Miner, T.W.; Stanton-Geddes, Z. *Building Urban Resilience: Principles, Tools, and Practice*; World Bank Publications: Washington, DC, USA, 2013.
2. McPhearson, T.; Haase, D.; Kabisch, N.; Gren, Å. Advancing understanding of the complex nature of urban systems. *Ecol. Indic.* **2016**, *70*, 566–573. [CrossRef]
3. Godschalk, D.R. Urban hazard mitigation: Creating resilient cities. *Nat. Hazards Rev.* **2003**, *4*, 136–143. [CrossRef]
4. Desouza, K.C.; Flanery, T.H. Designing, planning, and managing resilient cities: A conceptual framework. *Cities* **2013**, *35*, 89–99. [CrossRef]
5. Jabareen, Y. Planning the resilient city: Concepts and strategies for coping with climate change and environmental risk. *Cities* **2013**, *31*, 220–229. [CrossRef]
6. Vale, L.J. *The Resilient City: How Modern Cities Recover from Disaster*; Oxford University Press: Oxford, UK, 2005.
7. Roberts, D. Prioritizing climate change adaptation and local level resilience in Durban, South Africa. *Environ. Urban.* **2010**, *22*, 397–413. [CrossRef]
8. Therrien, M.-C.; Normandin, J.-M.; Paterson, S.; Pelling, M. Mapping and weaving for urban resilience implementation: A tale of two cities. *Cities* **2021**, *108*, 102931. [CrossRef]
9. Lu, P.; Stead, D. Understanding the notion of resilience in spatial planning: A case study of Rotterdam, The Netherlands. *Cities* **2013**, *35*, 200–212. [CrossRef]
10. Martín, C.; McTarnaghan, S. *Institutionalizing Urban Resilience: A Midterm Monitoring and Evaluation Report of 100 Resilient Cities*; Urban Institute and The Rockefeller Foundation: New York, NY, USA, 2018.
11. Galderisi, A.; Limongi, G.; Salata, K.-D. Strengths and weaknesses of the 100 Resilient Cities Initiative in Southern Europe: Rome and Athens' experiences. *City Territ. Archit.* **2020**, *7*, 16. [CrossRef]
12. Spaans, M.; Waterhout, B. Building up resilience in cities worldwide—Rotterdam as participant in the 100 Resilient Cities Programme. *Cities* **2017**, *61*, 109–116. [CrossRef]
13. Roberts, D.; Douwes, J.; Sutherland, C.; Sim, V. Durban's 100 Resilient Cities journey: Governing resilience from within. *Environ. Urban.* **2020**, *32*, 547–568. [CrossRef]
14. Vale, L.J. The politics of resilient cities: Whose resilience and whose city? *Build. Res. Inf.* **2014**, *42*, 191–201. [CrossRef]
15. Kaika, M. 'Don't call me resilient again!': The New Urban Agenda as immunology ... or ... what happens when communities refuse to be vaccinated with 'smart cities' and indicators. *Environ. Urban.* **2017**, *29*, 89–102. [CrossRef]
16. Pizzo, B. Problematizing resilience: Implications for planning theory and practice. *Cities* **2015**, *43*, 133–140. [CrossRef]
17. Meerow, S.; Newell, J.P. Urban resilience for whom, what, when, where, and why? *Urban Geogr.* **2019**, *40*, 309–329. [CrossRef]
18. De Jong, M.; Joss, S.; Schraven, D.; Zhan, C.; Weijnen, M. Sustainable–smart–resilient–low carbon–eco–knowledge cities; making sense of a multitude of concepts promoting sustainable urbanization. *J. Clean. Prod.* **2015**, *109*, 25–38. [CrossRef]
19. Shamsuddin, S. Resilience resistance: The challenges and implications of urban resilience implementation. *Cities* **2020**, *103*, 102763. [CrossRef] [PubMed]
20. Sanchez, A.X.; van der Heijden, J.; Osmond, P. The city politics of an urban age: Urban resilience conceptualisations and policies. *Palgrave Commun.* **2018**, *4*, 25. [CrossRef]
21. Boano, C. Pathways towards the Resilient City: Presupposition of Equality and Active Justice in Bangkok, Thailand. In *Environmental Justice and Urban Resilience in the Global South*; Allen, A., Griffin, L., Johnson, C., Eds.; Palgrave Macmillan US: New York, NY, USA, 2017; pp. 137–152.
22. Huck, A.; Monstadt, J.; Driessen, P.P.J.; Rudolph-Cleff, A. Towards Resilient Rotterdam? Key conditions for a networked approach to managing urban infrastructure risks. *J. Contingencies Crisis Manag.* **2021**, *29*, 12–22. [CrossRef]
23. Szpakowska-Loranc, E.; Matusik, A. Łódź—Towards a resilient city. *Cities* **2020**, *107*, 102936. [CrossRef]
24. Ilgen, S.; Sengers, F.; Wardekker, A. City-To-City Learning for Urban Resilience: The Case of Water Squares in Rotterdam and Mexico City. *Water* **2019**, *11*, 983. [CrossRef]
25. DuPuis, E.M.; Greenberg, M. The right to the resilient city: Progressive politics and the green growth machine in New York City. *J. Environ. Stud. Sci.* **2019**, *9*, 352–363. [CrossRef]
26. Holling, C.S. Resilience and Stability of Ecological Systems. *Annu. Rev. Ecol. Syst.* **1973**, *4*, 1–23. [CrossRef]
27. Holling, C.S.; Gunderson, L.H. *Panarchy: Understanding Transformations in Human and Natural Systems*; Island Press: Washington, DC, USA, 2002.
28. Adger, W.N. Social and ecological resilience: Are they related? *Prog. Hum. Geogr.* **2000**, *24*, 347–364. [CrossRef]
29. Pickett, S.T.; Cadenasso, M.L.; Grove, J.M. Resilient cities: Meaning, models, and metaphor for integrating the ecological, socio-economic, and planning realms. *Landsc. Urban Plan.* **2004**, *69*, 369–384. [CrossRef]
30. Ernstson, H.; van der Leeuw, S.E.; Redman, C.L.; Meffert, D.J.; Davis, G.; Alfsen, C.; Elmqvist, T. Urban Transitions: On Urban Resilience and Human-Dominated Ecosystems. *AMBIO—A J. Hum. Environ.* **2010**, *39*, 531–545. [CrossRef]
31. Ziervogel, G.; Pelling, M.; Cartwright, A.; Chu, E.; Deshpande, T.; Harris, L.; Hyams, K.; Kaunda, J.; Klaus, B.; Michael, K.; et al. Inserting rights and justice into urban resilience: A focus on everyday risk. *Environ. Urban.* **2017**, *29*, 123–138. [CrossRef]
32. Collier, M.J.; Nedović-Budić, Z.; Aerts, J.; Connop, S.; Foley, D.; Foley, K.; Newport, D.; McQuaid, S.; Slaev, A.; Verburg, P.J.C. Transitioning to resilience and sustainability in urban communities. *Cities* **2013**, *32*, S21–S28. [CrossRef]

33. Connolly, J.J.T. From Jacobs to the Just City: A foundation for challenging the green planning orthodoxy. *Cities* **2019**, *91*, 64–70. [CrossRef]
34. Ribeiro, P.J.G.; Goncalves, L. Urban resilience: A conceptual framework. *Sustain. Cities Soc.* **2019**, *50*, 11. [CrossRef]
35. Webber, S.; Leitner, H.; Sheppard, E. Wheeling out Urban Resilience: Philanthrocapitalism, Marketization, and Local Practice. *Ann. Am. Assoc. Geogr.* **2021**, *111*, 343–363. [CrossRef]
36. Leitner, H.; Sheppard, E.; Webber, S.; Colven, E. Globalizing urban resilience. *Urban Geogr.* **2018**, *39*, 1276–1284. [CrossRef]
37. Wang, L.; Xue, X.L.; Zhang, Y.X.; Luo, X.W. Exploring the Emerging Evolution Trends of Urban Resilience Research by Scientometric Analysis. *Int. J. Environ. Res. Public Health* **2018**, *16*, 29. [CrossRef]
38. Fitzgerald, A. Querying the resilient local authority: The question of 'resilience for whom'? *Local Gov. Stud.* **2018**, *44*, 788–806. [CrossRef]
39. Rodrigues, S. Speech to Resilience First. Available online: https://www.dropbox.com/s/ekh3m3lmoae359m/Speech%20to%20Resilience%20First%20by%20Shirley%20Rodrigues%20-%20120121.pdf?dl=0 (accessed on 2 September 2021).
40. Assefa, M. Urban Resilience in Addis Ketama and Lideta Sub Cities of Addis Ababa: The Case Tekilehaimanot Area Addis Ababa, Ethiopia. *Urban Reg. Plan.* **2018**, *3*, 44–54. [CrossRef]
41. Baron, N.; Cherenet, Z. Perceptions and pathways of resilience in Addis Ababa. *Int. J. Build. Pathol. Adapt.* **2018**, *36*, 337–353. [CrossRef]
42. Baron, N.; Donath, D. Bridging the Western Gaze—A Research Framework for Spatial Resilience in Addis Ababa. In Proceedings of the Transformation and Resilience in Urban Development. Presented at the 23rd International Seminar on Urban Form, International Seminar on Urban Form, Nanjing, China, 8–10 July 2016.
43. Birhanu, D.; Kim, H.; Jang, C.; Park, S. Flood Risk and Vulnerability of Addis Ababa City Due to Climate Change and Urbanization. *Procedia Eng.* **2016**, *154*, 696–702. [CrossRef]
44. Hagos, K.G.; Adnan, M.; Yasar, A.-u.-H. Effect of sidewalk vendors on pedestrian movement characteristics: A microscopic simulation study of Addis Ababa, Ethiopia. *Cities* **2020**, *103*, 102769. [CrossRef]
45. Herslund, L.; Backhaus, A.; Fryd, O.; Jørgensen, G.; Jensen, M.B.; Limbumba, T.M.; Liu, L.; Mguni, P.; Mkupasi, M.; Workalemahu, L.; et al. Conditions and opportunities for green infrastructure—Aiming for green, water-resilient cities in Addis Ababa and Dar es Salaam. *Landsc. Urban Plan.* **2018**, *180*, 319–327. [CrossRef]
46. Mohamed, A.; Worku, H.; Lika, T. Urban and regional planning approaches for sustainable governance: The case of Addis Ababa and the surrounding area changing landscape. *City Environ. Interact.* **2020**, *8*, 100050. [CrossRef]
47. Alem, G. Urban Plans and Conflicting Interests in Sustainable Cross-Boundary Land Governance, the Case of National Urban and Regional Plans in Ethiopia. *Sustainability* **2021**, *13*, 3081. [CrossRef]
48. RC100. A Milestone in Addis Ababa's Resilience Journey: A Conversation with Addis Ababa's First Chief Resilience Officer. Available online: https://medium.com/@100ResilientCities/a-milestone-in-addis-ababas-resilience-journey-8c70a8e15e83 (accessed on 15 May 2021).
49. Huck, A.; Monstadt, J.; Driessen, P. Mainstreaming resilience in urban policy making? Insights from Christchurch and Rotterdam. *Geoforum* **2020**, *117*, 194–205. [CrossRef]
50. Wakefield, S. Urban resilience as critique: Problematizing infrastructure in post-Sandy New York City. *Polit. Geogr.* **2020**, *79*, 102148. [CrossRef]
51. Mahadevia, D.; Bhatia, N.; Verma, P.; Raj, V.; Patel, S. Affordable Housing as Flood Resilient for Low-Income Households: Case of Ahmedabad and Surat. 2018. Available online: https://cept.ac.in/UserFiles/File/CUE/Working%20Papers/Revised%20New/CUE%20WP40%20-%20Affordable%20Housing%20as%20Flood%20Resilient%20for%20Low%20Income%20Households%20Case%20of%20Ahmedabad%20and%20Surat%20-%20Sept%202018%20-%20RR.pdf (accessed on 12 September 2021).
52. Mavalankar, D. Lessons from Massive Floods of 2006 in Surat City: A Framework for Application of MS/OR Techniques to Improve Dam Management to Prevent Flood. Available online: https://web.iima.ac.in/assets/snippets/workingpaperpdf/2008-07-06Mavalankar.pdf (accessed on 12 September 2021).
53. Raj, V. Affordable Housing as Flood Resilient for Low Income Households: A Case of Surat. Master's Thesis, CEPT University, Ahmedabad, India, 2018.
54. World Resources Institute India. Implementing the Urban Community Resilience Assessment UCRA in Three Communities in Surat City, 2018. Available online: https://projects.citiesalliance.org/projects/documents/Implementing_the_Urban_Community_Resilience_Assessment__UCRA__in_Three_Communities_in_Surat_City.pdf (accessed on 17 October 2021).
55. United Nations. 11. Sustainable Cities and Communities. Available online: https://www.globalgoals.org/11-sustainable-cities-and-communities (accessed on 20 November 2021).
56. United Nations Office for Disaster Risk Reduction (UNDRR). Disaster Resilience Scorecard for Cities. 2017. Available online: https://mcr2030.undrr.org/disaster-resilience-scorecard-cities (accessed on 12 November 2021).
57. Cutter, S.L.; Barnes, L.; Berry, M.; Burton, C.; Evans, E.; Tate, E.; Webb, J. A place-based model for understanding community resilience to natural disasters. *Glob. Environ. Chang.* **2008**, *18*, 598–606. [CrossRef]
58. Bryant, A.; Charmaz, K. *The Sage Handbook of Grounded Theory*; Sage: London, UK, 2007.
59. Allen, N.; Davey, M. The Value of Constructivist Grounded Theory for Built Environment Researchers. *J. Plan. Educ. Res.* **2018**, *38*, 222–232. [CrossRef]

60. Ostadtaghizadeh, A.; Ardalan, A.; Paton, D.; Jabbari, H.; Khankeh, H.R. Community disaster resilience: A systematic review on assessment models and tools. *PLoS Curr.* **2015**, *7*. [CrossRef] [PubMed]
61. Ahern, J. From fail-safe to safe-to-fail: Sustainability and resilience in the new urban world. *Landsc. Urban Plan.* **2011**, *100*, 341–343. [CrossRef]
62. Arup. *City Resilience Framework*; Arup and Rockefeller Foundation: London, UK, 2014. Available online: https://www.rockefellerfoundation.org/wp-content/uploads/City-Resilience-Framework-2015.pdf (accessed on 8 August 2021).
63. Patel, R.; Nosal, L. *Defining the Resilient City*; United Nations University Centre for Policy Research: New York, NY, USA, 2016.
64. Dey, I. Grounding Categories. In *The Sage Handbook of Grounded Theory*; Bryant, A., Charmaz, K., Eds.; Sage: London, UK, 2007; pp. 167–190.
65. Oke, C.; Bekessy, S.A.; Frantzeskaki, N.; Bush, J.; Fitzsimons, J.A.; Garrard, G.E.; Grenfell, M.; Harrison, L.; Hartigan, M.; Callow, D.; et al. Cities should respond to the biodiversity extinction crisis. *NPJ Urban Sustain.* **2021**, *1*, 11. [CrossRef]
66. Greener City Fund. Available online: https://www.london.gov.uk/what-we-do/environment/parks-green-spaces-and-biodiversity/greener-city-fund (accessed on 9 September 2021).
67. Davenport, S.; Whicheloe, R.; Frith, M.; Massini, P. Urban Greening for Biodiversity Net Gain: A Design Guide. 2021. Available online: https://www.london.gov.uk/sites/default/files/urban_greening_and_bng_design_guide_march_2021.pdf (accessed on 12 October 2021).
68. Koert, M.v.; Brandt, T. *Public Transporter RET: Taking a New and Sustainable Route in Rotterdam*; RSM Case Development Centre, Erasmus University: Rotterdam, The Netherlands, 2020.
69. World Bank Group; GFDRR. *Addis Ababa—Enhancing Urban Resilience: City Strength Resilient Cities Program (English)*; The World Bank Group: Washington, DC, USA, 2015.
70. RC100. Addis Ababa Resilience Strategy. Available online: https://resilientcitiesnetwork.org/networks/addis-ababa/ (accessed on 15 March 2021).
71. Khader, M. Rotterdam Resilience Strategy, Rotterdam. In *Urban Planning for Transitions*; Douay, N., Minja, M., Eds.; ISTE: London, UK, 2021; pp. 1–18.
72. London Wins Membership of Prestigious Network of Resilient Cities. Available online: https://www.london.gov.uk/press-releases/mayoral/london-wins-prestigious-membership (accessed on 8 August 2021).
73. Naef, P. Resilience as a City Brand: The Cases of the Comuna 13 and Moravia in Medellin, Colombia. *Sustainability* **2020**, *12*, 8469. [CrossRef]
74. Campelo, A. Global city branding. In *A Research Agenda for Place Branding*; Medway, D., Warnaby, G., Byrom, J., Eds.; Edward Elgar Publishing: Cheltenham, UK, 2021; pp. 101–116.
75. Lavers, T. Responding to land-based conflict in Ethiopia: The land rights of ethnic minorities under federalism. *Afr. Aff.* **2018**, *117*, 462–484. [CrossRef]
76. Wang, C. Do planning concepts matter? A Lacanian interpretation of the urban village in a British context. *Plan. Theory* **2021**. [CrossRef]
77. Rosenzweig, C.; Solecki, W. Hurricane Sandy and adaptation pathways in New York: Lessons from a first-responder city. *Glob. Environ. Chang.* **2014**, *28*, 395–408. [CrossRef]
78. Grove, K. *Resilience*; Routledge: London, UK, 2018.
79. Brown, T. Sustainability as empty signifier: Its rise, fall, and radical potential. *Antipode* **2016**, *48*, 115–133. [CrossRef]
80. Khan, S.; Zaman, A.U. Future cities: Conceptualizing the future based on a critical examination of existing notions of cities. *Cities* **2018**, *72*, 217–225. [CrossRef]
81. Davidson, M. Sustainability as ideological praxis: The acting out of planning's master-signifier. *City* **2010**, *14*, 390–405. [CrossRef]

land

Article

Ecological Embeddedness in the Maya Built Environment: Inspiration for Contemporary Cities

Naji Akbar [1], Ismaila Rimi Abubakar [2,*], Ayesha Agha Shah [3] and Wafa Al-Madani [3]

1 Department of Interior Architecture, College of Architecture and Planning, Imam Abdulrahman Bin Faisal University, Dammam 31451, Saudi Arabia; nakbar@iau.edu.sa
2 College of Architecture and Planning, Imam Abdulrahman Bin Faisal University, Dammam 31451, Saudi Arabia
3 Department of Architecture and Interior Design, College of Engineering, University of Bahrain, P.O. Box 32038, Zallaq, Bahrain; aashah@uob.edu.bh (A.A.S.); walmadani@uob.edu.bh (W.A.-M.)
* Correspondence: irabubakar@iau.edu.sa

Abstract: Cities nowadays are the most significant source of environmental degradation threatening local and global ecosystems. Interestingly, many ancient settlements present excellent lessons and inspiration for addressing our current urban predicaments, given their environmental stewardship. This research explores how the ecologically embedded settlement patterns, building configurations, urban agriculture and home gardening, and water conservation of the Maya-built environment can offer insights about mitigating contemporary urban sustainability challenges. Mayans' respect for nature not only guaranteed sustainable habitats but also engendered one of the most remarkable civilizations in a region that did not offer generous support for human accommodation. The Mayan world view promoted the idea of one spirit dwelling in all humans and other-than-human entities in an environment, making everything sacred and kin to everything else. The regional climate was kept under control by protecting the vegetation that also provided other ecological benefits. Land use was mixed, and residences were constructed with native and recyclable materials utilizing natural light and ventilation. The Mayan civilization inspires us to manage and protect plants, not cut them down; conserve water, not waste it; listen to the environmental feedback, not reject it; and, most importantly, it begs us to embrace nature as our own mother, not disown it as something dispensable. These principles have significant implications on urban land-use planning and policies today.

Keywords: ecological embeddedness; environmental sustainability; cities; home gardens; Maya; settlement pattern; urban agriculture; water conservation

Citation: Akbar, N.; Abubakar, I.R.; Shah, A.A.; Al-Madani, W. Ecological Embeddedness in the Maya Built Environment: Inspiration for Contemporary Cities. *Land* **2021**, *10*, 1360. https://doi.org/10.3390/land10121360

Academic Editors: Elizelle Juanee Cilliers and Sarel Cilliers

Received: 18 November 2021
Accepted: 7 December 2021
Published: 9 December 2021

1. Introduction

The literature strongly suggests that cities today are the major cause of adverse environmental impacts globally, which negatively affect them and the wider landscapes [1–4]. Cities are expected to account for 68% of the global population by 2050, with land consumption outpacing population growth by as much as 50% [5]. This sprawling development pattern has deleterious impacts on the ecosystem, especially biodiversity. The ecological dis-embeddedness of cities, which has divorced people and their activities from the local and regional environments, is among the causes of environmental degradation [6–10]. This has left many societies without concrete guidance that could be drawn from the context of which they are a part. Many scholars have studied this grave situation regarding modern cities and proposed the idea of a different and a better kind of habitat [11–15]. It is now becoming evident that healthy habitats can only be reformulated through developing respect for nature, which can be accomplished by learning lessons from the ecological embeddedness of ancient settlements [16–18].

To explore the past with an intention to seek inspiration for current and future cities, this paper surveys the Mayan built environment, investigating how the settlement pattern, building configurations, urban agriculture and home gardening, and water conservation were ecologically embedded. The paper then discusses some important lessons for contemporary-built environments to help mitigate their environmental sustainability challenges. Can we look at the past to reconceive our present and future cities? While some scholars, such as Campbell [13], argue that it would be less relevant, several others opine that exemplary instances of ecological embeddedness and sustainable practices regarding the built environment can be found in past civilizations for us to think about and possibly adapt for the benefit of the present and future generations [6,7,19,20].

Ecological embeddedness is the concept of identifying with the ecosystem of any given region and adhering to the principles of conservation, respect, and reciprocity implemented for the betterment of all stakeholders [21]. The concept consists of an evolving knowledge that is "transgenerational, transcultural, trans-philosophical, and transdisciplinary", gathered through experimentation with and monitoring of ecosystems [22]; it is much more than just a 'know-how' [17]. This understanding of the intricate relationships between the different components of an environment and their effects over longer periods is vital for taking effective sustainability decisions [23]. Turner et al. [24] enumerate six principles of ecological embeddedness: (a) interconnecting and rotating all entities in an environment; (b) environmental feedback and sustainable resource harvesting and reuse; (c) knowledge accumulation and transfer; (d) respectful and collaborative approaches to human activities; (e) associating oneself as a part of one's ancestral land; and (f) acknowledging spirituality in nature.

Dis-embeddedness, on the other hand, is a short-term gains (mainly economic) over long-term benefits (sustainability) approach, discounting the multiplicity of perspectives, looking at humans and nature as a binary, and ignoring spirituality and the call for meaning behind life [10]. The drive towards limitless material accumulation and the disregard for the environment is informed by the idea that humans and other-than-human stakeholders do not need to co-evolve. This fact is exemplified by the current situation in which cities, which cover less than 2% of the earth's surface, consume 78% of the energy generated and produce 60% of greenhouse gas (GHG) emissions [5]. Buildings alone consume over 40% of generated energy and contribute nearly one-third of GHG emissions in most countries [3]. Consequently, this situation has led not only to environmental degradation, but also to decontextualization and ecological dis-embeddedness of contemporary cities. The following are among the major environmental sustainability challenges faced by contemporary cities.

Air pollution and GHG emissions from buildings, the transportation sector, and industries are important factors undermining urban sustainability. The release of GHG emissions links directly to poor air quality with attendant health risks [3,5]. Likewise, over-dependence on unsustainable fossil-based energy and technologies contributes to environmental pollution. Burning fossil fuels releases carbon dioxide and other greenhouse gases, contributing to global warming and climate change [2,14]. Urban heat islands, formed due to increased energy consumption from buildings, hard structures, and concentrated pavements, threaten the sustainable development of urban communities. Moreover, the lack of blue-green infrastructure leads to higher temperatures and undesired climate change [25,26].

Similarly, sprawling development away from the city center leads to a serious challenge of increased mobility, congestion, environmental pollution, and accidents. In addition, the increasing scale of contemporary cities changes the dynamic of the carbon cycle and earth radiation and leads to high infrastructure demands and capital costs. Habitat fragmentation is another prominent issue of extensive urban development that weakens the ecosystem [4,15,25]. Furthermore, unsustainable resource consumption and lifestyles are increasingly exerting pressure on the environment, leading to the loss of prime agricultural land, biodiversity, and water resources [1,9,27,28], and unsustainable solid and liquid waste management practices and rising urbanization levels lead to water and land contamination [29,30]. Moreover, urban floods are triggered by ineffective land-use patterns and rainfall management [31,32].

Additionally, segregated and mono-functional land uses are sustainability challenges that encourage mobility to other zones for urban activities and discourage street-level socioeconomic interactions due to a lack of diversity in land-use planning and overreliance on cars for mobility [11]. Moreover, inequality and poverty are widespread in contemporary cities. These socioeconomic challenges have arisen due to unemployment and spatial mismatch, because of the detachment of homes from workplaces [13,16]. Lastly, urban development in most cities is devoid of public participation, and cities lack public realm and a sense of community, which results in feelings of alienation from the land and community [31].

A review of similar literature indicates that there is dearth of studies on lessons Mayan settlements can offer to help with addressing these urban sustainability challenges. For example, Akbar et al. [16] investigated how the ecological wisdom of Asian traditional settlements can help foster environmental sustainability by addressing contemporary urban challenges, such as disassociation with the land, overcrowding, over dependence on fossil-based energy and technology, GHG emissions, costly and resources-intensive projects, import of foreign ideas and materials, mono-functional urbanism, one-size-fits-all designs, flooding, and the urban heat island (UHI) effect. Possible solutions to these problems can be found in the ecological embeddedness of the Mayan-built environment. For instance, Ashmore [33] studied physically and spiritually meaningful site-planning concepts and principles among the ancient Maya. Munson and Pinzón [34] researched the way early Mayans constructed environmentally sustainable buildings. Murtha [35] demonstrated how low-urban density fostered sustainability in the Mayan-built environment. While Davis-Salazar [36] studied the Mayan-built environment's drainage and flood control, Dine et al. [37] explored the modern lessons of Mayan agriculture towards food security. So far, the present study is the first to extract the ecological embeddedness of Mayans and their settlements and apply their principles toward fostering environmental sustainability in contemporary cities.

This study can help cities mainstream ecological considerations into land-use planning and policymaking. It can also improve our knowledge of the evolving body of literature about the role of and lessons to be learned from the ecological wisdom of ancient-built environment in addressing urban sustainability challenges facing cities today.

2. Materials and Methods

2.1. Maya Background

The Maya occupied the region which is currently El Salvador, Belize, Guatemala, Honduras, and the southern part of Mexico [38]. This region lies in the tropics, 23.5 degrees north of the equator within the Meso-American region (Figure 1). The ecological understanding of the Maya is well documented. Their civilization faced many challenges, such as unpredictable drought, excessive rainfall, high humidity, a profusion of pests, a shortage of potable water, the inadequacy of irrigation systems, unfavorable karst land [39–41], and a lack of metals during the Classic era [42]. Despite these daunting challenges, the Maya still effectively utilized opportunities present in their land through close observation of and association with nature [43,44].

Large expanses of land occupied by the Maya comprised southern highlands, central tropical lowlands, and less-vegetated northern areas (Figure 1). These are noted by scholars as centers of influence during the great Mayan eras. The highlands, lowlands, and northern areas correspond to the Pre-Classic (1200 BC–250 AD), Classic (250–950 AD), and Post-Classic (950–1500 AD) ages, respectively [45]. A considerable population settled in the Maya region during the Late Pre-Classic Period (400 BC–250 AD), requiring them to adapt to the topographic variation as well as the microclimate. Most of the populations were employed in agriculture. By applying a trial-and-error method over years, people were able to modify the landscape to achieve exceptional agricultural production while enhancing the natural carrying capacity of their land at the same time.

Figure 1. Location of the land occupied by the Maya (inside green boundary) within the Meso-American region [21] [Akbar, 2021].

Time and its passage were significant for the Maya. The movement of heavenly bodies over time indicated when farming and rituals should take place [46]. For this, city elevations and vertical markers in the built environment played a key role. These aligned with certain movements and locations of various stars, which signified important turns in the temporal dimension. The observation of the heavens also designated proper time for crop plantation [47], failure to do so would affect agricultural yield.

The Maya were a society with complex urban systems and activities, as signified by their house mounds, agricultural terraces, raised fields, home gardens, and water reservoirs. Their cities were also places for public gatherings, specialized labor, recreation, and water conservation (especially the lowland locations) for the benefit of urban and rural populations [48]. The kings controlled the water that was available in limited amounts [49]. The urbanites were supported by their rural neighbors, who supplied them with food and various natural resources. Due to this reciprocation, a top-down and bottom-up balance was achieved in decision making regarding resource management and environmental conservation [50,51].

Urbanism has not been thought of as a feature of the Pre-Columbian New World, which is a trend in need of reconsideration. This is attributed to the focus of research on religion, leaving out other factors such as economy, politics, environment, and settlement patterns. While Aztec settlements such as Teotihuacan and Tenochtitlan have been termed urban centers, it is surprising to see that smaller and dispersed Mayan settlements, such as Tikal, Chan Chan, Calakmul, Cahokia, Tiwanaku, and Paquime, are left out as not satisfying the general definition of 'urbanism' according to some scholars. The Mayan-built environment can be considered as urban based on the social and spatial heterogeneity of its settlements, as well as its economic activities, not to mention its complex water conservation systems [52,53]. Furthermore, a multidisciplinary, multivariable, comparative, and cross-cultural study may bring out the characteristics of urbanism in Mayan environments and society further [54].

The functionality of the Mayan-built environment conforms to the people's religious thoughts. Even heavenly bodies and their spiritual and physical impacts on earth are part of this conformity. Mayans regarded the earth as a god where plants, animals, land, caves, and mountains were considered sacred and invaluable [55] That is why many Maya settlements were placed around *Cenotes* (underground water bodies) and caves, providing a meaningful link between people, land, and water. This level of connectedness can provide rich lessons for contemporary settlements threatened by a lack of spirituality and varied sustainability challenges. The following are the main reasons for choosing to study Maya culture: first, Mayan-occupied land appears to be one of the most intriguing historical regions, with ecological wisdom that survives to the present; second, the Mayan-built environment has often been likened to the 'Garden Cities', which remain famous amongst the public, scholars, and planners today; and lastly, significant literature, published over several decades of research and archaeological works, is available to carry out this research.

2.2. Data Collection and Analysis

This study employs the desktop research method of reviewing relevant literature to achieve its aim. The method consists of three iterative phases: the problem definition, literature survey, and data analysis [16]. The first phase consisted of identifying and defining the research problem to be investigated. In this study, the focus is on how lessons from the ecological embeddedness of the Mayan-built environment can help to mitigate sustainability challenges facing contemporary cities. The second phase consisted of surveying and gathering relevant literature and secondary data. Using the Web of Science, Scopus, and Google Scholar search engines (globally recognized comprehensive databases), literature related to the study domain were manually selected and downloaded. The identified relevant literature included published articles and books (academic literature) as well as official reports, statistics, urban plans, and maps (grey literature).

Table 1 shows an example of search terms used to identify and download resources using Google Scholar on 10 November 2021. Documents that satisfied the following four inclusion criteria were used. It should be (a) written in the English language; (b) full text; (c) related to the objective of the study; and (d) published after 1990, which was the time when interest in the legacy of Mayan settlements became heightened. After removing duplicates, 86 documents met the criteria and were included in the study.

Table 1. Result of manual search for relevant literature.

Search Terms	No. of Papers Identified	No. of Papers Downloaded
"Maya built environment"	170	32
"Sustainable architecture in Maya"	7	5
"Maya water management"	324	27
"Maya ecology"	66	24
"Maya home gardens"	66	13
"Maya + forest"	73	8
"ancient Maya settlement patterns"	245	52
Maya urban form	7	2
"Agriculture + Maya"	89	17
"Maya home gardens"	11	6
"Ancient Maya roads"	15	7

The last phase of the analysis involved organizing the gathered documents based on similarity to the topics set out in Table 1, albeit with some document belonging to more than one group. After this, each document was thoroughly analyzed to extract themes related to the Mayan-built environment that satisfied at least one of the six principles of ecological embeddedness identified earlier [24] or uncovered new ones. The extracted themes were finally collated, synthesized, and harmonized, which resulted in four categories: settlement pattern, building configuration, urban agriculture and home gardening, and water conservation. The analysis also involved extracting some quantitative data, such as building floor area and spacing, road width, and the ratio of the built-up areas to green areas from city maps. The quantitative data can help determine the environmental sustainability of settlement patterns. Literature review research has been employed in several studies due to the availability of secondary data online, its low costs, and suitability in exploratory research [16,29].

3. Results and Discussion

This section presents and discusses the study findings, divided along the ecological embeddedness of the Mayan-built environment according to the four themes that emerged from data analyses: (1) settlement pattern, (2) building configuration and construction, (3) urban agriculture and home gardening, and (4) water conservation. The paper then discusses some important lessons for contemporary-built environments to help alleviate their environmental sustainability challenges.

3.1. Settlement Patterns in Maya

Mayan settlements are famous for their residential clusters, home gardens, ritual and regal centers, and socioeconomic and political activities. Based on this, a city acquired an identity according to its dominant features [56]. It took Maya continuous and intensive procedures of agronomy and landscape treatment to adapt various locations for habitation [44]. As a result, adequate urban characteristics seem to emerge from these habitats such as refined construction, residential clusters [57], home gardens [38,58,59], land-use variation [60], neighborhoods, districts, complex institutions [15,61], market exchanges [62], and developed land and water resource management systems [63]. Often, even small settlements such as Joya de Ceren exhibited the civic and social order found in large settlements [64].

The Mayan world view was manifest in the design and arrangement of their habitat [65]. Their built environment was a patchwork of different and scattered agro-systems. Their land use included terraces, raised fields, kitchen gardens, orchids, slash-and-burn agriculture areas, and natural and created forests. Although land was cleared for urban developments, complete deforestation is not evidenced [43]. As shown in Table 2, only approximately 32% of central Tikkal was covered by buildings while the remaining 68% was open spaces and green areas. In contrast, central Chunchucmil is denser with 42% of the area occupied by buildings. In both settlements, building-floor areas ranged from 300–360 m^2 with 0.27–1.38 dwelling units per hectare, while major roads were 60–70 m wide and minor roads measured around 20–40 m wide, which compares well with many contemporary cities.

Table 2. Average road width, development density, and building floor area and spacing in central Tikal and Chunchucmil.

Settlement	Size (Hectare)			Buildings				Road Width (m)	
	Developed	Green and Open Areas	Total	Floor Area	Spacing	Total Units	Density (Units/Total Area)	Major Road	Minor Road
Tikal city center (Figure 2b)	240 ha (32%)	510 ha (68%)	750 ha	360 m^2	60–100 m	200	0.27 units/ha	60 m	20–30 m
Chunchucmil center (Figure 3a)	62 ha (42%)	83 ha (58%)	145 ha	300 m^2	40–60 m	187	1.38 units/ha	60–70 m	30–40 m

Mayan settlements did not display use of a grid, but this does not mean the dispersed layout was unplanned. Buildings were arranged to align with natural topography and contours, which enhanced the visual aesthetics of the landscape, increased access to sunlight and ventilation, minimized the need for excavation and hence was cost saving, reduced the risk of erosion, and supported protection of biodiversity. Runoff was dammed by causeways and canals and diverted towards low-lying areas for irrigation purposes where the landscape was marked with all kinds of natural and human-made reservoirs [40].

(a) Spatial Logic of Mayan Settlements

Urban form refers to the organization of all functional features of an urban area, including streets, buildings, open spaces, and natural areas. The arrangement of these features in Mayan-built areas was shaped by Mayan belief systems and daily activities that in turn transformed and influenced settlement structures [16]. According to Mayan cosmology, the earth is surrounded by supernatural worlds, which can be accessed through the mountains above (arboreal ecosystem), water bodies below the earth's surface (aquatic ecosystem), and the land in between (terrestrial ecosystem). In this respect, the Maya specially regarded the *Ceiba* tree as symbolizing this 3-part division of the cosmos, representing the underworld, earth, and heavens through its deep roots, high trunk, and spread-out branches, respectively, bringing these worlds together [46]. In other words, the *Ceiba* tree also represents the unity and interdependence of different ecological systems that make life possible.

Given the importance of cardinal directions, buildings were oriented to maximize the use of sunlight and natural ventilation and sacred buildings were located in such a way as to be visible to the general public [66]. Since north is believed to be where the ancestors and their wisdom reside, important structures are found in the northern zone that was not only suitable for sustainable habitation but also most visible to the surrounding commoner residences [67]. A good example of this orientation is the Twin Pyramid Complex in Tikal (Figure 2a) where the structures are north-south oriented. Mayan settlements displayed the above-mentioned characteristics to create a built environment with the highest level of harmonized living, interconnectivity of all entities with the environment, and acknowledgment of spirituality in nature.

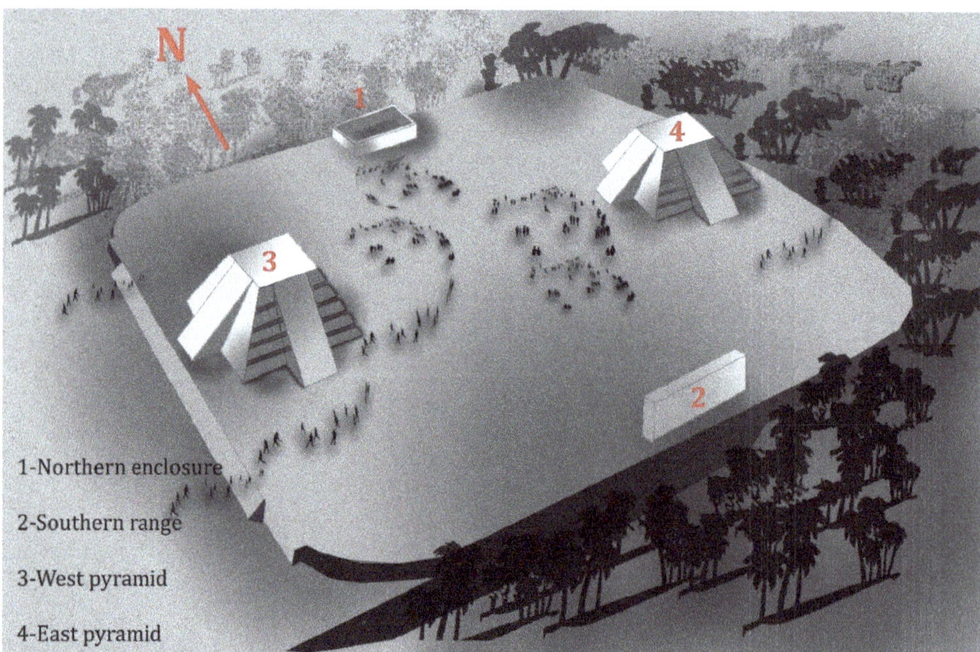

(a)

Figure 2. *Cont.*

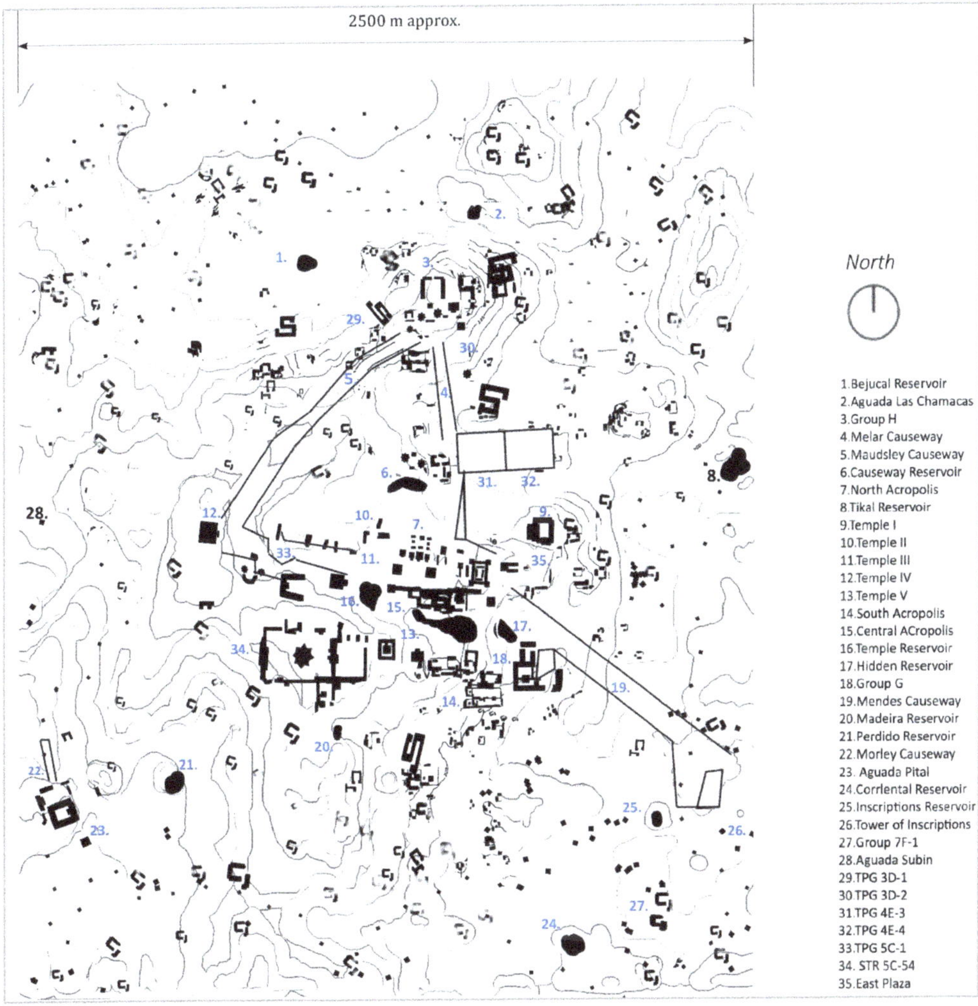

(b)

Figure 2. (**a**) Twin Pyramid Complex (group 4E-4) at Tikal (Guatemala) having cardinal placement (adapted from Ashmore [33] (p. 201)); (**b**) Cardinal alignment of the city center acting as an axis mundi connecting spread-out residential units with the underworld as well as the heavens at Tikal, Guatemala (adapted from Ashmore [33] (p. 204)).

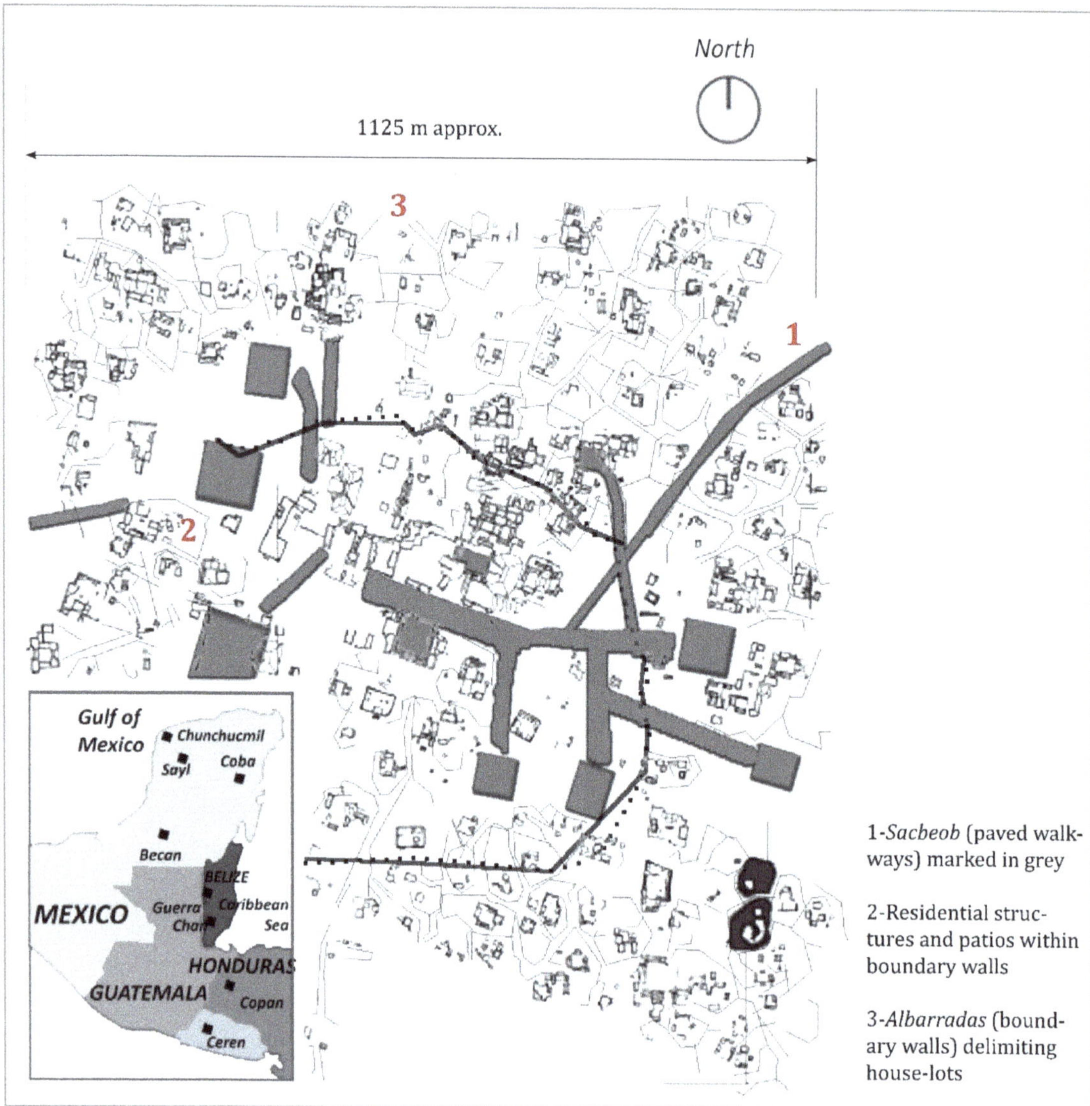

(a)

Figure 3. *Cont.*

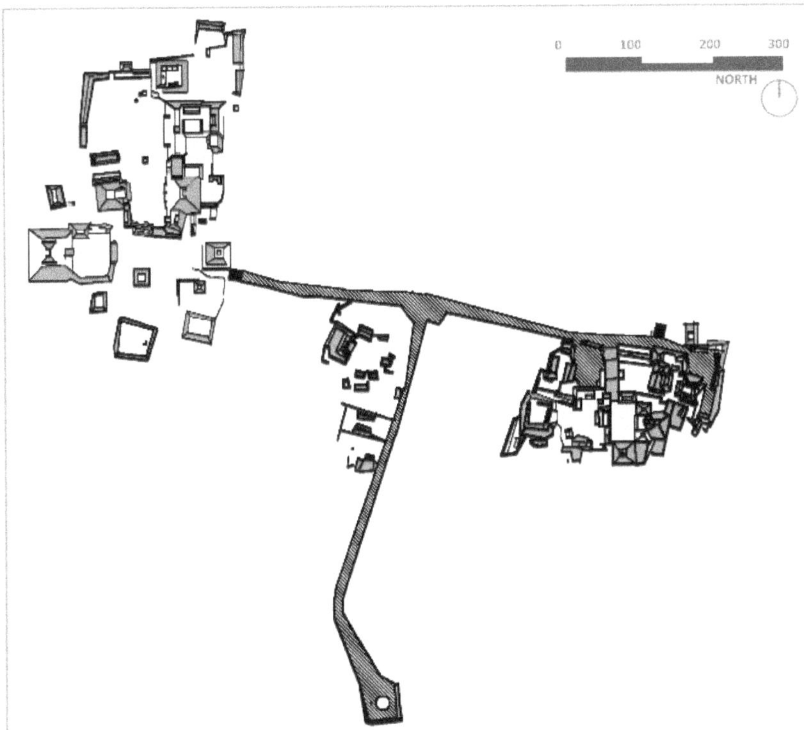

(**b**)

Figure 3. (**a**) *Sacbeob* (marked as a colored grey area) in the center of Chunchucmil connecting house-lots (adapted from Hutson et al., 2007 [68] (p. 466)). (**b**) *Sacbeob* in Seibal (adapted from Ashmore & Sabloff [69] (p. 209)).

The forms of Maya settlements could be informal, but they largely display sensitive harmonization of patio groups (residential and home garden areas), agricultural terraces, raised fields, streets, open spaces, standardization of features according to the locale, mixed and diverse functions, shared interests, and communal uses of facilities which promoted substantial face-to-face interactions [64]. The functional features of a settlement were incorporated into the public realm, including arenas, courts, and accessible markets, thereby fostering inclusion, and increasing social cohesion [61]. In Sayil, for instance, the settlement pattern is more nucleated than some other Maya settlements. Agricultural terraces follow the existing topography while raised fields are an appropriate adaptation of the human habitat under the climatic and geological conditions. The neighborhood open spaces and streets nurture a great sense of community. Since most production, consumption, and recycling took place within or close to the home, walkability is pronounced in Mayan-built environments.

The Maya practiced living, working, trading, and gardening within the same locale, giving rise to a better empathy with the environment. Their work and social lives were intertwined with the surrounding nature. Based on the logic discussed above, settlements were modelled as multi-layered (the sky, the earth, and the underworld), connected through time (as represented by movement cycle of Venus, the moon, and the sun), supported by a vertical axis (axis mundi represented by the *Ceiba* tree-trunk), and divided into four cardinal parts (with an emphasis on a north-south orientation (Figure 2b).

(b) Street Networks

Various buildings in a Mayan city were connected through *Sacbeob* which were elevated streets constructed for many transportation-related and socio-economic purposes

(Figure 4a,b). Three types of *Sacbeob* are commonly found: local, intra-site, and inter-site. Most *Sacbeob* were roughly between 500 m and 20 km long, indicating human-scale and enhanced walkability of a city [70]. Earlier designs did not include parapets, but they were later adopted in various *Sacbeob*. Their structure was similar to the foundation of any Mayan building, comprising large stones on the edges and grading to smaller gravel towards the upper surface which was finished with powdered limestone. Depending on the location and availability, the backfill for *Sacbeob* contained organic materials such as snail shells, which highlights waste recycling in Mayan societies.

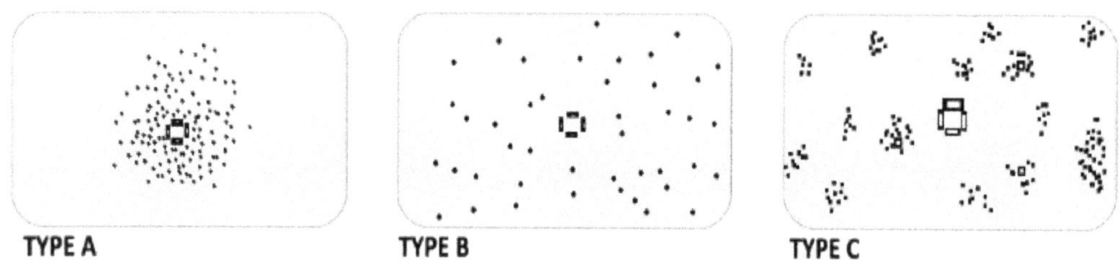

(**a**)

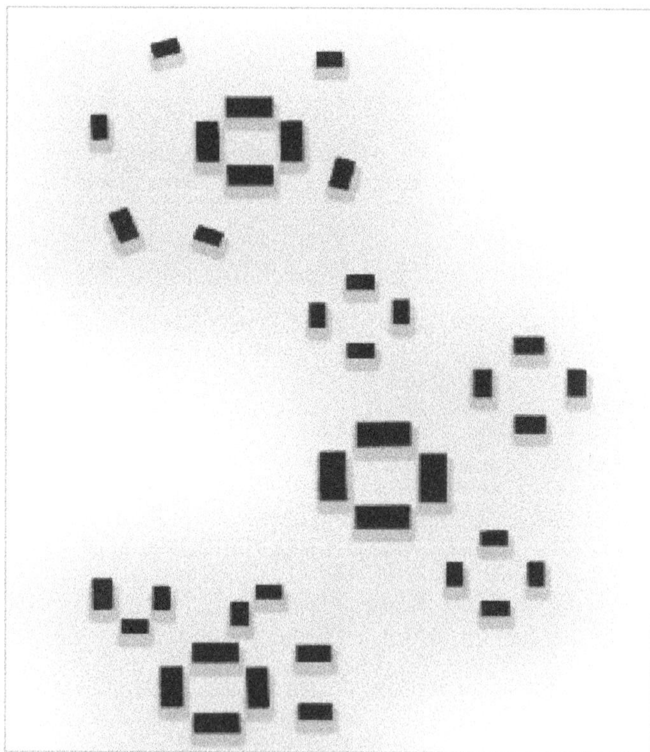

(**b**)

Figure 4. (**a**) Schematic drawings for three types of structure clusters in Mayan settlements; dots are house-lots surrounding the religious settlement center (axis mundi) (adapted from Smith [61] (p. 56)). (**b**) Mayan patio groups showing open patios within built up residences (adapted from Smith [61] (p. 56)).

Sacbeob served many functions, such as transporting people, animals, and goods. They promoted commerce, multi-level interaction, and processions representing political, socio-economic, and religious significance [71] through the architectural spaces. *Sacbeob* also formed the water-retaining edges of reservoirs, directing runoff away from inhabited areas towards natural water collection ponds [70]. By virtue of these *Sacbeob* and their orientation, a city center acquired an axial position (axis mundi) symbolizing a World Tree connecting the underworld, earth, and heavens in seamless continuity. Collaborative maintenance of *Sacbeob* (based on green human labor) brought people together and reminded them of a shared memory, sense of place, and religious, political, and environmental order.

3.2. Building Configuration and Construction

In several present-day cities, buildings have little regard for local climatic conditions, vernacular construction materials, and environmental sustainability [31]. However, among the hallmarks of Maya settlements is the embeddedness of environmental sustainability in architecture and building arrangement through the appreciation of indigenous knowledge, local climates, and the ecosystem. This Mayan trademark was a result of the prolonged and deep observations and experiences of the society [72]. Climatic conditions and cardinal orientations determined building configuration, architectural form, identity, and functional efficacy, while indigenous building materials lowered building costs. Building construction was based on centuries-old, revered, and adopted techniques that were passed from the ancestors to the current and the future generations.

Space available for housing differs in various Mayan settlements depending on land availability, climate, vegetation, topography, soil condition, and socioeconomic structures. Maya were outdoor people due to the nature of their locale. The resultant lifestyle, daily activities, and architecture, especially in the residential areas, are considered in designing the open spaces within a house-lot and the immediate surroundings. These were important factors in building configuration [73]. Given the nature of a house-lot, a household functioned as a small social group that, in turn, was connected and contributed to the larger population it was a part of. In Tikal, for instance, house clusters are closely spaced but there are large amounts of space between clusters. Continuous causeways may not have been needed; instead, a raised pathway in the landscape might have sufficed for movement [35]. On the other hand, the city of Waka had a nucleated pattern of dense residential quarters, unlike other built environments that have surrounding dispersed settlements.

Various types of building clusters are found in the Classic Mayan settlements (Figure 4a). It was initially thought that each one of them was a residence. However, scholars are now convinced that a certain number of structures were part of a house-lot (Figure 4b). For instance, 2280 structures were studied in Tikal that were later believed to be belonging to 691 groups, each group housing extended families comprising around 25 members [74]. Made of perishable materials and following vernacular style, these buildings fulfilled many purposes, such as sleeping, cooking, washing, processing, and storage [75]. One main difference between a Mayan house-lot and a modern residence today is the former had separate buildings, whereas the later holds all activities under one roof.

The simplicity and sustainability of buildings acquired through vernacular design is reported by Abrams [76]. While discussing the city of Copan during the Classic Mayan Period, the author highlights that new structures were mostly built over the foundations of the old ones. Designs were tried and tested to save time and labor efforts. A structure could be constructed in 100 days or less if it were not elaborate, making sure there was no shortage of labor required for other tasks such as agricultural work. Construction was aligned with the climate; given this, the dry season was the time during which most construction took place. Many other scholars found that construction was accretional, using old structures and available materials; water usage was kept to a minimum; basic design requirements, such as solar and wind movement, line of sight, and utility of the enclosure, fulfilled functional needs; construction details were well known, therefore, unskilled labor sufficed for building, reducing time and money [73,76,77]. This resulted in the energy saving required for material procurement.

Religious structures were different in design and construction from the residences of the common people. These were highly symbolic, offering spiritual nourishment to the citizens [78]. As discussed above, the ecological and environmental understanding of the Maya was not limited to the earth only; it included the skies and the heavenly bodies. This is corroborated by growing evidence that the orientation of religious buildings is astronomical. Architectural elements used in certain sacred buildings were aligned with the movement of important stars that predicted initiation of certain rituals [67–79]. Apart from fulfilling symbolic and religious needs, these buildings, by virtue of their orientation, helped in measuring time and predicting weather, which was significant for agricultural purposes.

(c) Examples of Ecological Embeddedness of Mayan Settlements

- **Coba:** Coba was a large urban center during the Classic Mayan era, spread over 30 miles^2 with a population of approximately 50,000 [58]. Green human labor that comprised thinking and sensitive persons was adopted for all tasks that garnered patience and work ethics. Given this, it is believed that the great 120-foot-tall pyramid of Coba was built without using wheels, pulleys, and draft animals. Residential buildings were mainly pole and thatch construction; stone was used with care. Terraces marked the residential landscape as the soil was thin and could not retain moisture and nutrients. Fish farming ponds diversified the food supply [80].
- **Cuello:** Cuello was a Late Pre-Classic settlement famous for its architectural characteristics. For instance, discussing an excavated site in the city, Hammond and Gerhardt [81] found that it consisted of a group of buildings intentionally planned around a patio. Multiple activities are believed to have taken place there, ranging from sleeping, cooking, and storage to burials. This indicates that the site provided self-sufficiency in many respects without having the need to acquire more land to obtain life essentials. The buildings were simple and reparable as local recyclable materials, such as cobble stones, clay, and lime plaster were used for building and repair work. The importance of the patio is signified by the buildings that are oriented towards it, meaningfully marrying the built with the open.
- **Caobal:** Architecture and construction appear to have had a social and political effect on the population of Caobal, such as building a spirit of community and maintaining connections with past generations and their experiences. The amount of domestic refuse found at different sites hints at the collaborative aspect of construction that, among many other factors, encouraged social sustainability. It appears that involvement in the construction process was as important as the output. The built environment also displays ecological knowledge and environmental stewardship. This is evident from practices such as adoption of reusing, recycling, and selecting earthen and local materials and keeping technology simple. Given the above, the built environment in Caobal became a container of social and environmental memory for future generations to follow [34].
- **Chan Noohol:** Located in the upland of the Mopan and Macal Rivers, this settlement comprises seven farmsteads that are 50–70 m apart. Every farmstead has a dedicated area for cultivation created on a non-orthogonal terrace. Both the terrace and the delimiting low-height retaining wall follow the existing topography that points towards site familiarity and the respect that the residents had towards what was already present. Existing bedrock was used as the foundation for structures, reducing foundation work and material required. This also helped to raise the level of the structures naturally. Local materials were used, such as chert cobbles and unshaped stones. New landscape features were created where the previous inhabitants had indicated they ought to be placed. This ensured cultural continuity and adherence to the site's demands [82].
- **Copan:** Abrams [76] shares interesting information about the construction of a religious site and its structures in Copan. The construction work appears to be a single

event requiring less labor and time because it was completed via the communal efforts of both men and women. The capability of the residents made them efficient if not highly skilled laborers. The author estimated that the construction was most probably completed with a special workforce of 40 persons employed on a yearly or half-yearly cycle, and about 370 non-specialized workers for about 60 days.

- **Chunchucmil:** Located in north-western Yucatan, Chunchucmil had a population of about 35,000 spread over 20–25 km^2 during the Classic Period. Dense house-lots delimited by *albarradas* (shared boundary walls) and *chichbes* (winding causeways) were distinguishing features here (Figure 5). The average house-lot size was approximately 4200 m^2 containing open spaces of approximately 3595 m^2. Although the city's location does not support intensive agriculture, the open house-lots still contained areas for climatically and geologically suitable cultivation that needed little tending. The trees in the house-lots imitated regional forests and provided economic benefits. Apart from residences, open spaces within house-lots were also used for other activities, including handicrafts, animal keeping, apiaries, processing, corn storage, rituals, burial, washing, toilets, and dumping organic waste (Figure 6b). The low height of the *albarradas* suggests strong social networks and sustainability. Existing materials and past platforms were used for the foundations, whereas the superstructures were made from perishable materials [68,83].

Figure 5. Schematic drawing showing homes, clear area, intermediate area, and garden area in Chunchucmil (adapted from Hutson et al. [68] (p. 444)).

(a)

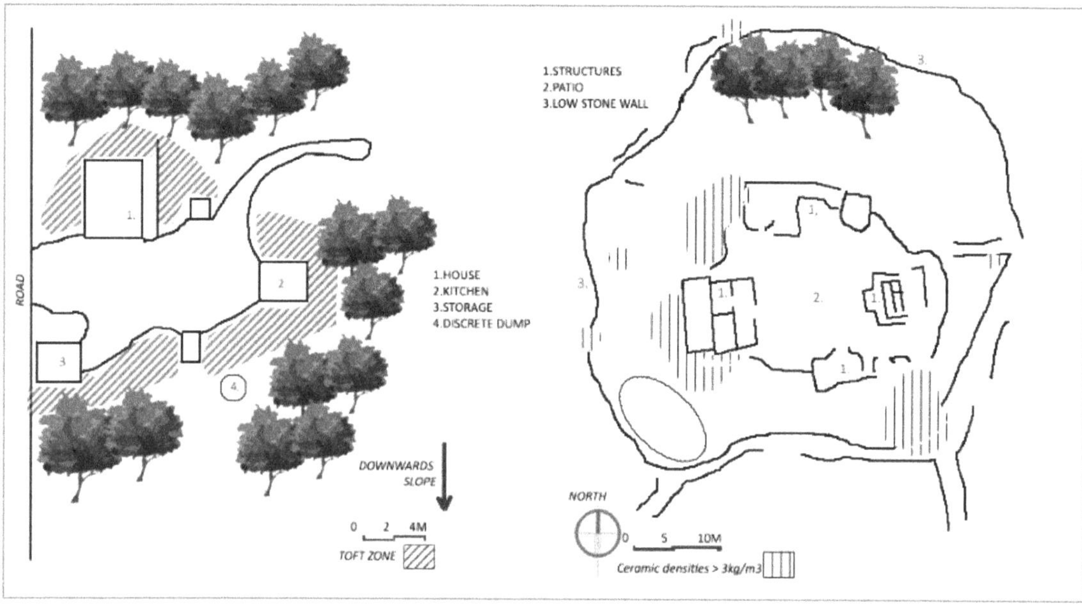

(b)

Figure 6. (**a**) Forest garden and the Maya House: Mesoamerican Research Center today following ancient Maya house garden techniques with a variety of plants and trees offering shade and nutrition (adapted from Ford [84]). (**b**) Map showing residential groups in Chunchucmil with home gardens and other miscellaneous functions (adapted from Halperin and Foias [85] (p. 13).

- **El Palmar:** The abundant presence of the Golden Section in nature and its recognition and application by the Maya points towards their connection with nature and all its constituent ecosystems. One of the examples of this fact is given by Doyle [86] who, after studying selected buildings in El Palmar, makes the case of the use of the Golden Section by the Maya in urban and architectural works. The author contends that, even though historical continuity and cardinal alignments were factors in determining the spread of the group under discussion, the usage of 48.5 m by 78.5 m Golden Section squares is also evident while sizing the building plans. The Golden Section is also applied to work out platforms for other buildings and open spaces; for instance, the size of the Triadic Group in El Palmar is 100 m by 160 m which, again, is a Golden Section square. Interestingly, the Golden Section was also used to determine lot sizes for Milpa farming.

- **Joya de Ceren:** Living in sustainable houses constructed with wattle-daub walls and thatch roof, the residents of this village also exhibited remarkable knowledge of the climate, soil, and plant biology. Intensively cultivated gardens surrounding the houses contained a variety of plants and trees, which provided food as well as shade and demarcation for the house-lots. The agricultural yield from the house gardens as well as the infield maize crop not only ensured self-sufficiency, but also provided enough surpluses to be exchanged for valuable goods [64].

3.3. *Urban Agriculture and Home Gardening*

The basic unit of Mayan settlements comprised not only residential compounds and neighborhoods, but also autonomously managed gardens, along with their surrounding farmlands. The amount of tree cover helped control soil erosion, while urban green areas connected the civic and ceremonial functional zones that linked to the residential areas [87]. Urban agriculture depended on diverse farming practices based on the local ecological, socioeconomic, and political situations. The fact that the agricultural landscape was a part of the Mayan-built environment is archaeologically evidenced by the variations in the identity of settlements and the people that lived in various regions.

For instance, even though infield agriculture is quite a common feature, nuances exit in the way it is made part of the built environment in different communities [88]. This distinction is a result of adaption required for the semitropical climate, varying terrain, and soil conditions [50], and the presence of diverse species [89]. Environmental feedback was instrumental in the decision-making process regarding how to modify the land for agriculture, habitation, and gardening. In this respect, terraces embedded in the landscape were particularly helpful in retaining moisture in the ground as well as preventing erosion of soil that was already thin [51]. For example, in Tikal, agricultural terraces were present in almost every household that had access to two to four acres of land [35].

Apart from water retention, terracing also provided a suitable flat surface for cultivation, nutrient penetration, and residential structures. As evidenced in Tikal, sometimes berms were also constructed to delimit terraces, provide raised surfaces for walking if needed, divert water to low-lying *bajos* (seasonal wetlands), and prevent silting [46]. Since the maintenance of these terraces required collaborative intensive labor, population growth supplied an essential human resource [90]. Thus, settlement patterns integrated with the agricultural land became a strong norm unifying the Maya in their common social and economic undertakings.

The favorable climate encouraged outfield agriculture in settlements such as Coba, therefore, house-lots relied less on home gardens. In comparison, Chunchucmil was dependent on infield agriculture as demonstrated by the presence of larger house-lots that had *rejolladas* and *sascaberas* (laterally dug chambers on the dry side of a sinkhole in the karst ground within the lot boundary) [91]. The combination of *infield* and *outfield* areas is another land-use pattern found in Mayan settlements. This assisted the Maya in dealing with unpredictable weather and crop yield. *Infield* areas were close to home and occupied by households as a primary base. Distant *outfield* areas, on the other hand, complimented the *infield* by becoming secondary cultivation and residential areas in case there was a need, depending upon the weather conditions [44].

Milpa farming is another instance of sustainable agricultural practice where the landscape setting within a built environment was purposefully and periodically changed to ensure sustainable agricultural yield while maintaining forest cover. This is a method where certain crops are grown on a rotating basis through transitional stages, providing control and soil enrichment. The Milpa cycle, shown in Figure 7a has three main phases:

- Looking for forest clearance mainly caused by cyclones and preparing the land for cultivation.
- Introducing crops that are rotational, diverse, and receive enough organic fertilizer as well as soil nutrients. This gives rise to a landscape which is a mix of varied crops, wild plants, and carefully selected trees.
- Leaving the land fallow for several years after a fixed cycle of cultivation to regain its original vegetation and nutrients. This is conducted by replanting seeds from the mature neighboring trees and crops.

Figure 7. *Cont.*

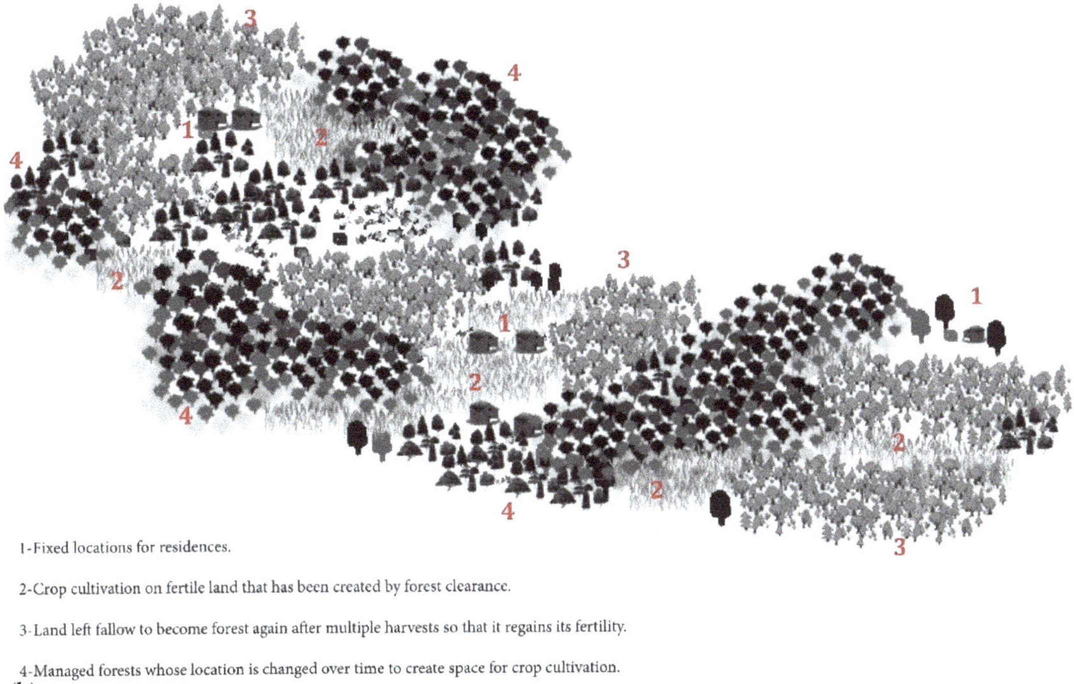

(**b**)

Figure 7. (**a**) Schematic drawing for Milpa cycle showing mature forest to agricultural field (labels 6, 1, 2, and 3 counter-clockwise) to long perennial/ecological succession (labels 3, 4, and 5) and back to forest again (labels 5 and 6) (adapted from Ford & Nigh [63] (p. 185). (**b**) Landscape defined by Milpa cycle (adapted from Bates [92]).

In short, the Milpa cycle is moving from forest to Milpa, followed by Milpa to forest garden, and finally, from forest garden back to the original forest (Figure 7b). Given these ecologically thought-out rotations, the built environment retained what already existed in the form of landscape [90]. Furthermore, diverse interactions with their given locales helped the Maya to spread out risks through house gardens, infields/outfields, Milpa farming, and forest management [90].

Home gardens were also significant green infrastructure that enlivened workplaces and residences. These lent meaning to such spaces and stored memories for future generations to relate and add to. Given the collaborative nature of the work required through human labor, home gardens advanced social sustainability [82]. Apart from being a symbol, home gardens were also utilitarian in nature, providing food and medicinal herbs, functioning as sunshades and windbreaks, and capturing and storing carbon [60–82]. According to the ethnographic research, lot sizes for intensively maintained home gardens were approximately 2.28 ha [50]. This diverse and environmentally responsive land use was complemented by less intensive outfields cultivated by the same family. Sayil, located in the Puuc, region presents a classic example of a garden city, owing to favorable soil conditions [93]. These gardens were dedicated to growing domestic, edible, and cover crops such as palms and cocoa plants requiring frequent and intensive tending [94]. Other plants were nitrogen-fixing and ornamental-shading legumes.

Home gardens are a noticeable feature of the Mayan settlements that exist even today [37,88,95–99]. In this respect, archaeological site 272–025 (also known as Tzunu'un) in the Mayan city of El Pilar presents an illuminating example of recreating a Mayan home garden in the present time. Maya forest gardening was interpreted by inviting traditional Mayan farmers and gardeners to implement their practices on site. Figure 6a shows the nature, profile, and location of vegetation on this site that provides food, climate control, and soil conservation.

(d) Examples of Home Gardens in Maya

- **Guijarral:** Home gardens that are close to the residences formed one component of the settlement pattern in Guijarral. This kind of garden is complemented by distant infields and outfields where major crops were cultivated. This purposefully embedded the population spatially, environmentally, economically, and spiritually in the built environment [87].
- **Papaloapan River:** This urban settlement was dispersed over large area. Engagement in intensive and extensive home gardens and contiguous infields made this settlement look like a 'Garden City'. Extensive distant outfields were cultivated rotationally and added resilience and risk mitigation [60].
- **Sayil:** The population in Sayil practiced intensive home gardening within city's bounds, making it another precedent of a 'Garden City'. Agricultural input was provided through green human labor that was essentially collaborative. Organic refuse, mulch, and human and animal waste were used as fertilizer that did not damage the soil. The presence of small garden lots indicates autonomy at the lower social levels.

3.4. Water Conservation

Water and its conservation were critical to the Maya given its scarcity, especially during the dry season. Its conservation and safe discharge to other areas throughout the year hints at their advanced knowledge [43]. As such, features such as *aguadas* (natural or human-made land depressions), *bajos*, *chultans* (underground water storage tanks in residential lots), berms, and dams formed essential elements of the Mayan-built environment [43], providing enough water for household and agricultural functions over the course of the year, especially the long annual drought [89]. Water ponds were multifunctional, offering ecological, spiritual, social, and economic benefits.

During the Late Pre-Classic Period (400 B.C–250 A.D.), some settlements were located close to existing natural depressions that were carefully reformed to collect the runoff from natural watersheds. For instance, in Sayil, underground water tanks known as *chultunes* form a noticeable feature of the built environment. The ensuing Classic Period (250 A.D.–950 A.D.), on the other hand, marks a change in water conservation in many sites such as Calakmul and Tikal. At those sites, important buildings and reservoirs were placed on higher altitudes, for instance hill summits. The water conservation profile during that period was convex, contrasting with the concave one adopted in the Late Pre-Classic Period (Figure 8). The impermeable pavement around sacred buildings on the summit guided rainwater towards large reservoirs. Canals from these reservoirs then transferred water to the dispersed farmsteads and their peripheral *bajo* reservoirs. However, this strategy differed for locations where there was abundant rainfall [40,89,100]. For instance, in settlements such as Copan and Palenque, water was not collected at the summit but diverted away to avoid damage due to flooding [36].

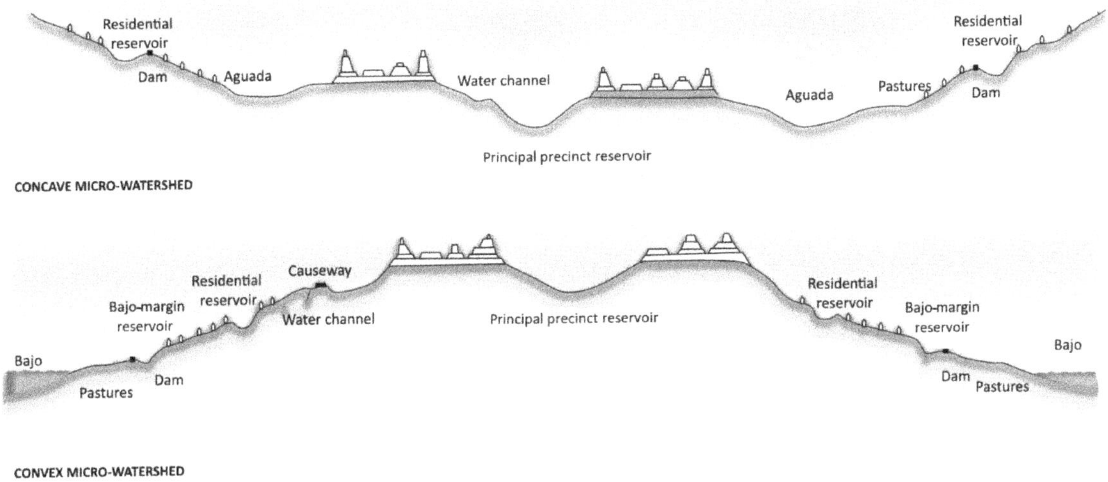

Figure 8. A change from concave micro-watershed (top: Late Pre-Classic Period) to convex micro-watershed (bottom: Late Classic Period) (adapted from Scarborough [101] (p. 140)).

(e) Examples of Water Conservation by the Maya

- **Tikal**: The central district in Tikal was placed on the summit of hills where watersheds were artificially modified to manage runoff. Six catchment reservoirs are located on this summit on an area of 6 to 9 ha that eventually led the runoff to the low-lying *bajo* and *aguada* reservoirs. The summit reservoirs could store approximately 900,000 m^3 of water, which was adequate for daily needs as well as replenishing the *bajos* (Figure 9). These reservoirs had raised causeways on their edges to dam water and direct it towards naturally occurring swamps [48] (p. 6). Floodgates were provided under the causeways to release water into the *bajo*-margin reservoirs for irrigation and drinking purposes. *Pozas* are another form of small-sized ponds within house-lots that were filled by the central water reservoirs [90].
- **Yanxohcah:** Located in the Central Uplands of the Yucatan Peninsula, this city is another example of skillful water conservation by the Maya. Here, many land depressions of around 20 m^3 have been identified through LiDar analysis [102] (p. 293). Since the soil was porous, the local reservoirs were lined with lime plaster to prevent water seeping into the ground. Water was received in these reservoirs through runoff flow under gravity and used for religious and agricultural purposes [100].
- **Cayo:** Apart from storing water for domestic purposes, the dam in Cayo also appears to have been used for breeding freshwater invertebrates that were a source of protein. This fact is confirmed by the presence of snail shells. The dam was strategically placed between converging hills and has a capacity of around 300 m^3. The retaining wall for the dam is well anchored to the bottom by being thick at the lower end while tapering to a slimmer section towards the top side [103] (p. 151). This indicates the developed understanding of the Maya about the way water exerts pressure on the enclosing surfaces of a dam.
- **Copan:** It was ecologically challenging for people to live in Copan due to the high rate of rainfall. Given this, water collection and dispersal were equally important to avoid flooding and soil erosion. To address this challenge, exceptional familiarity with the climate and hydrology played a key role. Water conservation undertaken at Copan further considered local hydrology, population density, site planning and architecture.

Effective water dispersal was thus achieved through the design and implementation of causeways, splashboards, roof drains, paved patios, stucco channels, substructure, and subterranean conduits [36].

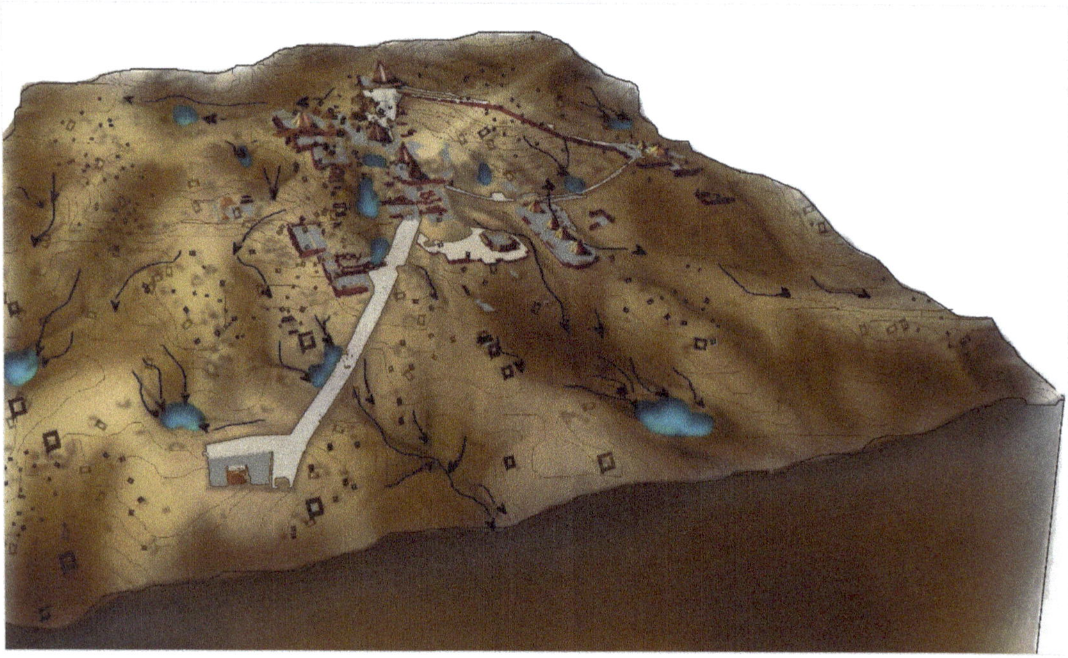

Figure 9. Water management in Tikal: The image shows the flow of water under gravity from the watershed created on the hill's summit. The water is then stored in the low-lying *bajos* (natural and human-made water reservoirs) that dot the landscape for use by the common people especially during the dry season. This shows an increased understanding by the Maya of their climate and terrain (adapted from Lucero & Gonzalez Cruz [48] (p. 6)).

4. Discussion: Lessons toward Mitigating Current Urban Sustainability Challenges

This study of the Classic Mayan-built environment highlights the ecological understanding and sustainable practices of the people that shaped their settlements and helped them survive for many millennia. Table 3 demonstrates that the environmental sustainability practices of the Maya are strongly related to the six principles of ecological embeddedness identified by Turner et al. [24]. The next sub-section discuses some key lessons that can be drawn from the ecological wisdom of the Maya that can improve the planning and building of contemporary cities.

4.1. Settlement Patterns

The Maya understood well environmental components, their interdependencies, and how to work with them for mutual benefit. The results of this are manifest at all levels, such as laying out buildings and connecting various areas within a city. The societies identified with and were embedded in the land they occupied, coevolving with all the components of their context. The people lived, traded, worked, farmed, played, and socially interacted in the same place. These activities, plus the culture and history, sustainably functioned and coexisted with the various components of the ecosystem, including the land, plants, animals, and sustainable lifestyle. The Mayan-built environment comprising residences, agricultural terraces, and managed forests was much more than just a means of getting food, clothing, and shelter. It was a concentration of diverse activities such as occupational specialization, production in surplus, individual and community interdependence, and art and culture. Settlement patterns in many Classic Mayan settlements corresponded with diverse activities that were ecologically friendly: the density fostered walkability, causeways are multi-functional, and site development was incremental, ensuring reuse of existing structural elements.

Table 3. The extent to which the six principles of ecological embeddedness were met by the Mayan-built environment.

#	Six Principles of Ecological Embeddedness [24]	Examples of Ecological Embeddedness of the Mayan-Built Environment
1	Interconnecting and rotating all entities in an environment	• Ceiba tree symbolizes connection of all entities present in an area. Therefore, interdependence of different ecological systems is recognized. Another instance of this is the cosmic interconnections with earthly undertakings, such as the beginnings of rituals and cultivation periods. These connections are of demonstrated practical value. One more interesting example of interconnectivity of different elements is purification of water to make it drinkable. This is conducted through the interdependent working of algae, lime mortar, fish, and water lilies living together in a water reservoir. • Time is perceived as cyclical/rotational. Phases in personal and practical life are predicted based on this perception, preparing one for future events. In everyday life, this is reflected by the rotation of crops and forested areas through Milpa farming, ensuring reliable subsistence. Another instance is basing the urban land-use concept of infield and outfield areas on rotation of intensive and extensive crop cultivation that provides sufficient agricultural yield. This also helps spread out risks and develop resilience in the built environment.
2	Environmental feedback and sustainable resource harvesting and reuse	• The degree to which foreign plants can be cultivated is based on environmental feedback. This fact is also visible in the settlement patterns that reflect the surrounding landscape. Carrying capacity is increased, working in collaboration with all elements present in an environment and their response to human intervention. Another example of this principle is that complete deforestation is not observed. The local climate is controlled through maintaining forest cover. Information for its success or failure is competently received through environmental feedback and responded to on time. • Water conservation through thoughtful management is vital and a permanent feature of the built environment. The same thought is applied to waste that is reused or recycled; for instance, human and animal waste is reused as fertilizer, broken pots and other items help in backfilling foundations for residences and *sacbeob*. Sustainable use of resources is also evidenced in houses for commoners that are made from perishable and recyclable materials.
3	Knowledge accumulation and transfer	• Occurrence of environmental feedback ranges from a few weeks to centuries. Therefore, knowledge acquired by the previous generations is extremely important, as it contains information about previous feedback and their cycles. Given this, knowledge is accretional and not only enriches the present generation, but also becomes a valuable source for future generations. • Human labor allows for a direct and effective knowledge transfer.
4	Respectful and collaborative approaches to human activities	• The Mayan-built environment is an anthropogenic landscape located in its geography. However, this is far from colonization of nature; the resultant Mayan habitat is a collaboration between humans and nature marked by reciprocity and respect. • Since environmental feedback is received by the common people present on ground, their role is crucial if the society is to survive. This results in the promotion of bottom-up policy decision making. Therefore, creation of the built environment is based on green human labor, which brings people together and encourages social sustainability.
5	Associating oneself as a part of their ancestral land	• There is one spirit that dwells in all human and other-than-human entities. This means the land and its inhabitants share one soul that bind them together. Given this, culture develops from the local physical environment, its constraints, and opportunities, as all share one soul. Meaningfulness found in the built environment is due to Mayans' association with their context. • Ecological knowledge varies from region to region. Migrating to a different land would mean reinvesting centuries to acquire new knowledge. Therefore, survival in the ancestral land, even during the worst environmental events such as droughts is more likely than elsewhere, given the Mayans' association with their land. Amongst other factors, burials provided a strong emotional and spiritual tie to the land. The Mayans gathered around the graves to seek inspiration and guidance from the spirits of the deceased.
6	Acknowledging spirituality in nature	• The way an environment and all its constituents come together to make life possible was awe-inspiring for Mayans. The spirituality derived from this is that there is a supreme soul that oversees and guides life. One example of this is the pronounced visibility of religious centers in the Mayan-built environment. • The idea that there is one supreme spirit that powers all human and other-than-human entities lends reverence and sanctity to everything present in a physical environment. For instance, unnecessarily cutting down trees is killing oneself as the soul that dwells in both the human and the tree is the same. The same idea of a shared soul converts the built environment from anthropocentric to cosmos-centric. This encourages environmental sustainability.

A residential unit was a place of habitation as well as work. In other words, home was where a possibility of sustenance could be created through interaction with nature. In this regard, terracing was adopted that corresponded to the terrain, provided levelled ground for residences, and prevented soil erosion and nutrient loss. Terrain informed the settlement style, which is non-orthogonal, spread out, and interspersed with areas for urban agriculture [12]. Common houses were made of local building materials, including perishable materials. Agricultural by-products and organic materials were used in layers for the ground floor construction. Mayan landscapes were lively, resulting from the interactions between social organization and the ecology of the surrounding locale, mutually benefitting all stakeholders living in an environment.

4.2. Urban Agriculture

The notion that urban living can be detached from rural still exists in some contemporary urban studies. The Maya, however, present a different case. Their settlements promote organic food production and ensure food security. Urban agriculture ensured crop yield while, at the same time, it did not rely on artificial fertilizer or pesticides. Human and animal waste was recycled and used as fertilizer that provided natural phosphorus to crops [48]. Taking environmental feedback into consideration, farming retained the natural features of the land without disturbing the local ecology. Given the rotation of crops, forest locations, and fallow land, the Mayan-built environment was dynamic, perfectly responding to the climate and the ecological allowances of a locale. Introduction of new plants was performed carefully, which, together with sustainable soil management, enabled native and new plants to thrive. There were primary and secondary locations for farming and living to be occupied for the purpose of acquiring additional production when needed.

Through maintaining native plants and domesticating foreign plants, the Maya were able to create an environment that offered shade and fruits. This, along with terrace construction and raised fields in certain built environments, resulted in a unique landscape pattern that is distinct from the deforestation and mono-cropping of the present time [104]. an increase in soil fertility is noted in these aggraded Mayan landscapes. It appears that the Mayans' impact on ecological systems helped many plant species to thrive, for instance cacao and pine groves in the northern Yucatan and Peten, respectively. Cultivation method, crop selection, and rotation did not deplete the soil of naturally occurring nutrients.

The Maya possessed highly developed ecological knowledge acquired over generations [41]. The knowledge of plant biology and soil type converted the less fertile lands into productive ones. For instance, maize was successfully grown in large quantities on land that was mostly unfavorable. This ability of the Maya formed anthropogenic landscapes, which were full of life and extended benefits, even on infertile terrain [50]. Crop cultivation was climate determined and frequently conducted near homes to facilitate harvesting and conservation, which proved successful as it imitated the way the ecosystems of the area functioned [101]. In the event of dry season, there was a shift from extensive to intensive cropping, swidden agriculture was reduced, and appropriate crops were selected to grow well in dry weather. In addition, mulching was introduced to retain moisture in the ground [43]. Moreover, protecting the forest cover controls the climate and provides other ecological benefits. Ancient forests that were cared for by the Maya survive even today, presenting another instance of the Mayan ecological legacy [43].

4.3. Water Conservation

Water conservation was essential during the Classic Maya Period. Water supply under gravity, construction of a variety of artificial channels and reservoirs, and capitalizing on naturally occurring water bodies made the Mayan-built environment contextual, climatically responsive, and resilient, especially during the dry months. The way the natural wetland biosphere functioned was mimicked to keep water clean in various reservoirs [41]. The blue-green infrastructure constructed by the Maya was sustainable. This included terrace walls that retained moisture and nutrients, artificial reservoirs, and modified natural reservoirs. The green infrastructure consisted of systematized green areas providing various ecosystem services such as gardening, leisure, flood protection, improving air quality, and reducing UHI effect.

Given the scarcity of water during the dry season, familiarity with the terrain, drainage routes, and moisture absorption of the soil, the Maya utilized natural and human-made water reservoirs. Potable water was obtained through imitating the wetland biosphere functions. Terrace walls retained moisture and prevented soil erosion. These also provided space for human residences that were made from perishable and recyclable materials. Through multiplicity of green cover and water reservoirs, the Maya were able to build resilience into their built environment and prevent flood risks.

4.4. Summary of Lessons for Addressing Contemporary Urban Sustainability Challenges

A summary of how the Mayans' ecological awareness can help mitigate the contemporary urban sustainability challenges is given in Table 4 by comparing the challenges [Column A] with lessons drawn from the Classic Mayan Period [Column B]. Briefly, emissions of GHG can be reduced if both production and consumption are localized. Emissions will further come down if production is linked with the carrying capacity of a region not overshooting the ecosystems. Connected to this is the provision of urban agriculture that will not only control UHI effect but, along with forestation, also function as a sponge to absorb torrential rains. If they are based on human labor, urban agriculture and its associated professions can provide self-sufficiency leading to availability of work and controlling the outflow of the local population looking for work. It will also cut down imports. Recycling is yet another lesson that can be learned from the Classic Mayan Period. When coupled with downscaled production and green human labor, recycling can significantly control all kinds of pollution. The end result of all of these actions is a built environment where the ecosystems are respected, where the natural setting of a locale is not altered and retain its identity, and where humans are part of their surroundings, all giving rise to a built environment that is ecologically embedded.

Table 4. Summary of lessons the Mayan-built environments can offer toward addressing current urban sustainability challenges.

Urban Sustainability Challenges	Solutions from the Ecological Embeddedness of the Mayan-Built Environment
High rate of GHG emissions and air pollution [3,5]	There is no mass scale production and consumption. This situation lowers emissions and maintains a normal climate [21].
Extensive transportation infrastructure adversely affects the natural landscape [16]	Both production and consumption are localized, and land-use is mixed, significantly reducing mobility and other infrastructure needs [64,105]
Mass production and unsustainable consumption lifestyle [27]	Role played by 'intense market' is limited. There is a communal and symbiotic way of life [41,43]. Conservation is a religious duty [48].
High levels of air, land, and water pollution [4,30]	Settlement density is appropriate and coupled with downscaled production, the environment is clean [35,58,61]. All waste is recyclable and biodegradable [106].
Alienation from one's land and food insecurity [107]	People are embedded in their locale. Food security is a high priority achieved through urban agriculture and gardening [44,50].
Cities overly dependent on unsustainable energy and technologies [2,14,108]	Green human labor defines the work style. Society respects the ecological allowances and ensured a fair outcome [109]. Buildings utilize natural lighting and ventilation [66].
Urban areas depend on imports that are susceptible to disruptions and have enormous carbon footprints [4]	Production areas lie within the built environment, imports are limited to mainly non-essential items, and risks are spread out [40,93,110]. This makes availability of life essentials predictable without any significant carbon footprint.
Urban flooding [31,32]	Maya settlements act as sponge cities. Blue-green infrastructure absorbs run-off and significantly reduces flooding [36,40,77,89,102].
Urban heat islands and sprawl [15,25,26,111]	Towns are medium to low density, mixed uses. Landscaping, water reservoirs, and urban agriculture are widespread, which capture carbon and absorb heat [12,105,112].
Inequality and unemployment [13,29]	There is fair guarantee of production [113]. Due to intensive labor requirement, individuals are not without work [107].

The principle of mixed land uses in the Mayan-built environment can help contemporary cities reduce urban sprawl and rising temperatures by minimizing the need to travel long distances. Building configuration that allows the use of natural lighting and ventilation is another lesson for creating more sustainable settlements. As in the Mayan-built environment, there is the need to promote indigenous building materials because they are low cost, reusable and recyclable, thus lowering energy use and increasing housing affordability. Participatory maintenance of the built environment is yet another Mayan feature than can bring humans and their environment together through interactions on multiple levels. This can strike a balance between top-down and bottom-up decisions and ensure timely reception of environmental feedback.

5. Concluding Remarks

The Mayan civilization is one of those eras of human history known for its environmental legacy and ecological wisdom. In the face of growing environmental and societal challenges, taking inspiration from Mayan culture may help us, in part at least, to shed light on possible solutions regarding contemporary urban sustainability issues. This civilization inspires us to achieve environmental sustainability through settlement patterns that promote mixed land use, walkability, native and recyclable construction materials, natural light and ventilation in buildings, and complete waste recycling. Urban agriculture and gardening capture carbon improve air quality, serve as windbreaks, reduce soil erosion, and ensure food security via organic food production. Water conservation through canals and reservoirs provides sufficient water for household and agricultural purposes, even during droughts. Water reservoirs are multifunctional, offering ecological, spiritual, and socioeconomic benefits, including food production, recreation, and curbing UHI effect and floods.

Mayans conducted their socioeconomic activities within the ecological allowances of a given locality and listened to and addressed the environmental feedback. All components of the ecosystem, such as land, vegetation, water, air, and animals, were revered and protected. Human activities harmoniously co-existed with the ecosystem and were linked to the culture and history of the society. As a result, the Mayans lived with and adapted to the favorable and unfavorable environmental changes within their context. These inspirations of Maya culture and their relevance are a significant and ethical frame worthy of adoption today, especially in the face of increasing urban sustainability challenges.

This study has some important implications for towns and cities in developing countries, which are the future of global urbanization because of their rapid population growth and expansion. The present study contributed to the recent search for sustainable urban solutions to the existing environmental problems using the values of Mayans' ecological wisdom accumulated through long historical experience. The study applied a holistic approach to improving land-use planning of contemporary cities based on the principles of the Mayans' settlement patterns, building configuration and construction, urban agriculture, home gardening, and water conservation. In addition, the heritage left by the ancient Maya civilization provides remarkable inspiration for this and future generations.

This study raises a fundamental question for the scholarship on contemporary cities. What are the challenges to theoretically and practically interpret the ecological wisdom of ancient civilizations when planning and managing overpopulated cities with the enormously increased complexity of interactions, high resource consumption, climate change, and pollution? From a strategic urban policy and planning perspective, future studies are needed to further operationalize the idea of ecological embeddedness. The ecological embeddedness of this ancient civilization proffers invaluable inspirations for addressing the environmental challenges of contemporary cities towards developing more sustainable built environments.

Author Contributions: Conceptualization: N.A.; methodology: I.R.A.; investigation: N.A., I.R.A., A.A.S., and W.A.-M.; data curation: N.A., I.R.A., A.A.S., and W.A.-M.; writing—original draft preparation: N.A., A.A.S., and W.A.-M.; writing—review and editing: I.R.A.; supervision: I.R.A.; project administration: N.A.; funding acquisition: I.R.A., A.A.S., and W.A.-M. All authors have read and agreed to the published version of the manuscript.

Funding: This research received no external funding.

Acknowledgments: Naji Akbar thanks Noel G. Keough, David P. Monteyne, Elizabeth H. Paris, Mohammed Alhefnawi, and Engr. Rehan Jamil for commenting on the draft manuscript, as well as Engr. Yazeed Alshiddi for assistance with selected images.

Conflicts of Interest: The authors declare no conflict of interest in conducting this study.

References

1. Harvey, D. The urban process under capitalism: A framework for analysis. In *Urbanization and Urban Planning in Capitalist Society*; Routledge: New York, NY, USA, 2020; pp. 91–121.
2. Moran, D.; Kanemoto, K.; Jiborn, M.; Wood, R.; Többen, J.; Seto, K.C. Carbon footprints of 13 000 cities. *Environ. Res. Lett.* **2018**, *13*, 064041. [CrossRef]
3. UN-Habitat. *Global Report on Human Settlements 2011: Cities and Climate Change*; Earthscan: London, UK, 2011.
4. Newman, P. The environmental impact of cities. *Environ. Urbaniz.* **2006**, *18*, 275–295. [CrossRef]
5. United Nations. *Cities and Pollution. Cities and Local Action*; United Nations: New York, NY, USA, 2021. Available online: https://www.un.org/en/climatechange/climate-solutions/cities-pollution (accessed on 16 October 2021).
6. Gail, W.; Cooper, W.H. Ecological embeddedness. *Acad. Manag. J.* **2000**, *43*, 1265–1282.
7. Guttmann-Bond, E. Sustainability out of the past: How archaeology can save the planet. *World Archaeol.* **2010**, *42*, 355–366. [CrossRef]
8. Hornborg, A. Ecological embeddedness and personhood. *Anthropol. Today* **1998**, *14*, 3. [CrossRef]
9. Mu'azu, N.D.; Abubakar, I.R.; Blaisi, N.I. Public acceptability of treated wastewater reuse in Saudi Arabia: Implications for water management policy. *Sci. Total Environ.* **2020**, *721*, 137659. [CrossRef] [PubMed]
10. Wang, X.; Palazzo, D.; Carper, M. Ecological wisdom as an emerging field of scholarly inquiry in urban planning and design. *Landsc. Urban Plan.* **2016**, *155*, 100–107. [CrossRef]
11. Abubakar, I.R.; Bununu, Y.A. Low carbon city: Strategies and case studies. In *Sustainable Cities and Communities. Encyclopedia of Sustainability in Higher Education*; Springer: Cham, Switzerland, 2020; pp. 366–376.
12. Barthel, S.; Isendahl, C. Urban gardens, agriculture, and water management: Sources of resilience for long-Term food security in cities. *Ecol. Econ.* **2013**, *86*, 224–234. [CrossRef]
13. Campbell, S. Green Cities, Growing Cities, Just Cities? Urban Planning and the Contradictions of Sustainable Development. *J. Am. Plan. Assoc.* **2007**, *62*, 296–312.
14. Watson, V. The planned city sweeps the poor away: Urban planning and 21st century urbanisation. *Prog. Plan.* **2009**, *72*, 151–193.
15. Smith, M.E. The archaeological study of neighborhoods and districts in ancient cities. *J. Anthropol. Archaeol.* **2010**, *29*, 137–154. [CrossRef]
16. Akbar, N.; Abubakar, I.R.; Bouregh, A.S. Fostering Urban Sustainability through the Ecological Wisdom of Traditional Settlements. *Sustainability* **2020**, *12*, 10033. [CrossRef]
17. Ardelt, M. Wisdom as expert knowledge system: A critical review of a contemporary operationalization of an ancient concept. *Hum. Dev.* **2004**, *47*, 257–285. [CrossRef]
18. Yoffee, N. Making ancient cities plausible. *Rev. Anthropol.* **2009**, *38*, 264–289. [CrossRef]
19. Pérez Rodríguez, V.; Anderson, K.C.; Neff, M.K. The Cerro Jazmín Archaeological Project: Investigating prehispanic urbanism and its environmental impact in the Mixteca Alta, Oaxaca, Mexico. *J. Field Archaeol.* **2011**, *36*, 83–99. [CrossRef]
20. Roberts, P.; Hunt, C.; Arroyo-Kalin, M.; Evans, D.; Boivin, N. The deep human prehistory of global tropical forests and its relevance for modern conservation. *Nat. Plants* **2017**, *3*, 1–9. [CrossRef] [PubMed]
21. Akbar, N. Surveying the Ecologically Embedded Built Environments of the Preindustrial Era with a Focus on the Classic Maya. Unpublished Master's Thesis, University of Calgary, Calgary, Canada, 2021.
22. Yang, B.; Young, R.F. (Eds.) *Ecological Wisdom: Theory and Practice*; Springer: Berlin/Heildberg, Germany, 2019.
23. Kakoty, S. Ecology, sustainability and traditional wisdom. *J. Clean. Prod.* **2018**, *172*, 3215–3224. [CrossRef]
24. Turner, N.J.; Ignace, M.B.; Ignace, R. Traditional ecological knowledge and wisdom of aboriginal peoples in British Columbia. *Ecol. Appl.* **2000**, *10*, 1275–1287. [CrossRef]
25. Buo, I.; Sagris, V.; Burdun, I.; Uuemaa, E. Estimating the expansion of urban areas and urban heat islands (UHI) in Ghana: A case study. *Nat. Hazards* **2021**, *105*, 1299–1321. [CrossRef]
26. Aina, Y.A.; Parvez, I.M.; Balogun, A.L.; Adam, E. Urban Heat Island Effects and Mitigation Strategies in Saudi Arabian Cities. In *Urban Heat Island (UHI) Mitigation*; Springer: Singapore, 2021; pp. 235–248.
27. Almulhim, A.I.; Abubakar, I.R. Understanding Public Environmental Awareness and Attitudes toward Circular Economy Transition in Saudi Arabia. *Sustainability* **2021**, *13*, 10157. [CrossRef]
28. Ferreira, C.S.; Walsh, R.P.; Ferreira, A.J. Degradation in urban areas. *Curr. Opin. Environ. Sci. Health* **2018**, *5*, 19–25. [CrossRef]
29. Abubakar, I.R. Strategies for coping with inadequate domestic water supply in Abuja, Nigeria. *Water Internat.* **2018**, *43*, 570–590. [CrossRef]
30. Joshi, R.; Ahmed, S. Status and challenges of municipal solid waste management in India: A review. *Cogent Environ. Sci.* **2016**, *2*, 1139434. [CrossRef]
31. Bulkeley, H. Urban sustainability: Learning from best practice? *Environ. Plan. A* **2006**, *38*, 1029–1044. [CrossRef]

32. Dano, U.L. An AHP-Based assessment of flood triggering factors to enhance resiliency in Dammam, Saudi Arabia. *Geo J.* **2021**, 1–16. [CrossRef]
33. Ashmore, W. Site-Planning principles and concepts of directionality among the ancient Maya. *Lat. Am. Antiq.* **2017**, *2*, 199–226. [CrossRef]
34. Munson, J.; Pinzón, F. Building an Early Maya Community: Archaeological Investigations at CAOBAL, Guatemala. *Anc. Mesoam.* **2017**, *28*, 265–278. [CrossRef]
35. Murtha, T. Rethinking Urban Density: Archaeology, Low Density Urbanism and Sustainability Lowland Maya Landscape Regionalism. In Proceedings of the Architecture, Archaeology and Contemporary City Planning, Valencia, Spain, 18–20 May 2015; pp. 92–100.
36. Davis-Salazar, K.L. Late Classic Maya drainage and flood control at Copan, Honduras. *Anc. Mesoam.* **2006**, *17*, 125–138. [CrossRef]
37. Dine, H.; Ardren, T.; Bascopé, G.; Báez, C.G. Famine Foods and Food Security in the Northern Maya Lowlands: Modern Lessons for Ancient Reconstructions. *Anc. Mesoam.* **2019**, *30*, 517–534. [CrossRef]
38. Turner, B.L.; Sabloff, J.A. Classic Period collapse of the Central Maya Lowlands: Insights about human-Environment relationships for sustainability. *Proc. Natl. Acad. Sci. USA* **2012**, *109*, 13908–13914. [CrossRef]
39. Hansen, R.D.; Bozarth, S.; Jacob, J.; Wahl, D.; Schreiner, T. Climatic and Environmental Variability in the Rise of Maya Civilization: A preliminary perspective from northern Peten. *Anc. Mesoam.* **2002**, *13*, 273–295. [CrossRef]
40. Lucero, L.J.; Fedick, S.L.; Dunning, N.P.; Lentz, D.L.; Scarborough, V.L. Water and landscape: Ancient Maya settlement decisions. *Archeol. Pap. Am. Anthropol. Assoc.* **2014**, *24*, 30–42. [CrossRef]
41. Lucero, L.J.; Gunn, J.D.; Scarborough, V.L. Climate Change and Classic Maya Water Management. *Water* **2011**, *3*, 479–494. [CrossRef]
42. Meanwell, J.L.; Paris, E.H.; Cruz Alvarado, W.; Peraza Lope, C. Metallurgical ceramics from Mayapán, Yucatán, Mexico. *J. Archaeol. Sci.* **2013**, *40*, 4306–4318. [CrossRef]
43. Beach, T.; Luzzadder-Beach, S.; Cook, D.; Dunning, N.; Kennett, D.J.; Krause, S.; Terry, R.; Trein, D.; Valdez, F. Ancient Maya impacts on the Earth's surface: An Early Anthropocene Analog? *Quat. Sci. Rev.* **2015**, *124*, 1–30. [CrossRef]
44. Ford, A. Valuing the Maya Forest as a Garden. In *Tropical Forest Conservation: Long-Term Processes of Human Evolution, Cultural Adaptations and Consumption Patterns*; Lewis, R.C., Mata, J.P., Connaughton, C., Eds.; United Nations Educational, Scientific and Cultural Organization: Paris, France, 2016; pp. 207–226.
45. Hammond, N. Maya: Revelation and re-Evaluation. *Antiquity* **2015**, *89*, 1503–1510. [CrossRef]
46. Ashmore, W. City and Town Planning of the Maya. In *Encyclopaedia of the History of Science, Technology, and Medicine in Non-Western Cultures*; Selin, H., Ed.; Springer: Dordrecht, Germany, 2014; pp. 1–9.
47. Zaro, G.; Lohse, J.C. Agricultural Rhythms and Rituals: Ancient Maya Solar Observation in Hinterland Blue Creek, Northwestern Belize. *Lat. Am. Antiq.* **2005**, *16*, 81–98. [CrossRef]
48. Lucero, L.J.; Gonzalez Cruz, J. Reconceptualizing Urbanism: Insights from Maya Cosmology. *Front. Sustain. Cities* **2020**, *2*, 1–15. [CrossRef]
49. Garrison, T.G.; Houston, S.D.; Scherer, A.K.; del Cid, D.; Garrido López, J.L.; Czapiewska-Halliday, E.; Román, E. A royal Maya country house: Archaeology at Bejucal, Guatemala. *J. Field Archaeol.* **2016**, *41*, 532–549. [CrossRef]
50. Fedick, S.L. Land Evaluation and Ancient Maya Land Use. In *The Upper Belize River Area, Belize, Central America*; Cambridge University Press: Cambridge, UK, 2018.
51. McCane, C.A.; Macrae, S.A.; Iannone, G.A. Consideration of the Spatial Arrangement of Settlement Groups and Terraces in Contreras, Minanha, Belize. *Res. Rep. Belizean Archaeol.* **2010**, *7*, 141–151.
52. Chase, D.Z.; Chase, A.F. Caracol, Belize, and Changing Perceptions of Ancient Maya Society. *J. Archaeol. Res.* **2017**, *25*, 185–249. [CrossRef]
53. Paris, E.H.; Taladoire, E.; Whiting, T.A.L. Return to Moxviquil: Form and Function in a Small Maya City. *Anc. Mesoam.* **2015**, *26*, 81–112. [CrossRef]
54. Peuramaki-Brown, M. The Integration and Disintegration of Ancient Maya Urban Centres: Charting Households and Community at Buenavista del Cayo, Belize. Unpublished Ph.D. Thesis, University of Calgary, Calgary, AB, Canada, 2012.
55. Brady, E.J.; Ashmore, W. Mountains, Caves, Water: Ideational Landscapes of the Ancient Maya. In *Archaeologies of Landscape: Contemporary Perspectives*; Blackwell: Hoboken, NJ, USA, 1999; pp. 124–145.
56. Eppich, K. Book Review: Maya Cities and Villages. *Lat. Am. Antiq.* **2016**, *27*, 132–134.
57. Restall, M. Maya Ethnogenesis. *J. Lat. Am. Anthropol.* **2008**, *9*, 64–89. [CrossRef]
58. Isendahl, C.; Smith, M.E. Sustainable agrarian urbanism: The low-Density cities of the Mayas and Aztecs. *Cities* **2013**, *31*, 132–143. [CrossRef]
59. Tsukamoto, K.; Camacho, J.L.; Valenzuela, L.E.C.; Kotegawa, H.; Esparza Olguín, O.Q. Political Interactions among Social Actors: Spatial Organization at the Classic Maya Polity of El Palmar, Campeche, Mexicó. *Lat. Am. Antiq.* **2015**, *26*, 200–220. [CrossRef]
60. Stark, B.L.; Ossa, A. *Ancient Settlement, Urban Gardening, and Environment in the Gulf Lowlands of Mexico*; Cambridge University Press: Cambridge, UK, 2019.
61. Smith, M.E. Classic Maya Settlement Clusters as Urban Neighborhoods: A Comparative Perspective on Low-Density Urbanism. *J. De La Société Des Américanistes* **2011**, *97*, 51–73. [CrossRef]

62. Masson, M.A.; Freidel, D.A. An argument for Classic era Maya market exchange. *J. Anthropol. Archaeol.* **2012**, *31*, 455–484. [CrossRef]
63. Ford, A.; Nigh, R. The Milpa Cycle and the making of the Maya forest garden. *Res. Rep. Belizean Archaeol.* **2010**, *7*, 183–190.
64. Farahani, A.; Chiou, K.L.; Harkey, A.; Hastorf, C.A.; Lentz, D.L.; Sheets, P. Identifying 'plantscapes' at the Classic Maya village of Joya de Cerén, El Salvador. *Antiquity* **2017**, *91*, 980–997. [CrossRef]
65. Pugh, T.W.; Rice, P.M. Early urban planning, Spatial strategies, and the Maya gridded city of nixtun-Ch'ich', Petén, Guatemala. *Curr. Anthropol.* **2017**, *58*, 576–603. [CrossRef]
66. Richards-Rissetto, H. An iterative 3D GIS analysis of the role of visibility in ancient Maya landscapes: A case study from Copan, Honduras. *Digit. Scholarsh. Humanit.* **2017**, *32*, 195–212. [CrossRef]
67. Landau, K. Spatial logic and Maya city planning: The case for cosmology. *Camb. Archaeol. J.* **2015**, *25*, 275–292. [CrossRef]
68. Hutson, S.R.; Stanton, T.W.; Magnoni, A.; Terry, R.; Craner, J. Beyond the buildings: Formation processes of ancient Maya houselots and methods for the study of non-Architectural space. *J. Anthropol. Archaeol.* **2007**, *26*, 442–473. [CrossRef]
69. Ashmore, W.; Sabloff, J.A. Spatial Orders in Maya Civic Plans. Latin American Antiquity. *Soc. Am. Archaeol.* **2002**, *13*, 201–215.
70. Shaw, M.J. 17-Maya Sacbeob: Form and function. *Anc. Mesoam.* **2001**, *12*, 261–272. [CrossRef]
71. Stanton, T.W.; Freidel, D.A. Placing the centre, centring the place: The influence of formative Sacbeob in classic site design at Yaxuná, Yucatán. *Camb. Archaeol. J.* **2005**, *15*, 225–249. [CrossRef]
72. Grauer, K.C. Active environments: Relational ontologies of landscape at the ancient Maya city of Aventura, Belize. *J. Soc. Archaeol.* **2020**, *20*, 74–94. [CrossRef]
73. Wernecke, D.C. A Stone Canvas: Interpreting Maya Building Materials and Construction. Unpublished Ph.D. Thesis, The University of Texas, Austin, TX, USA, 2005.
74. Becker, M. Ancient Maya houses and their identification: An evaluation of architectural groups at Tikal and inferences regarding their functions. *Rev. Española De Antropol. Am.* **1982**, *12*, 111–130.
75. Graham, E. Stone Cities, Green Cities. *Archeol. Pap. Am. Anthropol. Assoc.* **2008**, *9*, 185–194. [CrossRef]
76. Abrams, E.M. *Economic Specialization and Construction Personnel in Classic Period Copan, Honduras*; Cambridge University Press: Cambridge, UK, 2019; pp. 485–499.
77. French, K.D.; Duffy, C.J.; Bhatt, G. The urban hydrology and hydraulic engineering at the classic Maya site of Palenque. *Water Hist.* **2013**, *5*, 43–69. [CrossRef]
78. Stuart, D.S. The Hills Are Alive: Sacred Mountains in the Maya Cosmos. *Symbols* **1997**, *13*, 13–17.
79. Sprajc, I. Astronomical and Cosmological Aspects of Maya Architecture and Urbanism. *Cosmol. Across Cult.* **2009**, *409*, 303–314.
80. Barr, V.A. Mayan Engineering Legacy: Coba. *Mech. Eng.* **1990**, *112*, 66–71.
81. Hammond, N.; Gerhardt, J.C. Early Maya architectural innovation at Cuello, Belize. *World Archaeol.* **1990**, *21*, 461–481. [CrossRef]
82. Robin, C. Outside of houses. *J. Soc. Archaeol.* **2002**, *2*, 245–268. [CrossRef]
83. Magnoni, A.; Hutson, S.R.; Dahlin, B.H. Living in the city: Settlement patterns and the urban experience at classic period Chunchucmil, Yucatan, Mexico. *Anc. Mesoam.* **2012**, *23*, 313–343. [CrossRef]
84. Ford, A. Maya Forest Garden. In *Encyclopedia of Global Archaeology*; Springer: Cham, Switzerland, 2020; pp. 1–11.
85. Halperin, C.T.; Foias, A. Household Garbage: Classic period (ca. 300–900 CE) Maya Practices of Discard. *J. Field Archaeol.* **2016**, *8*, 45. [CrossRef]
86. Doyle, J.A. Early Maya geometric planning conventions at El Palmar, Guatemala. *J. Archaeol. Sci.* **2013**, *40*, 793–798. [CrossRef]
87. Hughbanks, P.J. Settlement and Land Use at Guijarral, Northwest Belize. *Cult. Agric.* **2008**, *20*, 107–120. [CrossRef]
88. Fisher, C. The role of infield agriculture in Maya cities. *J. Anthropol. Archaeol.* **2014**, *36*, 196–210. [CrossRef]
89. Scarborough, V.L.; Gallopin, G.G. A Water Storage Adaptation in the Maya Lowlands. *Science* **1991**, *251*, 658–662. [CrossRef] [PubMed]
90. Lucero, L.J. A Cosmology of Conservation in the Ancient Maya World. *J. Anthropol. Res.* **2018**, *July*, 327–359. [CrossRef]
91. Dahlin, B.H.; Bastamow, M.; Beach, T.; Hruby, Z.X.; Hutson, S.R.; Mazeau, D. Phantom Lithics at Chunchucmil, Yucatán, Mexico. *Technol. Maya Civiliz. Politic. Econ. Beyond Lithic Stud.* **2009**, 156–178.
92. Bates, A. The Gift of the Maya. The Great Change. 2015. Available online: https://peaksurfer.blogspot.com/2015/07/the-gift-of-maya.html (accessed on 2 October 2021).
93. Dunning, N.P.; Beach, T. Farms and Forests: Spatial and Temporal Perspectives on Ancient Maya Landscapes. In *Landscapes and Societies*; Martini, I.P., Chesworth, W., Eds.; Springer: Dordrecht/Heidelberg, Germany; London, UK; New York, NY, USA, 2010; pp. 369–390.
94. Lohse, J.C.; Findlay, N.P. A Classic Maya House-Lot Drainage System in Northwestern Belize. *Lat. Am. Antiq.* **2000**, *11*, 175–185. [CrossRef]
95. Castro, A.; Lascurain-Rangel, M.; Gómez-Díaz, J.A.; Sosa, V. Mayan Homegardens in Decline: The Case of the Pitahaya (Hylocereus Undatus), a Vine Cactus with Edible Fruit. *Trop. Conserv. Sci.* **2018**, *11*, 1–10. [CrossRef]
96. Dedrick, M.; Webb, E.A.; McAnany, P.A.; Kanxoc Kumul, J.M.; Jones, J.G.; Batún Alpuche, A.I.; Pope, C.; Russell, M. Influential landscapes: Temporal trends in the agricultural use of Rejolladas at Tahcabo, Yucatán, Mexico. *J. Anthropol. Archaeol.* **2020**, *59*, 101175. [CrossRef]
97. Martínez-Ballesté, A.; Martorell, C.; Caballero, J. Cultural or ecological sustainability? The effect of cultural change on Sabal palm management among the lowland Maya of Mexico. *Ecol. Soc.* **2006**, *11*, 27. [CrossRef]

98. Poot-Pool, W.S.; van der Wal, H.; Flores-Guido, S.; Pat-Fernández, J.M.; Esparza-Olguín, L. Economic Stratification Differentiates Home Gardens in the Maya Village of Pomuch, Mexico. *Econ. Bot.* **2012**, *66*, 264–275. [CrossRef]
99. Rojas, M.O.; Collins, S.; Cal, V.; Caal, F.; Knight, K.; Arnason, J.; Poveda, L.; Sanches-Vindas, P.; Pesek, T. Sustaining Rainforest Plants, People and Global Health: A Model for Learning from Traditions in Holistic Health Promotion and Community Based Conservation as Implemented by Q'eqchi' Maya Healers, Maya Mountains, Belize. *Sustainability* **2010**, *2*, 3383–3398. [CrossRef]
100. Brewer, J.L.; Carr, C.; Dunning, N.P.; Walker, D.S.; Anaya Hernández, A.; Peuramaki-Brown, M.; Reese-Taylor, K. Employing airborne lidar and archaeological testing to determine the role of small depressions in water management at the ancient Maya site of Yaxnohcah, Campeche, Mexico. *J. Archaeol. Sci.* **2017**, *13*, 291–302. [CrossRef]
101. Scarborough, V.L. Ecology and Ritual: Water Management and the Maya. *Lat. Am. Antiq.* **1998**, *9*, 135–159. [CrossRef]
102. Scarborough, V.L.; Dunning, N.P.; Tankersley, K.B.; Carr, C.; Weaver, E.; Grazioso, L.; Lane, B.; Jones, J.G.; Buttles, P.; Valdez, F.; et al. Water and sustainable land use at the ancient tropical city of Tikal, Guatemala. *Proc. Natl. Acad. Sci. USA* **2012**, *109*, 12408–12413. [CrossRef] [PubMed]
103. Healy, P.F. An Ancient Maya Dam in the Cayo District, Belize. *J. Field Archaeol.* **1983**, *10*, 147.
104. Wilk, R.R. Little house in the jungle: The causes of variation in house size among modern Kekchi Maya. *J. Anthropol. Archaeol.* **1983**, *2*, 99–116. [CrossRef]
105. Marken, D.; Murtha, T. Maya Cities, People and Place: Comparative Perspectives from El Peru and Tikal. Paper Presented at the 14th Belize Archeology Symposium, San Ignacio, Belize, 1 July 2016.
106. Smyth, M.P.; Dore, C.D.; Dunning, N.P. Interpreting prehistoric settlement patterns: Lessons from the Maya center of Sayil, Yucatan. *J. Field Archaeol.* **1995**, *22*, 321–347.
107. Williams, D.R.; Stewart, S.I. Sense of Place: An Elusive Concept That Is Finding a Home in Ecosystem Management. *J. For.* **1998**, *5*, 18–23.
108. Alshuwaikhat, H.M.; Abubakar, I. Towards a sustainable urban environmental management approach (SUEMA): Incorporating environmental management with strategic environmental assessment (SEA). *J. Environ. Plan. Manag.* **2007**, *50*, 257–270. [CrossRef]
109. Scarborough, V.L.; Burnside, W.R. Complexity and Sustainability: Perspectives from the Ancient Maya and the Modern Balinese. *Am. Antiq.* **2010**, *75*, 327–363. [CrossRef]
110. Chase, D.Z.; Chase, A.F. Path dependency in the rise and denouement of a classic Maya City: The case of Caracol, Belize. *Archeol. Pap. Am. Anthropol. Assoc.* **2014**, *24*, 142–154. [CrossRef]
111. Rees, W.; Wackernagel, M. Urban Ecological Footprint: Why Cities Cannot Be Sustainable and Why They Are a Key to Sustainability. *Environ. Impact Assess. Rev.* **1996**, *16*, 223–248. [CrossRef]
112. Robinson, M.E.; McKillop, H.I. Ancient Maya wood selection and forest exploitation: A view from the Paynes Creek salt works, Belize. *J. Archaeol. Sci.* **2013**, *40*, 3584–3595. [CrossRef]
113. Patten, D.T. The role of ecological wisdom in managing for sustainable interdependent urban and natural ecosystems. *Landsc. Urban Plan.* **2016**, *155*, 3–10. [CrossRef]

land MDPI

Article

Challenges and Opportunities for Public Participation in Urban and Regional Planning during the COVID-19 Pandemic—Lessons Learned for the Future

Marijana Pantić [1,*], Juaneé Cilliers [2], Guido Cimadomo [3], Fernando Montaño [4], Olusola Olufemi [5], Sally Torres Mallma [6] and Johan van den Berg [7]

1 Institute of Architecture and Urban & Spatial Planning of Serbia, 11000 Belgrade, Serbia
2 Faculty of Design, Architecture and Building, University of Technology Sydney, Sydney 2007, Australia; jua.cilliers@uts.edu.au
3 Department of Art and Architecture, Escuela Técnica Superior de Arquitectura, Universidad de Málaga, 29016 Málaga, Spain; cimadomo@uma.es
4 Digital City Science, HafenCity Universität, 20457 Hamburg, Germany; fermontz8@gmail.com
5 Independent Researcher, Tanner Court, Oakville, ON 1497, Canada; solaoluf@yahoo.com
6 Faculty of Architecture and Urbanism, Universidad Ricardo Palma, Lima 15039, Peru; sally.torres@urp.edu.pe
7 Urbanissimo bv, 3650 Dilsen-Stokkem, Belgium; info@urbanissimo.be
* Correspondence: marijana.d.pantic@gmail.com

Citation: Pantić, M.; Cilliers, J.; Cimadomo, G.; Montaño, F.; Olufemi, O.; Torres Mallma, S.; van den Berg, J. Challenges and Opportunities for Public Participation in Urban and Regional Planning during the COVID-19 Pandemic—Lessons Learned for the Future. *Land* **2021**, *10*, 1379. https://doi.org/10.3390/land10121379

Academic Editor: Alexandru-Ionuţ Petrişor

Received: 5 November 2021
Accepted: 8 December 2021
Published: 13 December 2021

Publisher's Note: MDPI stays neutral with regard to jurisdictional claims in published maps and institutional affiliations.

Abstract: The COVID-19 pandemic has spurred significant changes in the fields of economic development, social issues, everyday life, etc. Activities that used to depend on face-to-face communication were firstly suspended and then shifted to new forms of communication. This includes the public participation process in urban and spatial planning. Therefore, this study explores the new domain developed in urban and spatial planning with regard to public participation and surmises future realms in the post-pandemic era. On the occasion of the virtual collaboration platform Cyber Agora organized by the ISOCARP (International Society of City and Regional Planners), chosen participants got together virtually to share, discuss, and compare their practical knowledge in public participation before and during COVID-19. In addition, they addressed the potential benefits of shifting from traditional to virtual participation and potential benefits in the post-COVID-19 era. Considering the collected data and understanding them in the light of the available literature, this study concludes that the application of a combined approach (using both traditional and virtual modes of participation) is recommended because it would enable a larger number and higher diversity of participants. The study also elaborates particular modes of virtual participation with the pros and cons of their use in a particular context.

Keywords: collaborative decision making; urbanism; spatial planning; online platforms; lockdown

1. Introduction

Public participation in urban and regional planning increases the chances that decisions correspond to the diverse interests and needs of different stakeholders and citizen groups [1–5]. It also brings implementation closer to cost-effective investments and builds the trust and satisfaction of citizens toward the government [5]. However, at the beginning of its development, participatory planning was reserved solely for the elite [6], which changed in some countries in the second half of the 20th century [3]. The relevance of citizen participation has increased and become necessary in planning processes [7].

Since the COVID-19 pandemic prompted a lockdown in almost all countries in the world [8], governments and planners were forced to improvise or develop a new functionality for public spaces [9] and methods of participation [10–13]. Many techniques characteristic of the traditional approach to participation, which required face-to-face interaction, have become unviable with the COVID-19 outbreak [14]. Therefore, the use of technologies and virtual space has become a focal point in the development of new participation approaches [12] and conducting studies dealing with the new roles of information and communication technologies (e.g., [15]).

Although the use of the internet was seen as anarchic at the beginning of its existence [16], it has significantly evolved, and its use exploded [17]. Together with new technologies, the internet transformed the possibilities of collaboration [18], networking [19], and participation itself [20,21]. Thus, it has developed into a "democratizing force" that facilitates in particular the involvement of young people and individuals with disabilities in participative processes [21–23]. About a decade ago, the use of the internet and technologies was seen as an opportunity to upgrade participation, but it was also open to criticism that the same people would always be interested in participation regardless of the methods used [24]. Some cities have developed efficient virtual systems of data collection, and [25,26] expressed concern that citizens are being reduced to the role of a sensor, with no meaningful exchange with planners and government. The point might be, as [27] notes, that the internet still needs to be improved in order to evolve from a "public space" into a "public sphere".

Authors such as [10,28,29] examined the most plausible scenarios in the post-COVID-19 era regarding participatory planning. All of them noted that a shift of paradigms took place during the COVID-19 pandemic, inevitably leading toward new shifts in the post-pandemic era. The authors of [23,30–32] proposed that the introduction of technology in the participatory process during the COVID-19 pandemic will turn into a "new normal" in the period to come.

The purpose of participation should not only be voting for the offered solution, but also debate with a vivid communication exchange through which an idea will be developed [33,34], especially in the early stages of any planning process. This is one of the reasons for the skepticism related to certain forms of virtual participation in urban and regional planning, such as online questionnaires or data collection via mobile phone applications. Regardless of the compensatory aspects of virtual participation during the COVID-19 pandemic, [34] noticed that the current legal context needs to be changed in order to equalize participants' decisional power to its extent in an institutionalized traditional approach. According to [34,35], there is no solid evaluation of the e-planning and virtual approach; therefore, its effectiveness in participation is not clear. The COVID-19 pandemic has created a specific context and rapidly prompted both the ad hoc and planned development of new virtual approaches. Their challenges and opportunities have yet to be explored and determined. Therefore, this paper evaluates these aspects of public participation in urban and regional planning, additionally discussing lessons learned for the post-COVID-19 era.

This paper first presents short histories and the contexts in which public participation and virtual public participation have developed in the case study countries. Subsequently, it presents results from the conducted survey. Lastly, the results are discussed and conclusions are noted in light of other academic literature.

1.1. The Practice of Public Participation

The idea of public participation in urban planning was born with Jean Jack Rousseau in the 18th century. However, its application and institutionalization took another two centuries [36]. The concept was first adopted nationally; however, with the Rio Declaration on Environment and Development in 1992 [37], which promoted participation at the global level, it has become a constitutional part of other international documents such as the Leipzig Charter on Sustainable European Cities (Leipzig Charter 2007) [38], Treaty of Lisbon [39], and the New Leipzig Charter [40].

The time frame in which public participation was formalized in the planning of analyzed countries spans a long period and features different evolution threads. Starting in the 1920s, Canada was the first to introduce the institution of participation in planning, which has increased significantly since then [41]. Unlike other socialist countries at that time, the involvement of the public in planning in Serbia (formerly part of Yugoslavia) was established in 1949 by the "Basic Decree on the Master Urban Plan" (Osnovna uredba o generalnom urbanističkom planu) [3]. Unlike the other countries analyzed in this paper, the development of public participation fluctuated significantly in accordance with socio-political changes; the highest level of participation was recorded in the 1970s, then subsided during the 1990s, was quite restricted in 2003, and improved slightly in 2014 compared to 2003, but not compared to the 1970s [3]. The first federal regulations for the participation of citizens in urban planning in Germany date from 1960 [42], and they were later modified and improved (e.g., German Building Law of 1976 [43], European Construction Law Adaptation Act [44,45]). In Belgium, public participation arose in the 1960s from resistance to decisions and was unregulated, before being formed into a mandatory procedure; since the 21st century, it involves mandatory dialogue and testing. In Spain, public consultation, as a first step in the development of urban and territorial planning, was implemented only with the promulgation of the Planning Regulation for the development and application of the Law on Land Regime and Urban Planning of 1978, art. 116.1 [46], which is, among others, a result of the first democratic elections in 1977 after a long dictatorship period. It was nevertheless an indirect kind of participation, fostered only by those public institutions that considered it necessary in the preliminary phases of planning to collect all kinds of stakeholders' opinions. These ideas were implemented in very few cases [47]. Public consultation in different stages of urban development was regulated only with the promulgation of the Land Regime and Valorization Law [48,49]. In Peru, mandatory participation was introduced by the Rights of Citizen Participation and Control Law-Law No. 26,300 in 1994 [50] to strengthen the process of the country's democratic acculturation. In addition, the UNDP and the National Elections Jury published a guide in 2008 to establish mechanisms for citizen participation that do not generate abuse or chaos [51]. Lastly, public participation in South Africa was pushed forward by the "Report on the Assessment of Public Participation Practices in the Public Service" in 2008 [52]. In the report, the Public Service Commission categorically urges government departments to institutionalize the practice of public participation by, inter alia, developing their policies, guidelines, and structures such as dedicated public participation units to drive public participation. Thus, the introduction of public participation into formal procedures in the analyzed countries lasted for almost an entire century, mostly expanding the rights and modes of participation, but also with examples of hindrance and degradation due to political changes (Table 1).

Table 1. Public participation in urban and regional planning.

Country	Formalization of the Participation	Participation Models		
		Consulting before a Plan Is Drafted	**Commenting on the First Draft**	**Commenting on the Final Draft**
Belgium	1960s	Obligatory	Obligatory	Obligatory
Canada	1920s	Obligatory	Obligatory	Obligatory
Germany	1960s	Non-binding but commonly practiced	Obligatory	Obligatory
Peru	2010s	Obligatory	Obligatory	Obligatory
Serbia	1940s	Non-binding but occasionally practiced	Obligatory	Obligatory
South Africa	2000s	Non-binding but occasionally practiced	Obligatory	Obligatory
Spain	2000s	Non-binding but occasionally practiced	Obligatory	Obligatory

Except in countries where formalized public participation was established before the 2000s, a combination of formal elements with nonbinding approaches of participation is common along with obligatory terms of public participation. In Belgium, progressive cities and companies go beyond formal regulations and open explicit dialogue with residents and the population with scenarios. Under the repealed Canada's Environmental Assessment Act (CEAA), a public participation guide for environmental assessments required obligatory and meaningful participation, but there was also voluntary and nonbinding participation in other aspects that required public input in other planning processes. The Guide to Canada's Impact Assessment Act, 2020 (formerly the CEAA) introduced substantive changes that broaden the scope of all federal assessments, and the new Impact Assessment Act (IAA), which went into effect in August 2019, represents a fundamental shift in federal environmental assessment in Canada [53]. The IAA [54] has a clear Public Participation Plan which helps to identify and address the potential impact of a designated project. Due to Germany's federal system, there are guidelines in the different German states (Länder) that derive their basis from federal law and stipulate specific regulations on citizen participation in addition to regulations at the federal level. Second-level regulations have been applied in all states except one during the 1990s [45]. As in Germany, urban and territorial regulations in Spain are delegated to regions (comunidades). "Indirect democracy" intended as the right of neighbors to be informed and to participate in public affairs has been regulated since 1985 by the Law Regulating the Bases of Local Governments [55] and was extended to the specific planning framework in the Land Regime and Valorization Law [48]. "Direct democracy" or full participation in defining the transformation of urban models is nonbinding and implemented by planners and several municipalities that understand the benefits of citizens' involvement. True participation is only possible when municipalities or planners foster strategies that include "novel" approaches.

1.2. *The Practice of Virtual Public Participation*

In contrast to traditional forms of participation, virtual public participation is not legally binding in all the analyzed countries. In Belgium, Germany, Peru, and Serbia, the practice of virtual public participation is partially defined by regulations, whereas nonbinding forms were developed along the way. In other countries, the COVID-19 pandemic was the trigger for the use of virtual participation (Table 2).

Table 2. Virtual public participation in urban and regional planning.

Country	When Consideration of the Practice Began	Formalization of Virtual Participation
Belgium	2020 due to the COVID-19 pandemic	Nonbinding but intensified due to the COVID-19 pandemic
Canada	The 2010s	Obligatory, changes made due to the COVID-19 pandemic
Germany	The late 1990s	Obligatory
Peru	2020 due to the COVID-19 pandemic	Nonbinding but intensified due to the COVID-19 pandemic
Serbia	The late 2000s	Obligatory, but only regarding public information; new nonbinding forms developed due to the COVID-19 pandemic
South Africa	The 2010s	Nonbinding but intensified due to the COVID-19 pandemic
Spain	The 2000s	Obligatory, but only regarding public information

Independent of COVID-19, Germany was the first to consider virtual participation in planning in the late 1990s and early 2000s. The first large-scale study of E-Citizen Participation in German Cities was launched in 2004: "Digital Citizens' Participation in Germany's Large Cities 2004" ("Elektronische Bürgerbeteiligung in deutschen Großstädten 2004") [56]. With the most recent amendment to the Building Code (BauGB), the additional use of the internet has become mandatory for public participation in building land-use planning since May 2017 (Section 4a (4) BauGB). The documents must be displayed and accessed by the public, for example, via the municipality's internet portal [57].

The Revised Text of the Land and Urban Rehabilitation Law 7/2015 [58] in Spain made virtual participation obligatory, but only in terms of publishing current urban plans, public consultation calls, and information on the internet. Due to the delegation of administrative management, some regions applied similar rules earlier, as well as fostered the presentation of queries in a digital format (Andalucia since 2013 with the Transparency, Access to

Public Information and Good Governance Law 9/2013 [59]). By the end of the 2000s, the municipality of Rivas Vaciamadrid (Spain) fostered real virtual participatory processes spurred by the ambition to reach all community members [60].

In Serbia, the Law on Planning and Construction adopted in 2009 [61] required the public display of plans on the internet, thus establishing the first obligatory aspect of virtual participation, which has not developed further since then. However, nonbinding forms started to emerge increasingly with the COVID-19 pandemic. COVID-19 realities also boosted the need for virtual participation in South Africa, although virtual public participation is not a common feature due to technology constraints and limited infrastructure. Similarly, the pandemic hastened alternative ways of participation (meaning virtual participation) in Peru and Belgium.

Pandemic-inspired actions led to official regulation changes in Canada. Canada Bill 189 [62] was an Act to amend various Acts (including the Planning Act) to address COVID-19. Moreover, Ontario Regulation 149/20 [63] stipulates special rules related to the declared emergency. The "Canada Digital Charter in Action" initiative was created earlier. However, specifically since 2020, virtual public participation and engagement have become evident due to pandemic measures such as lockdown and social distancing.

In Peru, the Public Participation and Digital Government Law [64] is mandatory, but work with virtual mechanisms in urban planning began as a result of the pandemic. There is no specific related law, and it is not mentioned in the new Sustainable Urban Development Law of 2021 [65]. There was only one Manual of Metropolitan Development Plans in 2020, where it is recommended to adopt virtual measures for public participation [66]. This is because the development of two metropolitan plans for Lima and El Callao provinces began at the end of 2019 and the beginning of 2020, respectively.

The COVID-19 pandemic is the reason why online meetings have become one of the inevitable tools for public participation in all the analyzed countries. In South Africa, it even became legally binding. Another means of virtual public participation inspired by the pandemic is online voting, which was introduced as legally binding in South Africa and as legally nonbinding in Canada, Germany, Serbia, and Spain. Other forms of virtual public participation were introduced in only a few countries (Table 3). In Ontario, the City of Toronto approved virtual public hearings and participation due to COVID-19 via telephone, tablets, computers, and smartphones.

Table 3. Forms of virtual public participation in urban and regional planning.

Forms of Virtual Participation		Belgium	Canada	Germany	Peru	Serbia	South Africa	Spain
Documents available for public inspection on a website	Legally nonbinding	/	/	/	/	/	/	/
	Legally binding	Yes	Yes	Yes	Yes	Yes	Yes	Yes
	COVID-19 inspired	/	/	/	/	/	Yes	/
Online voting	Legally nonbinding	/	Yes	Yes	Yes	Yes	/	Yes
	Legally binding	/	/	/	/	/	Yes	/
	COVID-19 inspired	/	Yes	Yes	Yes	Yes	Yes	Yes
Online meetings	Legally nonbinding	Yes	Yes	Yes	Yes	Yes	/	Yes
	Legally binding	/	/	/	/	/	Yes	/
	COVID-19 inspired	No	Yes	Yes	Yes	Yes	Yes	Yes
Mobile phone applications	Legally nonbinding	Yes	/	Yes	/	/	/	/
	Legally binding	/	/	/	/	/	/	/
	COVID-19 inspired	Yes	/	/	/	/	/	/
Submission of remarks and suggestions by e-mail/digital registry	Legally nonbinding	/	/	/	/	Yes	/	/
	Legally binding	/	/	/	/	/	/	Yes
	COVID-19 inspired	/	/	/	Yes	Yes	/	/
Simplified procedure to consult planners by phone/mail	Legally nonbinding	/	/	/	/	/	/	Yes
	Legally binding	Yes	/	/	/	/	/	/
	COVID-19 inspired	/	/	/	Yes	/	/	Yes

Regardless of the transition to virtual forms of participation, stakeholders have mainly remained the same. Virtual meetings, information on web pages, and online voting were accessible to all stakeholders—from citizens to nongovernmental organizations, the civil sector, public institutions, etc. In some cases, the range of stakeholders depends on the type of project (plan), the governmental body responsible for the plan, and other factors. For example, in Germany, there is no exact participation procedure or defined tools, but they could vary depending on the state/local government. In the case of Serbia, alternative (legally nonbinding) documents such as Green City Action Plan (GCAP) were financed by the EBRD and the Japanese government; therefore, the documents went through a different procedure than the Spatial Plan of the Republic of Serbia that was conducted by the state government (both were created during the pandemic). The action plan procedure involved a wider range of virtual participation tools (e.g., online voting, online meetings, and e-mail communication), whereas, for the Spatial Plan, only remarks and suggestions could be sent via e-mail. However, the Spatial Plan involved all possible stakeholders willing to participate, whereas GCAP meetings were mainly accessible to invited City Government representatives.

One of the successful case finalizations of an urban project appears to be the Europaplatz in Tübingen (Germany) (open public space), whose renovation was discussed for 25 years before the COVID-19 pandemic. The last public planning workshop was canceled due to pandemic restrictions tightening in autumn 2020. Instead, a digital format was developed. The planning was explained by various experts in videos. All the information was processed textually and graphically on the website and presented by the daily newspapers. Participation was intensively advertised via various social networks and mailing lists. Valuable suggestions were received such that the planning was modified in parts. The positive vote with 70% approval and only 10% clear rejection showed that the planning is based on broad urban society consensus. A comparatively large number of younger people and families were reached. Overall, the media and public considered this to be a success, especially regarding the fact that it was one of the first digital formats for citizen participation in Tübingen and, thus, still in the trial stage [67].

2. Materials and Methods

The onset of the COVID-19 pandemic hindered traditional and prompted new forms of public participation in urban and regional planning. Many activities, including planning, were relegated to a position of secondary importance compared to the health (and economic) emergency. This forced shift resulted in a range of ad hoc solutions, simultaneously showing the drawbacks of previously never-tested tools and new opportunities to join the participatory process. The virtual experiences produced under the pandemic will be even more relevant in the future (of planning) than their chance to serve as participation processes during the lockdown. The objective of this research was to investigate the problems of and opportunities for virtual public participation tools and evaluate them with a view to their integration into the official public participation system in urban and regional planning.

This research is, therefore, based on the following hypotheses:

Hypothesis 1 (H1). *The ad hoc development and application of virtual participatory tools, which could not be previously tested, highlight problems and obstacles in the public participation process.*

Hypothesis 2 (H2). *Regardless of their ad hoc development and application, virtual participatory tools show certain advantages that make them worthy of becoming standardized tools in public participation that can even be implemented beyond the present COVID-19 times.*

Concerning research design, this research is based on case-controlled studies evaluated by the expert assessment method. Seven countries were analyzed as the case-controlled studies in order to identify the possible predictors of positive outcomes of using virtual participation tools in public participation in urban and regional planning. Case-controlled studies are also referred to as observational studies where, in this particular research, seven experts in urban and/or regional planning (authors of the paper) assessed the problems of and opportunities for virtual participation and the prospects of its application in the post-COVID era.

The experts and, hence, the cases were selected by the convenience method with reasoning. Specifically, due to the new pandemic circumstances, the ISOCARP (International Society of City and Regional Planners) organized a virtual collaboration platform (Cyber Agora) for its members. At the point of the COVID-19 outbreak, the gathering of urban and regional planers from all over the globe was considered the best opportunity to recruit professionals in this field, all affiliated with the ISOCARP and being interested and experienced in dealing with participation before and during the COVID-19. On the occasion of the second meeting devoted to "Virtual Public Planning Participation: Hype or New Normal?", the participants were invited to collaborate on the preparation of this study. The authors of this paper are those who accepted and went through the preparation processes. In addition to the convenience of the selection, the choice of case studies also corresponded to the basic reliability criteria: (1) encompassing different sociopolitical backgrounds, (2) involving cases with different lengths of experience in public participation, and (3) addressing experts from both academia and practice.

As a result, the subjects of this research are Belgium, Canada, Germany, Peru, Serbia, South Africa, and Spain. The study is designed as exploratory research in which data collected for the analysis include expert observations and assessments by the authors of the paper, as well as assessments reported in the academic literature and media on understanding the implications for public participation in urban and regional planning during the COVID-19 pandemic. Primary data (the authors' observations and assessments) were collected by survey, technically supported by the Google Form platform. Survey-based research is conducted without predetermined notions of the expected responses [68]. Such surveys are common in social and psychological research and are often used to describe and explore human behavior [69]. The questionnaire in this study consisted of 42 questions divided into four major topics: (1) main characteristics of the planning systems, (2) formal/informal occurrence of virtual public participation, (3) public participation challenges occurring during the COVID-19 pandemic, and (4) public participation opportunities developed during the COVID-19 pandemic. More than half of the questions were open-ended, whereas the remaining questions were multiple-choice with the possibility of entering original answers ("Other: . . . "). This way, a balance was enabled between new knowledge and the comparability of the answers.

A qualitative evaluation was made of the collected data. Despite the possibility of quantifying some of the answers collected by the survey, the fact that the research involves only seven countries influenced the decision to evaluate the results qualitatively. The responses were interpreted narratively, with the frequent use of phrases characteristic for qualitative research (e.g., "predominantly", "mostly", "hardly", "on average", "majority"). In addition, the categorization method was used to identify the main groups of challenges to and opportunities for virtual participation.

3. Results

3.1. Challenges to Virtual Public Participation

The challenges to virtual participation in urban and regional planning are not necessarily related to the shift from traditional to virtual platforms. In general terms, COVID-19 caused timeline, deadline, and process delays; however, despite a certain level of adaptation to the new context, virtual participation also suffered from the insufficiencies inherited from traditional forms of participation. This included the challenge of reaching all relevant stakeholders (particularly vulnerable groups), as well as participants' knowledge and understanding of planning procedures and aims.

However, experiences from the analysed countries indicate that virtual public participation during the COVID-19 pandemic has met specific challenges. These challenges can be categorised as follows: (1) accessibility, (2) reliability, and (3) trustworthiness.

This section may be divided by subheadings. It should provide a concise and precise description of the experimental results, their interpretation, as well as the experimental conclusions that can be drawn.

3.1.1. Accessibility

The accessibility issue refers to inequalities in terms of participation from the moment when the internet and technical equipment (computers and mobile phones) became an essential prerequisite for participation in planning. Even though it was not intentional, experiences from the analyzed countries show that accessibility to virtual participation was hindered by (a) an urban–rural divide, (b) a wealthy–poor divide, (c) a young–elderly divide, and (d) a gender divide. According to the survey, Canada, South Africa, and Peru were challenged the most, while Belgium successfully diminished the problem.

Since the pandemic, it should be noted that rural municipalities have had difficulty participating because of poor access to the internet. To address this inequity, postcards of notification and hard copies of documents were sent to be picked up at certain locations in Canada. The significant difference in infrastructure levels between urban and rural areas in Serbia includes the road infrastructure, water supply, sewage system, and accessibility to the Internet [70]. As rural areas are usually populated by an older population, the accessibility problem increases. In Spain, internet access of the elderly population (65–74 years old) is 69.7%, which is significantly smaller compared to the average of 93.2% (16–74 years old) [71]. As in many world cities, the COVID-19 pandemic also highlighted the digital divide and inequalities within Canadian communities. Wi-Fi hotspots and public spaces, public libraries, and restaurants providing Wi-Fi access became unavailable during COVID-19 lockdown periods, whereby a large number of residents were left without internet access.

In Canada, online inequality and poor access to the internet were problematic not only for seniors but also for people with disabilities and BIPOC (Black Indigenous People of Color) in low-income neighborhoods. Limited access to digital platforms is, thus, a common feature. In most cases, it was the vulnerable and disadvantaged communities that were most affected by internet access during the lockdown. In rural communities, there are often fewer internet services and lower internet service speeds compared to urban centers [72]. In Spain, it was determined that 91.4% of Spanish households have access to the internet, but the COVID-19 pandemic highlighted that low-income families in this country have access mostly through mobile devices. Although this might seem adequate, it was found that this situation leads to a wider divide.

The inclusiveness of marginalized groups in some societies remains a global challenge. There are cultures where not all community members can speak up during public meetings or have access to mobile phones or the internet. Therefore, their attempts to reach out to these groups digitally are more challenging.

Inequality in terms of accessibility is further magnified by uneven digital literacy, where elderly and rural populations in particular show lower levels of computer literacy. In this sense, digital infrastructure, access, and literacy became very relevant for virtual public participation during the COVID-19 lockdown. In Serbia, for example, more than half of the population (51.0%) over 15 years of age is computer-illiterate, with substantial regional differences [73]. The Government of Ontario (Canada), therefore, allocated 150 million CAD to expand broadband internet across the province, as part of a previous 315 million CAD plan developed to improve digital connectivity in rural communities, emphasizing the high costs associated with technology advances, but also the importance of leadership and political will in addressing inequality in technology accessibility.

3.1.2. Reliability

Due to its ad hoc development and considerable improvisation, virtual participation still has not had the opportunity to solve technical and methodological issues. Therefore, a range of challenges were recognized, starting with (a) the gap between traditional and virtual participation, (b) the lack of virtual participation methodologies, (c) poorly targeted participants, (d) the rigid discussion structure, (e) anonymity, (f) the lack of informal discussion, (g) "self-service", and (h) the lack of reliance on body language. Some countries, such as Belgium, found a new efficient way to involve citizens, which works well despite the initial skepticism. In contrast, South Africa and Peru reported the greatest difficulties.

There is limited knowledge of virtual participation methodologies. Virtual engagement can be negatively and automatically characterized as being inequitable. Judging by the experience in South Africa, virtual communication is emphasized more than virtual participation. Large Zoom meetings are possible but expensive; therefore, not all interested parties can participate at once. This diminishes the value and the purpose of the discussion. This is often viewed as a lesser investment of resources and is, thus, equated with a lesser effort to engage diverse and representative participants within a community. In Serbia, citizens were deprived of the public presentation of the Spatial Plan of the Republic of Serbia. This and similar deficiencies in supportive documentation and data are a barrier to proper presentation of the plans and, hence, meaningful participation.

Due to the sudden shift from traditional and well-known methods of participation materials and participation itself, some stakeholder groups and citizens were left out. The narrower scope of participants was risky in terms of the involvement of all target groups, which led to the provision of false information.

Online voting (survey) on a specific aspect of a plan provides easy access, but it is also usually anonymous. Thus, it is a safe place to express opinions, but the accountability of the contributors should be carefully considered, since the outcomes of "self-service" further guide public spending and affect others. The City of Belgrade (Serbia) also tested the possibilities of such virtual participation through the Green City Action Plan for the City of Belgrade, which was envisioned as a participatory procedure to define a vision, mission, and main objectives. The primary idea was to gather various parties in a workshop, which was dismissed after the COVID-19 crisis accelerated. To enable the timely completion of the project, the consultant and the City decided to provide public online voting where citizens could suggest various proposals. The submitted proposals showed a lack of understanding of the term "vision". This example pointed out the great relevance of well-trained moderators in leading the participation process, since they are excluded by default in forms of virtual participation such as online voting (survey).

Similar findings were recorded in Spain with a preliminary participatory process conducted in Santander in July 2020 during the lockdown period. The process constituted in-person and online activities, but the latter lacked explanations. Moreover, limited space was provided for participants to include personal opinions and suggestions, thus limiting participation and failing the objective of being citizen-driven.

Furthermore, the apathetic nature of virtual participation means that it is not possible (in most cases) to read body language, further limiting meaningful conversation and the reliability of the participation results. Virtual discussions tend to have more rigidly structured questions compared to in-person discussions; thus, they do not lead to discussions outside their realm that may be important depending on the sociocultural circumstances. In digital media, there may be a degree of abstraction and social isolation that can reduce the joy associated with a group effort, which discourages potential participants from taking part.

In Lima, different results were achieved when the participatory processes of the Metropolitan Development Plan (which started in late 2019) were taken online. There were very limited discussions, and this created further uncertainty for the stakeholders. In Serbia, especially in rural areas, it was also found that stakeholders were bewildered by the unexpected transition to virtual means of discussion. In Munich (Germany), on the other hand, strict planning regulations and formal processes lacked the freedom of participatory and cocreation formats. As mentioned in the ongoing Smart Together program "not all local problems can be solved with the smart city project, which can lead to frustrations of residents" [74]. Other cities, however, do embrace the benefits of participatory planning. For example, in Antwerp (Belgium), the enormous resistance to finishing the city ring grew into a collaboration with Ringland authorities, which resulted in the ring being covered in strategic places, such that the city is connected to its spacious surroundings by parks.

3.1.3. Trustworthiness

Remote communication and "self-service" participation only accentuate the issue of trust between governments and experts, on the one hand, and participants, on the other. In the survey conducted for this analysis, the experts identified the following aspects that harm trustworthiness in virtual public participation: (a) depersonalized participation, (b) give-up rate, (c) mistrust in the corrupted system, (d) fear of misuse of personal data, (e) interference of hackers, and (f) appearance of fewer efforts. The most concern regarding these aspects was shown in South Africa, whereas it was very little in Serbia and Spain, and completely insignificantly in Belgium.

Even when public administration offers online services, some citizens might still prefer to use traditional (in-person) ways to communicate with the local administration, arguing that it might provide an opportunity for more interaction and clarification regarding intentions and attitudes. There is often the lack of a "human touch" when face-to-face engagements are replaced by virtual encounters. Members of some communities feel more confident sharing details with people of the same gender or ethnicity. Virtual participation does not seem as inviting as face-to-face collaborations, and a preference for tête-à-tête communication deters some participants from involvement in virtual participation. Video call meetings might make it easier to be present, but they do not necessarily enhance participation, especially since some participants tend to keep their microphone muted and video off, thus depersonalizing their participation to the maximum. A facilitator would tend to skip such participants during the discussion and would, thus, limit the meaningfulness of the participation.

Inclusive virtual participation could face more challenges in cities where there is poor trust in governance. For example, in Lima (Peru), significant omissions are evident in city planning due to corruption, weak coordination mechanisms by the Metropolitan Governance, and the lack of an official term for planning in the Peruvian legislation. Research on the biosphere reserve in Serbia has shown that trustworthiness and overall willingness to participate in planning processes highly depend on the implementation success of previous projects based on a participatory approach. The perception that a complex matter is difficult to do online, along with mistrust in the use of personal data on the internet, plays an important role when attracting people to take part in virtual participation. When it comes to the use of technologies, especially applications, citizens showed skepticism regarding personal data manipulation, while participation hosts encountered problems caused by hackers.

3.2. Opportunities for Virtual Public Participation

Virtual participatory planning approaches do, however, provide various opportunities to improve public participation. Some of the opportunities are even universally applicable to all three groups of challenges—accessibility, reliability, and trustworthiness. The case of the town of Hildesheim (Germany) has shown that online participation in times of a pandemic can certainly work and also lead to good and usable results. The number of clicks and participants proved that a high number of people could be reached [75].

Virtual platforms were also found to make it easier to engage with larger audiences. In Lima, for example, given the COVID-19 pandemic, citizens' and architects' participation in virtual conferences, seminars, and forums developed by the College of Architects of Peru greatly increased. This was seemingly due to the elimination of commuting to the venue, as well as growing interest in topics about the city's future.

In 2019, in Rivas Vaciamadrid, a small city in Spain with 88,150 inhabitants, a digital platform for participatory budgeting attracted approximately 200 proposals, followed by 3000 votes. The platform enabled an easy evaluation process, resulting in 14 actions being duly selected for funding, worth 300,000 EUR. In Munich, it became evident that a combination of working offline and online can lead to better results, for example, optimizing workshops using the MunichApp in combination with existing events and local bottom-up initiatives. However, the overall impression in Spain is to be cautious, since the opportunities for virtual participation during COVID-19 could lose their relevance in post-COVID times. After all, the lockdown unencumbered many people who had more time to participate in online meetings. This does not necessarily mean that, in the future, virtual activities will not have the same number of participants.

Therefore, Spain is the only example that does not see shifts during the COVID-19 pandemic as a sound opportunity to improve accessibility to participation through lower costs in organizing (no expenses for the venue and catering) and participating (no need to travel), as well as a significantly smaller time requirement when participating from a personal computer/mobile phone. Even though the reduced costs and lower time requirement were identified as opportunities in the majority of the analyzed countries, they did not significantly influence the number of participants in Serbia and Spain. However, if a virtual meeting had taken place in the case of Serbia's National Spatial Plan, the results would have been different, because an opportunity would have been provided for citizens outside of the capital city to participate, thus having a much better inclusion of rural and remote areas.

Similarly, it was noticed that young people are more comfortable using new technologies than older people. Therefore, virtual participation balanced out young people with experienced citizens in Canada, Germany, Peru, and South Africa, but not in Belgium, Serbia, and Spain. In the latter three countries, a misconception was noted about the digital knowledge of the elderly who are eager to learn, especially since COVID-19, and who wish to improve their information and gain inspiration. In addition, the youngest generations were not included, since having technological skills without experience does not necessarily mean an improvement in participation. The target group of participants was aged 30–50 years.

Representatives of all the analyzed cases were nevertheless unified on the matter that some applications (e.g., Zoom) are an opportunity to improve trustworthiness by running polls and presenting the results during meetings. Virtual participation processes assist with the facilitation of meetings, as the platform ensures that all voices are equal, rather than the frequent scenario where the "loudest person is heard". Six out of seven experts estimated that user-friendly interface applications could enhance participatory approaches and quality control, as well as present information effectively and transparently via virtual platforms, thus contributing to accessibility, reliability, and trustworthiness. Virtual platforms could ensure that reachability is extended even further, especially through asynchronous processes and by providing the option to view the recording of events at a convenient time and day. Virtual platforms make it possible to extend inclusiveness and overcome many of the traditional barriers to participatory processes.

The research identified that virtual platforms often lead to promptly obtained results. Voting for scenarios can be done virtually (graphically, with videos telling the story and explaining the scenarios). Virtual participatory platforms enable measurable results from the input of all participants, and these inputs are formally acknowledged through the digital structures. The city of Munich, for example, has created a smart data platform for the project Smarter Together known as the Transparency Dashboard, which gives interested people a clear and comprehensive overview of what kind of data are being collected in the project, how they are processed, and the measures taken to protect data and privacy, among others [74]. These digital advances improve transparency and, as a direct result, enhance trustworthiness. Two experts agreed that virtual platforms provide an opportunity to build trust in their countries, four of them thought that virtual platforms help to overcome language barriers, and three of them thought that COVID-19 is an opportunity to rethink "business as usual".

3.3. Legacy for Post-Pandemic Times

The pandemic has forced many municipalities to move toward digital participatory methods to continue developing their ongoing urban projects. Some of them have had the opportunity to try different digital tools and assess them on the basis of their own experience. There have been some difficulties given the novelty of these methods, but a considerable number of municipalities are satisfied with implementing online methods. They were able to reach more people with good levels of participation. At the same time, they have seen the multiple possibilities offered by online tools and how to combine them with well-known offline participatory methods.

They have become more aware of the importance and need for virtual participation, as experienced in Belgium, Germany, Peru, South Africa, and Serbia. The advantages, such as the easier organization of online discussions, as well as data collection (opinions) via websites, e-mails, and digital registries, have made governments realize that methods developed to deal with the specific conditions of the pandemic are also practical for post-pandemic times. For example, in Germany, public participation, which used to be primarily analog, is now inevitably becoming largely digital. All over Germany, cities such as Hildesheim, Duisburg, Dortmund, Munich, Frankfurt, and Lindau are calling for digital participation formats for the development of planning areas [76]. Numerous cities have developed online participation formats such as online meetings, online voting, and increasing communication on social media.

Governmental actions in Canada went beyond awareness and the idea of new possibilities, since amendments were made to the legislation. For example, Ontario's Bill 197 (Chapter 18 of the Statutes of Ontario) [77] is an Act to amend various statutes in response to COVID-19 and to enact, amend, and repeal various statutes, Bill 189–Chapter 6 [64] is an Act to amend various Acts to address COVID-19, and Ontario Regulation 149/20 Special Rules Relating to Declared Emergency [65] is the provincial legislation that sets the rules for land-use planning in Ontario.

In any case, there is still a long way to go in most other countries when it comes to the implementation of these new tools for urban planning purposes. In South Africa, for example, the issue is limited funds and willingness to invest in such infrastructure, whereas, in Serbia, there is a fear that the government will use the inclusion of online meetings as justification for the exclusion of in-person dialogue between experts and citizens in public inspections. In Spain, a different understanding of the value of public participation in the whole process of planning (empowerment of neighbors) would be the first step toward the later implementation of all kinds of virtual participatory processes. Several experiences, such as in the cities of Rivas Vaciamadrid, Santander, Oviedo, or Albacete, despite the focus on smaller city areas, show the worthiness of the participatory approach.

All experts nevertheless agreed that the most successful aspects of public participation created during the COVID-19 pandemic should remain part of regular planning practice. The main arguments in favor of this were as follows: they do not contravene the application of a traditional approach; online meetings are an opportunity for those who cannot participate due to time constraints to see the recorded video; new virtual participation and public discussion can reach a greater number of citizens along with a different target of stakeholders; digital submissions of remarks and suggestions might be easier for citizens, governments, and consultants; a digital record of entries is easier to sort and integrate into public reports. The development of a hybrid model of public participation is seen as a win–win situation for all.

Out of 13 opportunities recognized in the analysis, experts from Canada, Germany, and South Africa estimated that more than 10 of these would be suitable as part of traditional planning in the future, and experts from the other countries chose between six and 10. There are two opportunities that all experts considered relevant to keep as part of regular practice: (1) recorded online meetings because they enable their content to be seen by groups of interested stakeholders that could not participate in the meeting, and (2) the use of applications that allow the creation of polls during the meetings so that results can be shared with participants immediately. The majority of the experts also considered that virtual participation is an excellent opportunity to enlarge the number of participants willing to get involved due to (1) lower costs for the participants (no need to travel), (2) smaller time demands (no traveling to the venue and back), and (3) an easier evaluation process enabled by the use of online platforms.

Less popular, but still chosen by the majority of experts, were opportunities to (1) balance out young people with experienced citizens because they are more comfortable using new technologies, (2) reduce costs for the organizers (no venue, catering, etc.), (3) enhance participatory approaches and quality control with the help of user-friendly interface applications, (4) rethink "business as usual", and (5) be more transparent by effectively presenting information via virtual platforms. Only a few experts considered that virtual platforms help to overcome language barriers and build trust between participants and government/consultants.

4. Discussion and Conclusions

This study showed that virtual participation in urban and regional planning faces similar drawbacks to those in traditional planning participation, such as the participants' knowledge of planning principles and goals, as well as the involvement of a large number of stakeholders from diverse sectors and spheres. As [78,79] noted, the majority of the population is not interested in active participation in decision making, but they often get involved in order to protect their private interests. They do not realize the benefits or the important contribution of citizen participation [80]. Compared to traditional ways of participation, participants' knowledge must include how to use the technology (computer/mobile phone) and applications installed on electronic devices. Although [81] stated that technological possibilities and skills no longer represent an obstacle, [82] noted that technological literacy is developing gradually, and that part of the population is still not familiar with or comfortable using ICT.

On the one hand, the use of technologies in the participation process opens the possibility of involving a larger number of people and, hence, a broader scope of stakeholders. This issue has also been broadly addressed in other sources. The authors of [2,11,15,19–21,83] considered that virtual participation enables more flexible involvement of stakeholders and saves time and money for both organizers and attendees. If these are the reasons why participants should take part in the processes, technology and accompanying software make it possible. Forms of participation that do not require discussion also help; potential participants can share their opinions at a time that suits them best via e-mail, online questionnaire, survey, online platform, etc. [15]. Virtual participation eliminates the dimension of physical distance [84] and reduces the CO_2 footprint due to reduced travel [83,85].

On the other hand, it has been noted by other researchers that using a solely virtual approach eliminates some participation from the public participation process [82], especially in the case of the elderly [7]. Those who are used to traditional participation forms, do not have a computer/mobile phone, or are not sufficiently familiar with the use of technology could choose or be forced to stay out of the participation process. Additionally, part of the population does not have internet access, which implies a digital divide when combined with the lack of technical equipment and knowledge [13,21].

Technical knowledge and internet access are closely related to the local context: individual and group access to the infrastructure and their social status. Thus, the digital divide results from the urban–rural [86], wealthy–poor [12,13], young–elderly [21,87], and gender divides [12,82,87–89], as recognized in this study and by other authors. It was also noted, as confirmed by [11], that the divide based on social status is the foundation for the divide based on skin color. The authors of [90] also registered the problem of disabled persons who might be included in the participation more easily with the introduction of technological means, but only if the interface is adapted to their requirements.

Therefore, overcoming technical obstacles is relevant to improving the possibility of marginalized groups' participation and overcoming difficulties in communication and understanding between participants from different milieux and social groups [2,7,19,33]. A serious level of improvement is also expected in the sphere of data protection, anonymity, and transparency on the technical side of the processes [21,91,92]. New and more visually attractive forms of plan presentation also increase responsiveness in the participation process and, to a certain extent, eliminate the absence of face-to-face communication between plan creators and other stakeholders [15,87]. This is crucial for target groups reluctant to enter virtual (nonpersonal) exchange without informal social interactions [7,19,82]. Since virtual exchange reduces the ability to ask questions [19], it leaves the virtual participants (if the meetings are mixed) on the margin of the conversation [93,94].

Other studies stated that some participants might withdraw from participation because they do not feel competent enough to share their opinion with professionals [95]. In contrast, some forms of virtual participation (e.g., voting, mobile applications) do not require a public display, thereby easing participation for those who feel embarrassed [96]. A simplification of planned design and its outcomes might be clearer to a wider range of participants through virtual reality (VR) and augmented reality (AR). As technology is an essential part of a smart city, it is natural that it plays a significant role in planning itself. The authors of [97,98] emphasized that VR and AR greatly assist in good quality planning because they enable authentic simulation and visualization of planned interventions in space, thus allowing participants to make the right decision when deciding to either accept or decline the offered design. VR and AR are, nevertheless, new tools in planning; therefore, they require further improvements to present the scenario more realistically and engage a wider range of stakeholders by improving easy access, simple understanding, level of amusement, etc. [99,100].

Some authors (e.g., [3,101–103]) argued that public participation suffers from formality, which does not fulfil its purpose of the true inclusion of stakeholders and citizens in the process. As noted by this study and confirmed by [12], this might significantly hinder participants' trust in the process and their motivation to get involved. New and virtual forms of presenting public discussion results (e.g., graphs, polls) increase transparency and, hence, trust [7]. In addition to the use of presentation forms, the technology enables video and audio recording [82], which greatly increases the dissemination, transparency, understanding of the discussed document, and the quality of the stakeholders' response in the ensuing participation rounds.

In conclusion, the COVID-19 pandemic has served as a change initiator in public participation, including participation in urban and regional planning. The changes were mainly directed toward the introduction of technology, with improvised and planned new forms of online participation. According to the results of this and other studies that dealt with the consequences of COVID-19 (e.g., [23,30,104]), new approaches and a shift toward virtual participation have already been considered the "new normal". In this sense, it is underlined that face-to-face communication and traditional forms of participation cannot be simply replaced by virtual forms of participation [2], but virtual participation offers new opportunities that were not possible in the traditional approach. Therefore, other studies (e.g., [9,12,13,17,81]) advocate a combined approach as the best solution for future participative processes, which is emphasized by the authors of this study, particularly for the field of urban and regional planning.

The shift toward a combined approach should flow in parallel with technological advances and their application in participation, along with the awareness-raising and education of all potential participants in the use of technology and the purpose of urban and regional planning in order to maximize the meaningfulness of the shift [4,15,87]. A combined approach in the post-COVID-19 era would spur a larger number of stakeholders and their diversity, which is considered the ultimate precondition for a successful and meaningful participatory process [2,3,7,12,82,95,105]. Before COVID-19, [2] explored technical and methodologically detailed solutions to improve virtual participation, which [81] conducted in the specific COVID-19 context. Their results can be used as a starting point when adjusting to a combined approach. Additionally, [7] noted that the practice of at least one face-to-face meeting should be kept, following the recommendations by [15] to hold the participation process in several rounds and by [87] to involve participation in the earliest stage of planning and keep it through the entire process.

Author Contributions: Conceptualization, M.P. and J.C.; methodology, M.P. and J.C.; software, M.P., J.C., G.C., F.M., O.O., S.T.M. and J.v.d.B.; validation, M.P., J.C., G.C., F.M., O.O., S.T.M. and J.v.d.B.; formal analysis, M.P.; investigation, M.P., J.C., G.C., F.M., O.O., S.T.M. and J.v.d.B.; resources, M.P., J.C., G.C., F.M., O.O., S.T.M. and J.v.d.B.; data curation, M.P.; writing—original draft preparation, M.P.; writing—review and editing, M.P., J.C., G.C., F.M., O.O., S.T.M. and J.v.d.B.; visualization, M.P.; supervision, M.P.; project administration, M.P. and J.C.; funding acquisition, M.P., J.C., G.C., F.M., O.O., S.T.M. and J.v.d.B. All authors read and agreed to the published version of the manuscript.

Funding: This research received no external funding.

Institutional Review Board Statement: Not applicable.

Informed Consent Statement: Not applicable.

Data Availability Statement: Data are available in a publicly accessible repository that does not issue DOIs. Publicly available datasets were analyzed in this study. These data sources can be found in the references.

Conflicts of Interest: The authors declare no conflict of interest. The funders had no role in the design of the study; in the collection, analyses, or interpretation of data; in the writing of the manuscript, or in the decision to publish the results.

References

1. Healey, P. The Communicative Turn in Planning Theory and its Implications for Spatial Strategy Formation. *Environ. Plan. B Plan. Des.* **1996**, *23*, 217–234. [CrossRef]
2. Hampton, S.E.; Halpern, B.S.; Winter, M.; Balch, J.K.; Parker, J.N.; Baron, J.S.; Palmer, M.; Schildhauer, M.P.; Bishop, P.; Meagher, T.R.; et al. Best practices for virtual participation in meetings: Experiences from synthesis centers. *Bull. Ecol. Soc. Am.* **2017**, *98*, 57–63. [CrossRef]
3. Maričić, T.; Cvetinović, M.; Bolay, J.C. *Participatory Planning in the Urban Development of Post-Socialist Serbia. A Support to Urban Development Process*, 1st ed.; École Polytechnique Fédérale de Lausanne (EPFL) Cooperation and Development Center (CODEV): Lausanne, Switzerland; Institute of Architecture and Urban & Spatial Planning of Serbia (IAUS): Belgrade, Serbia, 2018.
4. Cairns, G.; Wright, G. A reflection on the mass production of scenarios in response to COVID-19. *Futures Foresights Sci.* **2020**, *2*, 1–5. [CrossRef]
5. Gohari, S.; Baer, D.; Nielsen, B.F.; Glicher, E.; Situmorang, W.Z. Prevailing approaches and practices of citizen participation in smart city projects: Lessons from Trondheim, Norway. *Infrastructures* **2020**, *5*, 36. [CrossRef]
6. Hall, P. *Cities of Tomorrow: An Intellectual History of Urban Planning and Design in the 20th Century*, updated edition; Blackwell: London, UK, 2002.
7. Jaeweon, Y.; Sae-Woom, J.; Dongoh, H.; Seung-Won, K.; Ju-Chul, J. Post COVID-19 Visioning of Urban Comprehensive Plan through Citizen Participation: Focusing on the Citizen Participation of Busan Metropolitan City. *J. Korea Plan. Assoc.* **2021**, *56*, 156–168.
8. World Health Organization. Available online: www.who.int/emergencies/diseases/novel-coronavirus-2019 (accessed on 25 May 2020).
9. Rooij, R.; Aalbers, K.; Hausleitner, B.; Newton, C.; Rocco, R. Education for the resilient city—Teaching and learning urban design and planning in COVID-19 times. *Urban Des. Plan.* **2020**, *173*, 119–124. [CrossRef]
10. Azizi, M.; Azar, A.; Dehghan Nayeri, M. Participatory decision making in the post COVID-19 period. *Mod. Res. Decis. Mak.* **2020**, *5*, 165–192.
11. Valdez, E.S.; Gubrium, A. Shifting to Virtual CBPR Protocols in the Time of Corona Virus/COVID-19. *Int. J. Qual. Methods* **2020**, *19*, 1–9. [CrossRef]
12. Bricout, J.; Baker, M.A.P.; Moon, W.N.; Sharma, B. Exploring the Smart Future of Participation: Community, Inclusivity, and People with Disabilities. *Int. J. E-Plan. Res.* **2021**, *10*, 94–108. [CrossRef]
13. Graziano, T. Smart Technologies, Back-to-the-Village Rhetoric, and Tactical Urbanism: Post-COVID Planning Scenarios in Italy. *Int. J. E-Plan. Res. (IJEPR)* **2021**, *10*, 80–93. [CrossRef]
14. Liang, Z.C.; Ooi, S.B.S.; Wang, W. Pandemics and their impact on medical training: Lessons from Singapore. *Acad. Med.* **2020**, *95*, 1359–1361. [CrossRef]
15. Rajhans, V.; Rege, S.; Memon, U.; Shinde, A. Adopting a modified Delphi technique for revisiting the curriculum: A useful approach during the COVID-19 pandemic. *Qual. Res. J.* **2020**, *20*, 373–382. [CrossRef]
16. Norris, P. Virtual democracy. *Harv. Int. J. Press Politics* **1998**, *3*, 1–8. [CrossRef]

17. Mehmood, A.; Imran, M. Digital social innovation and civic participation: Toward responsible and inclusive transport planning. *Eur. Plan. Stud.* **2021**, *29*, 1870–1885. [CrossRef]
18. Olson, G.M.; Zimmerman, A. (Eds.) *Scientific Collaboration on the Internet*, 1st ed.; MIT Press: Cambridge, MA, USA, 2008.
19. Shirmohammadi, S.; Hu, S.; Ooi, W.T.; Schiele, G.; Wacker, A. Mixing virtual and physical participation: The future of conference attendance? In Proceedings of the IEEE International Workshop on Haptic Audio Visual Environments and Games (HAVE 2012) Proceedings, Munich, Germany, 8–9 October 2012; pp. 150–155.
20. Stanyer, J. The British Public and Political Attitude Expression: The Emergence of a Self-expressive Political Culture? *Contemp. Politics* **2005**, *11*, 19–32. [CrossRef]
21. Vromen, A. Building virtual spaces: Young people, participation and the Internet. *Aust. J. Political Sci.* **2008**, *43*, 79–97. [CrossRef]
22. Dubov, A.; Shoptaw, S. The value and ethics of using technology to contain the COVID-19 epidemic (Letter to the Editor). *Am. J. Bioeth.* **2020**, *20*, W7–W11. [CrossRef]
23. Torous, J.; Myrick, K.J.; Rauseo-Ripcupero, N.; Firth, J. Digital mental health and COVID-19: Using technology today to accelerate the curve on access and quality tomorrow. *JMIR Ment. Health* **2020**, *7*, 38848. [CrossRef] [PubMed]
24. Norris, P. *Digital Divide: Civic Engagement, Information Poverty and the Internet Worldwide*; Cambridge University Press: Cambridge, MA, USA, 2001.
25. Goodchild, M.F. Citizens as sensors: The world of volunteered geography. *GeoJournal* **2007**, *69*, 211–221. [CrossRef]
26. Gabrys, J. Programming Environments: Environmentality and citizen sensing in the smart city. *Environ. Planning. D Soc. Space* **2014**, *32*, 30–48. [CrossRef]
27. Papacharissi, Z. The virtual sphere: The internet as a public sphere. *New Media Soc.* **2002**, *4*, 9–27. [CrossRef]
28. Buheji, M.; Buheji, A. Planning Competency in the New Normal—Employability Competency in Post- COVID-19 Pandemic. *Int. J. Hum. Resour. Stud.* **2020**, *10*, 237–251. [CrossRef]
29. Park, H.-J. Urban Planning in Post-COVID-19 Era: Humanist Perspective Revisited. *J. Econ. Geogr. Soc. Korea* **2020**, *23*, 247–261.
30. Castka, P.; Searcy, C.; Fisher, S. Technology-enhanced auditing in voluntary sustainability standards: The impact of COVID-19. *Sustainability* **2020**, *12*, 4740. [CrossRef]
31. Mazzoleni, S.; Turchetti, G.; Ambrosini, N. The COVID-19 outbreak: From 'black swan' to global challenges and opportunities [Editorial]. *Pulmonology* **2020**, *26*, 117–118. [CrossRef] [PubMed]
32. Ting, D.S.W.; Carin, L.; Dzau, V.; Wong, T.Y. Digital technology and COVID-19. *Nat. Med.* **2020**, *26*, 458–464. [CrossRef] [PubMed]
33. Hudson-Smith, A.; Evans, S.; Batty, M. Building the virtual city: Public participation through e-democracy. *Know. Techn. Pol.* **2005**, *18*, 62–85. [CrossRef]
34. Thoneick, R. Integrating Online and Onsite Participation in Urban Planning: Assessment of a Digital Participation System. *Int. J. E-Plan. Res.* **2021**, *10*, 1–20. [CrossRef]
35. Anttiroiko, A.-V. Digital Urban Planning Platforms: The Interplay of Digital and Local Embeddedness in Urban Planning. *Int. J. E-Plan. Res.* **2021**, *10*, 35–49. [CrossRef]
36. Sameh, H.M.M. Public Participation in Urban Development Process through Information and Communication Technologies. Master's Thesis, Department of Urban Planning, Faculty of Engineering, Ain Shams University, Cairo, Egypt, 2011. Available online: www.cpas-egypt.com/pdf/Hend_Magdy/M.Sc/002Public-Participation-in-Urban-Planning.pdf (accessed on 26 November 2021).
37. Rio Declaration on Environment and Development. 1992. Available online: www.jus.uio.no/lm/environmental.development.rio.declaration.1992/portrait.a4.pdf (accessed on 26 November 2021).
38. Leipzig Charter on Sustainable European Cities (Leipzig Charter 2007). Available online: https://ec.europa.eu/regional_policy/sources/activity/urban/leipzig_charter.pdf (accessed on 26 November 2021).
39. Treaty of Lisbon. Available online: https://eur-lex.europa.eu/legal-content/EN/TXT/?uri=CELEX%3A12007L%2FTXT (accessed on 26 November 2021).
40. The New Leipzig Charter–the Transformative Power of Cities for the Common Good. Available online: https://ec.europa.eu/regional_policy/sources/docgener/brochure/new_leipzig_charter/new_leipzig_charter_en.pdf (accessed on 26 November 2021).
41. Hodge, G.; Gordon, L.A.D. *Planning Canadian Communities: An Introduction to the Principles, Practice, and Participants*, 5th ed.; Thomson Nelson Publishers: Toronto, ON, Canada, 2008.
42. Bundesbaugesetz (German Federal Building Act) vom 23. Juni 1960 Erster Teil: Bauleitplanung §2 Aufstellung der Bauleitpläne (6). Available online: www.stadtgrenze.de/s/bbg/1960/bbaug1960.htm (accessed on 12 July 2021).
43. Bundesbaugesetz (German Building Law) vom 18.08.1976 BGBI. I S. 2256 §2a Beteiligung der Bürger an der Bauleitplanung. Available online: www.bgbl.de/xaver/bgbl/start.xav?start=//*%5B@attr_id=%27bgbl176s2256.pdf%27%5D#__bgbl__%2F%2F*%5B%40attr_id%3D%27bgbl176s2256.pdf%27%5D__1628501890392 (accessed on 13 July 2021).
44. Europarechtsanpassungsgesetz Bau–EAG Bau (European Construction Law Adaptation Act) vom 24. Juni 2004, Stadtumbau §§171a-d Soziale Stadt §171e. Available online: www.bgbl.de/xaver/bgbl/start.xav?start=//*[@attr_id=%27bgbl104s1359.pdf%27]#__bgbl__%2F%2F*%5B%40attr_id%3D%27bgbl104s1359.pdf%27%5D__1628502700349 (accessed on 12 July 2021).
45. Busch, L. Bürgerbeteiligung in der Städtebaulichen Planung-das Beispiel der Kreisangehörigen Städte Schleswig-Holsteins. Ph.D. Dissertation, HafenCity Universität, Hamburg, Germany, 2009. Available online: https://repos.hcu-hamburg.de/bitstream/hcu/413/1/Diss_Busch.pdf (accessed on 7 July 2021).

46. Reglamento 2159/1978 de Planeamiento para el desarrollo y aplicación de la Ley sobre Régimen del Suelo y Ordenación Urbana (Planning Regulation for the development and application of the Law on Land Regime and Urban Planning). 1978. Available online: www.boe.es/eli/es/rd/1978/06/23/2159 (accessed on 12 July 2021).
47. Delgado Pérez, G. La participación ciudadana en el planeamiento urbanístico y el fenómeno de la teledirección: Análisis crítico sobre el proceso de formación de los planes urbanísticos en España y la incidencia de la opinión de los ciudadanos en su redacción, tramitación y gestión posterior. *Scr. Nova. Rev. Electrónica De Geogr. Y Cienc. Soc.* **2007**, *14*, 245.
48. Ley 6/1998 sobre régimen del suelo y valoraciones (Land Regime and Valorization Law 6/1998). Boletín Oficial del Estado No. 89, 14 April 1998. Available online: www.boe.es/eli/es/l/1998/04/13/6 (accessed on 5 July 2021).
49. Mozo Amo, J. El urbanismo y la participación ciudadana. *Encuentros Multidiscip.* **2015**, *50*, 1–8.
50. Ley de los Derechos de Participación y Control Ciudadanos. Ley 26300, Diario Oficial 'El Peruano (Rights of Citizen Participation and Control Law-Law No. 26300). Available online: https://www2.congreso.gob.pe/sicr/cendocbib/con5_uibd.nsf/76AA4A37BC523EA8052586DB00226C55/$FILE/LEY-26300.pdf (accessed on 15 July 2021).
51. JNE & UNDP. *Guia de Participación Ciudadana del Perú. Lima*; Jurado Nacional de Elecciones: Lima, Peru, 2008; Available online: www2.congreso.gob.pe/sicr/cendocbib/con4_uibd.nsf/B3465D6F1868627205257CD7005DE4B8/$FILE/1_pdfsam_Guia_de_participacion_ciudadana.pdf (accessed on 3 August 2021).
52. Guide on Public Participation in the Public Service; Department: Public Service and Administration—Republic of South Africa (DPSA). 2008. Available online: www.dpsa.gov.za/dpsa2g/documents/cdw/2014/citizenengagement.pdf (accessed on 13 July 2021).
53. Vitello, C. Updated Guide to Canada's Impact Assessment Act. 2020. Available online: https://environmentaljournal.ca (accessed on 15 July 2021).
54. Impact Assessment Act [Enacted by section 1 of chapter 28 of the Statutes of Canada 2019, in force 28 August 2019, see SI/2019-86]. Available online: www.laws-lois.justice.gc.ca/PDF.1-2.75.pdf (accessed on 15 July 2021).
55. Ley 7/1985 Reguladora de las Bases del Régimen Local (Law Regulating the Bases of Local Governments). 1985. Available online: www.boe.es/eli/es/l/1985/04/02/7/con (accessed on 12 July 2021).
56. Bräuer, M.; Biewendt, T. Elektronische Bürgerbeteiligung in deutschen Großstädten 2004. Website-Ranking Herausgegeben von der Initiative eParticipation. Available online: www.initiative-eparticipation.de (accessed on 7 July 2021).
57. Wiegandt, C.-C.; Lobeck, M.; Märker, O.; Wolf, K.; Häußler, J.; Christ, S. Webbasierte Medien in der Stadtentwicklung: Bürgerbeteiligung und Bürgerengagement in der Digitalen Gesellschaft; BBSR-Online-Publikation Nr. 28/2017. 2018. Available online: https://www.bbsr.bund.de/BBSR/DE/veroeffentlichungen/bbsr-online/2017/bbsr-online-28-2017-dl.html?__blob=publicationFile%3D1 (accessed on 10 October 2021).
58. Real Decreto Legislativo 7/2015 por el que se Aprueba el texto Refundido de la Ley de Suelo y Rehabilitación Urbana (Urban Rehabilitation Law 7/2015). Available online: www.boe.es/eli/es/rdlg/2015/10/30/7/con (accessed on 3 July 2021).
59. Ley 19/2013 de Transparencia, Acceso a la Información Pública y buen Gobierno (Transparency, Access to Public Information and Good Governance Law). Available online: www.boe.es/eli/es/l/2013/12/09/19/con (accessed on 15 July 2021).
60. Rubio, G.; González, C.; López, L.; Pau, I.; Anes, S. Experiencia de e-participación en Rivas Vaciamadrid. In *Proceedings CollECTeR Iberoamérica 2008*; Universidad Politécnica de Madrid: Madrid, Spain, 2008.
61. Law on Planning and Construction. 2009. Available online: www.pravno-informacioni-sistem.rs/SlGlasnikPortal/eli/rep/sgrs/skupstina/zakon/2009/72/11/reg (accessed on 2 July 2021).
62. Bill 189 An Act to Amend Various Acts to Address the Coronavirus (COVID-19), Coronavirus (COVID-19) Support and Protection Act, 2020, Chapter 6 of the Statutes of Ontario. 2020. Available online: https://www.ola.org/en/legislative-business/bills/parliament-42/session-1/bill-189 (accessed on 15 July 2021).
63. Special Rules Relating to Declared Emergency, Ontario Regulation 149/20 made under the Planning Act. Available online: www.omtario.ca/laws/regulation/r20149 (accessed on 15 July 2021).
64. Decreto Legislativo No.1412 que aprueba la Ley de Gobierno Digital (Public Participation and the Digital Government Law), 2018, pp. 4–8. Available online: https://cdn.www.gob.pe/uploads/document/file/353216/decreto-legislativo-que-aprueba-la-ley-de-gobierno-digital-decreto-legislativo-n-1412-1691026-1.pdf (accessed on 15 July 2021).
65. Ley de Desarrollo Urbano Sostenible (Sustainable Urban Development Law of 2021), 2018, pp. 2008–2010. Available online: https://busquedas.elperuano.pe/normaslegales/decreto-legislativo-que-aprueba-el-codigo-de-responsabilidad-decreto-legislativo-n-1348-1471548-8/ (accessed on 3 July 2021).
66. Manual para la Elaboracion de Planes de Desarrollo Metropolitano—PDM 2020 (Manual of Metropolitan Development Plans in 2020), 2020, p. 370. Available online: https://cdn.www.gob.pe/uploads/document/file/1481389/Manual%20para%20la%20Elaboraci%C3%B3n%20de%20Planes%20de%20Desarrollo%20Metropolitano%20-%20PDM.pdf (accessed on 3 August 2021).
67. Stadt + Grün. Available online: https://stadtundgruen.de/artikel/buergerbeteiligung-nicht-nur-in-zeiten-von-corona-15492.html (accessed on 3 July 2021).
68. Ponto, J. Understanding and Evaluating Survey Research. *J. Adv. Pract. Oncol.* **2015**, *6*, 168–171.
69. Singleton, R.A.; Straits, B.C. *Approaches to Social Research*, 5th ed.; Oxford University Press: New York, NY, USA, 2009.
70. Pantić, M. Sustainable Development Perspectives for Serbian Mountain Areas: Lessons from the European Context. Ph.D. Thesis, TU-Dresden, Dresden, Germany, 2014. Available online: https://nbn-resolving.org/urn:nbn:de:bsz:14-qucosa-144339 (accessed on 1 July 2021).

71. Instituto Nacional de Estadística Encuesta sobre Equipamiento y Uso de Tecnologías de Información y Comunicación en los Hogares, Notas de Prensa. 2020. Available online: www.ine.es/prensa/tich_2020.pdf (accessed on 3 July 2021).
72. Canadian Radio-Television and Telecommunications Commission (CRTC). Available online: https://crtc.gc.ca/pubs/cmr2019-en.pdf (accessed on 7 July 2020).
73. Statistical Office of the Republic of Serbia (SORS). *Educational Attainment, Literacy and Computer Literacy—Data by Municipalities and Cities*; SORS: Belgrade, Serbia, 2013.
74. Smarter Together. Available online: www.smarter-together.eu/projects/citizen-stakeholder-engagement-munich (accessed on 12 September 2020).
75. Stadt Hildesheim-Online-Bürgerbeteiligung in der Stadtplanung. Available online: www.hildesheim.de/rathaus/pressemeldungen/2021-03-09/online-buergerbeteiligung-in-der-stadtplanung.html?type=2# (accessed on 23 May 2021).
76. Urban Digital. Available online: https://urban-digital.de/digitalisierung-stadt-in-zeiten-covid-19-teil-2/ (accessed on 3 July 2021).
77. Bill 197 An Act to Amend Various Statutes in Response to COVID-19 and to Enact, Amend and Repeal Various Statutes, Chapter 18 of the Statutes of Ontario. Available online: https://www.ola.org/en/legislative-business/bills/parliament-42/session-1/bill-197 (accessed on 15 July 2021).
78. Putnam, R.D. Bowling alone: America's declining social capital. *J. Democr.* **1995**, *6*, 65–78. [CrossRef]
79. Hahn, J. *Soviet Grassroots: Citizen Participation in Local Soviet Government*; IB Tauris: London, UK, 1988.
80. Lee, J.J.; Kim, Y.S.; Kim, H.; Kim, D.Y. The Present Status and Government Officials Recognition on the Citizen Participation in Comprehensive Plan–Focused on Municipalities of Gyeonggi-do. *J. Urban Des. Inst. Korea Urban Des.* **2015**, *16*, 5–16.
81. Henry, S.M.; Bazilian, D.M.; Markuson, C. Just Transitions: Histories and Futures in a Post-COVID World. *Energy Res. Soc. Sci.* **2020**, *68*, 101668. Available online: www.sciencedirect.com/science/article/pii/S2214629620302437 (accessed on 17 September 2021). [CrossRef]
82. Osafo, E. Engaging Communities in Challenging Times: Lessons Learned from the Master Gardener Program during the COVID-19 Pandemic. *Adv. Dev. Hum. Resour.* **2021**, *23*, 75–87. [CrossRef]
83. Fox, H.E.; Kareiva, P.; Silliman, B.; Hitt, J.; Lytle, A.D.; Halpern, S.B.; Hawkes, V.C.; Lawler, J.; Neel, M.; Olden, D.J.; et al. Why do we fly? Ecologists' sins of emission. *Front. Ecol. Environ.* **2009**, *7*, 294–296. [CrossRef]
84. Jiminez-Zarco, A.; Gonzalez-Gonzalez, I.; Sagio-Rubio, F.; Torrent-Sellens, J. The co-learning process in healthcare professionals: Assessing user satisfaction in virtual communities of practice. *Comput. Hum. Behav.* **2014**, *51*, 1303–1313. [CrossRef]
85. Fraser, H.; Soanes, K.; Jones, S.A.; Jones, C.S.; Malishev, M. The value of virtual conferencing for ecology and conservation. *Conserv. Biol.* **2016**, *31*, 540–546. [CrossRef]
86. De Filippi, F.; Coscia, C.; Guido, R. From smart-cities to smart-communities: How can we evaluate the impacts of innovation and inclusive processes in urban context? *Int. J. E-Plan. Res.* **2019**, *8*, 24–44. [CrossRef]
87. Rikanović, M.; Zindović, M.; Đurić, N.; Lalić, M. *Analiza Urbanističkog Planiranja i Projektovanja u Srbiji sa Stanovišta Rodne Ravnopravnosti (Urban Planning and Design Analysis from the Gender Equity Perspective in Serbia)*; Žensko Arhitektonsko Društvo & UN Women: Belgrade, Serbia, 2020.
88. Goldberg, G. Rethinking the public/virtual sphere: The problem with participation. *New Media Soc.* **2011**, *13*, 739–754. [CrossRef]
89. Jon, I. Reframing postmodern planning with feminist social theory: Toward "anti-essentialist norms". *Plan. Theory* **2020**, *19*, 147–171. [CrossRef]
90. Kupper, H.; Banks, L.; Bright, T.; Davey, C.; Shakespeare, T. Disability-inclusive COVID-19 response: What it is, why it is important and what we can learn from the United Kingdom's response. *Wellcome Open Res.* **2020**, *5*, 79. [CrossRef] [PubMed]
91. De Wijis, L.; Witte, P.; Geertman, S. How smart is smart? Theoretical and empirical considerations on implementing smart city objectives–a case study of Dutch railway station areas. Innovation. *Eur. J. Soil Sci.* **2016**, *29*, 424–441. [CrossRef]
92. Yigitcanlar, T.; Desouza, K.C.; Butler, L.; Roozkhosh, F. Contributions and risks of Artificial Intelligence (AI) in building smarter cities: Insights from a systematic review of the literature. *Energies* **2020**, *13*, 1473. [CrossRef]
93. Tang, A.; Neustaedter, C.; Greenberg, S. VideoArms: Embodiments for Mixed Presence Groupware. In *People and Computers XX—Engage*; Bryan-Kinns, N., Blanford, A., Curzon, P., Nigay, L., Eds.; Springer: London, UK, 2007; pp. 85–102. Available online: https://link.springer.com/chapter/10.1007%2F978-1-84628-664-3_8 (accessed on 12 September 2021).
94. Harry, D.; Green, J.; Donath, J. Backchan.nl: Integrating Backchannels in Physical Space. In *Proceedings of the SIGCHI Conference on Human Factors in Computing Systems, Florence, Italy, 5–10 April 2008*; Association for Computing Machinery: New York, NY, USA, 2009; pp. 1361–1370.
95. Green, R.A. The Delphi technique in educational research. *SAGE Open* **2014**, *4*, 1–8. [CrossRef]
96. McKenna, K.Y.A.; Bargh, J.A. Coming out in the age of the Internet: Identity "demarginalization" through virtual group participation. *J. Personal. Soc. Psychol.* **1998**, *75*, 681–694. [CrossRef]
97. Jamei, E.; Mortimer, M.; Seyedmahmoudian, M.; Horan, B.; Stojcevski, A. Investigating the Role of Virtual Reality in Planning for Sustainable Smart Cities. *Sustainability* **2017**, *9*, 2006. [CrossRef]
98. Olszewski, R.; Gnat, M.; Trojanowska, H.; Turek, A.; Wieladek, A. Towards social fuzzy geoparticipation stimulated by gamification and augmented reality. In Proceedings of the 13th International Conference on Natural Computation, Fuzzy Systems and Knowledge Discovery (ICNC-FSKD), Guilin, China, 29–31 July 2017; pp. 1363–1370.

99. Imottesjo, H.; Kain, J.-H. The Urban CoBuilder—A mobile augmented reality tool for crowd-sourced simulation of emergent urban development patterns: Requirements, prototyping and assessment. *Comput. Environ. Urban Syst.* **2018**, *71*, 120–130. [CrossRef]
100. Fegert, J.; Pfeiffer, J.; Peukert, C.; Weinhardt, C. Enriching E-Participation through Augmented Reality: First Results of a Qualitative Study. In *Changing Landscapes—WI 2020—Entwicklungen, Chancen und Herausforderungen der Digitalisierung, Proceedings of the 15th International Conference on Wirtschaftsinformatik, Potsdam, Germany, 8–11 March 2020*; Gronau, N., Heine, M., Krasnova, H., Pousttchi, K., Eds.; GITO Verlag: Berlin, Germany, 2020; pp. 560–566.
101. Fraser, N. Transnationalizing the public sphere: On the legitimacy and efficacy of public opinion in a post-westphalian world. *Theory Cult. Soc.* **2007**, *24*, 7–30. [CrossRef]
102. Park, J.J.; Sung, H.J.; Park, C.H. Development and Application of Collective Spatial Decision Support System for Participatory Urban Planning. *Seoul Stud.* **2015**, *16*, 1–16.
103. Jeon, C.M.; Lee, H.C. The Trends and Characteristics of Engaged Urban Planning in Korea: Focused on 2030 Master Plans for Cheongju, Suwon, and Seoul. *Seoul Stud.* **2016**, *17*, 1–16.
104. Cilliers, J.; Simarmata, H.A.; Taufiqurrahman, T.; Olufemi, O.; Cimadomo, G.; Pantić, M.; Cap, C.; Treske, E.; Kumar, R.; Nakil, S.; et al. Virtual public participation during the COVID-19 crises. Hype or new normal for city planning? In Proceedings of the 56th ISOCARP World Planning Congress "Post-Oil City: Planning for Urban Green Deals", Doha, Qatar, 8–12 November 2020; pp. 1513–1525.
105. Nambisan, S.; Baron, R.A. Virtual customer environments: Testing a model of voluntary participation in value co-creation activities. *J. Prod. Innov. Manag.* **2009**, *26*, 388–406. [CrossRef]

land

Article

A Novel Approach for the Assessment of Cities through Ecosystem Integrity

Ian MacGregor-Fors [1,2,*], Ina Falfán [1,3], Michelle García-Arroyo [1,2], Richard Lemoine-Rodríguez [4], Miguel A. Gómez-Martínez [5], Oscar H. Marín-Gómez [1,6], Octavio Pérez-Maqueo [1] and Miguel Equihua [1]

1 Red de Ambiente y Sustentabilidad, Instituto de Ecología, A.C. (INECOL), Xalapa 91073, Mexico; ina.falfan@st.ib.unam.mx (I.F.); michelle.garciaarroyo@helsinki.fi (M.G.-A.); oschumar@iztacala.unam.mx (O.H.M.-G.); octavio.maqueo@inecol.mx (O.P.-M.); miguel.equihua@inecol.mx (M.E.)
2 Ecosystems and Environment Research Programme, Faculty of Biological and Environmental Sciences, University of Helsinki, 15140 Lahti, Finland
3 Laboratorio de Restauración Ecológica, Instituto de Biología, Universidad Nacional Autónoma de México, Ciudad de Mexico 04510, Mexico
4 Institute of Geography, Ruhr University Bochum, 44801 Bochum, Germany; Richard.LemoineRodriguez@rub.de
5 Instituto de Biotecnología y Ecología Aplicada (INBIOTECA), Universidad Veracruzana, Xalapa 91090, Mexico; zS18016054@estudiantes.uv.mx
6 Laboratorio de Ecología, Facultad de Estudios Superiores Iztacala, Universidad Nacional Autónoma de México, Tlalnepantla 05490, Estado de México, Mexico
* Correspondence: ian.macgregor@helsinki.fi

Abstract: To tackle urban heterogeneity and complexity, several indices have been proposed, commonly aiming to provide information for decision-makers. In this study, we propose a novel and customizable procedure for quantifying urban ecosystem integrity. Based on a citywide approach, we developed an easy-to-use index that contrasts physical and biological variables of urban ecosystems with a given reference system. The Urban Ecosystem Integrity Index (UEII) is the sum of the averages from the variables that make up its intensity of urbanization and biological components. We applied the UEII in a Mexican tropical city using land surface temperature, built cover, and the richness of native plants and birds. The overall ecosystem integrity of the city, having montane cloud, tropical dry, and temperate forests as reference systems, was low ($-0.34 \pm$ SD 0.32), showing that, beyond its biodiverse greenspace network, the built-up structure highly differs from the ecosystems of reference. The UEII showed to be a flexible and easy-to-calculate tool to evaluate ecosystem integrity for cities, allowing for comparisons between or among cities, as well as the sectors/regions within cities. If used properly, the index could become a useful tool for decision making and resource allocation at a city level.

Keywords: citywide; ecological integrity; integrity index; species retention; urban ecology; urbanization intensity

Citation: MacGregor-Fors, I.; Falfán, I.; García-Arroyo, M.; Lemoine-Rodríguez, R.; Gómez-Martínez, M.A.; Marín-Gómez, O.H.; Pérez-Maqueo, O.; Equihua, M. A Novel Approach for the Assessment of Cities through Ecosystem Integrity. *Land* **2022**, *11*, 3. https://doi.org/10.3390/land11010003

Academic Editor: Elizelle Juanee Cilliers

Received: 25 October 2021
Accepted: 9 December 2021
Published: 21 December 2021

1. Introduction

Urbanization leaves some of the most impressive imprints of the human population and activities on ecosystem structure and function [1,2]. Its effects, assessed from an urban-metabolism abstraction, extend far beyond city limits, posing pressing social, environmental, and ecological challenges [1,3]. Undoubtedly, cities have been proven to provide an important set of advantages in terms of economy, governance, and access to services (e.g., electricity, drinking water, health care, education [4]). Yet, cities are highly heterogeneous in many dimensions. Urban lifestyle not only has a profound impact on ecosystems [5–7], but it also extensively affects human health [8]. All the above makes it imperative to revisit the way in which we urbanize landscapes and manage our cities to build more biodiverse and livable cities, where efforts toward more sustainable practices result in resilient systems.

Urban heterogeneity and complexity have prompted research interest across many disciplines, from architecture, the humanities, and social sciences, to the natural sciences [2,9–12]. To tackle urban complexity, several indices have been proposed [13,14]. Such metrics most commonly aim to provide components to decision makers for them to have solid evidence upon which they can base their actions [15–17]. Some indices focus on different aspects of the urban environment and population, such as lifestyle, economy, ecosystem "health", and biodiversity, aiming to assess urban environmental conditions and resilience [18]. Among those indices, two are the most frequently used: the Ecological Footprint and the City Biodiversity Index (also known as the Singapore Index on Cities' Biodiversity). The Ecological Footprint index evaluates "urban sustainability" based on ecological and economic principles and concepts (such as carrying capacity and natural capital) by accounting for the area of biologically productive land and sea necessary to produce the renewable resources the urban population demands and to assimilate the generated waste [7]. On the other hand, the Singapore Index assesses biodiversity at a local level and provides cities with the ability to monitor their conservation efforts over time [19]; it consists of 23 indicators focused on three key components: native biodiversity, ecosystem-services provision, and governance and management [20,21]. However, an important area of opportunity in the approach to urban complexity has been overlooked, which is based on the ecosystem integrity concept [22,23].

Ecosystem integrity (also referred to as ecological integrity, biotic integrity, biological integrity, natural integrity, integrity of the landscape, or integrity of habitats) was initially defined as "the capability of supporting and maintaining a balanced, integrated, adaptive community of organisms having a species composition, diversity, and functional organization comparable to that of natural habitat of the region" [24] (p. 56). This vision encompasses elements such as the health, biodiversity, stability, sustainability, naturalness, wildness, and beauty of the ecosystems [15] in such a way that a system with integrity can withstand and recover from most perturbations, both natural and induced [24]. As an attribute of ecosystems [25], integrity can be measured through the sum of their chemical, physical, and biological integrities [15,24]. It is also considered as a tool for the management and restoration of ecosystems, for species and environmental protection, for providing services [15,25,26] (see Rohwer and Marris [27]), and for being a bridge among scientists, managers, policymakers, and the general audience for decision making [28,29].

Ecosystem integrity can be used from a local to a global level, in a wide array of systems [30–36], although its use has been higher in natural ecosystems than in human-dominated systems [36]. Considering this wide spectrum, it can actually be even more informative in highly disturbed systems if the reference systems are correctly identified [15,26]. In 2004, Noss [23] introduced the first adaptation of the concept of ecosystem integrity and its evaluation for urban systems, conceiving it as a comparative measure between the built-up areas and the surrounding "natural" systems taken as a reference within a biogeographic region. He clearly recognized that "although urban areas will never have the biodiversity, naturalness, and ecological resilience of pristine wilderness areas, there are reasonable standards they can meet" [23] (p. 4).

Bringing ecosystem integrity into the assessment of urban systems provides an opportunity to push our understanding of cities forward in comparison to previous urban metrics. As such, in this study, we propose a novel and customizable approach to quantify the integrity of urban ecosystems. Briefly, based on a citywide approach, we developed an easy-to-use index that takes as many variables as desired from the physical and biological dimensions of urban ecosystems, which, upon proper scaling and standardization, are contrasted with a given reference system. As citywide assessments are required to populate the index, ecosystem integrity values can be retrieved for entire cities—which can be contrasted to that of other cities—as well as sectors/regions within cities. To show how the index performs, we applied it to a Neotropical green city (Xalapa, Mexico), a widely studied urban system, with a set of variables that allowed us to test its use and performance; yet, as alerted above, the index is customizable and can be calculated with a different set of variables as long as they meet the criteria reviewed in Section 2.1.

2. Materials and Methods

2.1. Urban Ecosystem Integrity Index

To fully understand urban ecosystems, both positive and negative aspects, as well as the trade-offs between them, they have to be rigorously assessed [18]. The Urban Ecosystem Integrity Index (UEII) we propose here was conceived as a balance between two urban dimensions: the physical (i.e., urbanization intensity component) and the biological (i.e., species richness).

In the UEII, the urban intensity component subtracts from the ecosystem integrity, while the biological component adds to it (Figure 1). Regarding the urban intensity component, variables should summarize urban encroachment over a given site, such as built cover or human activities (measured directly and/or indirectly; e.g., vehicle traffic, anthropogenic noise, and urban heat islands). For the biological component, it is important to consider only the biodiversity that was recorded in the city and is part of the biota of the reference system or systems [23]. Thus, sites that are more similar in terms of reference biodiversity will indicate, at least partially, higher ecological integrity.

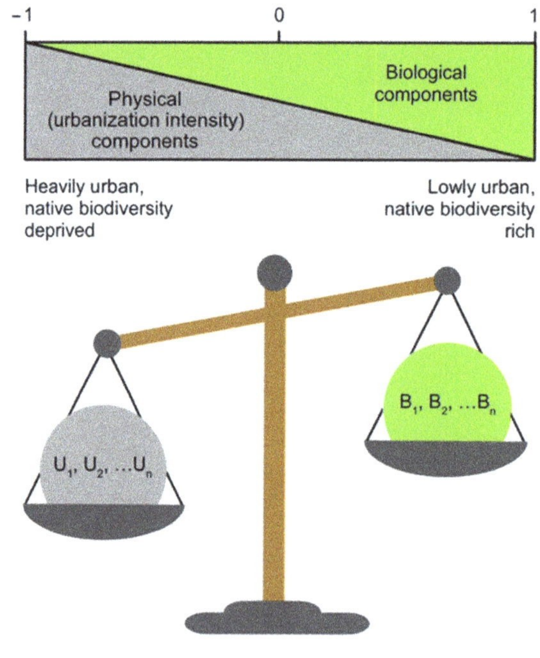

Figure 1. Graphical representation of the rationale behind the Urban Ecosystem Integrity Index (UEII), where U_1–U_n represent the variables that describe the physical (urbanization intensity) component and pull towards the negative values of the UEII and B_1–B_n are the variables describing the biological dimension of the urban continuum, drawing into the positive values of the index, which, in their totality, range from −1 to 1.

As a reference framework, a polygon of the study area is needed as a boundary to perform field surveys and to retrieve remote-sensing data. Although any polygon can be used, we strongly recommend considering the urban continuum, particularly the continuous built-up area of an urban system (see Falfán et al. [37] for an applied example). Mathematically, the UEII is the sum of the averages from all the variables that make up each dimension's component (whose amounts and magnitudes can vary between both components). Given that the nature of variables can differ greatly, not only conceptually but also mathematically, all variables used in the index are required to be standardized and re-scaled [38].

Finally, the UEII is computed at the pixel level (which needs to be determined a priori, most commonly given by methodological or technological constraints) for an entire urban continuum as follows:

$$\text{UEII} = \{[(U_1 + U_2 + \ldots U_n)/n_U]\,(-1)\} + [(B_1 + B_2 + \ldots B_n)/n_B], \quad (1)$$

where U_1–U_n represent the variables describing the physical (urbanization intensity) dimension, B_1–B_n describe the biological component, n_U is the number of variables used to describe urbanization intensity, and n_B represents that of the biological component. As urbanization components are considered features that erode ecological integrity, they are represented as negative values. On the other hand, biodiversity assets are deemed to be positive contributions. Hence, the range of values of the sum of both components is from −1 to 1 (Figure 2). It is of the utmost importance that the set of variables included to describe the physical and biological dimensions of cities are meaningful and comparable in case that comparative approaches are sought. In the variable selection process, we advise users to consider the potential redundancy (correlation), variance, and previous knowledge of the specific set of chosen variables, to assess crucial aspects of urban systems and to increase the value of UEII results. Although we do not have a suggested list of variables to include for the calculation of the index, we advise that all variables associated with the urbanization intensity component can be used as a proxy (e.g., built cover, temperature, noise, traffic) and those from the biological component follow the "bioindicator" profile (see Moreno et al. [39] for further details). Index values can be used for a single city or for multiple cities. In the case of applying the index in a comparative approach, the variables used to calculate the index ought to be the same across cities, as different proxies could lead to more differences in index values than those actually occurring between (or among) cities.

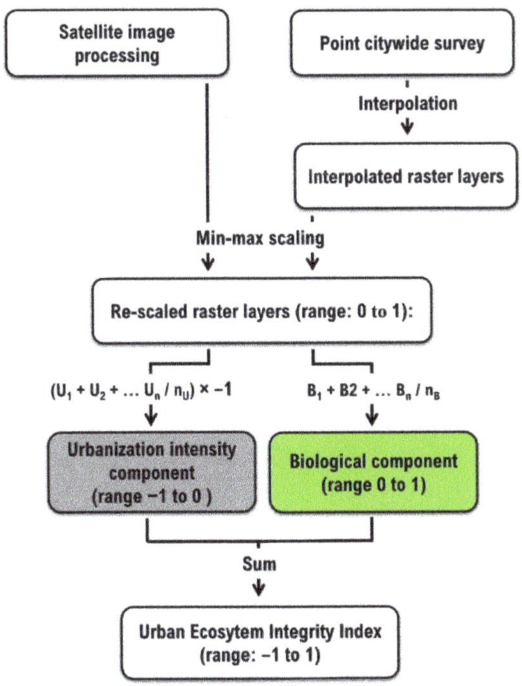

Figure 2. Flowchart for the quantification of the Urban Ecosystem Integrity Index (UEII).

2.2. Case Study: Xalapa

2.2.1. Study Site

We tested the applicability of the UEII in the Neotropical city of Xalapa, which is located in central Veracruz, Mexico (19°32′38″ N, 96°54′36″ W). The urban continuum of Xalapa (66.53 km^2; ~548,282 inhabitants [40]) settles on a transitional climatic region in such a way that it has two main climate types: temperate humid to the northwest and semi-warm humid to the southeast pp. 81-98 in [41].The vegetation of the city was originally comprised of montane cloud, tropical dry, and temperate forests [42], which were used as reference systems for the quantification of the UEII in Xalapa. The city has an important system of public and private vegetation that makes up almost 40% of the city's surface (Falfán et al. [37]). Urban vegetation is composed of local species from these ecosystem types, as well as non-native species from other vegetation types and regions [43,44]. Within its greenspaces and other land uses, Xalapa has been shown to be a city that shelters a sizeable number of wildlife species [45–47].

2.2.2. UEII for Xalapa

To apply the UEII to Xalapa, we firstly delineated its urban continuum. For this, we generated an updated polygon that encompassed the aggregated built-up infrastructure of the city in 2020, following the clustering and communication criteria specified by MacGregor-Fors [48] and Lemoine-Rodríguez et al. [49]. Then, we generated spatially explicit data for the four components that would be used to compute the UEII: land surface temperature (LST) and built cover for the urbanization intensity component and native plant and bird species richness for the biological component.

We retrieved LST (°C) from a Landsat 8 OLI/TIR satellite image with a spatial resolution of 30 m and no cloud cover over the city area (12 April 2019). We pre-processed the image by applying the dark-object subtraction (DOS) image-based atmospheric correction to the visible and infrared bands in order to obtain surface-reflectance values [50], employing the RStoolbox R package [51]. We conducted further processing steps to obtain LST by means of the R package LSTtools [52], used to process thermal data. We converted the digital numbers (DN) of the thermal infrared band 10 (TIR) to at-sensor spectral radiance (L_λ), employing the multiplicative and re-scaling factors contained in the image metadata, as suggested by the USGS [53]. We applied the modified normalized difference vegetation index (NDVI) threshold method [54] to estimate the per-pixel ground emissivity. According to this output, we defined fixed emissivity values for soil, vegetated, and mixed pixels. We derived the brightness temperature (BT) from the at-sensor radiance of the TIR band. After this, we corrected the BT values by using the Planck function to add the influence of the computed per-pixel emissivity and then converted the values from degrees Kelvin to degrees Celsius.

With the aim of extracting the built-up extent inside the city polygon, we employed an atmospherically corrected Sentinel 2 L2A image of 10 m spatial resolution (19 May 2019). We geographically adjusted the Sentinel 2 bands to the Landsat 8 image to make them spatially comparable and snapped the pixel grids based on their upper left corners. We computed the NDVI based on the red and near-infrared (NIR) bands [55]. Applying an NDVI threshold of 0.40, we removed the vegetation pixels. We manually excluded bare-soil and water pixels in order to retain only built-up features. Finally, we computed the total area in m^2 of the 10 m built-up pixels inside each 30 m LST pixel.

In the case of the biological components, we followed a standardized citywide sampling scheme [56] to describe native woody plants (i.e., trees and shrubs) and bird species richness across the city (114 sampling sites within 50 m radii of the sites). For the bird surveys, we performed one 10 min point count per site, which occurred from sunrise to four hours later, recording all birds seen or heard at each survey site (except overflying individuals; following Ralph et al. [57]). Surveys took place in April and May 2019 and again in January and February 2020 to encompass the bird species richness of both the breeding and wintering seasons. For the plant surveys, we recorded all woody species in street verges, median strips, and residential yards (only when visible from the streets in the latter case) located within 50 m of sampling sites during April and May 2019. We recorded as many species as possible in a period of ten minutes. When we were not able to identify species at the site, we collected a sample to later identify in the XAL Herbarium of Instituto de Ecología, A. C. We confirmed species identification, nomenclature, and distribution in Plants of the World Online (POWO [58]). Importantly, not all plants and birds were considered, but only those that would be found in the original regional ecosystems. Given that the biological data were based on a site-specific sampling scheme, we generated interpolated raster layers for both groups through the inverse distance weighting (IDW) interpolation method [59,60].

In order to have comparable data to populate the index, we re-scaled all four raster layers so that values ranged between 0 and 1. For this, we used the Min–Max transformation formula [38] as follows:

$$X_{\text{re-scaled}} = [(X - X_{\min})/(X_{\max} - X_{\min})] \tag{2}$$

Afterwards, we simply computed Formula (1) provided in Section 2.1 for the index, which was calculated for each 30 × 30 m pixel, as follows:

$$\text{UEII}_{\text{Xalapa}} = \{[(\text{LST}_{\text{re-scaled}} + \text{Built cover}_{\text{re-scaled}})/2]\,(-1)\} + [(\text{Native plant richness}_{\text{re-scaled}} + \text{Native bird richnes}_{\text{sre-scaled}})/2] \tag{3}$$

As aforementioned, pixels can be of any size for the UEII to work, but, in this case, the restriction was the LST, for which we could only retrieve values at that resolution. Thus, we geographically adjusted all layers to be spatially consistent (aligned) at that spatial resolution. For the computation of the input variables, as well as the UEII index, we used R [61] and QGIS 3.4 [62].

3. Results and Discussion

The computation of the UEII for Xalapa was informative, highlighting the urban ecology and biodiversity knowledge available to date [41,44,63–66]. The data gathered documented few variables, but were enough to assess the ecosystem integrity of the city. The average UEII for the city was −0.34 (±SD 0.32), showing that, although the city has been considered green and biodiverse outside its greenspace network, an important proportion of the city's sprawl highly differs from the reference ecosystem, as it is heavily urbanized (Figure 3). This can be clearly seen when contrasted with the vegetation cover in Xalapa (previously published by Falfán et al. [37] (p. 16)), where not all greenspaces have the same UEII value. Simultaneously, many of the streetscapes in the south are not fully covered by vegetation, but their UEII values are higher than many streetscapes in the northeastern region of the city.

The graphical representation of the UEII of Xalapa clearly shows higher ecosystem integrity along the greenspace network of the city, while the intermediate values mainly corresponded to built-up areas mixed with public and private greenspaces, including tree-lined streets. Finally, the lower ecosystem integrity values corresponded to densely built zones with sparse and scarce presence of vegetated areas and some exotic and/or invasive birds (Figure 3).

These results concur with previous findings for the city regarding its "gray" (highly urban) and "green" spaces. Actually, the greenspace network of Xalapa has been shown to retain an important part of the native plants and birds from montane cloud, tropical dry, and temperate forests, as well as other groups, such as butterflies, ants, and fungi, which could be further integrated into the UEII calculation [46,47,65,67]. Spatially, the south and the west of Xalapa showed higher values of ecosystem integrity since these zones have more, larger, and higher-quality greenspaces; this region of the city is considered to be the wealthiest one. The center and northeast showed lower integrity values, albeit with some dispersed spaces and streetscapes of higher integrity (Figure 3).

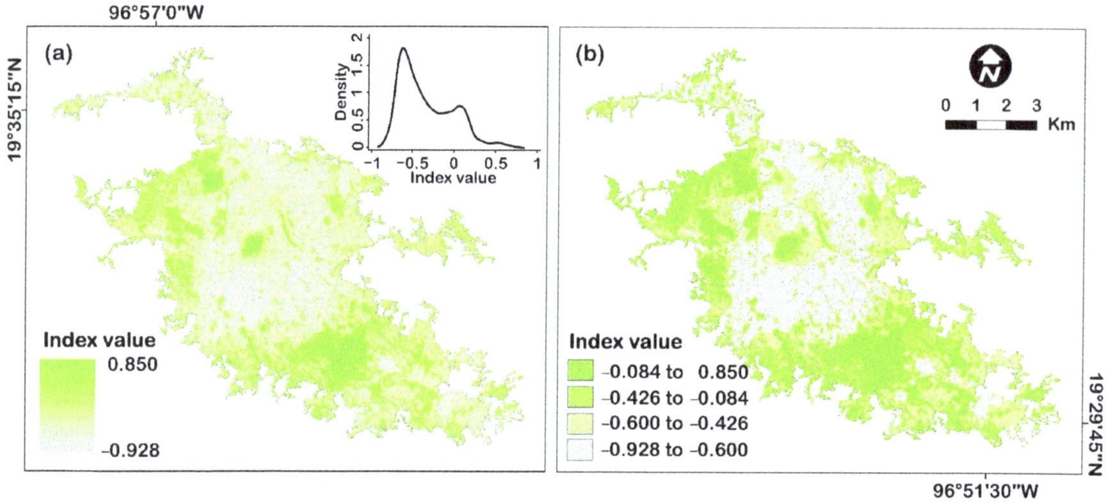

Figure 3. Spatially explicit representation of the Urban Ecosystem Integrity Index (UEII) in Xalapa (30 m × 30 m pixel resolution), showing (**a**) continuous results illustrating the distribution of the index values in a density plot and (**b**) its zoning by quartiles.

In the process of computing the index for Xalapa, we identified advantages and limitations, some of which stand out in relation to existent indices, including those focused on urban biodiversity [15,18,21,29]. Among the novel advantages of the UEII is its potential of assessing urban continuum without the constraints of administrative limits. Furthermore, as it is calculated in a spatially explicit way (at the pixel level), the UEII (1) deems cities to be heterogeneous in land cover and land use [5] and (2) can be obtained for all the administrative zones through which the urban continuum extends. Thus, it allows for intra-city comparisons that cannot be achieved with most existing indices since they are calculated considering the city as one homogeneous indivisible entity (e.g., Shaty and Reza [29]). Because the UEII is context-dependent (i.e., the index is a comparative measure with ecosystems of reference), it can be applied to any urban area worldwide, and the number of variables integrated into the index is decided by the researchers based on their knowledge of the cities under study and their intra-environment and surrounding ecosystems. Although previous studies have shown the type of variables suitable for the evaluation of the integrity of urban areas [29], for this index, there is not a fixed number of variables to fulfill. Hence, users can include as many variables as they consider appropriate for their city's conditions. In this study we used four variables—two for each component—which were sufficient for a first evaluation of the index itself and the ecosystem integrity of the urban continuum of Xalapa. When following comparative citywide approaches, UEII values can be contrasted by comparing their corresponding histograms (or density plots; as shown in Figure 3), as long as all variables are the same and were retrieved using similar standardized procedures.

One potential limitation for the calculation of the index is that input information must be spatially explicit. Nonetheless, in most cases, suitable data (apart from the biological component) exist at municipal, regional, or national scales. This aspect could represent a drawback if the biological component information is not available and needs to be generated, and if satellite imagery needs to be generated as well. Yet, there are increasing free options for both cases, such as citizen-based data and free imagery (e.g., remote-sensing information for academic purposes is now available at low or no cost provided by sources such as Google Earth Engine or the United States Geological Survey).

Furthermore, although we only used terrestrial variables in this assessment, variables regarding urban waterbodies (e.g., lakes, ponds, rivers, and dams, both natural and artificial) may also be included in the UEII since many cities worldwide have an important number of "bluespaces" within their limits. Such urban bluespaces can also be used as, or compared to, reference systems and inform on ecosystem integrity [24,30,34,68,69]. Notably, we did not integrate chemical, socioeconomic, or other physical variables as in other studies (e.g., Shaty and Reza [29]). However, these variables could be included within the urbanization intensity component. The addition of variables to either of the components of the UEII must be clearly specified to express the addition to or subtraction of ecosystem integrity. Additionally, since the UEII was conceived as a measure of ecosystem integrity across the urban continuum, it does not offer information about the effect of the city's metabolism on external ecosystems [7,18], and, therefore, is limited to its boundaries.

Other emerging features of the use of the UEII remain to be explored, such as its contribution to the evaluation of other aspects, such as: ecosystem services, management, conservation, or sustainability measures [15,18,22]; the pertinence of including other wildlife groups, physical variables, or abiotic measures [15,70]; or the validity and usefulness of comparing UEII results with either the same or different reference ecosystems, as has also been recommended for other indices [15,71].

4. Conclusions

By computing the UEII with the associated physical variables for the urbanization intensity dimension and biological component, we showed that the ecological integrity of Xalapa could be successfully measured when contrasted with its reference ecosystems. In this example, overall ecosystem integrity, considering montane cloud, tropical dry, and temperate forests as reference systems, was low. Nonetheless, the main greenspaces showed the highest values of ecosystem integrity, as expected.

In general, the UEII showed to be a flexible and easy-to-calculate tool to evaluate urban ecosystem integrity. Thus, the index can provide both basic research and applied management and decision-making uses. On the one hand, colleagues interested in the "ecology of the city" paradigm—aiming at understanding the complexity of entire cities—could use this approach to test hypotheses that require spatially explicit data. Index results can also be used as a metric of urban ecosystem integrity, interpreted according to the input variables. This quantitative metric can then be used as a predictor of dependent ecological variables of interest, falling within the "ecology in the city" paradigm. On the other hand, results of the index could be used for city-scale decision making and resource allocation in developing more sustainable, biodiverse, resilient, and livable cities, following the "ecology for the city" paradigm, seeking to apply the available evidence-based knowledge to action [72,73].

Author Contributions: Conceptualization, I.M.-F., O.P.-M. and M.E.; methodology, I.M.-F., I.F. and M.G.-A.; software, R.L.-R.; validation, I.M.-F. and R.L.-R.; formal analysis, I.F.; investigation, I.F., M.G.-A., R.L.-R., M.A.G.-M. and O.H.M.-G.; data curation, I.F. and R.L.-R.; writing—original draft preparation, I.M.-F., I.F., M.G.-A., R.L.-R., M.A.G.-M. and O.H.M.-G.; writing—review and editing, I.M.-F., I.F., M.G.-A., R.L.-R., M.A.G.-M., O.H.M.-G., O.P.-M. and M.E.; visualization, I.M.-F. and I.F.; supervision, I.M.-F.; project administration, I.M.-F.; funding acquisition, O.P.-M. and M.E. All authors have read and agreed to the published version of the manuscript.

Funding: This study was supported by Consejo Nacional de Ciencia y Tecnología (CONACYT) through a FORDECYT project: 296842.

Data Availability Statement: Data available on request.

Acknowledgments: We are most grateful to Eleanor Diamant for her comments and for proofreading the English grammar and wording. We also thank Miguel Á. Domínguez-López for his support with the plant surveys and plant identification, Carlos Durán-Espinosa for plant identification in Herbarium, and Rafael Rueda-Hernández for leading one of the bird survey seasons.

Conflicts of Interest: The authors declare no conflict of interest.

References

1. Elmqvist, T.; Andersson, E.; McPhearson, T.; Bai, X.; Bettencourt, L.; Brondizio, E.; Colding, J.; Daily, G.; Folke, C.; Grimm, N.; et al. Urbanization in and for the Anthropocene. *NPJ Urban Sustain.* **2021**, *1*, 6. [CrossRef]
2. Grimm, N.B.; Faeth, S.H.; Golubiewski, N.E.; Redman, C.L.; Wu, J.; Bai, X.; Briggs, J.M. Global change and the ecology of cities. *Science* **2008**, *319*, 756–760. [CrossRef] [PubMed]
3. Rees, W.E. Urban ecosystems: The human dimension. *Urban Ecosyst.* **1997**, *1*, 63–75. [CrossRef]
4. United Nations. *World Urbanization Prospects: The 2014 Revision. Highlights*; UN: New York, NY, USA, 2014.
5. Cadenasso, M.L.; Pickett, S.T.A.; Schwarz, K. Spatial heterogeneity in urban ecosystems: Reconceptualizing land cover and a framework for classification. *Front. Ecol. Environ.* **2007**, *5*, 80–88. [CrossRef]
6. Pickett, S.T.A.; Cadenasso, M.L.; Grove, J.M.; Nilon, C.H.; Pouyat, R.V.; Zipperer, W.C.; Costanza, R. Urban ecological systems: Linking terrestrial ecological, physical, and socioeconomic components of metropolitan areas. In *Urban Ecology: An International Perspective on the Interaction between Humans and Nature*; Marzluff, J.M., Shulenberger, E., Endlicher, W., Alberti, M., Bradley, G., Ryan, C., Simon, U., ZumBrunnen, C., Eds.; Springer: Boston, MA, USA, 2008; pp. 99–122.
7. Wackernagel, M.; Kitzes, J.; Moran, D.; Goldfinger, S.; Thomas, M. The ecological footprint of cities and regions: Comparing resource availability with resource demand. *Environ. Urban* **2006**, *18*, 103–112. [CrossRef]
8. Vlahov, D.; Galea, S. Urbanization, urbanicity, and health. *J. Urban Health* **2002**, *79*, S1–S12. [CrossRef]
9. Batty, M. The size, scale and shape of cities. *Science* **2008**, *319*, 769–771. [CrossRef]
10. Castells, M. *La Cuestión Urbana*, 14th ed.; Siglo XXI: Distrito Federal, Mexico, 1997.
11. Grimm, N.B.; Grove, J.G.; Pickett, S.T.; Redman, C.L. Integrated approaches to long-term studies of urban ecological systems. *BioScience* **2000**, *50*, 571–584. [CrossRef]
12. Rydin, Y.; Bleahu, A.; Davies, M.; Dávila, J.D.; Friel, S.; De Grandis, G.; Groce, N.; Hallal, P.C.; Hamilton, I.; Howden-Chapman, P.; et a Shaping cities for health: Complexity and the planning of urban environments in the 21st century. *Lancet* **2012**, *379*, 2079–2108. [CrossRef]
13. Klopp, J.M.; Petretta, D.L. The urban sustainable development goal: Indicators, complexity and the politics of measuring cities. *Cities* **2017**, *63*, 92–97. [CrossRef]
14. Westfall, M.S.; De Villa, V.A. (Eds.) *Urban Indicators for Managing Cities*; Asian Development Bank: Manila, Philippines, 2001.
15. Andreasen, J.K.; O'Neill, R.V.; Noss, R.; Slosser, N.C. Considerations for the development of a terrestrial index of ecological integrity. *Ecol. Indic.* **2001**, *1*, 21–35. [CrossRef]
16. Ferreira, V.; Barreira, A.P.; Loures, L.; Antunes, D.; Panagopoulos, T. Stakeholders' engagement on nature-based solutions: A systematic literature review. *Sustainability* **2020**, *12*, 640. [CrossRef]
17. Kabisch, N.; Frantzeskaki, N.; Pauleit, S.; Naumann, S.; Davis, M.; Artmann, M.; Haase, D.; Knapp, S.; Korn, H.; Stadler, J.; et al. Nature-based solutions to climate change mitigation and adaptation in urban areas: Perspectives on indicators, knowledge gaps, barriers, and opportunities for action. *Ecol. Soc.* **2016**, *21*, 39. [CrossRef]
18. Mori, K.; Christodoulou, A. Review of sustainability indices and indicators: Towards a New City Sustainability Index (CSI). *Environ. Impact Assess. Rev.* **2012**, *32*, 94–106. [CrossRef]
19. Niemelä, J. Ecology of Urban Green Spaces: The way forward in answering major research questions. *Landsc. Urban Plan* **2014**, *125*, 298–303. [CrossRef]
20. Chan, L.; Djoghlaf, A. Invitation to help compile an index of biodiversity in cities. *Nature* **2009**, *460*, 33. [CrossRef]
21. Pierce, J.R.; Barton, M.A.; Tan, M.M.J.; Oertel, G.; Halder, M.D.; Lopez-Guijosa, P.A.; Nuttall, R. Actions, indicators, and outputs in urban biodiversity plans: A multinational analysis of city practice. *PLoS ONE* **2020**, *15*, e0235773. [CrossRef]
22. Equihua Zamora, M.; García Alaniz, N.; Pérez-Maqueo, O.; Benítez Badillo, G.; Kolb, M.; Schmidt, M.; Equihua, J.; Maeda, P.; Álvarez Palacios, J.L. Integridad ecológica como indicador de la calidad ambiental. In *Bioindicadores: Guardianes de Nuestro Futuro Ambiental*; González Zuarth, C.A., Vallarino, A., Pérez Jiménez, J.C., Low Pfeng, A.M., Eds.; El Colegio de la Frontera Sur, Instituto Nacional de Ecología y Cambio Climático: Distrito Federal, Mexico, 2014; pp. 687–710.
23. Noss, R.F. Can urban areas have ecological integrity? In Proceedings of the 4th International Urban Wildlife Symposium on Urban Wildlife Conservation, Tucson, AZ, USA, 1–5 May 1999; University of Arizona: Tucson, AZ, USA, 2004; pp. 3–8.
24. Karr, J.R.; Dudley, D.R. Ecological perspective on water quality goals. *Environ. Manag.* **1981**, *5*, 55–68. [CrossRef]
25. Conway, T.M.; Almas, A.D.; Coore, D. Ecosystem services, ecological integrity, and native species planting: How to balance these ideas in urban forest management? *Urban For. Urban Green.* **2019**, *41*, 1–5. [CrossRef]
26. Ordóñez, C.; Duinker, P.N. Ecological integrity in urban forests. *Urban Ecosyst.* **2012**, *15*, 863–877. [CrossRef]
27. Rohwer, Y.; Marris, E. Ecosystem integrity is neither real nor valuable. *Conserv. Sci. Pract.* **2021**, *3*, e411. [CrossRef]
28. Karr, J.R.; Larson, E.R.; Chu, E.W. Ecological integrity is both real and valuable. *Conserv. Sci. Pract.* **2021**, e583. [CrossRef]
29. Shathy, S.T.; Reza, M.I.H. Sustainable cities: A proposed Environmental Integrity Index (EII) for decision making. *Front. Environ. Sci.* **2016**, *4*, 82. [CrossRef]
30. Beyene, A.; Legesse, W.; Triest, L.; Kloos, H. Urban impact on ecological integrity of nearby rivers in developing countries: The borkena river in highland Ethiopia. *Environ. Monit. Assess.* **2009**, *153*, 461–476. [CrossRef]
31. Blumetto, O.; Castagna, A.; Cardozo, G.; García, F.; Tiscornia, G.; Ruggia, A.; Scarlato, S.; Albicette, M.M.; Aguerre, V.; Albin, A. Ecosystem integrity index, an innovative environmental evaluation tool for agricultural production systems. *Ecol. Indic.* **2019**, *101*, 725–733. [CrossRef]

32. Hansen, A.; Barnett, K.; Jantz, P.; Phillips, L.; Goetz, S.J.; Hansen, M.; Venter, O.; Watson, J.E.M.; Burns, P.; Atkinson, S.; et al. Global humid tropics forest structural condition and forest structural integrity maps. *Sci. Data* **2019**, *6*, 232. [CrossRef]
33. Grantham, H.S.; Duncan, A.; Evans, T.D.; Jones, K.R.; Beyer, H.L.; Schuster, R.; Walston, J.; Ray, J.C.; Robinson, J.G.; Callow, M.; et al. Anthropogenic modification of forests means only 40% of remaining forests have high ecosystem integrity. *Nat. Commun.* **2020**, *11*, 5978. [CrossRef]
34. Li, P.; Zhang, Y.; Lu, W.; Zhao, M.; Zhu, M. Identification of priority conservation areas for protected rivers based on ecosystem integrity and authenticity: A case study of the Qingzhu River, Southwest China. *Sustainability* **2021**, *13*, 323. [CrossRef]
35. Hansen, A.J.; Noble, B.P.; Veneros, J.; East, A.; Goetz, S.J.; Supples, C.; Watson, J.E.M.; Jantz, P.A.; Pillay, R.; Jetz, W.; et al. Toward monitoring forest ecosystem integrity within the post-2020 global biodiversity framework. *Conserv. Lett.* **2021**, *14*, e12822. [CrossRef]
36. Zelený, J.; Mercado-Bettín, D.; Müller, F. Towards the evaluation of regional ecosystem integrity using NDVI, brightness temperature and surface heterogeneity. *Sci. Total Environ.* **2021**, *796*, 148994. [CrossRef]
37. Falfán, I.; Muñoz-Robles, C.A.; Bonilla-Moheno, M.; MacGregor-Fors, I. Can you really see 'green'? assessing physical and self-reported measurements of urban greenery. *Urban For. Urban Green.* **2018**, *36*, 13–21. [CrossRef]
38. So, A.; Joseph, T.V.; John, R.T.; Worsley, A.; Asare, D.S. *The Data Science Workshop: A New, Interactive Approach to Learning Data Science*; Packt Publishing Ltd.: Birmingham, UK, 2020.
39. Moreno, C.E.; Rojas, G.S.; Pineda, E.; Escobar, F. Shortcuts for biodiversity evaluation: A review of terminology and recommendations for the use of target groups, bioindicators and surrogates. *Int. J. Environ. Health* **2007**, *1*, 71–86. [CrossRef]
40. INEGI (Instituto Nacional de Estadística y Geografía) Censo de Población y Vivienda 2020. Available online: https://www.inegi.org.mx/programas/ccpv/2020/#Microdatos (accessed on 22 February 2021).
41. López-Moreno, I.R. (Ed.) *Ecología Urbana Aplicada a La Ciudad de Xalapa*; Instituto de Ecología, A.C., Man and the Biosphere Programme—United Nations Educational, Scientific and Cultural Organization (MAB UNESCO), H. Ayuntamiento de Xalapa, Veracruz: Veracruz, Mexico, 1993.
42. Castillo-Campos, G. *Vegetación y Flora Del Municipio de Xalapa, Veracruz*; Instituto de Ecología, A.C., Man and the Biosphere Programme—United Nations Educational, Scientific and Cultural Organization (MAB UNESCO), H, Ayuntamiento de Xalapa, Veracruz: Veracruz, Mexico, 1991.
43. Capitanachi, M.C.; Amante, H.S. *Las Áreas Verdes Urbanas En Xalapa, Veracruz. Catálogo de Flora Urbana*; Universidad Veracruzana, Instituto de Ecología A.C.: Veracruz, Mexico, 1995; Volume 2.
44. Falfán, I.; MacGregor-Fors, I. Woody Neotropical streetscapes: A case study of tree and shrub species richness and composition in Xalapa. *Madera Bosques* **2016**, *22*, 95–110. [CrossRef]
45. González-García, F.; Straub, R.; García, J.A.L.; MacGregor-Fors, I. Birds of a neotropical green city: An up-to-date review of the avifauna of the city of Xalapa with additional unpublished records. *Urban Ecosyst.* **2014**, *17*, 991–1012. [CrossRef]
46. MacGregor-Fors, I.; Avendaño-Reyes, S.; Bandala, V.M.; Chacón-Zapata, S.; Díaz-Toribio, M.H.; González-García, F.; Lorea-Hernández, F.; Martínez-Gómez, J.; Montes de Oca, E.; Montoya, L.; et al. Multi-taxonomic diversity patterns in a neotropical green city: A rapid biological assessment. *Urban Ecosyst.* **2015**, *18*, 633–647. [CrossRef]
47. MacGregor-Fors, I.; Escobar, F.; Rueda-Hernández, R.; Avendaño-Reyes, S.; Baena, M.; Bandala, V.; Chacón-Zapata, S.; Guillén-Servent, A.; González-García, F.; Lorea-Hernández, F.; et al. City "green" contributions: The role of urban greenspaces as reservoirs for biodiversity. *Forests* **2016**, *7*, 146. [CrossRef]
48. MacGregor-Fors, I. How to measure the urban-wildland ecotone: Redefining 'Peri-Urban' areas. *Ecol. Res.* **2010**, *25*, 883–887. [CrossRef]
49. Lemoine-Rodríguez, R.; MacGregor-Fors, I.; Muñoz-Robles, C. Six Decades of urban green change in a Neotropical city: A case study of Xalapa, Veracruz, Mexico. *Urban Ecosyst.* **2019**, *22*, 609–618. [CrossRef]
50. Chavez, P.S. Image-based atmospheric corrections—Revisited and improved. *Photogramm. Eng. Remote Sens.* **1996**, *62*, 1025–1036.
51. Leutner, B.; Horning, N.; Schwalb-Willmann, J.; Hijmans, R.J. RStoolbox: Tools for Remote Sensing Data Analysis (R Package Version 0.2.6). 2019. Available online: https://github.com/bleutner/RStoolbox (accessed on 15 May 2020).
52. Lemoine-Rodríguez, R.; Mas, J.-F. LSTtools: An R Package to Process Thermal Data Derived from Landsat and MODIS Images (v0.0.2). 2020. Available online: https://doi.org/10.5281/zenodo.4010732 (accessed on 18 May 2020). [CrossRef]
53. USGS (United States Geological Service). *Landsat 8 Data Users Handbook*; USGS Earth Resources Observation and Science: Sioux Falls, SD, USA, 2019.
54. Sobrino, J.A.; Jiménez-Muñoz, J.C.; Sòria, G.; Romaguera, M.; Guanter, L.; Moreno, J.; Plaza, A.; Martínez, P. Land surface emissivity retrieval from different VNIR and TIR sensors. *IEEE Trans. Geosci. Remote Sens.* **2008**, *46*, 316–327. [CrossRef]
55. Morawitz, D.F.; Blewett, T.M.; Cohen, A.; Alberti, M. Using NDVI to assess vegetative land cover change in central puget sound. *Environ. Monit. Assess.* **2006**, *114*, 85–106. [CrossRef]
56. MacGregor-Fors, I. De mitos a hitos urbanos: ¿Cómo hacer ecología en selvas de Asfalto. In *Manual de Técnicas Para el Estudio de Fauna Nativa en Ambientes Urbanos*; Zuria, I., Olvera-Ramírez, A.M., Ramírez Bastida, P., Eds.; REFAMA, Universidad Autónoma de Querétaro: Querétaro, Mexico, 2019; pp. 19–38.
57. Ralph, C.J.; Geupel, G.R.; Pyle, P.; Martin, T.E.; DeSante, D.F.; Milá, B. *Handbook of Field Methods for Monitoring Landbirds*; U.S.D.A., Forest Service, Pacific Southwest Research Station: Albany, CA, USA, 1996; p. 41.

58. POWO Plants of the World Online Kew Science. Available online: http://www.plantsoftheworldonline.org/ (accessed on 20 May 2019).
59. Kennedy, H. *Data in Three Dimensions: A Guide to ArcGIS 3D Analyst*; Thomson Delmar Learning: Clifton Park, NY, USA, 2004.
60. Longley, P.A.; Goodchild, M.F.; Maguire, D.J.; Rhind, D.W. *Geographic Information Systems and Science*, 2nd ed.; John Wiley & Sons: New York, NY, USA, 2005.
61. *R Core Team R: A Language and Environment for Statistical Computing*; R Foundation for Statistical Computing: Vienna, Austria, 2020.
62. QGIS Development Team. *QGIS Geographic Information System*; Open Source Geospatial Foundation Project; 2020; Available online: http://qgis.osgeo.org (accessed on 25 November 2021).
63. Dáttilo, W.; MacGregor-Fors, I. Ant social foraging strategies along a neotropical gradient of urbanization. *Sci. Rep.* **2021**, *11*, 6119. [CrossRef] [PubMed]
64. Gómez-Martínez, M.A.; Klem, D.; Rojas-Soto, O.; González-García, F.; MacGregor-Fors, I. Window strikes: Bird collisions in a neotropical green city. *Urban Ecosyst.* **2019**, *22*, 699–708. [CrossRef]
65. Escobar-Ibáñez, J.F.; Rueda-Hernández, R.; MacGregor-Fors, I. The greener the better! Avian communities across a neotropical gradient of urbanization density. *Front. Ecol. Evol.* **2020**, *8*, 500791. [CrossRef]
66. Ramírez Restrepo, L.; Halffter, G. Butterfly diversity in a regional urbanization mosaic in two Mexican Cities. *Landsc. Urban Plan* **2013**, *115*, 39–48. [CrossRef]
67. Méndez Romero, E.A. Alteraciones Térmicas Derivadas de La Urbanización En La Ciudad de Xalapa, Veracruz. Análisis Espacial y Temporal: 1982–2015. Master's Thesis, El Colegio de Veracruz, Xalapa, Mexico, 2017.
68. Li, J.; Li, Y.; Qian, B.; Niu, L.; Zhang, W.; Cai, W.; Wu, H.; Wang, P.; Wang, C. Development and validation of a bacteria-based index of biotic integrity for assessing the ecological status of urban rivers: A case study of Qinhuai River basin in Nanjing, China. *J. Environ. Manag.* **2017**, *196*, 161–167. [CrossRef] [PubMed]
69. Kim, J.-J.; Atique, U.; An, K.-G. Long-term ecological health assessment of a restored urban stream based on chemical water quality, physical habitat conditions and biological integrity. *Water* **2019**, *11*, 114. [CrossRef]
70. Carignan, V.; Villard, M.-A. Selecting indicator species to monitor ecological integrity: A review. *Environ. Monit. Assess.* **2002**, *78*, 45–61. [CrossRef]
71. Singh, R.K.; Murty, H.R.; Gupta, S.K.; Dikshit, A.K. An overview of sustainability assessment methodologies. *Ecol. Indic.* **2009**, *9*, 189–212. [CrossRef]
72. Pickett, S.T.A.; Cadenasso, M.L.; Childers, D.L.; McDonnell, M.J.; Zhou, W. Evolution and future of urban ecological science: Ecology *in*, *of*, and *for* the city. *Ecosyst. Health Sustain.* **2016**, *2*, e01229. [CrossRef]
73. McDonnell, M.J.; MacGregor-Fors, I. The Ecological Future of Cities. *Science* **2016**, *352*, 936–938. [CrossRef] [PubMed]

land

MDPI

Article

Using "Live" Public Sector Projects in Design Teaching to Transform Urban Green Infrastructure in South Africa

Christina Breed [1,*] and Helge Mehrtens [2]

[1] Department of Architecture, School of the Built Environment, Faculty of Engineering, Built Environment and Information Technology, University of Pretoria, Hatfield Campus, Pretoria, Hatfield 0028, South Africa
[2] eThekwini Municipality, Durban 4001, South Africa; h.mehrtens@gmx.de
* Correspondence: ida.breed@up.ac.za

Abstract: Urban green infrastructure is not acknowledged in the Global South for the critical social and ecological functions it can provide. Contextual design solutions and innovative approaches are urgently needed to transform the status quo. University-local government collaboration could be a way to encourage new thinking, new roles and design skills to develop solutions to these complex problems. This paper presents a case study analysis of such a collaboration. Qualitative research was conducted to establish the degree to which the exposure to real-life projects stimulates postgraduate design students' transformative learning. The researchers also inquired into the benefits of the collaboration for the municipality. The participants' reflections were recorded by means of anonymous questionnaires. The findings show that the live project created a municipal setting for seeking alternative solutions in design processes and outcomes. For the students, the project created rich social dynamics and an interplay of familiarity and uncertainty, which aided transformative learning. The students' deeper learning indicates greater social empathy, reconsidering the role of the profession, greater design process flexibility, and learning and valuing skills across disciplines. The findings hold promise for a more just and sustainable future built environment through collaborations that transform the design professionals involved, the outcomes they pursue, and the processes they follow.

Keywords: Global South; local municipality; urban green infrastructure; ecosystem services; students; live projects; design; experiential learning; spatial transformation; built environment

Citation: Breed, C.; Mehrtens, H. Using "Live" Public Sector Projects in Design Teaching to Transform Urban Green Infrastructure in South Africa. *Land* **2022**, *11*, 45. https://doi.org/10.3390/land11010045

Academic Editors: Sarel Cilliers, Thomas Panagopoulos and Elizelle Juanee Cilliers

Received: 19 November 2021
Accepted: 17 December 2021
Published: 29 December 2021

Publisher's Note: MDPI stays neutral with regard to jurisdictional claims in published maps and institutional affiliations.

1. Introduction

In several parts of the Global South, high biodiversity, alongside rapid urbanization and a lack of basic infrastructure, dominates [1–3]. The nature of spatial expansion, the prevalence of informal economies and settlements, combined with the demographic profiles of cities, increase risks [4,5]. These social and environmental challenges will worsen as the effects of climate change progressively manifest [5–7]. Built environment practitioners need to be trained to engage with informality, urban poverty, infrastructure and service provision, land management and local governance [5]. Staying within planetary boundaries [8] (which entails net-zero emissions, avoiding biodiversity loss and stopping land-system change), while meeting basic needs for all, will require some profound shifts in how cities are built, maintained, and operated. This calls for a reimagining of service and infrastructure provision, including harnessing green infrastructural solutions instead of "grey" solutions that contain embedded emissions and other ecological costs. Design skills are ever more critical to develop innovative responses to the current multifaceted challenges [9].

The inclusion of urban green infrastructure (UGI), as part of the infrastructural services of urban development, remains incidental in the Global South [6,10]. The UGI requires purposeful planning, design, and managed interventions to provide a network of green spaces that are accessible, multifunctional, and of high quality [11,12].

African cities are still shaped by historic (colonial) planning traditions [13]. For example, African Anglophone cities tend toward urban sprawl, which contains newer developments with less intense land use and a more irregular layout [13]. This is because colonial and native sections developed without an overall plan and coordination [13]. These legacies result in the perpetuation of township layouts, plot sizes, and building typologies that are often disconnected from current urban realities and how people actually live [14], including informal economies and settlements [5]. Similarly, building forms that are promoted internationally seldom relate to the local needs, climatic conditions, plot size and layout, land tenure, and rules [15].

In a neoliberal market, developers, planners, and municipalities often favor land-use types that provide direct economic benefits for some [16,17]. In so doing, they often disregard the welfare of the poor and the fragmentation of ecological areas [18]. While planners in general show a low to moderate knowledge and awareness of ecological aspects [19], the public sector is additionally tested by a lack of capacity and governance [20]. Due to competing interests and a shortage of funds, green areas and conservation plans that are included in open space planning frameworks do not always influence site development plan decisions on the ground [3]. Combined with a lack of service delivery, informality becomes the norm. Green spaces are therefore often unlawfully occupied for living [21] or used for illegal dumping. Development or densification almost inevitably happens on available, well-located land. When recreational green spaces form part of these formal developments, the requirements for high use, low cost and low maintenance result in a small palette of tried and tested design solutions and a lack of innovation [22].

Several authors discuss the importance of partnerships between different stakeholders (including transdisciplinary work) in contributing to improved UGI planning [23–25] to ease institutional and capacity constraints, overcome the divide between science and practice, and bring about systemic change [26–28]. Roux et al. [28] offer an excellent case study analysis for transdisciplinary researchers that provide lessons to improve the chances of project success for systemic change. Their examples are of academic research collaboration with the community and the public sector, and include postgraduate research students [28]. Their study highlights the role of students as "bridging agents" due to their neutrality and genuine interest in local issues [28], but does not consider how the engagement shapes the professional formation of the students as part of the required systemic change. Another study, by Jabeen et al. [29], focuses particularly on approaches to build the skills of architectural design students. It considers community collaboration as a way to improve the students' design ability to be rational, creative, and more socially responsive. This study describes the design outcomes of students in dealing with a challenging socio-ecological context as an explicit process of critical reflection, but does not inquire into the students' extrarational development. A study by Sara [30] considers live projects for architectural design students. It considers both rational and extrarational learning, but without contemplating the relation of the findings to the context for systemic change. Our study combines ambitions from these former studies by focusing on postgraduate planning and design students in collaborative partnerships as future professionals that need to learn the skills to bring about the systemic and intuitional changes required to manage particular risks and operate within planetary resources.

Design faculties increasingly aim to broaden design relevance beyond aesthetic self-expression and originality [29,31] and prepare students to deal with global and contextual challenges [9]. Although good planning and design "value" standards exist, designers do not always have the opportunity or inclination to utilize them [32]. Furthermore, in the Global South, design challenges are more complex, involving poor communities located in ecologically vulnerable areas such as flood plains and steep slopes [3]. Case study evidence of UGI integration in mainstream African urban planning and design practice exist [33,34], yet empirical evidence in the literature remains scarce [19]. Education, training, and professional development initiatives must help forge a common language of risk-conscious, greener alternatives for future spatial planners and designers [6,35,36].

To co-evolve their understanding of a socio-ecological issue [26], built environment designers could benefit from participation in live projects in the public sector during their education [29–31]. Specifically advocated for vocational education in architecture, live projects have been used as a successful way to bridge education and practice [37]. The real-world conditions of live projects foster experiential learning that could be deep and transformative [30,38,39]. The educational benefits include adapting to a fast-changing world [40], bridging the science-implementation gap [27] and improving participants' understanding of themselves and their professional aspirations [29,31]. However, planning and design guidelines do not manifest during experience; they must be inferred through questioning preconceptions and critical reflection on the consequences of action [30]. For this reason, live projects and experiential learning for designers have been criticized for diminishing creativity when a reflective process is absent [41,42]. Live projects can also hinder problem-solving abilities if insufficient guidance is given to participants [43], while the project and outcomes are harder to control and predict, entailing greater risks for educators [30].

The need for a more experimental approach through "learning-by-doing" has been newly emphasized by UGI designers [44,45] and the public planning sector [7]. This approach requires a willingness for risk taking [7], adapting to change, uncertainty, and ambiguity [45], exposing participants to diverse values and insights, and establishing environments where learning is directed at relevant and meaningful problems [29,31]. Adult learning consists of modifying, transforming, and integrating knowledge and skills [46]. For adults, experiential learning could foster transformative learning if learners are led to self-examination to become contextually aware [30,47]. This psycho-social approach to learning replaces technical content with personal transformation as the driver of learning [48]. Experiential learning entails experiences in the context of others and reflection [40], while transformative learning is a shift in consciousness in terms of beliefs and practices [49]. This dual process could lead to the development of much-required new roles, relationships, or actions [49] to transform the current way cities are planned and built.

This paper focuses on the combined quest of local municipalities and future professionals (students) to improve UGI planning and design. Design students at postgraduate level are trained to scrutinize value standards and explore creative solutions. As one possible, partial solution to institutional inertia, we bring together current and future professionals—with the support of academics—to explore creative solutions for UGI. We believe this can open the horizons of current municipal staff who are exposed to new ways of thinking and have access to creative alternatives. It can also equip students to deal with 21st-century systemic challenges, including those involved in navigating municipal bureaucracies. This paper responds to two research questions: Is there value for municipalities in a Global South context to involve postgraduate students in live projects for UGI design? To what degree does the specific learning environment created by the live project and its context stimulate transformative learning processes for the students (as future professionals)?

To answer these questions, we present a case study analysis of a collaborative live project in eThekwini, South Africa. The case study context presents many of the typical African and Global South UGI challenges, along with a municipal ambition to improve both the state of social livelihoods and ecological infrastructure. We then present the live project objectives, the student design project and the research design. We continue by conveying and discussing the findings and offering some reflections to assist future collaborative partnerships.

2. Method

2.1. Study Context and Challenges

South Africa's urban population of 66.4% in 2018 is projected to reach 79.8% by 2050 [2]. Most of the urbanization is taking place in the form of informal settlements where people are living in poverty and where basic facilities and services are missing [50]. Added to these are biodiversity hotspots [51] and projected climate change, manifesting as extremes in temperature and flooding patterns [52]. The task of professionals and local government to manage trade-offs between social livelihoods and ecological infrastructure is complex. The legacy of former Apartheid planning still dominates urban areas in South Africa [18], manifesting as an unjust distribution of green assets across race and income [12]. Planned as dormitories, former "townships" still need mobility and green infrastructure to turn them into well-connected, healthy mixed-use neighborhoods [53]. Township areas under Apartheid planning had few developed green spaces, while vacant space operated as a barrier to planned racial and economic opportunity [53]. Open spaces are therefore often still perceived and experienced negatively by the communities in which they are situated, and green spaces do not feature greatly in academic, policy, and civil society discourse [54]. The lack of substantial change in the inefficient and inequitable spatial layout of cities is largely due to institutional inertia, economic forces and weak spatial management [55]. Municipal management responsibilities for UGI are not well coordinated, while development funding is scant [20]. Better integrated city-level decision-making, municipal leadership, and improved technical capabilities could lead to much-required change [55].

This situation persists, while the socio-economic importance that quality open spaces offer could be significant for surrounding developments and could be catalysts for economic development [4]. In addition to a great need for well-designed, multifunctional UGI [23], existing recreational green spaces are not meeting community needs [56]. Recent calls appeal to private planning professionals and authorities alike to address UGI injustices by promoting activities and representations of an urban nature that reflect residents' world views through more inclusive planning and design processes [12,18,53].

Case Study: KwaMashu, Durban, eThekwini Municipality

The eThekwini Municipality, which encompasses the city of Durban, lies within the KwaZulu-Natal Coastal Belt vegetation type, with high rainfall (up to 1200 mm) and temperatures (mean maximum 32 °C and minimum 5 °C) [57]. Originally occupying subtropical forested areas, the vegetation today is affected by an intricate mosaic of sugar cane fields, timber plantations, and urban development, with interspersed secondary grasslands, thickets, and patches of coastal thornveld [57]. Durban was the first city in South Africa to apply a Metropolitan Open Space System (MOSS) approach in city planning [58]. Subsequently, the city has built a reputation as a leader in developing a climate adaptation strategy [59].

The eThekwini Municipality's good reputation is based on a community and ecosystem-based adaptation (CEBA) approach [60]. The strong and early focus on adaptation sets the city apart from many other urban climate change initiatives globally [7]. In 2015, the Municipal Council approved the Durban Climate Change Strategy [7]. It aims to provide locally based solutions that not only reduce climate change vulnerability, but also address the socio-economic challenges faced by residents in developing green jobs and improving people's livelihoods [60]. Consequently, several CEBA-focused projects have been initiated. These have followed a "learning-by-doing" model of development and implementation, and show a "willingness to experiment" [7].

Nonetheless, the eThekwini Municipality, similar to other local municipalities, has had to admit that current policy, law, governance, and environmental management efforts have been inadequate to prevent the degradation of the UGI [61]. The municipality is facing various challenges around open space.

These first entail the open space legacy of Apartheid. Second, the present-day poor management and maintenance of open spaces encourage crime, littering, and dumping. Third, detailed accurate spatial information on open spaces is hardly available, which leads to vacant spaces without ownership. Furthermore, there is a lack of assigned responsibility within departments to assess and address service delivery gaps, synchronize open space service delivery across departments, and develop and activate open spaces [55,62]. Despite these challenges and shortfalls, dedicated individuals in the municipality remain solution driven and realize significant feats.

The Bridge City KwaMashu Open Space Project follows the CEBA approach and stresses the importance of UGI for sustainable township regeneration in the KwaMashu Urban Hub Regeneration Project, with the newly developed Bridge City in its heart. The project received significant funding from National Treasury with a grant that aims to stimulate private investment in former township areas [63]. The Open Space Project leverages the large built infrastructure projects related to the Bridge City development. The Open Space Project within the Regeneration Project aims to bring about social and environmental sustainability and resilience [60].

KwaMashu is one of the many township areas created during Apartheid through forced removals and the relocation of people [60]. The precinct faces several challenges, such as poverty and a high unemployment rate [60]. Moreover, it has been identified as a crime hotspot [60].

The KwaMashu Urban Hub Regeneration Project covers approximately 1500 ha with about 135,000 inhabitants (see Figure 1). The population size of the Regeneration Project is similar to urban centers in sub-Saharan Africa (less than 250,000) that accounts for 140.7 million urban dwellers [64], thereby providing scope for learning by other African urban centers of similar size and deprivation levels. The greater project area houses nearly one million inhabitants, which is almost a quarter of the greater eThekwini's marginalized resident population [65]. The Regeneration Project includes visions for a consolidated commercial node and enhanced public transport interchange zone, along with existing residential and industrial precincts [61]. Officially launched in 2007, the Bridge City development is still on its way. Key grey infrastructure, such as a train station, as well as a mall and a hospital, have been built and are meant to attract further investment.

Figure 1. An aerial photograph showing the Bridge City KwaMashu Open Space Project as a "regional park" that is located within the KwaMashu Urban Hub Project. Source: [61].

The Bridge City development is located on the ridge above the adjacent Piesang's River valley within the historical Apartheid "buffer zone" [60]. Apartheid planning took advantage of the natural separation of areas through topography and riverine systems to enforce racial segregation. The name "Bridge City" was chosen to stress the concept of bridging this divide (see Figure 2).

Figure 2. View from the KwaMashu residential area towards the new Bridge City development taken across the Piesang's River flood plain. Only the train bridge crosses the vast wetland that has become an insurmountable barrier to pedestrians. Source: Gesellschaft fur Zusammenarbeit (GIZ), 2018.

One of the challenges for the Bridge City development has been precisely the crossing of the Piesang's River flood plain [61]. Over a length of nearly 2 km, there is no pathway or pedestrian bridge over the Piesang's River to allow the population of old-town KwaMashu to safely cross the river and steep slopes to reach the facilities of Bridge City [65]. Additionally, part of the protected biodiversity of the Durban MOSS green space is in a state of neglect and is perceived to be a dangerous, no-go area (see Figure 3).

Figure 3. The Piesang's River flood plain, viewed from KwaMashu towards Bridge City. Urban agriculture can be seen in the foreground, with the cemetery in the middle left. Source: GIZ, 2018.

As in other sub-Saharan cities with urban expansion into flood-prone areas [3], the Piesang's River valley is one of the largest of a few remaining patches of open space in the KwaMashu area. Nearly the entire valley falls within the 100-year flood line [66], which needs to be respected in any design solution for recreational use. A delineation and assessment of the wetland found the flood plain wetland system to be critically modified as a result of the transformation of catchment hydrology [66]. These impacts have resulted in major losses in ecosystem services, and riverine and wetland habitat. The majority of the flood plain currently comprises dryland alien thicket [66]. The flood plain's current potential is mainly the provision of regulating services of flood storage capacity, as well as enhanced downstream water quality, but it needs intervention [66]. As carbon offset, the open space has the potential to provide a range of additional ecosystem services if it is properly designed.

The Open Space Project aims to develop the Piesang's River valley into a major park, which needs to respond to regular flooding and has a reinstated and protected wetland in its center (see Figure 4). The rehabilitated wetland would improve the above-mentioned regulating services and add additional supporting services of habitat and increased biodiversity. New and sometimes conflicting uses are required. Cultural services include safe spaces for sports, rest and recreation for the surrounding communities, and infrastructure for pedestrians and cyclists, such as pathways and pedestrian bridges to reach Bridge City. The design would require intangible aspects such as respect for the cultural heritage (e.g., cemeteries) to retain the urban identity and foster a sense of place [67]. Provisioning services required include designated allotments for urban agriculture and food security. A Spatial Framework Plan (Figure 4) and subsequent spatial designs were developed to encourage sustainable and affordable non-motorized transport and support economic opportunities and green jobs.

Figure 4. The Spatial Framework Plan for the proposed Piesang's River Park. Pedestrian connections are essential for the functioning of the system. Source: [61].

To steer the planning process, the eThekwini Municipality set up an interdisciplinary team with members from various departments, assisted by private consultants. Significant experimental CEBA approaches were developed and tested within the planning process. This included a partnership with the Durban Green Corridor (DGC), a registered non-profit organization that aims to see communities thrive in balance with the habitats around them for execution [68]. It includes the development of plans and innovative concepts, such as a process of rethinking green space management in cooperation with the Sihlanzimvelo community-based stream management program. This program receives significant funding through the Cities Finance Facility (CFF) [69] and identifies various international alternative funding sources, such as the offset strategy [70], for the open space project. The team established one of the first Go! Durban bike parks.

An opportunity for collaboration between the University of Pretoria and the eThekwini Municipality on the conceptual design of the development of the Piesang's River Park was an additional way to bring a fresh perspective to the project's design.

2.2. *Live Project Aim and Objectives*

A case study analysis is presented on the pedagogical practice of a live project, which fits into the transformative pedagogy model [30]. The case study was selected because it is representative of the challenges of UGI design in the developing world. The research aim was to extract important insights on the value of collaborative real-life projects in transforming how young and established professionals deal with UGI design challenges.

Live projects characteristically engage external collaborators. Students leave the classroom for this engagement and produce something of value to these collaborators [30]. In 2018, at a national conference, the eThekwini Municipality presented the framework development of the Piesang's River Park in KwaMashu. Due to previous experience in engaging students in service learning and live projects, the university lecturer identified the Piesang's River as a potential learning site for engaging the fourth-year design studio in ecological design. In 2019, the Department of Architecture at the eThekwini Municipality and the Landscape Architecture program at the University of Pretoria agreed to collaborate on the conceptual design of the Piesang's River Park.

Typical of a live project, the remit of the project was worked out cooperatively [30] to gain joint benefit for the municipality and the University's students. In line with the two research questions stated in the introduction, the live project objectives were to expose students to a real-life project to comprehend the complexities of ecological design; to include municipal staff to provide guidance to the students to gain an understanding of the municipal agencies and processes; to obtain feedback from the students on the value of real-life projects for their learning and professional formation; and to receive report from the municipality on the benefits of the engagement for them.

2.2.1. Live Project Activities

As part of an existing design studio course, students could elect to form part of the eThekwini ecological design studio. As a design project, they had to choose one of six earmarked areas within the Piesang's River Park framework on which to focus (see Figure 5). In line with a typical live project development [30], two municipal staff members actively engaged with the lecturer in writing the student brief. The student brief identified key considerations for the Piesang's River Park based on the Open Space Framework (described above). Each site came with a brief explanation of the "design program" or intended uses and amenities to include.

Figure 5. The Piesang's River flood plain area, indicating the six sites for the students' design projects. Source: Author, 2019.

The municipal leaders for the project and the University's landscape design studio staff managed the students' design stimulus activities and provided continuous guidance. Experiential learning requires action and reflection, which can be stimulated through participation in activities, dialogue and written feedback [40]. To open up the likelihood of transformative learning experiences for students, the lecturer purposefully allowed the project to embrace aspects of risk by introducing the unexpected, spontaneous and unknown [71]. These elements were enhanced by traveling to an unfamiliar context. With funding from the University, the students were able to visit eThekwini and KwaMashu—an 800 km drive from Pretoria—for a six-day field trip. The first day and the evenings were for socializing and reconnaissance. The coastal subtropical environment, along with the urban social setting, offered an exciting prospect of new experiences for the students.

To assist students to identify key design factors to respond to, the eThekwini Municipality arranged a five-day technical workshop. On the first day, several consultants and municipal staff presented environmental and demographic information that gave direction to the framework they had developed. For example, the main routes and transport networks were explained and how people typically commute to work opportunities. The assessment of the ecological state of the wetlands was presented, as well as proposals for its phased rehabilitation. This information, including digital aerial photographs and specialist reports from consultants, was made available to the students for their use.

As part of the workshop, two site visits to the Piesang's River and KwaMashu (see Figure 6), on-site engagement with the residents under the guidance of the municipal team (see Figure 7) and guided visits to two examples of best-practice urban parks in Durban were conducted. Private and public consultants explained the challenges of each example to the students and discussed the design decisions they had selected to respond to them and how successful or unsuccessful these decisions were. Students could then physically inspect the project and take note of how activities and amenities were grouped and what materials and construction details were used. The students spent the last two days working on conceptual design ideas and presented these to the municipal staff for initial feedback.

Figure 6. Meaningful field trip activities, such as site and case study visits, arranged by the municipal staff for the students during their visit to eThekwini. Source: Author, 2019.

Figure 7. Workshop activities and discussions between the community, municipal staff, and students during the student visit to eThekwini (top) and back in Pretoria (bottom). Source: Author, 2019.

The Department of Architecture at the University of Pretoria operates within a studio-based design learning paradigm where critical discussions are used to point out the strengths and weaknesses in the students' work. Precedent studies are also used as a way to find possible solutions that must then be adjusted to the local social and environmental context. On return to Pretoria, the responsible lecturers gave weekly critique to the students on their design progress. Midway through the design process, staff from eThekwini came to Pretoria to provide their input to the students' design developments (see Figure 7).

Being more aware of local contextual realities and municipal processes and frameworks, the municipal feedback—representing the client—would focus on different aspects to that of the lecturers. The two municipal staff members were also present during the project's final examination (see Figure 8). They selected and awarded the project that was most responsive to their expectations. All the final projects were made available to the municipality digitally for their use and further development. Extracts from two of the final design projects can be seen in Figure 9.

2.2.2. Research Design

The authors made use of a participatory action research approach during the live project. Action research involves the investigation of complex, real-world questions, and makes use of cycles of action followed by critical reflection. It typically uses qualitative data and focuses on clarification for specific situations [72]. Action research allows for participant observation [73], involving post-action discussions and critical reflection between the authors. Together with the written feedback questionnaires, these methods allowed for triangulation.

The ten postgraduate architecture and landscape architecture students that formed part of the studio volunteered to participate in anonymous feedback questionnaires. The sample was representative of students from different cultural backgrounds, and the gender divide was equal. The two staff members involved in the studio from eThekwini were also asked for their feedback on the benefit of the intervention to the municipality. The lecturer used anonymous questionnaires to reduce the potential influence of personal contact on the responses provided. The students' written feedback is a method to stimulate the reflective processes required in design thinking [30,74].

Figure 8. Students, questioned by external examiners, defend their design proposals on the Piesang's River Park in the presence of the lecturer and municipal staff during the final examination for the quarter-year project. Source: Author, 2019.

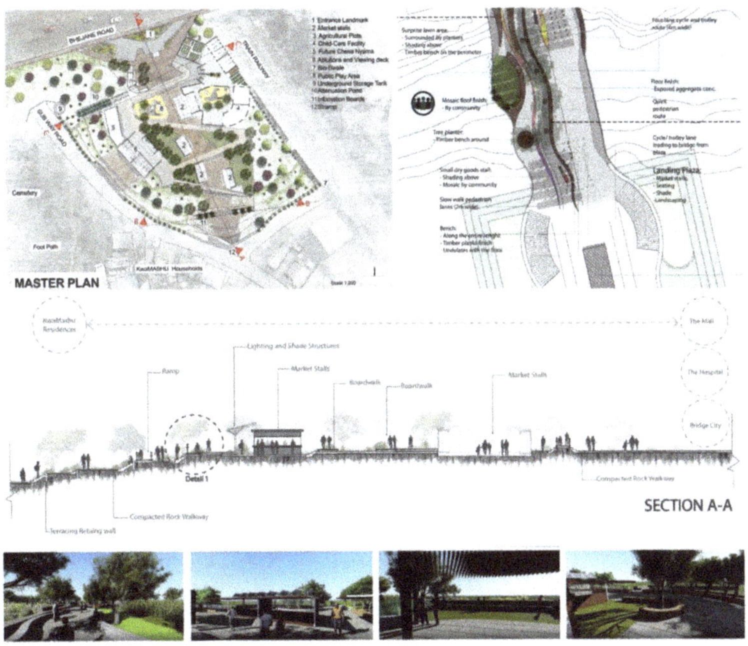

Figure 9. A collage with selected design work from the students that was shared with the eThekwini Municipality. The top left and middle images feature a linking market space on a steep slope to Bridge City. The top right and bottom images feature a pedestrian bridge crossing the Piesang's River with recreational lookout points. Source: University of Pretoria, Department of Architecture, 2019.

Students completed questionnaires manually on four occasions during the seven-week project period. The students were prompted with open-ended statements they had to complete in their own words to reflect on their experiences. The survey questions inquired broadly about the physical environment, social engagement, and the students' learning process. Working within a psycho-socialist school of thought, the questionnaire sought to identify extrarational sources that are more intuitively or emotionally based [48]. Therefore, specific prompts were included to identify memorable aspects, challenges, emotions experienced, and any changes in perspectives that would indicate areas of transformative learning [47,75]. However, the research was undertaken inductively to develop understanding rather than test existing theory. The initial questions are outlined in Table 1. The questionnaires were adjusted slightly throughout the seven weeks to stay in tune with the design process and types of activities, for example, leaving out the initial three questions about the reasons for choosing the studio and asking about the field trip or the studio engagement activities more specifically.

Table 1. Guiding questions contained in the research questionnaires to the students.

Tick the applicable option:
1. This studio is my first choice.
2. I had to take this discipline-specific studio.
If you chose this studio, please explain why:
Please complete the following statements (if they apply to you):
1. What made a great impact on me in this project (so far) was ...
2. What I really enjoyed/liked was ...
3. I really disliked the fact that ...
4. What I found memorable about the week was ...
5. (Cross out what does not apply): Something that surprised/angered/excited me was ...
6. The following experience(s) influenced my learning process in the following way ...
7. I feel/I don't feel ... that I now think differently about things (because ...)

Source: Author, 2019.

An open-ended questionnaire, inquiring about the benefits derived from the student engagement, was sent to the participating municipal staff at the end of the project. With a low response rate, the municipal feedback was mainly collated through the participant observation of the relevant author. This entailed the observation of all project activities and discussions with the municipal team members [75]. However, the limited direct feedback from municipal staff members is a limitation of the study that would require further research to substantiate the findings.

The 40 questionnaire answers from the students were transcribed and qualitatively analyzed and coded with Atlas TI (Version 7) software [76], using the interview question themes as initial deductive coding categories [77]. The codes were grouped into six families: reason for selection of the studio, learning/learning experience, memorable aspects, challenges, emotional experiences, and transformation/change. In line with the transformative pedagogy within which the questionnaire was conceived, the analysis focused on identifying extrarational sources and personal attributes [48] rather than technical competencies in learning. Saliency regarding the number of times themes were articulated in the total body of the data was recorded [78]. This included the repeated articulation of themes in response to different questions by the same person on the same feedback form. The saliency is indicated in brackets in the findings as a reference.

The ten students involved in the case study had a fair representation of culture and gender. The sample size is comparable to similar studies, such as a study [30] that involved 15 students per case study, interviewed as three focus groups (not individually), giving a total of 30 participants and six focus group interviews. In our study, the questioning of the same participants throughout the four different phases of the studio project provided four responses per person. The inquiry resulted in rich personal descriptions, which are considered a valuable basis from which to draw findings concerned with deep learning. In line with the psycho-socialist school of thought, qualitative research is mainly concerned with the quality and depth of the feedback received [76]. Similar to Sara [30], our study included evaluative feedback from those managing the project in the dataset, and made use of educational literature to further interpret and contextualize the findings.

3. Findings

3.1. UGI Design Project Outcomes and Municipal Benefits

In response to the first research question, we present a summary of the observations and discussions with the author as part of the municipal team. The benefits from the engagement were evident in two areas: the planning and design process, and the design work outcomes.

3.1.1. Critical Evaluation of the Planning and Design Process

Reviewing, Questioning and Adapting Municipal Ideas and Solutions

Within the eThekwini Municipality, the collaboration with the University of Pretoria was done as part of the municipal project evaluation process. The municipal project team used the five-day technical workshop with the students to reflect on the work process and the results achieved after a four-year planning period. Working in an interdisciplinary team representing roughly 10 different line departments within eThekwini Municipality meant rotating staff, and dealing with conflicting interests, and posed interpersonal and logistical challenges. Under these circumstances, the technical workshop allowed the members of the steering committee to reflect together on the project status. It also allowed the interdisciplinary team to focus on and rethink the approach taken thus far, and to explain and review the project scope and outcomes.

Following an experimental approach, the municipal team of professionals needed to be prepared to think beyond the conventional, consider unexpected problems, and constantly evaluate and adjust their actions. There is no international best practice to duplicate. Specific local community-driven answers needed to be developed within the specific neighborhood. The municipal team showed an eagerness to share their "on-the-ground experience". At the same time, they were challenged by the students' honest critical questions. By sharing, they assisted the students to understand the various challenges experienced in the real-life eThekwini environment. As mentioned by the staff, "the students were able to provide some strong design solutions for the nodes and came up with fresh ideas".

Creativity, Experimentation and Inspirational Ideas and Solutions from the Students

Students are typically free to experiment and are slightly naïve. They lack the firm inhibitions and value biases that are often found in long-standing professionals and can therefore bring creative and inspiring ideas to the planning process. The feedback from the municipal staff was that "it is important for departments to be open to innovation and academic views". This comment indicates a desire to embrace theoretical, conceptual, and more speculative thinking. The students' presentations provoked an in-depth discussion among the different departments. Based on the students' design results, conflicting views were expressed and deliberated in a solution-driven way. The profound discussion about the design challenges that the students faced helped reveal weaknesses in the Piesang's River Park's aims and ambitions, and forced the team to consider alternatives.

3.1.2. Macro-Level Design Outcomes

At the time of the student engagement, the planning process for the Piesang's River Park was still on a generic framework planning level. For most of the sites, the students' design work was the first exploratory design to be undertaken and captured in drawings. Based on the Piesang's River framework, the students' proposals included functional UGI aspects, such as agroecology and wetland rehabilitation, while also creating opportunities for social interaction and economic activities. Other students provided the necessary infrastructure associated with the UGI such as recycling centers, pedestrian bridges, and markets with day care facilities. The students were ambitious to include various considerations of user comfort and wellbeing into the functional programs they were provided with, along with sustainable design initiatives. For example, the residents' social and physical connection to the river environment and views towards it were exploited in the river boardwalks and recreation areas. The proposals allowed for nature experiences and didactical aspects, along with sustainable practices such as stormwater capture for infiltration or re-use.

The project team was impressed with the students' well-reasoned "strong design solutions". Creative and innovative thinking resulted in "fresh ideas", which is a key component of a successful experimental CEBA-focused approach. The students' design proposals were more conceptual, speculative, and risky than the typical proposals received from consultants that were well versed with practical constraints, such as cost, durability, and maintenance. Although the students' proposals lacked technical refinement, they presented conceptual ideas that went beyond mere problem solving. They combined various social and ecological design functions with the creation of places for social encounters and environmental experiences that were not impossible to execute. In that way, the proposals challenged the status quo in both process and outcomes—which are very necessary steps in transforming the current state of the built environment.

The University's collaboration process was included in the intermediate project report. All design work delivered by the students was incorporated into the status quo project sheets. The students' work will serve the practical role of stimulating the design scope beyond the conventional when it is included in the municipality's future design briefs. Professional consultants will develop these conceptual design proposals when funding becomes available for implementation. The feedback received recommended packaging the students' work in A4 booklets, rather than physical pin-up or digital presentations, to make it more accessible for future use.

3.2. Students' Transformative Learning Ascribed to the Specific Learning Environment

In response to the second research question, we identify the transformative learning that could be ascribed to the specific learning environment created by the live project, as well as the context for the students, with saliency indicated in brackets. In line with the research design, we gave preference to the findings that were more intuitive or emotionally based.

3.2.1. Learning Environment and Context

There is ample evidence that the environment created by the project stimulated the students' learning. The project's real-life nature (33) was mentioned most frequently by students, as illustrated by this response: "dealing with a real-life scenario hindered my creative process a lot [...] because the real constraints actually pushed me to new ways of thinking creatively".

Other aspects mentioned that influenced learning were the new/different context (19), the interdisciplinary nature of the studio (18), social engagements (14), and combined environmental and social engagements (14). Specific activities mostly referred to social engagement and people-aided learning, such as the feedback from the municipality (13), engagement with the community (12), the examples given by the municipality (12), the lecturer (10), theoretical work reviewed (9) and the "expert" consultants involved (9). Examples of specific references to the environment are: "going to site and experiencing real-life issues practically," and social aspects such as thinking differently about "community engagement because ... it is the core of design and should always be a priority."

The real-life project specifically impacted on design thinking: "I need to link my solutions more closely to reality and community" and "It helped me understand aspects of the design better than I could really grasp before."

3.2.2. Memorable Aspects of the Project Environment and Learning

The students were specifically questioned about what they found to be memorable. In their responses, they most recurrently mentioned social interactions (19), for example: What I find memorable/meaningful in terms of my learning: "The contact with potential users of the space helped me better understand who I would be designing for."

Other aspects that were mentioned were the context (7), elements of the environment (5), and a combination of social and environmental engagement (4). The students mentioned memorable aspects of their learning as their own skills development (3) and thought development (3). Thought development is indicative of transformative learning.

3.2.3. Emotional Experiences that Shaped the Learning Environment

The students were prompted to express the emotions they experienced and the reasons for these emotions. The most frequently mentioned emotions were degree of uncertainty (13), as illustrated by the response to the question: How did you experience the site visits? "I was nervous to go into the wetland, but I never felt unsafe or in danger."

The second-most frequently mentioned emotion was personal passion (5). Students also expressed empathy towards the social/human condition (4) of others, as well as feeling challenged (3). There were also expressions of feeling enabled, having fun or being inspired (3). Emotions was also indicated as part of social experiences.

3.2.4. Perception of Inner Change or Transformative Learning

Students were asked whether they felt or thought differently about anything. The response was most recurrently that their basis for design decisions (13) had changed, as expressed by the response to the question: I feel that I think differently about: "Material choices in design projects because ... there are so many other factors that can be considered in the choice of materials other than aesthetics."

They also expressed changed thoughts on the role of the profession (10) and the design process (6). Some confirmed the importance of aspects of prior learning (5), the importance of an interdisciplinary approach to design (4) and exposure to new things (4).

Examples of reported change in students' thinking are in response to the question: I now think differently about things because: "I can understand that we have a role to play in making tangible differences to people's lives as opposed to existing in an 'ethereal artistic, unattainable' realm"; I now think differently about design because: "Architecture does not necessarily have to be glamorous and flashy. Simple, practical architecture impacts real people's lives more, especially in the township context."

3.3. *Other Dimensions and Challenges of the Student Learning*

The data were analyzed for *what* students reported they had learned. This included how to focus design (7), the importance of ecology (6), new design approaches (5), and the role of the professional (4). The experience thus seems to have made students think specifically about design and their professional role, as expressed in this response: "Architecture is not building/a physical building–it's about space and people and the environment and how this all comes together."

The students indicated several reasons for their selection of the studio: the environmental focus (5) and the fact that the studio was different from previous studios (3), as expressed in the response: "I don't know anything or much about ecology and I think it's a necessary field to study within the context of architecture, given the upheaval that our craft tends to cause in the natural environment." The unfamiliar context was an important reason why students selected the studio. The new context was strongly articulated as part of the learning experience and was the second-most memorable aspect expressed. The context was often responsible for students' emotional experiences, such as uncertainty, challenges, fun, and excitement. The students also expressed exposure to a new context as a reason for changes in the way they thought and felt, as can be seen in the response to the question: What I really enjoyed/liked: "Experiencing a different context, which opened my eyes and perspective. We went to various sites to understand the overall context"; What I found memorable was: "Community interaction and a new context experience. New thinking processes came about."

The students were asked to express "challenges" they experienced. Among these, time constraints (22) featured greatly, which is not uncommon for design studios. A student revealed how time limitations had perceived effects on their learning: "The speed of the quarter made for fast learning processes ... this was both good and bad. Retention may not be great, but design decisions were made quickly." Another recurring "challenge" expressed was uncertainty (13) at different moments of the process. Some of this was due to the different expectations (4) they perceived from role players, as stated in the response: "There was confusion between what is expected of us by the eThekwini Municipality and our academic outcomes." Students felt challenged in retrospect because the community engagement (9) and site engagement (4) they had during the six-day visit to Durban was not enough, as expressed in the response: "The community interaction was great, but I feel that we needed to have much more interaction and representation from them."

4. Discussion

4.1. Improving Local UGI Planning and Design

The eThekwini Municipality follows an experimental CEBA approach in UGI planning [60]. This approach encourages the municipality to develop specific solutions within the local context. Although not everyone has bought into this approach, a culture has been created to think creatively and critically reflect on the outcomes during the entire planning process. In this context, university engagement resulted in being a valuable tool, since it assisted the municipality with the above ambitions that are part of the institutional culture. Not all local municipalities have a culture to work experimentally or creatively. We speculate that these municipalities might benefit more from student engagement, but the real impact and perceived value of engaging students might be lower. Roux et al. [28] warn that, in the public sector, successful collaboration is often hampered by prejudices or established beliefs, capacity limitations, and an inability to traverse power inequalities among those involved. We concur with former studies that there should be appropriate incentives for the different partners involved in collaborations [79]. It is further important to be cognizant of how the objectives of public sector partners differ from those of academics and to make sure that the student work is provided in a useful format.

Roux et al. [28] have indicated that an active agent should be embedded within an institution that has a primary interest in implementing the desired change. Our experience confirms that having active champions in both the university and the municipality that had similar ambitions determined the short-term operationalization of the learning ideals pursued in the collaborative project. Since the collaborating author left the eThekwini municipality three months after the completion of the student design studio, it was not possible to verify the long-term impact of the collaboration on the municipal processes or the Piesang's River project as such. The duration, timing, and continuation potential of the project limited the findings, confirming research lessons by Roux et al. [28] on successful systemic change. Longer-term monitoring is required to gain a more detailed understanding of the municipal processes and potential uptake of ideas as a result of student collaboration.

Cilliers [36] advocates that, in South Africa and Africa, UGI must be considered as an integral spatial issue. This compels public spaces to be designed to reflect context-specific needs and challenges in long-term planning strategies for the improved provision of UGI [36]. Our findings confirmed that, during the collaboration process, the students' difficulties with some of the site planning, conflicting instructions from different stakeholders and community interests revealed profound design challenges brought about by the proposed municipal framework. The municipal team understood the complexities of UGI design better, but is conditioned by simple solutions. Roux et al. [28], warn that the attraction of simple solutions often retains the status quo. This pitfall needs to be consciously avoided. The live project process and students' attitudes created a fertile environment for seeking solutions that went beyond the status quo. We concur with Roux et al. [28] that students are good bridging agents because they are truly interested and candid, while being perceived as neutral. Not all the design challenges could be fully resolved during the live project, but it is important to note that many questions and trade-offs were grappled with, which led to changes in how students and municipal staff discern the desired solutions to these design problems. The design proposals that attempted alternative UGI outcomes could serve as a good foundation and inspiration for the future detailed design of the selected sites. If and how the municipality will take these ideas forward was not verified as part of the research.

4.2. The Role of Real-Life Projects in Design Teaching

Although the students performed well in the studio, it could not be said that they reached above-average technical design outcomes. The complexity of the design problems they were faced with was high. Our experience has indicated the multidisciplinary nature of UGI design solutions that would greatly benefit from insights from stormwater engineering and urban design students or professionals, among others, in future studios. That being said, technically refined design resolution was not the main goal of the project, but rather responses that reflect deeper-seated value changes through confrontation with the diverse social and environmental challenges that form part of the complex system of design in the Global South. Cockburn et al. [27] point out that one of the reasons for the science-action gap in sustainability challenges is that technical knowledge management solutions receive far more attention than social learning-based approaches. The same has been argued in education, especially with regard to vocational professions that are focused on competencies. Newton [48] argues that, more than providing a set of required competencies, education should seek to create graduates who are mindful, able to deal with change and are socially responsible. A focus on knowledge transfer requires a process that differs from gaining the capacity to deal with adaptive challenges [48].

Dewey [80] emphasized the importance of the learning environment in encouraging and directing specific learning to take place. As can be seen from the findings, the "real-life" aspect of the project was by far the most articulated as part of the perceived changes in the learning process. This illustrates the stimulus of such an environment for transformative learning. The dynamic live project learning environment was responsible for the memorable aspects, challenges and emotional experiences mentioned. Emotions often reveal an underlaying process of transformation [42].

Our findings indicate that the primary way in which the real-life project stimulated transformative learning was through social dynamics. The interactions included a diversity of new role players with a real-life interest in the project that, at times, resulted in conflicting feedback and ideas. For the students, this came in the form of confrontation between the academic outcomes of the design brief, the needs of the community, the municipality's functional requirements and budget constraints. Sara [30], in her findings, indicates how the contact with a real client and the perceived value the client has for the work become a strong motivation for students. This coincides with our findings, where the students expressed time limits for meeting the municipality's requirements and fully understanding the needs of the community as project challenges. The pressure they felt is indicative of their inner motivation to meet the client's expectations. Sara [30] points out that the challenges students raise around communication, negotiation, and empathy are positive indications of the development of valuable skills for professionals. Designers need to make value judgments and learn to respond sensitively and sensibly to the diversity of values that exist in the design profession, in society and between them [28,29]. We found that the social dynamics assisted students to become more socially aware and make more sensitive (and by implication sensible) value judgements.

Historic examples will continue to influence expectations about well-functioning and orderly cities, while students are enticed by aspirational (technologically advanced) examples of acontextual "smart" and eco-cities [14]. These historic and futuristic ideals are hard to reconcile with the on-the-ground realities in a Global South context, to which students are exposed in a real-life project experience. For example, students were confronted with the poor, homeless persons, and potential migrants and drug users. This is considered "problematic". Historic and futuristic utopias formulate ideas of "appropriate public and proper behavior" in the interest of order and safety, which indirectly exclude people based on income, gender, age, ethnicity, and religion [81]. In South Africa, the desire for safety and order still conflicts with the ideals of accessibility and inclusion [81]. This paradox compels students to exercise greater social empathy, while actively considering local value and public safety in design solutions. In so doing, they define and translate green infrastructure in the local context [36]. Although limited community engagement was possible, the students still became aware that values should be captured and linked to beneficiaries, comprising both the community and the authorities [36]. This experience would make them cognizant of the importance of perception factors that are instrumental in the optimization and implementation of UGI in Africa [36].

The second way in which the real-life project stimulated learning was through emotions: creating a balance between the excitement and enthusiasm of the students, and uncertainty, the pressure of time constraints and the variety of contrasting expectations. Our findings confirm that of Sara's [30], who found that students were motivated and experienced the live project as exciting. As mentioned, the higher levels of motivation were strongly linked to the involvement of real clients and the perceived relevance of the work. In addition, our findings emphasize the value of open choices, and how students' own passion and identity were associated with these choices (to elect the studio and a specific site within the Piesang's River Framework), creating meaning for them during the learning process. This aligns with educational literature, which indicates that free choice connects learners to their sense of self [46]. This intrinsic motivation could lead to superior learning achievements [82].

Issues of time constraints, also mentioned by Sara [30], are typical of design projects at university or in practice, compelling learners to make swifter decisions about design trade-offs based on the information they have at hand. Sara [30] found that the students became better at self-organization during live projects due to their motivation to take greater ownership. The perceived pressure resulted in them gathering the required information more quickly in order to take the necessary design decisions. Although time constrains were a challenge, our findings do not contain evidence of students' improved self-organization.

In our findings, the students mentioned uncertainty numerous times during the learning experience. Ironically, this uncertainty was often created by the same ingredient that created motivation—the involvement of the client and the unfamiliar context. Tracey and Hutchinson [74] reason that uncertainty plays an important role in the design process. How designers experience states of uncertainty could be different, yet often lead to emotional experiences [74], which aid deeper transformative learning [47].

In sum, an interplay between the familiar and the unknown was created by ownership matched with a novel physical context and aspects of uncertainty. From this we propose that it is important to consider this balance as a factor that creates a "supportive yet charged learning environment" [49]. Based on our findings, we argue that these conditions also make the project more memorable, with the probability of longer-term transformation in students.

4.3. Professional Relevance of Self-Reported Transformation

The learning-by-doing educational approach has an implicit emphasis on informal learning [41], while transformative learning focuses on personal attributes [48]. From our findings, five areas were identified in which learning transformation occurred: the premises for design decisions, the role of the profession, confirmation of prior learning aspects, the design process, and interdisciplinary importance. Below, we consider their relevance for 21st-century designers who must develop the skills of integrated decision-making, leadership, and technical capabilities.

Students reported both thought and skill development in their feedback, which are important for professional development and making design decisions. From the municipal feedback, there is also evidence that the student engagement triggered new ways for the staff to think about design solutions. Design is about problem-solving [29,31]. A designer's main task in complex systems is to weigh up the many possible ways a space can function and balance social, environmental, and economic benefits with cognizance and transparency of the trade-offs [83]. What constitutes a fair trade-off is a value judgment, based on personal and professional formation, but also influenced by contextual factors [84]. Our findings concur with Jabeen et al. [29] that the real-life confrontation with a variety of values and insights allowed students to reflect and make decisions about their professional values, reflecting greater empathy. For the municipal staff, the confrontation rather purposefully challenged the status quo and prevented defaulting on design solutions that are too simple.

Professional roles and responsibilities have constantly been redefined as social values have changed over the last century [85]. Levels of global warming, poverty, biodiversity loss, and other detrimental consequences of human action compel the moral improvisation of knowledge for improved design choices [33]. Students need to be confronted during their education to help them discern the choices of what kind of a professional they would like to become [29,31]. Many professionals, while operating within ethical boundaries, do not desire to change the status quo that provides them with a living. Built environment professionals who are not confronted with the shortfalls of current planning systems and urban realities will not be inclined or enabled to create a different future with alternative UGI solutions. Professionals must not only learn to respect other disciplines and the knowledge of unschooled community members, but must willfully desire to break out of silos to make relevant systemic connections and thereby create alternative futures. Our research findings show that a real-life project confrontation in a challenging environment could assist with such a decision about the role an individual decides to play. Newton [48] argues that, when students undergo such transformational change, it provides them with essential tools to become effective agents of change.

The aim of the real-life project is also to experiment and test the application of theoretical concepts [38]. Similar to findings by Sara [30], students reported testing what they had learned before by applying previous skills in a real-world setting. The reflection needed for this process to be internalized was aided by the feedback questionnaires [74]. The contextual complexities [9] and the learning environment [29,31] challenged students' conventional ways of doing ("hindered the creative process"). Students consequently adapted their design process and learned to become more flexible in how they apply theory.

The complexity of urban problems and the shortcomings of solutions that aim to deal with these problems in isolation have led to greater emphasis on interdisciplinary and even transdisciplinary engagement in research, design, and planning [86]. Multiple perspectives can more comprehensively document the context and significance of UGI to respond more directly to social wellbeing [36], while providing solutions that are more sustainable and resilient to local conditions. Our findings correspond to those of Sara's [30], indicating learning across disciplines. The architecture students specifically reported a greater appreciation of landscape architecture skills. The proficiency to work across disciplines is an essential skill for future professionals to improve design solutions [9]. Technical know-how of stormwater engineers or community insight from social anthropologists would have further assisted the students on the project to build competencies across disciplines.

Interdisciplinary engagements come with challenges. Students paired in groups of two across architecture and landscape architecture disciplines did not all work in synergy. Only one student group was successful with integrated decision-making across their disciplines, coincidentally receiving the award for the best studio project. At least two other groups experienced intense conflict in making joint decisions they were unable to resolve, leading to poorer design outcomes. It is important to acknowledge that designers notoriously have strong egos, while individual creative conceptions are still highly valued within their disciplinary formation. This emphasizes the importance of the educational changes required, as advised by several authors [6,9,36,48]. The young professionals learned that all the solutions do not lie in one profession or one approach, but in different ways of knowing, which includes the knowhow of the community.

5. Conclusions

Collaboration between universities and local municipalities could improve UGI planning and design as far as it confronts preconceived ideas and acontextual applications with on-the-ground realities and conceptualizes alternative design processes and proposals. Collaborators can benefit from an experimental learning-by-doing approach to improve contextual, procedural, systemic, and transformative learning that could start altering conventional and institutionalized ways of approaching UGI challenges and cultivate innovative ideas for balancing trade-offs.

Our study has confirmed previous findings and elaborated on them in the following ways. We found that, for the municipal staff, the live project process and student attitudes and ideas created a setting for seeking solutions that went beyond the status quo. Familiar questions and trade-offs were reconsidered, with changes in how students and municipal staff formulated their ideas of "ideal" solutions to UGI design problems.

Live projects provided the exposure of young (and potentially adult) professionals to greater social dynamics, which developed their ability to more sensitive and sensible value judgements. Live projects created an interplay between the familiar and the uncertain, which is a fertile environment for transformative learning. This environment can be conceived by open choice and ownership, matched with a novel and/or challenging physical context. Transformative learning for designers through live projects was associated with greater social empathy in design decision, influencing choices about the role a young professional decides to play, creating greater design process flexibility and learning, and the acknowledgement of skills across disciplines.

Although collaboration and live projects cannot solve all the challenges of UGI in Global South countries, they do pose one possible way of transforming the people involved, the current processes followed and the outcomes pursued. When combined, these aspects hold promise for the radical shifts required to reimagine and create a more just and sustainable built environment. Our case study focused specifically on challenges for UGI in a Global South context, making the findings more applicable to these settings. However, we believe the findings remain equally valuable to all countries that aim to improve their ability to live inside planetary boundaries. Social empathy, self-activation, flexibility, and interdisciplinarity are valuable qualities for designers, regardless of their socio-ecological context.

Our study particularly had limitations in terms of the spectrum and depth of the feedback received from municipal staff. To gain a better understanding of the real uptake of ideas by municipalities and changes over time, a longer-term and more in-depth feedback process with a greater representation of built environment disciplines needs to be undertaken in future studies. Based on our experience, we offer some suggestions as guidance for future work.

Collaborative projects render benefits for those engaged in them through time spent either engaging with the site and local community or critically reflecting on and debating about fresh solutions to the status quo. Incentives for the academics and municipal staff must be aligned with the shared outcomes and deliverables expected from team members in the collaborative process. For example, educational outputs by students, if provided in the right format, could be taken further by the municipal staff or academic researchers involved to develop as institutional outputs.

Site visits are critical to assist in the comprehension of activities, users and environmental qualities that are present over time. Since there is mostly no guarantee that student projects will be implemented, caution should be exercised in creating expectations in the community, while transparently communicating the potential lack of tangible outcomes through key figures and existing social structures.

A remote context seems to intensify student learning. Initial site visit experiences are very acute, but a remote context and its users become progressively more abstract as time passes. If there is not enough time and funding for site visits and community consultation, this could potentially counteract the objective of live projects to confront students with the realities on the ground.

Although feedback from different role players is an important part of the experiential learning process, adequate guidance of students is also critical. Incorporating a wider pool of disciplines in projects could be logistically challenging, but could broaden the learning experience and quality of the design proposals. Existing or additional coursework and lectures should be provided in the studio to explain new interdisciplinary concepts and provide guidance to students in terms of existing knowledge on these subjects.

Author Contributions: Conceptualization, C.B.; methodology, C.B. and H.M.; formal analysis, C.B.; writing—original draft preparation, C.B.; writing—review and editing, C.B. and H.M.; visualization, C.B.; project administration, C.B and H.M.; funding acquisition, C.B. All authors have read and agreed to the published version of the manuscript.

Funding: University of Pretoria, Scholarship of Teaching and Learning (SoTL) grants 2019.

Institutional Review Board Statement: The study was conducted according to the guidelines of the Declaration of Helsinki, and approved by the Institutional Review Board (or Ethics Committee) of Faculty of Engineering, Built Environment and Information Technology at the University of Pretoria (reference number EBIT/117/2019 and date of approval 14 June 2019).

Informed Consent Statement: Informed consent was obtained from all subjects involved in the study.

Conflicts of Interest: The authors declare no conflict of interest.

References

1. Cincotta, R.P.; Wisnewski, J.; Engelman, R. Human population in the biodiversity hotspots. *Nature* **2000**, *404*, 990–992. [CrossRef]
2. United Nations, Department of Economic and Social Affairs, Population Division (UN/DESA/PD). World Urbanization Prospects: The 2018 Revision. Available online: https://population.un.org/wup/Download/ (accessed on 2 March 2020).
3. Cilliers, S.; Breed, C.; Cilliers, J.; Lategan, L. Urban ecological planning and design in the Global South. In *Urban Ecology of the Global South*; Shackleton, C., Cilliers, S., Davoren, E., Du Toit, M., Eds.; Springer Press: Berlin, Germany, 2021; pp. 365–400.
4. White, R.; Turpie, J.; Letley, G. *Greening Africa's Cities: Enhancing the Relationship between Urbanization, Environmental Assets and Ecosystem Services*; The World Bank: Washington, DC, USA, 2017; p. 50.
5. Dodman, D.; Leck, H.; Rusca, M.; Colenbrander, S. African urbanisation and urbanism: Implications for risk accumulation and reduction. *Int. J. Disaster Risk Reduct.* **2017**, *26*, 7–15. [CrossRef]
6. Dobbs, C.; Escobedo, F.J.; Clerici, N.; De la Barrera, F.; Eleuterio, A.A.; MacGregor-Fors, I.; Reyes-Paecke, S.; Vásquez, A.; Zea Camaño, J.D.; Hernández, H.J. Urban ecosystem services in Latin America: Mismatch between global concepts and regional realities? *Urban Ecosyst.* **2019**, *22*, 173–187. [CrossRef]
7. Roberts, D.; Boon, R.; Diederichs, N.; Douwes, E.; Govender, N.; McInnes, A.; Mclean, C.; O'Donoghue, S.; Spires, M. Exploring ecosystem-based adaptation in Durban, South Africa: 'Learning-by-doing' at the local government coal face. *Environ. Urban.* **2012**, *24*, 167–195. [CrossRef]
8. Steffen, W.; Richardson, K.; Rockström, J.; Cornell, S.E.; Fetzer, I.; Bennett, E.M.; Biggs, R.; Carpenter, S.R.; de Vries, W.; de Wit, C.A.; et al. Planetary boundaries: Guiding human development on a changing planet. *Science* **2015**, *347*, 1259855. [CrossRef] [PubMed]
9. Meyer, M.W.; Norman, D. Changing design education for the 21st century. *She Ji J. Des. Econ. Innov.* **2020**, *6*, 13–49. [CrossRef]
10. Du Toit, M.; Cilliers, S.; Dallimer, M.; Goddard, M.; Guenat, S.; Cornelius, S. Urban green infrastructure and ecosystem services in sub-Saharan Africa. *Landsc. Urban Plan.* **2018**, *180*, 249–261. [CrossRef]
11. Pauleit, S.; Liu, L.; Ahern, J.; Kaźmierczak, A. Multifunctional green infra-structure planning to promote ecological services in the city. In *Urban Ecology: Patterns, Processes, and Applications*; Niemelä, J., Ed.; Oxford University Press: New York, NY, USA, 2011; pp. 272–285.
12. Venter, Z.S.; Shackleton, C.M.; Van Staden, F.; Selomane, O.; Masterson, V.A. Green apartheid: Urban green infrastructure remains unequally distributed across income and race geographies in South Africa. *Landsc. Urban Plan.* **2020**, *203*, 103889. [CrossRef]
13. Baruah, N.G.; Henderson, J.V.; Peng, C. Colonial legacies: Shaping African cities. *J. Econ. Geogr.* **2020**, *21*, 29–65. [CrossRef]
14. De Satgé, R.; Watson, V. African cities: Planning ambitions and planning realities. In *Urban Planning in the Global South*; De Satgé, R., Watson, V., Eds.; Palgrave Macmillan: London, UK, 2018.

15. De Schiller, S. Supporting sustainability issues in urban transformation. *Urban Des. Int.* **2004**, *9*, 53–60. [CrossRef]
16. Dhakal, K.P.; Chevalier, L.R. Managing urban stormwater for urban sustainability: Barriers and policy solutions for green infrastructure application. *J. Environ. Manag.* **2017**, *203*, 71–181. [CrossRef]
17. Seeliger, L.; Turok, I. Green-sighted but city-blind: Developer attitudes to sustainable urban transformation. *Urban Forum* **2015**, *26*, 321–341. [CrossRef]
18. Moodley, S. Why do planners think that planning has failed post-apartheid? The case of eThekwini Municipality, Durban, South Africa. *Urban Forum* **2018**, *30*, 307–323. [CrossRef]
19. Van Zyl, B.; Cilliers, E.J.; Lategan, L.G.; Cilliers, S.S. Closing the gap between urban planning and urban ecology: A South African perspective. *Urban Plan.* **2021**, *6*, 122–134. [CrossRef]
20. Pasquini, L.; Enqvist, J.P. Green infrastructure in South African cities. In *Report for Cities Support Programme*; African Centre for Cities: Cape Town, South Africa, 2019.
21. Lategan, L.; Cilliers, J. Considering urban green space and informal backyard rentals in South Africa: Disproving the compensation hypothesis. *Town Region Plan* **2016**, *69*, 1–16. [CrossRef]
22. Wohlitz, E. *Bring Back Nature to the City*; Briza: Pretoria, South Africa, 2016; p. 402.
23. Lindley, S.; Pauleit, S.; Yeshitela, K.; Cilliers, S.; Shackleton, C. Rethinking urban green infrastructure and ecosystem services from the perspective of sub-Saharan African cities. *Landsc. Urban* **2018**, *180*, 328–338. [CrossRef]
24. Frischenbruder, M.T.M.; Pellegrino, P. Using greenways to reclaim nature in Brazilian cities. *Landsc. Urban Plan.* **2006**, *76*, 67–78. [CrossRef]
25. Cilliers, S.; Du Toit, M.; Cilliers, J.; Drewes, E.; Retief, F. Sustainable urban landscapes: South African perspectives on transdisciplinary possibilities. *Landsc. Urban Plan.* **2014**, *125*, 260–270. [CrossRef]
26. Carden, K.; Ellis, D.; Armitage, N.P. Water sensitive cities in South Africa: Developing a community of practice. In Proceedings of the 3rd International Conference on Design, Construction, Maintenance, Monitoring and Control of Urban Water Systems. *WIT Trans. Built Environ.* **2016**, *165*, 51–74.
27. Cockburn, J.; Rouget, M.; Slotow, R.; Roberts, D.; Boon, R.; Douwes, E.; O'Donoghue, S.; Downs, C.T.; Mukherjee, S.; Musakwa, W.; et al. How to build science-action partnerships for local land-use planning and management: Lessons from Durban, South Africa. *Ecol. Soc.* **2016**, *21*, 21. [CrossRef]
28. Roux, D.J.; Nel, J.L.; Cundill, G.; O'Farrell, P.; Fabricius, C. Transdisciplinary research for systemic change: Who to learn with, what to learn about and how to learn. *Sustain. Sci.* **2017**, *12*, 711–726. [CrossRef]
29. Jabeen, H.; Kabir, K.H.; Aziz, T. Balancing rationalism with creativity: An architectural studio's experience of responsive design solutions. *Environ. Urban.* **2021**, *33*, 63–82. [CrossRef]
30. Sara, R. Learning from life—Exploring the potential of live projects in higher education. *J. Educ. Built Environ.* **2011**, *6*, 8–25. [CrossRef]
31. Garrott, J.G. Facilitating experiential learning in environmental design. *Des. Stud.* **1983**, *4*, 115–123. [CrossRef]
32. Deming, M.E. (Ed.) *Values in Landscape Architecture and Environmental Design: Finding Center in Theory and Practice*; Louisiana State University Press: Baton Rouge, LA, USA, 2015; p. 250.
33. Breed, C.A. Incorporating the multiple benefits of urban nature into ecological design. In *The Routledge Handbook of Urban Ecology*, 2nd ed.; Douglas, I., Anderson, P.M.L., Goode, D., Houck, M.C., Maddox, D., Nagendra, H., Tan, P.Y., Eds.; Routledge: London, UK, 2020; pp. 784–798.
34. Botes, K.L.; Breed, C.A. Outdoor living wall systems in a developing economy: A prospect for supplementary food production in urban South Africa? *Acta Structilia* **2021**, *28*, 143–169. [CrossRef]
35. Wilhelm-Rechmann, A.; Cowling, R.M.; Difford, M. Responses of South African land-use planning conservation projects. *Biol. Conserv.* **2014**, *180*, 206–213. [CrossRef]
36. Cilliers, E.J. Reflecting on green infrastructure and spatial planning in Africa: The complexities, perceptions, and way forward. *Sustainability* **2019**, *11*, 455. [CrossRef]
37. Harriss, H.; Widder, L. (Eds.) *Architecture Live Projects Pedagogy into Practice*, 1st ed.; Routledge: London, UK, 2014; p. 230.
38. Kolb, D. *Experiential Learning: Experience as the Source of Learning and Development*; Prentice Hall: Englewood Cliffs, NJ, USA, 1984; p. 390.
39. Tzonis, A. Creativity real and imagined in architectural education. *Front. Archit. Res.* **2014**, *3*, 331–333. [CrossRef]
40. Richards, A. Adventure-based experiential learning. In *Empowerment Through Experiential Learning: Explorations of Good Practice*; Mulligan, J., Griffin, C., Eds.; Kogan Page: London, UK, 1992; pp. 118–123.
41. Taneri, B.; Dogan, F. How to learn to be creative in design: Architecture students' perceptions of design, design process, design learning, and their transformations throughout their education. *Think. Ski. Creat.* **2021**, *39*, 14. [CrossRef]
42. Schon, D.A. *The Reflective Practitioner. How Professionals Think in Action*, 1995 ed.; Basic Books: New York, NY, USA, 1983; p. 384.
43. Kirshner, P.; Sweller, J.; Clark, R. Why minimal guidance during instruction does not work: An analysis of the failure of constructivist, discovery, problem-based, experiential, and inquiry-based teaching. *Educ. Psychol.* **2006**, *41*, 75–86. [CrossRef]
44. Felson, A.J.; Bradford, M.A.; Merway, T.T. Promoting Earth Stewardship through urban design experiments. *Front. Ecol. Environ.* **2013**, *11*, 362–367. [CrossRef]
45. Kato, S.; Ahern, J. Learning by doing': Adaptive planning as a strategy to address uncertainty in planning. *J. Environ. Plan. Manag.* **2008**, *51*, 543–559. [CrossRef]

46. Mackeracher, D. *Making Sense of Adult Learning*; Culture Concepts: Toronto, ON, Canada, 1996; p. 288.
47. Mezirow, J. *Learning and Transformation: Critical Perspectives on a Theory in Progress*; Jossey-Bass: San Francisco, CA, USA, 2000; p. 371.
48. Newton, S. Transformational higher education in the built environment. *J. Educ. Built Environ.* **2009**, *4*, 100–112. [CrossRef]
49. Groen, J.; Jacob, J. Spiritual transformation in a secular context: A qualitative research study of transformative learning in a higher education setting. *Int. J. Teach. Learn. High. Educ.* **2006**, *18*, 75–88.
50. Pauleit, S.; Lindley, S.; Cilliers, S.; Shackleton, C. Urbanisation and ecosystem services in sub-Saharan Africa: Current status and senarios. *Landsc. Urban Plan.* **2018**, *180*, 247–248. [CrossRef]
51. Brooks, T.M.; Mittermeier, R.A.; Da Fonseca, G.A.B.; Gerlach, J.; Hoffmann, M.; Lamoreux, J.F.; Mittermeier, C.G.; Pilgrim, J.D.; Rodrigues, A.S.L. Global biodiversity conservation priorities. *Science* **2006**, *313*, 58–61. [CrossRef] [PubMed]
52. Maúre, G.; Pinto, I.; Ndebele-Murisa, M.; Muthige, M.; Lennard, C.; Nikulin, G.; Dosio, A.; Meque, A. The southern African climate under 1.5 °C and 2 °C of global warming as simulated by CORDEX regional climate models. *Environ. Res. Lett.* **2018**, *13*, 8. [CrossRef]
53. Shackleton, C.M.; Gwedla, N. The legacy effects of colonial and apartheid imprints on urban greening in South Africa: Spaces, species, and suitability. *Front. Ecol. Evol.* **2021**, *8*, 13. [CrossRef]
54. Schäffler, A.; Swilling, M. Valuing green infrastructure in an urban environment under pressure—The Johannesburg case. *Ecol. Econ.* **2013**, *86*, 246–257. [CrossRef]
55. Turok, I. Transforming South Africa's divided cities: Can devolution help? *Int. Plan. Stud.* **2013**, *18*, 68–187. [CrossRef]
56. Shackleton, C.M.; Blair, A.; De Lacy, P.; Kaoma, H.; Mugwagwa, N.; Dalu, M.T.; Walton, W. How important is green infrastructure in small and medium sized towns? Lessons from South Africa. *Landsc. Urban Plan.* **2017**, *180*, 273–281. [CrossRef]
57. Mucina, L.; Rutherford, M.C. *The Vegetation of South Africa, Lesotho and Swaziland. Strelitzia 19*; South African National Biodiversity Institute: Pretoria, South Africa, 2006.
58. Boon, R.; Cockburn, J.; Douwes, E.; Govender, N.; Ground, L.; Mclean, C.; Roberts, D.; Rouget, M.; Slotow, R. Managing a threatened savanna ecosystem, KwaZulu-Natal Sandstone Sourveld in an urban biodiversity hotspot: Durban, South Africa. *Bothalia* **2016**, *46*, 1–12. [CrossRef]
59. Roberts, D. Thinking globally, acting locally—Institutionalizing climate change at the local government level in Durban, South Africa. *Environ. Urban.* **2008**, *20*, 521–537. [CrossRef]
60. eThekwini Municipality. *Economic Development and Planning. Expanded Bridge City Precinct Green Spaces Programme Scope*; eThekwini Municipality: Durban, South Africa, 2016; p. 103.
61. eThekwini Municipality. *Integrated Development Plan—Five-Year Plan: 2017/18 to 2021/22*; eThekwini Municipality: Durban, South Africa, 2017; p. 670.
62. Morgan, D. *Management Model Report—New Management Model of Public Spaces: Increase Service Delivery in the Realm of Public (Open) Spaces—Case Study Bridge City KwaMashu Open Space Area*; eThekwini Municipality: Durban, South Africa, 2018; p. 25.
63. National Treasury, Republic of South Africa. *Neighbourhood Development Partnership Grant (Toolkit No 1 of 5): Grant Overview*; National Treasury: Pretoria, South Africa, 2007; p. 11.
64. Satterthwaite, D. The impact of urban development on risk in sub-Saharan Africa's cities with a focus on small and intermediate urban centres. *Int. J. Disaster Risk Reduct.* **2017**, *26*, 16–23. [CrossRef]
65. eThekwini Municipality. *The Kwamashu Urban Hub Precinct Plan: Status Quo and Conceptual Framework Development*; eThekwini Municipality: Durban, South Africa, 2015; p. 103.
66. Edwards, R.; Macfarlane, D.; Robinson, K.; Israel, A.; De Groen, M.; Dunsmore, S. *Wetland Management Guidelines: Building Capacity and Supporting Effective Management of Wetlands within South African*, Municipalities (ICLEI) Local Governments for Sustainability—Africa Secretariat. Annex A3: The Piesang River Floodplain Rehabilitation Project: Planning and Evaluation to Date. 2018; 210–221.
67. Funke, N.; Baker, A.; Mehrtens, H.; Buthelezi, N.; Edwards, R. Landscapes of integration: The Piesangs River Open Space Project—a case study. In Proceedings of the ILASA 2018 Conference, Drakensberg, South Africa, 13–14 August 2018; p. 6.
68. Green Corridors. Green Corridors. Available online: https://durbangreencorridor.co.za/ (accessed on 16 January 2021).
69. C40 Cities Finance Facility. Durban Flood Prevention Boosted with Support of C40 Cities Finance Facility 2017 (press release). Available online: https://www.c40.org/news/durban-flood-prevention-boosted-c40-finance-facility/ (accessed on 16 January 2021).
70. Bremen City. Climate partnership Bremen-Durban. Available online: https://www.nachhaltigkeitspartnerschaften.bremen.de/durban/urban_planning-9107 (accessed on 12 January 2021).
71. Cross, N. *Design Thinking: Understanding How Designers Think and Work*; Berg: Oxford, UK, 2011; p. 192.
72. Gray, K.; Chang, R.; Radloff, A. Enhancing the scholarship of teaching and learning: Evaluation of a scheme to improve teaching and learning through action research. *Int. J. Teach. Learn. High. Educ.* **2007**, *19*, 21–32.
73. Atkinson, P.; Hammersley, M. Ethnography and participant observation. In *Handbook of Qualitative Research*; Denzin, N.K., Lincoln, Y.S., Eds.; Sage: London, UK, 1994.
74. Tracey, M.W.; Hutchinson, A. Uncertainty, reflection, and designer identity development. *Des. Stud.* **2016**, *42*, 86–109. [CrossRef]
75. Kitchenham, A. The evolution of John Mezirow's transformative learning theory. *J. Transform. Educ.* **2008**, *6*, 104–123. [CrossRef]
76. Saldaňa, J. *The Coding Manual for Qualitative Researchers*; Sage: Los Angeles, CA, USA, 2013; p. 368.

77. Mayring, P. Qualitative Content Analysis Theoretical Foundation. Available online: https://d-nb.info/1191589749/34 (accessed on 16 January 2021).
78. Grbich, C. *Qualitative Data Analysis: An Introduction*, 2nd ed.; Sage: London, UK, 2013; p. 336.
79. Harris, F.; Lyon, F. Transdisciplinary environmental research: Building trust across professional cultures. *Environ. Sci. Policy* **2013**, *31*, 109–119. [CrossRef]
80. Dewey, J. *Experience and Education*; Kappa Delta Pi: New York, NY, USA, 1938; p. 116.
81. De Vries, L.A. Paradox of the commons? The planning and everyday management of Green Point Park. *Urban Forum* **2019**, *30*, 325–339. [CrossRef]
82. Urdan, T.C. (Ed.) *The Role of Context: Advances in Achievement and Motivation*; JAI Press: Stamford, CT, USA, 1999.
83. O'Farrell, P.J.; Anderson, P.M.L.; Le Maitre, D.C.; Holmes, P.M. Insights and opportunities offered by a rapid ecosystem service assessment in promoting a conservation agenda in an urban biodiversity hotspot. *Ecol. Soc.* **2012**, *17*, 27. [CrossRef]
84. Bengston, D.N. Changing forest values and ecosystem management. *Soc. Nat. Resour.* **1994**, *7*, 515–533. [CrossRef]
85. Breed, C.; Cilliers, S.; Fisher, R. Role of landscape designers in promoting a balanced approach to green infrastructure. *J. Urban Plan. Dev.* **2015**, *141*, 11. [CrossRef]
86. Ahern, J.; Cilliers, S.S.; Niemelä, J. The concept of ecosystem services in adaptive urban planning and design: A framework for supporting innovation. *Landsc. Urban Plan.* **2014**, *125*, 254–259. [CrossRef]

land

MDPI

Article

Enhancing Urban Biodiversity: A Theory of Planned Behavior Study of the Factors Influencing Real Estate Actors' Intention to Use Nature-Inclusive Design and Construction Concepts

Mariët A. van Haaster-de Winter *, Marijke W. C. Dijkshoorn-Dekker, Thomas J. M. Mattijssen and Nico B. P. Polman

Wageningen Economic Research, Wageningen University and Research, Prinses Beatrixlaan 582, 2595 BM Den Haag, The Netherlands; marijke.dijkshoorn@wur.nl (M.W.C.D.-D.); thomas.mattijssen@wur.nl (T.J.M.M.); nico.polman@wur.nl (N.B.P.P.)
* Correspondence: mariet.vanhaaster-dewinter@wur.nl; Tel.: +31-70-3358330

Abstract: This paper presents the results of an exploration into the adoption of biodiversity-friendly practices by the real estate sector, by researching which factors determine companies' intentions to use nature-inclusive design and construction concepts (NID). NID represents practices in which nature and building are inextricably linked when (re)designing building projects. We applied the Theory of Planned Behavior (TPB), a well-known framework for studying behavior. A telephone survey was carried out among 103 employees at different types of companies in the real estate sector. The findings showed that attitude, social norms, and perceived behavioral control are all significant explanatory variables for the intention of using NID. Perceived behavioral control is the strongest predictor. In addition, interest in NID strengthens the predictive value of perceived behavioral control. The empirical findings in this study serve as a first attempt to provide insights into the determinants of behavior in favor of using NID and, by extension, looking for drivers for change. The study was carried out in the Netherlands, but the results may be applicable or interesting to other countries as well when looking for opportunities to enhance biodiversity in urban areas or considering how the real estate sector could give substance to their vital role in spatial developments.

Keywords: sustainability; behavioral change; building sector; green spaces; urban areas

Citation: van Haaster-de Winter, M.A.; Dijkshoorn-Dekker, M.W.C.; Mattijssen, T.J.M.; Polman, N.B.P. Enhancing Urban Biodiversity: A Theory of Planned Behavior Study of the Factors Influencing Real Estate Actors' Intention to Use Nature-Inclusive Design and Construction Concepts. *Land* **2022**, *11*, 199. https://doi.org/10.3390/land11020199

Academic Editor: Elizelle Juanee Cilliers

Received: 16 December 2021
Accepted: 19 January 2022
Published: 27 January 2022

1. Introduction

More than half of the world's population lives in urban areas, and this number is expected to increase [1]. While cities' populations keep growing, cities are challenged by environmental issues and planetary boundaries [2]. Loss of biodiversity is one of these issues and is considered as a top-five global risk [3]. Urban green space provides important values for increasing urban biodiversity [4], as well as for climate resilience and the livability of cities [5,6]. City planners are encouraged to consider biodiversity conservation as a core value in urban planning [7]. Although the importance of urban green space for biodiversity and citizens is now widely recognized [8,9], green space in and around cities is often limited by competing spatial claims [10]. Continuing urban expansion and declining maintenance budgets for green spaces put pressure on the available quality and quantity of green space in many cities [11,12]. As a consequence, urban biodiversity and the quality of life of urban citizens are threatened across the globe.

1.1. The Role of the Real Estate Sector in Nature-Inclusive Construction and Design

A sector that h s a major influence on spatial developments in cities is the real estate sector [8,13,14]. In this paper, the term "real estate sector" is used in its broadest sense to refer to a sector that includes (business) actors who are active in building and planning—and thus shaping—the urban environment, including architects, project developers, urban planners, housing corporations, construction companies, landowners, and financial institutions. Despite the major role of the real estate sector in urban development, preserving or enhancing biodiversity is usually not one of its priority areas [8,15]. In fact, various studies indicate that the current real estate sector is not environmentally sustainable. It exerts great pressure on natural resources and contributes to the loss of biodiversity [15–17]. Nevertheless, urban biodiversity can recover and even improve when it is "built into the urban fabric", i.e., when it is truly and widely incorporated into construction projects and different urban infrastructures [18].

Over recent decades, scholars have come to realize that in the complexity of the modern city, strict managerial approaches to tackle urban issues often fail to achieve their desired outcomes because they lack recognition of the importance of stakeholder behavior [19] and of the fact that "firms or industry actors play critical roles in sustainability transitions" ([20]). In this regard, the real estate sector is no exception, especially because it has such a high environmental impact [15]. Therefore, the prominent role of the real estate sector in urban spatial development, combined with societal and environmental urgency for urban nature and biodiversity, demands a broader insight into the practices of real estate actors and opportunities for change.

Technical innovations are often necessary for the success of change, but what may be even more challenging is the social embedding of such changes in actual behavior. "Social" barriers are therefore as essential as "technical" ones [21,22]. From this perspective, promoting biodiversity-friendly practices by the real estate sector does not merely require the development of new forms of nature-inclusive design and construction (NID), but also an embedding of NID in actual behavior. Real estate companies need to change the way they are doing business to enhance (urban) biodiversity.

However, it is known that people do not change behaviors quickly. They often show little or no interest in new ideas or activities and are often unaware that their current behavior is problematic or will result in negative consequences. Behavioral change is a process that starts with recognizing and seeing things differently and encompasses moving through certain stages of change [23]. The adoption of NID shows similarities with circular economy [24–27], which currently emerges in the built environment [15]. Both concepts include the creation of new relationships with goods and materials and attempt to overcome the sustainable disadvantages of current behaviors.

Central to this study is the use of NID, in which nature conservation and construction activities are inextricably linked and reinforce each other. Ultimately, NID creates more biodiversity and a climate-adaptive and attractive living environment for residents and users. We understand NID as acting pro-actively in the interests of nature and biodiversity by integrating nature when building or renovating homes, offices, other buildings, or area development. When adopting such practices, flora and fauna are taken into account, preferably proactively at an early stage. Measures to be taken vary in ease of implementation and integration. Standard options include permeable pavements, insect hotels, or encouraging sparrows or bats to inhabit an area. However, creating a nesting box for a peregrine falcon or a green roof (or even a facade) requires a more specialized approach.

While there are many forms of NID that are already available in the urban environment, the uptake has often been slow. As we discussed above, this uptake will require conscious behavioral change among real estate stakeholders. Consider, for example, the behavioral changes required when using (new) roof tiles that are suitable for bats in buildings. These tiles cannot be inserted in buildings without prior thought about why and where to buy them, how to incorporate them in the building design, and/or where and how to install them. Thus, the first usage of such roof tiles requires significant modifications in existing procurement practices and procedures.

1.2. Research Objective and Scope

In the context of this background, we studied the development of new real estate practices for enhancing biodiversity in urban areas through the lens of behavioral change.

While the urgency for urban greening is clear, and it is also clear that the real estate sector has a big role to play, too little attention is currently being paid to the mechanisms that underpin behavioral change in this sector. The considerations of real estate stakeholders and the predictive factors for their engagement in NID are currently unknown. Although the literature highlights a number of innovative practices in which real estate actors incorporate biodiversity [28,29], insight into the perspectives of the sector as a whole is currently lacking. Addressing this knowledge gap can provide important insights for addressing the behavior of real estate stakeholders, with the aim of promoting NID and ultimately contributing to urban biodiversity.

Therefore, the aim of the present research was to explore factors determining companies' intentions to use NID. Through a survey, we gained insights into the considerations and practices of Dutch real estate actors in relation to NID. We did this by identifying the factors that determine companies' intentions to use NID.

This research was conducted in the Netherlands and specifically focused on those actors who are active in the real estate sector in a broad sense, i.e., the planning and construction of buildings and their (green or grey) surroundings. In the Netherlands, despite increasing attention, projects for nature-inclusive building are still developing only to a limited extent. The national government attempts to support a nature-inclusive development of urban regions, as highlighted by a letter to parliament from the responsible minister on 17 June 2020, in which she indicated her the ambition to make nature-inclusive building generally accepted [30]. In the recent coalition agreement by the new national government coalition, presented on 15 December 2021, the Minister of Housing, Spatial Planning and the Environment committed to focusing on solving bottlenecks to stimulate NID [31]. Several large and small cities have developed stimulating policies to further develop nature-inclusive building (e.g., Amsterdam, Arnhem, The Hague, and Zeist). There are also several initiatives at the provincial level, such as in the province of Zuid-Holland, to ensure that new building recovers or enhances biodiversity [32].

This paper starts with an explanation of the conceptual framework and then provides the methodology that was employed. It then provides the results, and finally presents and discusses the findings of the research.

2. Conceptual Framework

A framework that has been widely applied for predicting change in human behavior is the Theory of Planned Behavior (TPB), which was originally developed by Ajzen [33–35]. This framework has large predictive power [36] and has been used in various countries and behavioral domains, including health-related behavior, recycling behavior, physical activity, (organic) food purchasing, online purchasing behavior, and trade [37–41]. The TPB is based on the belief that behavioral intentions are the immediate predictors of behavior. Intentions, in turn, are influenced by three factors: a person's attitude towards the behavior, how much social pressure the person feels (i.e., social norms), and whether the person feels in control in performing the behavior (perceived behavioral control).

In general, the TPB views the intention to display a certain behavior as stronger when the three predictive factors are positive, making it more likely that a person will carry out this behavior [33]. Attitude is the first determinant that is assumed to capture people's overall evaluation of a specific behavior and its projected outcome [42]. The more positive an individual's evaluation regarding the outcome of using NID, the more favorable his or her attitude should be towards this behavior and, consequently, the stronger the intention to engage in NID. The second determinant is social norms, which are the perceived social pressures to perform a specific behavior. Different types of social norms are distinguished [43], and one is assumed to be more fruitful for behavior change [44]. This so-called "injunctive norm" refers to "what others think I should do" and represents perceived moral rules of so-called "significant others" (a given referent individual or a group or peer group) [45]. The last determinant is perceived behavioral control, which refers to the perceived ease or difficulty of performing a given behavior. It is concerned with beliefs about control factors that can facilitate or interfere with executing a specific behavior. The greater the perceived behavioral control over using NID, the stronger the individual's intention to engage in such activities.

In this study, we applied the TPB to the real estate sector and the use of NID. By doing so, we aimed to predict when real estate actors are willing (or not willing) to engage in NID, and what factors are of particular relevance in this respect. Our conceptual model is presented in Figure 1 and shows how we applied the TPB in the present study. We expected that all of the TPB variables (attitude, social norms, and perceived behavioral control) are positively associated with the intention to use NID (hypothesis 1).

The TPB is, in principle, open to the inclusion of additional predictors [35] and for the purpose of this study we have explored the role of two other variables. First, we were curious about the role of interest, because adopting a new behavior such as NID is not an easy, singular action to perform; people need to modify their behavior significantly. Quite often a lack of interest is an obstacle to a successful change in adopting green building [46]. Interest in NID might influence the strength of the relationship between attitude/social norms/perceived behavioral control and intention (i.e., the moderating effect of interest in NID on respective relationships, such as the attitude-intention relationship). Therefore, we hypothesized that interest moderates the factors that can influence actors' intention to use NID (hypothesis 2).

Second, as a business's decisions could be influenced by the nature of the business (for instance, architects or construction companies might have different interests related to NID), the context of a company was added as an additional variable, given the heterogeneity of the sector. Both additional variables were employed independently from the way in which we measured attitude, social norms, and perceived behavioral control. Therefore, we hypothesized that company context is associated with the intention to use NID (hypothesis 3).

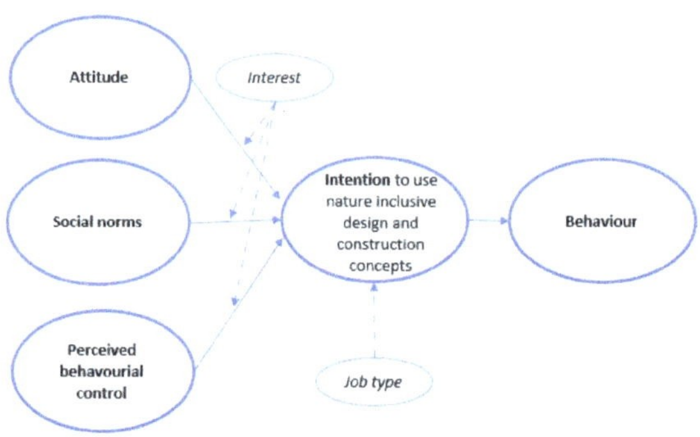

Figure 1. Conceptual model of this study based on the Theory of Planned Behavior (TPB). The solid arrows show the relationship between the TPB's variables and the dashed arrows show the effects of the additional variables.

3. Materials and Methods

3.1. Procedure and Participants

To obtain information on the relevant variables in this study, a survey method was adopted. For data collection, a structured questionnaire was developed and set out during June–August 2019 (before the outbreak of COVID-19). Respondents were invited to join the survey and, upon approval, were approached for an interview by phone. The interviews were held according to a fixed protocol, in Dutch, and consisted of closed-ended questions to prevent deviations from the central research questions. The protocol and the survey were developed by the authors of this article and slightly adapted after a pre-test among experts on either behavior or the real estate sector. The interviews were executed by a market research agency specializing in the real estate sector and their panel—the largest one in The Netherlands—was used for drawing a sample. Respondents were recruited based upon the following company characteristics: the segment of the real estate sector, company size, geographical distribution, and the respondent's involvement in the strategic decision-making process within the company. The survey was widely disseminated; a total of 103 company representatives were interviewed by telephone (minimum 1/2 h), of which 88 were used for the statistical analysis based on consistency and company size. Companies smaller than 1 full-time equivalent were left out because of their small size; they were considered to be self-employed.

As noted, the study sample consisted of 88 participants, who were all involved in the strategic decision processes within their companies, with a mean working experience of 10.5 years (SD = 9.5). The most common types of company contexts were contractors/builders (27%), followed by architects (21%) and real estate managers (21%). Architects plan, design, and oversee construction of buildings, whereas contractors are in charge of the everyday oversight of a construction site. Real estate managers represent companies that are charged with operations related to real estate property for a fee (e.g., housing corporations). Other respondents (31%) included investors, brokers, and project developers.

3.2. Measures

As NID is a relatively new and not yet well-defined concept, the interpretation of NID, as used in this study, was explained to the respondents at the start of each interview (see Appendix A). The study then began with some single-item questions that asked participants to indicate their familiarity with and i1nterest in NID using a 5-point Likert scale ranging from (1) "completely disagree" to (5) "completely agree". For example, the questions included "To what extent was your organization familiar with NID prior to this survey?", "To what extent is your company interested in using NID?", and "To what extent are you already using NID?". These questions were followed by several questions designed to assess the TPB's variables. The TPB variables were measured in accordance with the work of Ajzen (e.g., [34,35]) and the items used were adopted from previously validated instruments and modified so that the focus was on using NID. The variables were explained in the next section and Table 1 shows the items and their accompanying Cronbach's alphas. The items used show good or acceptable internal consistency [47]. Finally, the questionnaire asked participants to provide some information about their backgrounds and the company's background.

Attitude: to assess the attitude towards the use of nature-inclusive design, participants evaluated three items on 5-point bipolar adjective scales, such as "useless—useful", "bad—good." Responses were aggregated to yield a measure of attitude [33,48].

Social norms: (injunctive) social norms were measured using three Likert scale items that asked the respondents whether they believed that their customers/funders/government wanted them to use NID (ranging from 1 = "completely disagree" to 5 = "completely agree").

Perceived behavioral control: the perceived behavioral control refers to the expected simplicity or difficulty with which one can perform the intended behavior. The items were measured using two Likert scale items and assessed to what extent people consider themselves capable of using NID (ranging from 1 = "completely disagree" to 5 = "completely agree").

Intention: The respondents were asked to indicate their intention to perform NID behaviors in the upcoming year, with three items as set out in the table [33].

Table 1. Items of the TPB's variables.

Variable	Item	Cronbach's Alpha
Attitude	Negative—positive Bad—good Useless—useful	0.85
Social Norms	I believe that our customers want us to build nature-inclusive in the coming year. I believe that our funders want us to build nature-inclusive in the coming year. I believe that the government wants us to build nature-inclusive in the coming year.	0.66
Perceived behavior control	I have the feeling that we will be able to build nature-inclusive in the coming year. If we wanted to, we could build nature-inclusive in the coming year. It is usually our decision whether we are going to build nature-inclusive in the coming year.	0.74
Intention	We intend to build nature-inclusive in the coming year. We would like to build nature-inclusive in the coming year. We are sure that we will be building nature-inclusive in the coming year.	0.94

3.3. Method

All analyses were carried out using SPSS, version 20. To compare the means between two groups, we used the student's *t*-test. To assess whether the TBP's variables that have been adopted are fit for use, we estimated Cronbach's alphas, which showed the level of internal consistency within a group of items. The range was between 0 and 1 and a higher Cronbach's alpha indicated that a scale was more reliable. To measure the strength of association between two variables and the direction of the relationship, we calculated the Pearson's correlation, which showed correlation, not causation. If a correlation between two variables is present, it can be positive or negative.

To test the hypotheses and identify which behavioral variables predict companies' intentions for using NID, multiple linear regression analysis was conducted, which is a statistical technique that uses two or more independent variables to predict the outcome of a dependent variable. We included analysis of multicollinearity using variance inflation factors (VIF) to determine the independence of variables in our model. If the VIF was smaller than 10, multicollinearity was not a problem in this study. As the hypothesis indicated, we did not focus only on direct relationships between variables, but also focused on whether the relationship between two variables depends on (is moderated by) the value of a third variable and, therefore, we examined the moderating role of interest. For the moderating analysis, we used the PROCESS macro for SPSS and mean-centered the variable "interest".

4. Results

4.1. Current Situation

The interviews started with a discussion of the current NID practices among the respondents, as presented in Table 2. Of the respondents, 61% indicated that they were already somewhat involved in NID, while a further 36% indicated that they were not yet involved but expected to become involved in the future. Only 3% of the respondents indicated that they did not expect to become involved in NID. Furthermore, 33% of all respondents indicated that their company already has a vision with respect to NID.

Table 2. Involvement of companies in NID (n = 88).

	Percentage
1. Currently involved in NID	61%
2. Not yet involved in NID, expecting to become involved in the future	36%
3. Not expecting to become involved in NID	3%

4.2. Motivations for and Barriers to Engaging in NID

In Table 3, companies' main motivations for engaging in NID are highlighted, based on a Likert scale from 1 (strongly disagree) to 5 (strongly agree). Concerning the motivations (Table 3) to engage in NID, this shows some quite significant differences between companies that are already involved in NID and companies that are not yet involved. The most important motivation for engaging in NID was, "We want to do something good for society," but the contributions to the public image of the company were also a very important reason to engage in NID. The respondents who were already involved in NID ranked these motivations significantly higher than did those who were not yet involved.

Table 3. Main motivations for engaging in NID on a 1–5 Likert scale (n = 88).

	Companies Involved in NID	Companies Not Involved in NID
We want to do something good for society *	4.2	3.7
It is good for our public image *	4.0	3.4
NID sets us apart from others	3.4	3.3
It fits within our way of working **	3.3	2.3
The risk-reward ratio is acceptable	3.2	2.9
We have enough knowledge on NID **	3.1	2.3
NID is part of the customer demand *	2.9	2.3
Conservation or improvement of market position	2.8	2.7
NID is required for certification that we aim for	2.2	2.1

Notes: * significant at the 5% level and ** significant at the 1% level.

Table 4 highlights the main barriers that companies experience in relation to NID on a Likert scale from 1 (strongly disagree) to 5 (strongly agree). In this table, the differences between the two groups are generally quite small and not significant. The most important barrier was a perceived lack of willingness to pay among customers. Companies that are not yet involved in NID believed, more often than companies that were involved, that NID would involve too much uncertainty and would not fit with their way of working.

Table 4. Main barriers for engaging in NID on a 1–5 Likert scale (n = 88).

	Companies Involved in NID	Companies Not Involved in NID
Insufficient willingness to pay amongst customers	3.7	3.6
NID involves too much uncertainties *	2.8	3.3
Management of green is a problem	3.1	3.2
Not necessary for our market position	3.0	3.1
Complicated due to municipal rules or procedures	3.3	3.1
The risk-reward ratio is not acceptable	2.9	3.0
We have insufficient knowledge on NID	2.7	2.9
Technology has not yet been developed far enough	2.8	2.7
NID does not fit within our way of working *	2.1	2.6
Insufficient support for NID within our company	2.1	2.5

Notes: * significant at the 5% level.

4.3. Model Results

Descriptive statistics and Pearson's correlations between the study's variables are presented in Table 5. These figures show that on average, companies show a positive attitude, positive perceived behavioral control, and a positive intention to use NID. Respondents indicated that they perceived moderately low social pressure to use NID.

Table 5. Descriptive statistics and correlations for study variables.

	Mean	SD	1.	2.	3.	4.	5.
1. Attitude	3.96	0.86	1	0.165	0.370 *	0.543 *	0.450 *
2. Social norms	2.84	0.86		1	0.413 *	0.412 *	0.512 *
3. Perceived behavioral control	3.23	1.05				0.429 *	0.651 *
4. Interest	3.65	1.11				1	0.553 *
5. Intention	2.94	1.32					1

Notes: * significant at the 1% level.

Table 6 presents the results of regression analysis on the intention to use NID. The variables explained a moderate proportion of the variance (R2 = 0.55). Attitude, social norms, and perceived behavioral control were significant explanatory variables for the intention to build nature-inclusively. There were positive relationships between attitude, social norms, and perceived behavioral control in using NID, as hypothesized (H1). Therefore, if one of these variables increased, it had a positive effect on the intention to build nature-inclusively (i.e., it increased with it). Of the three variables, the perceived behavioral control had the largest coefficient and thus the greatest influence on intention. A coefficient indicates the degree to which the dependent variable increased (or decreased), with an increase in the associated variable of 1. In this case, when the behavioral control increased by 1, the intention to build nature-inclusively increased by 0.56, on a scale of 1 to 5.

Table 6. Predictors of behavioral intention to use NID.

	Coefficient	VIF
Constant	−1.56 *	
Attitude	0.41 *	1.17
Social norms	0.39 *	1.22
Perceived behavior control	0.56 *	1.36

Notes: * significant at the 1% level.

In the next stage of analysis, to test H2, the main model was further used to explore whether interest played a moderating role. The same model was applied, now including the moderating effect of interest in NID on the attitude-intention relationship, the social norms-intention relationship, and the perceived behavioral control-intention relationship. The results revealed that the relationship between perceived behavioral control and intention was moderated by interest in NID ($\beta = 0.15$, se = 0.08, $p < 0.05$). However, interest did not moderate significantly the relationship between either attitude or social norms and intention towards using NID ($\beta = 0.09$, se = 0.10, $p = 0.382$ and $\beta = 0.02$, se = 0.11, and $p = 0.885$). Hypothesis 2 was thus partly supported. Simple slope analyses showed that among those who had a relatively high interest in NID (1SD above the mean), the stronger the perceived behavioral control was, the higher the intention was to use NID ($\beta = 0.46$, se = 0.13, and $p < 0.001$). Among those who showed a relatively low interest (1SD below the mean), there was a positive but weaker relationship between perceived behavioral control and the intention to use NID ($\beta = 0.79$, se = 12, and $p < 0.000$). Thus, for individuals who experienced high levels of interest, the link between perceived behavioral control and intention to use NID was stronger, compared to individuals who experienced low levels of interest.

In the final part of the survey, the context of a company was added to the model, as a business's decisions could be influenced by the nature of the business. A regression analysis was run in the same manner as before. Table 7 indicates that by adding controls for the company context, the estimated coefficients for attitude and social norms decrease, whereas perceived behavioral control increases. Thus, the variables for architects and real estate development were positive for the intention to use NID, indicating, as hypothesized, the importance of company context (hypothesis 3).

Table 7. Predictors of behavioral intention to use NID, including company contexts.

	Coefficient
Constant	−1.42
Attitude	0.24 *
Social norms	0.41 **
Perceived behavioral control	0.61 **
Company context	
- Contractors/builders	0.33
- Architects	0.60 *
- Real estate development	0.72 *

Notes: * significant at the 5% level and ** significant at the 1% level.

5. Discussion and Conclusions

The aim of the present research was to explore factors determining companies' intentions to use NID. The results supported hypothesis 1. All key variables (attitude, social norms, and perceived behavioral control) were significant predictors of the intention to use NID. The findings of this study showed that interest partially moderates the relationships among the key variables and the intention to use NID. Thus, hypothesis 2 was partly confirmed. Interest only moderates the relationship between perceived behavior control and the intention to use NID. This means that the positive effect of perceived behavioral control on the intention to use NID is even stronger for companies with a high level of interest. Finally, company context directly influenced the intention to use NID (hypothesis 3), particularly for architects and real estate managers. This hypothesis could be explained by the fact that architects and real estate managers deal with NID on a more strategic level, as opposed to builders, who operate on an operational level. Thus, we provided empirical evidence that a company's intention to use NID is determined by attitude, social norms, perceived behavioral control, and company context. In addition, interest is an indirect aspect to consider when it comes to the adoption of NID. Targeted interventions could support the adoption of NID, and it is important to take interest into account in the development of future interventions.

As our analysis showed, the appeal of nature-inclusive construction for real estate stakeholders lies in social factors, as well as in business models. The most important motivation for incorporating a nature-friendly approach lies at the level of values—doing something good for society. However, the contribution to image and distinctiveness also has a motivating effect. Our analysis showed that Dutch real estate actors are aware of their corporate social responsibility [16], but they also engage in NID because they perceive it as beneficial for their business interests. Even so, while NID is perceived as beneficial for the image and distinctiveness of real estate companies, they do not engage in NID because of profit margins or potential extra income; respondents experienced a low willingness to pay among customers and perceived this as the most important barrier to engaging in NID. In this respect, NID might be good for the profile of companies, but many real estate stakeholders still need to develop adequate business models to profit financially from the engagement in NID [49].

A main difference between companies that are already engaged in NID and those that are not yet involved relates to how well NID fits within companies' ways of working and matches their knowledge and experience with NID. On average, companies that build nature-inclusively were more likely to find that such an approach is in line with the operational working method within the company and that there is sufficient knowledge about nature-inclusive building. Those that are less involved in NID felt that it involves too much uncertainty.

In addition to the main findings, we would like to highlight the role of social norms that could accelerate the adoption of NID-specific behavior, and sustainable behavior in general. A social norm is something that is developed by several individuals or groups together, but small incentives are needed to make the new behavior the standard or common practice. It is known [33,45] that people are more likely to perform a type of behavior if they feel that other people around them also do so or find it important to do so. Therefore, this awareness could be used in many interventions aimed at behavioral change [50]. Our study showed that moderately low social pressure to use NID was perceived. Those responsible within the companies did not feel a great deal of (dis)approval from their peers or significant others. Therefore, promoting social norms with respect to NID can help to motivate real estate stakeholders, as well as other actors, in preserving or enhancing urban biodiversity. To mobilize behavioral change in the real estate sector—and beyond—in favor of biodiversity, actions could be developed to normalize NID as desired behavior.

For example, communication activities such as developing a contest, structurally drawing attention to the subject in trade journals, or—more compellingly—including nature-inclusive buildings as a requirement in tender procedures or in administrative rules and regulations related to, for instance, permits for construction projects, could be considered.

Our main contribution to the literature is that we studied the enhancement of biodiversity in urban areas through the lens of behavioral change of real estate stakeholders/actors who play key roles in urban spatial development. It is recommended that behavior itself be understood before trying to change current behavior [50]. In addition, we provided empirical evidence for the mechanisms that underpin behavioral change in the real estate sector. Considering the important role that the real estate sector plays in urban development and the need to engage that sector in the field of urban biodiversity [8,15], the findings offer positive insights for greater involvement of the Dutch real estate sector in urban greening through NID. The majority of interviewed companies were already involved in NID, and from the perspective of growth it is important to highlight that almost all respondents that were not yet involved in NID expected to become involved in the future.

To conclude this article, some limitations of the present study need to be addressed. First, the research was undertaken in a specific setting of the Dutch building environment; hence, any generalizability of its findings for different contexts (e.g., other countries) should be treated with caution. The general approach to study the intention to adopt NID provides an opportunity for the analysis to be replicated in urban areas in other countries to ascertain whether companies in different subsectors within the building sector behave similarly. In this study, we focused on representatives of companies, as they act as agents of change. However, they can operate only within the constraints of a company, which can be either stimulating or restraining.

Further research could also be conducted to explore additional factors to enhance biodiversity by deploying the unconscious aspects of behavior. We examined leads for behavioral change in the conscious parts of behavior, as it is reasonable to expect that nature-inclusive building will require a certain degree of conscious thinking or planning. However, processes operate unconsciously as well, and as such influence behavior [51]. Unconscious processes, such as emotions, biases, and impulses, were not included in this survey. Exploring these processes and biases may provide further clarity and represent a promising extension of this study, and could bridge the failure to translate intentions into behavior (the so-called "intention-behavior gap") [52]. Finally, although this study was carried out in the Netherlands, the results may be applicable or interesting to other countries as well, when looking for opportunities for giving substance to the real estate sector's vital role in spatial developments or enhancing biodiversity in urban areas, and as such contribute to a more sustainable and circular world.

Author Contributions: Conceptualization, M.A.v.H.-d.W., M.W.C.D.-D., T.J.M.M. and N.B.P.P.; methodology, M.A.v.H.-d.W., M.W.C.D.-D., T.J.M.M. and N.B.P.P.; validation, M.A.v.H.-d.W., M.W.C.D.-D., T.J.M.M. and N.B.P.P.; formal analysis, M.A.v.H.-d.W. and N.B.P.P.; investigation, M.A.v.H.-d.W.,

M.W.C.D.-D. and T.J.M.M.; resources, M.A.v.H.-d.W., M.W.C.D.-D., T.J.M.M. and N.B.P.P.; data curation, M.A.v.H.-d.W. and N.B.P.P.; writing—original draft, M.A.v.H.-d.W., M.W.C.D.-D., T.J.M.M. and N.B.P.P.; writing—review and editing, M.A.v.H.-d.W., M.W.C.D.-D., T.J.M.M. and N.B.P.P.; supervision, M.W.C.D.-D.; project administration, M.W.C.D.-D.; funding acquisition, M.W.C.D.-D. All authors have read and agreed to the published version of the manuscript.

Funding: This research was funded by the Dutch Ministry of Agriculture, Nature and Food Quality, and Wageningen University and Research, with the theme "Nature-inclusive Transitions" [grant number KB36-005-004].

Institutional Review Board Statement: Ethical review and approval were waived for this study due no personal or sensitive information was gathered and no formal approval needed to be officially obtained.

Informed Consent Statement: Informed consent was obtained from all subjects involved in the study.

Data Availability Statement: The data collected for the purpose of this study are available in Dutch: see [53].

Acknowledgments: We would like to acknowledge Danny Taufik for his valuable support, and three anonymous reviewers for their valuable feedback.

Conflicts of Interest: The authors declare no conflict of interest.

Appendix A

This questionnaire is about nature-inclusive construction in (re)development of real estate, also referred to as "nature-inclusive design and construction concepts" (NID).

NID means taking nature and biodiversity into account in the process from planning to implementation and incorporating this in the building or the (public) environment so that more diverse plant and animal species can live there. This will give the living environment of both animals and people a boost.

- In new and rebuilt buildings across all kind of levels or phases, e.g., during the design phase or when decisions about investments are made, and in whichever field the company is active.
- For example, by creating green facades, roofs or indoor gardens, planting trees and bushes, integrating nest boxes or special roof tiles, or creating natural ponds.
- It is all about proactively acting for the benefit of biodiversity and integrating nature into current practices.

Real estate: (re)development, construction projects, and area development in the Netherlands.

References

1. United Nations. *World Urbanization Prospects: The 2018 Revision (ST/ESA/SER. A/420)*; United Nations: New York, NY, USA, 2019.
2. Rockstrom, J.; Steffen, W.; Noone, K.; Persson, A.; Chapin, F.S., III; Lambin, E.F.; Lenton, T.M.; Scheffer, M.; Folke, C.; Schellnhuber, H.J. A safe operating space for humanity: Identifying and quantifying planetary boundaries that must not be transgressed could help prevent human activities from causing unacceptable environmental change, argue Johan Rockstrom and colleagues. *Nature* **2009**, *461*, 472–476. [CrossRef] [PubMed]
3. World Economic Forum. *The Global Risks Report 2021*, 16th ed.; World Economic Forum: Geneva, Switzerland, 2021.
4. Fuller, R.A.; Irvine, K.N.; Devine-Wright, P.; Warren, P.H.; Gaston, K.J. Psychological benefits of greenspace increase with biodiversity. *Biol. Lett.* **2007**, *3*, 390–394. [CrossRef] [PubMed]
5. Raymond, C.M.; Frantzeskaki, N.; Kabisch, N.; Berry, P.; Breil, M.; Nita, M.R.; Geneletti, D.; Calfapietra, C. A framework for assessing and implementing the co-benefits of nature-based solutions in urban areas. *Environ. Sci. Policy* **2017**, *77*, 15–24. [CrossRef]
6. Lima, M.F.; Ward Thompson, C.; Aspinall, P. Friendly communities and outdoor spaces in contexts of urban population decline. *Land* **2020**, *9*, 439. [CrossRef]
7. Liu, J.; Slik, F. Are street trees friendly to biodiversity? *Landsc. Urban Plan.* **2022**, *218*, 104304. [CrossRef]
8. Opoku, A. Biodiversity and the built environment: Implications for the Sustainable Development Goals (SDGs). *Resour. Conserv. Recycl.* **2019**, *141*, 1–7. [CrossRef]

9. Lepczyk, C.A.; Aronson, M.F.; Evans, K.L.; Goddard, M.A.; Lerman, S.B.; MacIvor, J.S. Biodiversity in the city: Fundamental questions for understanding the ecology of urban green spaces for biodiversity conservation. *BioScience* **2017**, *67*, 799–807. [CrossRef]
10. Haase, D.; Frantzeskaki, N.; Elmqvist, T. Ecosystem services in urban landscapes: Practical applications and governance implications. *Ambio* **2014**, *43*, 407–412. [CrossRef]
11. Burton, M.; Dempsey, N.; Mathers, A. Connecting making and keeping: Design and management in place-keeping. In *Place-Keeping*; Routledge: London, UK, 2014; pp. 145–170.
12. Kabisch, N. Ecosystem service implementation and governance challenges in urban green space planning—The case of Berlin, Germany. *Land Use Policy* **2015**, *42*, 557–567. [CrossRef]
13. Roebeling, P.; Saraiva, M.; Palla, A.; Gnecco, I.; Teotónio, C.; Fidelis, T.; Martins, F.; Alves, H.; Rocha, J. Assessing the socio-economic impacts of green/blue space, urban residential and road infrastructure projects in the Confluence (Lyon): A hedonic pricing simulation approach. *J. Environ. Plan. Manag.* **2017**, *60*, 482–499. [CrossRef]
14. Sealey, K.S.; Binder, P.-M.; Burch, R.K. Financial credit drives urban land-use change in the United States. *Anthropocene* **2018**, *21*, 42–51. [CrossRef]
15. Munaro, M.R.; Tavares, S.F.; Bragança, L. Towards circular and more sustainable buildings: A systematic literature review on the circular economy in the built environment. *J. Clean. Prod.* **2020**, *260*, 121134. [CrossRef]
16. Battisti, F.; Guarini, M.R.; Chiovitti, A. The assessment of real estate initiatives to be included in the socially-responsible funds. *Sustainability* **2017**, *9*, 973. [CrossRef]
17. Vanags, J.; Butane, I. Major aspects of development of sustainable investment environment in real estate industry. *Procedia Eng.* **2013**, *57*, 1223–1229. [CrossRef]
18. Thomson, G.; Newman, P. Cities and the anthroposcene: Urban governance for the new era of regenerative cities. *Urban Studies.* **2020**, *57*, 1502–1519. [CrossRef]
19. Stubbs, M.; Lemon, M.; Longhurst, P. Intelligent urban management: Learning to manage and managing to learn together for a change. *Urban Stud.* **2000**, *37*, 1801–1811. [CrossRef]
20. Köhler, J.; Geels, F.W.; Kern, F.; Markard, J.; Onsongo, E.; Wieczorek, A.; Alkemade, F.; Avelino, F.; Bergek, A.; Boons, F. An agenda for sustainability transitions research: State of the art and future directions. *Environ. Innov. Soc. Transit.* **2019**, *31*, 1–32. [CrossRef]
21. Egbue, O.; Long, S. Barriers to widespread adoption of electric vehicles: An analysis of consumer attitudes and perceptions. *Energy Policy* **2012**, *48*, 717–729. [CrossRef]
22. Rohracher, H. Managing the technological transition to sustainable construction of buildings: A socio-technical perspective. *Technol. Anal. Strateg. Manag.* **2001**, *13*, 137–150. [CrossRef]
23. Prochaska, J.O.; Di Clemente, C.C. Transtheoretical therapy: Toward a more integrative model of change. *Psychother. Theory Res. Pract.* **1982**, *19*, 276. [CrossRef]
24. Chizaryfard, A.; Trucco, P.; Nuur, C. The transformation to a circular economy: Framing an evolutionary view. *J. Evol. Econ.* **2020**, *31*, 475–504. [CrossRef]
25. Merli, R.; Preziosi, M.; Acampora, A. How do scholars approach the circular economy? *A systematic literature review. J. Clean. Prod.* **2018**, *178*, 703–722.
26. Ellen MacArthur Foundation. Towards the circular economy. *J. Ind. Ecol.* **2013**, *2*, 23–44.
27. Stahel, W.R. The circular economy. *Nature* **2016**, *531*, 435–438. [CrossRef] [PubMed]
28. Cinderby, S.; Bagwell, S. Exploring the co-benefits of urban green infrastructure improvements for businesses and workers' wellbeing. *Area* **2018**, *50*, 126–135. [CrossRef]
29. Ziogou, I.; Michopoulos, A.; Voulgari, V.; Zachariadis, T. Implementation of green roof technology in residential buildings and neighborhoods of Cyprus. *Sustain. Cities Soc.* **2018**, *40*, 233–243. [CrossRef]
30. Dutch Government. *Letter to Parliament about Nature-Inclusive Building*; Ministry of Agriculture, Nature and Food Quality: The Hague, The Netherlands, 2020. (In Dutch)
31. Dutch Government. *Coalition Agreement "Looking out for Each Other, Looking Forward to the Future"*; Parliament: The Hague, The Netherlands, 2022. (In Dutch)
32. Province of South Holland. *Covenant on Climate-Adaptive Building in South Holland*; Province of South Holland: The Hague, The Netherlands, 2018. (In Dutch)
33. Ajzen, I. The theory of planned behavior. *Organ. Behav. Hum. Decis. Processes* **1991**, *50*, 179–211. [CrossRef]
34. Ajzen, I. *The Theory of Planned Behavior: Reactions and Reflections*; Taylor & Francis: Abingdon, UK, 2011.
35. Ajzen, I. The theory of planned behavior: Frequently asked questions. *Hum. Behav. Emerg. Technol.* **2020**, *2*, 314–324. [CrossRef]
36. Armitage, C.J.; Conner, M. Efficacy of the theory of planned behavior: A meta-analytic review. *Br. J. Soc. Psychol.* **2001**, *40*, 471–499. [CrossRef]
37. Fishbein, M.; Ajzen, I. *Predicting and Changing Behavior: The Reasoned Action Approach*; Taylor & Francis: Abingdon, UK, 2011.
38. Gopi, M.; Ramayah, T. Applicability of theory of planned behavior in predicting intention to trade online. *Int. J. Emerg. Mark.* **2007**, *2*, 348–360. [CrossRef]
39. Grazioli, V.S.; Lewis, M.A.; Fossos-Wong, N.; Larimer, M.E. Attitudes toward protective behavioral strategies: Do they predict use of strategies and alcohol outcomes over time? *Addict. Behav.* **2018**, *87*, 190–195. [CrossRef] [PubMed]

40. Yadav, R.; Pathak, G.S. Intention to purchase organic food among young consumers: Evidences from a developing nation. *Appetite* **2016**, *96*, 122–128. [CrossRef] [PubMed]
41. Hardeman, W.; Johnston, M.; Johnston, D.; Bonetti, D.; Wareham, N.; Kinmonth, A.L. Application of the theory of planned behavior in behavior change interventions: A systematic review. *Psychol. Health* **2002**, *17*, 123–158. [CrossRef]
42. Greaves, M.; Zibarras, L.D.; Stride, C. Using the theory of planned behavior to explore environmental behavioral intentions in the workplace. *J. Environ. Psychol.* **2013**, *34*, 109–120. [CrossRef]
43. Cialdini, R.B.; Kallgren, C.A.; Reno, R.R. A focus theory of normative conduct: A theoretical refinement and reevaluation of the role of norms in human behavior. In *Advances in Experimental Social Psychology*; Elsevier: Amsterdam, The Netherlands, 1991; Volume 24, pp. 201–234.
44. Smith, J.R.; Louis, W.R.; Terry, D.J.; Greenaway, K.H.; Clarke, M.R.; Cheng, X. Congruent or conflicted? The impact of injunctive and descriptive norms on environmental intentions. *J. Environ. Psychol.* **2012**, *32*, 353–361.
45. Minton, A.P.; Rose, R.L. The effects of environmental concern on environmentally friendly consumer behavior: An exploratory study. *J. Bus. Res.* **1997**, *40*, 37–48. [CrossRef]
46. Darko, A.; Chan, A.P. Review of barriers to green building adoption. *Sustain. Dev.* **2017**, *25*, 167–179. [CrossRef]
47. Taber, K.S. The use of Cronbach's alpha when developing and reporting research instruments in science education. *Res. Sci. Educ.* **2018**, *48*, 1273–1296. [CrossRef]
48. Petty, R.E.; Wegener, D.T.; Fabrigar, L.R. Attitudes and attitude change. *Annu. Rev. Psychol.* **1997**, *48*, 609–647. [CrossRef]
49. Mattijssen, T.J.M.; Dijkshoorn-Dekker, M.W.C.; Kortstee, H.; Polman, N.B.P. Nature-inclusive Real Estate: Lessons learned in three Real Estate Projects in Dutch Cities. *Int. J. Sustain. Urban Dev. in review.*
50. Michie, S.; Atkins, L.; West, R. The behavior change wheel. In *A Guide to Designing Interventions*, 1st ed.; Silverback Publishing: London, UK, 2014; pp. 1003–1010.
51. Kahneman, D. *Thinking, Fast and Slow*; Macmillan: New York, NY, USA, 2011.
52. Vermeir, I.; Verbeke, W. Sustainable food consumption: Exploring the consumer "attitude–behavioral intention" gap. *J. Agric. Environ. Ethics* **2006**, *19*, 169–194. [CrossRef]
53. Van Haaster-de Winter, M.A.; Dijkshoorn-Dekker, M.W.C.; Polman, N.B.P.; Mattijssen, T.J.M.; Kortstee, H. *Natuurinclusief Bouwen: Wat Beweegt de Nederlandse Vastgoedsector? Onderzoek Naar Motieven, Barrières en Gedrag in de Transitie Naar Meer Groen (NID: What Drives the Dutch Real Estate Sector? Research into Motives, Barriers and Behavior in the Transition towards More Green and Biodiversity)*; Wageningen Economic Research: The Hague, The Netherlands, 2020. (In Dutch)

land

MDPI

Article

Economic Valuation of Urban Green Spaces across a Socioeconomic Gradient: A South African Case Study

Louis Gerhardus Lategan [1,*], Zene Steynberg [1], Elizelle Juanee Cilliers [1,2] and Sarel Stephanus Cilliers [1]

[1] Unit for Environmental Sciences and Management, Faculty of Natural and Agricultural Sciences, North-West University, Potchefstroom 2520, South Africa; 27135349@nwu.ac.za (Z.S.); jua.cilliers@uts.edu.au (E.J.C.); sarel.cilliers@nwu.ac.za (S.S.C.)
[2] School of Built Environment, University of Technology Sydney, Ultimo, NSW 2007, Australia
* Correspondence: 21441480@nwu.ac.za

Abstract: Urban green spaces (UGSs) may present economic contributions through increases in proximate property values, encapsulated in the proximity principle (PP). More data on the PP is required from the Global South, where the quality and equitable distribution of UGSs are important considerations. This paper investigates the PP in Potchefstroom, South Africa following a quantitative approach, by statistically analyzing municipal property valuations in three districts differentiated according to their socioeconomic status (SES). Districts are divided into sample areas where three zones are demarcated according to their proximity to a UGS. The results show that property valuations are generally higher for properties in closer proximity to UGSs in lower- and higher-income samples, but are lower in middle-income areas. Neighborhood characteristics and SES, UGS amenity and maintenance, ecosystem services and disservices, domestic garden area and residential property size may be connected to the confirmation or rejection of the PP. The rejection of the PP in middle-income areas indicates a need to improve public UGSs as amenity destinations. The results confirming the PP in low-income areas could incentivize expenditures to improve UGS area and quality to increase the willingness to pay for proximity to such spaces and, reciprocally, increase revenue from municipal property taxes.

Keywords: green infrastructure; environmental justice; property values; proximity principle; socioeconomic status; Global South

Citation: Lategan, L.G.; Steynberg, Z.; Cilliers, E.J.; Cilliers, S.S. Economic Valuation of Urban Green Spaces across a Socioeconomic Gradient: A South African Case Study. *Land* **2022**, *11*, 413. https://doi.org/10.3390/land11030413

Academic Editors: Zhonghua Gou and Thomas Panagopoulos

Received: 4 January 2022
Accepted: 9 March 2022
Published: 11 March 2022

1. Introduction

Research under the broad umbrella of green urban environments is well established, following almost forty years of intensifying scholarship. The bulk of academic investigation has historically focused on the Global North, but interest has also increasingly turned to the Global South, highlighting the need to consider local scale, context and dynamics when developing principles and frameworks to understand the role green uses fulfil in the urban environment [1]. Such themes, once rooted in the biological sciences, have now permeated multiple fields, including disciplines such as urban planning and design and urban ecology [2,3]. Multiple research themes have emerged, initially focused on issues related to environmental degradation and protection, but also increasingly investigating social and economic considerations, often framed by the concept of environmental justice. Conceptualizations around environmental justice traditionally focused more on who receives the benefits of urban green spaces (UGSs) and who bears the brunt of potential disadvantages [4,5]. More recently, the emphasis has fallen on issues of distributive justice concerning environmental quality and equitable UGS distribution and access [5–8]. Whilst there have been exceptions [9], the bulk of international studies on the equitable distribution of UGSs have found that these spaces are frequently unevenly distributed [10–12], with lower-income residents often served by lower levels of access and reduced proximity to lower-quality public and private UGSs in comparison with higher income cohorts [13–18], often overlapping with their racial and ethnic minority status [19–24]. Differences in socioeconomic status have been directly and indirectly linked to the provision, access and use of public and private UGS at multiple levels. Spatiotemporal variations in socioeconomic factors and power relations within national and local structures can lead to significant variability in UGS supply and demand [25,26]. Socioeconomic status (SES) and associated political leverage may exert substantial influence on the planning, development and management of UGSs across the urban landscape [18,27,28]. UGSs are also perceived, visited and utilized in different ways by different socioeconomic groups [29,30] depending on the context, and influenced by, for example, socio-cultural conventions, amenity value, opportunities for social interaction, access and perceptions about inclusivity and safety [31–34]. Issues linked to safety have been cited specifically as concerns for the vulnerable, including socioeconomically disadvantaged factions, the elderly, women and certain ethnic groups internationally [10,29,35]. Such concerns are also often related to physical characteristics and facilities, for example, lush vegetation or a lack of lighting that may conceal criminal activity and limit surveillance [36–38].

The majority of research, and most case studies on UGS equality, quality and its associated positive and negative impacts, have been focused on the Global North and its highly urbanized contexts, which generally present higher incomes per capita and lower population growth figures than most of the Global South [10,13]. The Global South is forecasted to accommodate the largest share of future urbanization [17], establishing what Myers [39] calls a "southern urban world". The resulting appetite for land in contexts presenting more limited resources and capacities may jeopardize UGSs, compromising environmental quality [6] and raising concerns for environmental justice related to environmental degradation, loss of valuable ecosystems [10,17,34,40,41] and the further entrenchment of inequality in access to quality UGSs amongst the disenfranchised. The relationship between SES and planning for UGSs has also received limited attention to date [18,42]. There is a need for urban planners and authorities, who have often been shown to acknowledge the importance of UGSs to some extent [2,34,43], to be further sensitized to the value of these spaces based on research from and focused on the Global South related to the improved planning, development and management of UGSs across the socioeconomic gradient [3,33]. Evidence of the economic savings and contributions related to UGSs may prove fundamental to help justify the human and financial capital required to prioritize the equitable development and maintenance of UGSs to realize more sustainable human settlements [44].

As highlighted in Section 2.3, several studies have attempted to measure the economic benefits of UGSs internationally. Investigations have often employed hedonic price analyses to determine the value of residential property relative to their proximity to UGSs, thereby testing the proximity principle, which assumes increases in residential property prices as distance to UGSs decrease [45]. The principle has generally been confirmed in much of the Global North (for example Immergluck and Balan [46]; Daams et al. [47]), demonstrating that the market may value adjacency to UGSs, translating directly to increased municipal tax revenue. Conversely, South African cases (focused mainly on middle-income suburbs to date) have rejected the premise [48–50], indicating a reluctance to pay more for properties in proximity to UGSs. The limited research conducted in South Africa, providing contrasting results in comparison to international findings, in conjunction with the country's severe socioeconomic and UGS inequality (see Section 1.1), provide opportunities for further investigation.

In consideration of the points raised, this paper aims to test the proximity principle in the South African case study of Potchefstroom along a socioeconomic gradient, including neighborhoods in low-income townships, middle-income suburbs and an exclusive high-income gated community. This paper is the first, to the knowledge of the authors, to apply the proximity principle along a socioeconomic gradient in the Global South with the aim of delivering preliminary findings on the economic value attached to UGSs within differentiated socioeconomic contexts. This study employs hedonic price analysis and the proximity principle (see Section 3) to test the hypothesis that the economic value of UGSs may vary amongst low-income, middle-income and high-income areas in a South African case study. Results may help justify increased resource allocation and prioritization to inform tailored approaches to UGS development and management appropriate to each socioeconomic level.

1.1. Turning to the Global South: South Africa in Focus

Within the Global South, South Africa presents particularly fertile research potential owing to three features. Firstly, although research on multiple facets related to UGSs has predominantly focused on the Global North [51], neglecting sub-Saharan Africa in particular [34], South Africa has been especially well represented in scholarship [52], and the country provides a relatively rich literature base on urban ecology to inform future research. Secondly, South Africa provides an obverse case to conventions in the Global North, with racial minorities (of European descent) generally presenting as the wealthiest group in the country in contrast with the poorer Black majority, as a result of past institutionalized racial discrimination under colonialism and apartheid [8]. Thirdly, apartheid and its restrictions on Black urbanization resulted in suppressed levels of internal migration to urban centers by the majority of citizens for many decades. The fall of apartheid led to rapid urbanization, as those previously restricted then migrated to cities [10], resulting in immense service delivery challenges, while also often stressing existing UGSs. South Africa has continued to struggle with the legacies of its racially segregated development history [53] and remains one of the most unequal societies in the world [54], with disparities between groups of varying SES perhaps being more pronounced than anywhere else in the world.

The scars of the past are represented spatially and environmentally across the cityscape. Older leafy suburbs with low population densities, many modelled around the ideals of the "Garden City" [55], and new gated communities that increasingly boast eco-development credentials [56], are mostly inhabited by White residents and, increasingly, Black higher income earners as well [10]. These green enclaves contrast starkly against the apartheid era townships and new subsidized mega housing projects developed for lower-income beneficiaries [57,58] that have higher population densities and limited, underdeveloped and poorly maintained UGSs [10]. Within the context of socioeconomic inequality, a significant housing backlog, blooming informal settlement sectors and the failure to deliver basic services, UGSs have been considered luxury amenities [8,38,59]. As such, the quantity and quality of UGSs vary significantly along the socioeconomic gradient [49,57], in keeping with international precedents. For example, in a case study of nine South African towns in the Eastern Cape province, McConnachie and Shackleton [10] found that newer subsidized housing projects delivered five times less green area per capita than apartheid-era townships, and up to 15 times less than more affluent, older suburbs. These results were significant considering that, proportionally speaking, higher-income suburbs presented public UGS provisions comparable in size to more densely populated apartheid-era townships. Despite the subsidized housing projects being more recently developed, and in contradiction of the state's policy support for more sustainable and just outcomes and prescriptions on UGS standards [8], these developments have continued to deliver relatively poor public UGS provisions. In addition, the wealth and racial disparities between older townships and new subsidized housing projects were proven to be negligible in comparison with the vast differences these areas presented in comparison to more affluent suburbs [10]. Results were corroborated by Venter et al. at the national scale, presenting that public and private UGSs were "abundant, accessible, greener and more treed in high-income areas with predominantly White residents relative to low-income areas" [8]. More affluent suburbs presented incomes six times higher on average, with 11.7% more tree cover and 8.9% higher vegetation cover compared to areas accommodating predominantly Black, Indian and Colored citizens (an accepted term in South Africa) [8].

Despite inequities in the provision of and access to UGS along the socioeconomic gradient, international [60,61], African [34] and South African studies [33,62] have continuously shown local communities to appreciate and acknowledge the benefits and value of UGSs.

As discussed in Section 2.1 below, UGSs may constitute critical components of the urban fabric that deliver vital services and environmental, social and economic benefits. It is often much simpler to identify environmental and social benefits [44] in contrast with more obscure economic compensations [63] that may remain unclear and underestimated [64] amongst UGS users and even decision makers. Ignorance, misconceptions and pressures exerted by the challenges highlighted above often result in those in power prioritizing land uses with a more explicit monetary value or political cachet over UGSs [48]. Prescriptive targets on UGS quantity and spacing, often aimed at promoting equal access, may in fact further undermine [19] the actions needed to institute more equitable provisions while considering varying socioeconomic contexts [18].

In keeping with the aim of this paper, the next section discusses core concepts, starting with UGS framed within the paradigms of green infrastructure and ecosystem services and disservices. This is followed by Section 2.3, presenting a more detailed exploration of economic valuation methods, focusing on hedonic price analysis and the proximity principle, with an emphasis on South African examples. Section 3 explains the methodology followed in testing the proximity principle in the case study of Potchefstroom, followed by the presentation and discussion of the results. Concluding remarks and recommendations close the paper.

2. Literature Review

2.1. Urban Green Spaces and Green Infrastructure

UGS is a widely applied and interpreted concept used in reference to a variety of green spaces of varying scales and functions in the urban environment. These may include both natural or anthropogenically developed, formal or informal green spaces that accommodate or compliment a variety of land uses and services [65,66]. It falls beyond the scope of this paper to provide a detailed discussion of the various UGS typologies and examples presented in the literature, but van Zyl et al. [67] provide an apt synopsis tailored to the South African context.

UGSs are increasingly framed as elements of green infrastructure, especially when these spaces accommodate grey-green elements that combine vegetation with engineered technical structures or grey infrastructure [68]. The green infrastructure approach provides a useful lens for this discussion of UGSs, owing to its recognition of several services, benefits and disadvantages potentially levied by such spaces [65]. Green infrastructure, included under the broader banner of "nature-based solutions" [69], is not yet served by an all-encompassing and broadly agreed-upon definition [8,70,71]. The concept is often also supplemented by the term blue-green infrastructure, in recognition of the services provided by water bodies and wetlands when part of a network of UGSs [72]. This paper draws on various established definitions, incorporating elements of blue-green infrastructure, to describe green infrastructure as an interconnected, multifunctional network comprised of links and nodes constituted by natural, semi-natural and artificial blue and green spaces and systems that deliver benefits known as ecosystem services [51,67,73]. The term ecosystem services has evaded a general definition [74], but is defined in this paper as the benefits all living species (humans in particular) derive, directly or indirectly, from the capacity of ecosystems to deliver goods and services that satisfy their needs [75,76]. Ecosystem services may be classified as provisioning, regulating, cultural and habitat or supporting services [77,78], which are regarded as direct benefits. This is captured in Table 1, which also presents examples mainly from sub-Saharan Africa.

Table 1. Main classes of ecosystem services provided by urban green infrastructure, their definitions and some examples from sub-Saharan Africa.

Ecosystem Service Category	Definition	Examples
Provisioning ecosystem services	The capability of natural or semi-natural UGSs to contribute physical products, materials or goods consumed directly by humans.	Cultivated and wild food, water, raw materials such as wood and fuel or those for construction work and arts, crafts and tools, as well as for medicinal uses [79–82].
Regulating ecosystem services	The regulation and mitigation of various processes.	Regulating the climate; the removal of pollutants by air and water filtration; water storage, filtration, and drainage; protection from disasters such as landslides and storms; seed dispersal and pollination; and providing pest and human disease regulation [83–88].
Cultural ecosystem services	The non-material benefits humans obtain from ecosystems that meet cultural or spiritual needs.	Recreation; aesthetic appreciation and reflection; physical and psychological health, educational values, social interaction and social cohesion; spiritual and religious experiences; and sense of place [89–92].
Habitat/supporting ecosystem services	Necessary to facilitate the fulfilment of all other ecosystem services.	Providing habitats to encourage biodiversity and maintenance of genetic diversity [93,94].

Other literature sources used for all the categories: Cilliers et al. [57,95]; du Toit et al. [52]; and TEEB [96].

Economic contributions are regarded as indirect benefits owing to the potential income and savings generated in relation to the four categories of ecosystem services discussed above [63]. An important additional economic benefit relates to the potential for increased property values as a result of proximity or access to UGSs [49] (see Section 2.3 below), reflecting a demand for UGSs and their associated ecosystem services [57]. SES may influence demand for ecosystem services by dictating human needs and activities, and thus influence how UGSs are used and what is expected of them, as well as how the provision of an ecosystem service affects wellbeing, for example, through physical or psychological health [18]. It follows that a holistic view of ecosystem services is required that also considers less-positive impacts, or ecosystem disservices.

2.2. Considering Ecosystem Disservices

The direct benefits derived from ecosystem services are countered by "functions of ecosystems that are perceived as negative for human wellbeing" [35], termed ecosystem disservices [97]. Although no universal definition or typology has been devised, several categories of ecosystem disservices have been identified. Four general classifications that are globally applicable are captured in Table 2, as there have been limited studies focusing on ecosystem disservices in sub-Saharan Africa.

Table 2. Different categories of ecosystem disservices provided by green infrastructure, their origin and some examples.

Ecosystem Disservice Category	Ecosystem Origin	Examples
Environmental/Ecological	Biotic	Invasive species outcompete indigenous species; Changes in species interactions and species populations.
	Abiotic	Changes in environmental variables of species in terms of soil and climate; Maintenance and management expenditure;
Economic/Financial	Biotic	Damage to infrastructure (property, cables, sidewalks, roads) caused by vegetation and tree roots; Pest-disease control.
	Abiotic	Drought; Wildfires; Siltation; Leaching of nutrients; Pruning and planting (maintenance costs); Foregone land-use opportunities; Decreased property values.

Table 2. *Cont.*

Ecosystem Disservice Category	Ecosystem Origin	Examples
Health (Physical, Mental, Safety)	Biotic	Human diseases from pathogens; Negative health impacts due to volatile organic compounds (VOC's) and pollen; Poisonous plants and pests/venomous animals; Allergens; Disease vectors; Safety hazards due to tree falls and lack of maintenance.
	Abiotic	Security concerns (crime); Anxiety; Discomfort; Floods; Storms; Landslides; Changes in air quality due to the production of VOC's; Decreased property values.
Cultural (Aesthetic and Cultural)	Biotic	Bird excrement on stonework and sculptures; Damage to infrastructure (sidewalks/paving); Littering and dumping of waste; Unpleasant odors from rotting organic matter; Aesthetically unpleasing; High noise levels; Decreased property values.
	Abiotic	Soil erosion; Landslides; Exposure to the elements (winds) makes it an unpleasant experience; Blocked views; Obstruction of traffic infrastructure.

Source: Own composition, based on Lategan and Cilliers [62]; Cilliers et al. [57]; Lyytimäki and Sipilä [35]; Lyytimäki [98]; Gómez-Baggethun et al. [99]; von Dohren and Haas [100]; Shackleton et al. [101]; Davoren and Shackleton [97].

Ecosystem disservices may (i) occur at different spatial and temporal scales and (ii) require certain threshold conditions to be met, often interacting with other disservices, leading to cumulative negative effects [101]. These may be context-specific and vary from one socioeconomic group to another, with the poor and more vulnerable often being disproportionately affected [101]. It is critical that a balanced approach is followed and that ecosystem disservices are thoroughly considered when attempting to plan for UGSs [18]. It must be noted that whilst certain scholars, for example, Shackleton et al. [101], explicitly exclude impacts such as littering and crime from ecosystem disserves as they are not the direct outcome of an ecosystem process, these are included in Table 2, in recognition of the broad perspective required when evaluating and valuing UGSs and their associated effects. Such impacts have also been recognized by others, for example, Lyytimäki and Sipilä [35] and von Dohren and Haas [100]. Negative impacts related to the lack of a dedicated function or amenities, poor maintenance, nuisance and crime, as is often exhibited by public UGSs in South Africa [50], may sway public sentiment and affect the willingness of the market to pay for close proximity to certain UGSs [99]. Such impacts, together with the availability and size of private UGSs such as domestic gardens, may also result in a preference for green views and not necessarily for immediate proximity or access to public UGSs [102,103].

Accounting for both ecosystem services and disservices is crucial if the net benefits UGSs deliver in cities are to be demonstrated [104]. In light of the importance of demonstrating the obscure economic value of UGSs and the complex interrelationship between ecosystem services and disservices, several economic valuation methods have been applied. These include the stated preference approach, revealed preference approach, avoided cost, replacement cost, travel cost, contingent valuation and hedonic price analysis [57,99,105,106]. Hedonic price analysis is highlighted for its broad application internationally and in South Africa [50].

2.3. *Hedonic Price Analysis and the Proximity Principle*

Hedonic price analysis generally breaks down the price of an observed good into discrete marginal prices linked to separate characteristics, whilst also considering consumer choice [107,108]. Hedonic price analysis has been employed in the valuation of UGSs by investigating residential property prices to identify and quantify various factors that may exert an influence on property values, without considering the interactions between these factors [47,48,107,109]. The relative distance to UGSs has been particularly widely employed as a factor, translated via the proximity principle. As stated in the introduction, the proximity principle posits that property prices will increase as distance to UGSs decreases, thereby determining the value of amenities such as UGSs via the value of surrounding residential properties, taken as representative of consumers' willingness to pay for proximity to UGSs and their associated benefits [47,49]. Increased residential property values translate to increased residential property taxes for municipalities, which may present economic returns to account for public expenditures on the development and maintenance of these spaces [48,110].

Several studies have attempted to investigate the impacts of UGSs on residential property values, with the most focused-on case studies being located in the Global North. In general, the results have supported the proximity principle to varying degrees [50]. General trends seem to confirm the proximity principle where certain conditions are satisfied, for example, when residential properties have fairly direct access to a UGS or enjoy views of it; certain active or passive uses are accommodated; UGSs are perceived as safe and well-maintained public spaces; and surrounding residential properties are smaller in area, for example, by accommodating higher-density dwelling units or where socioeconomic conditions result in smaller or less developed domestic gardens [110–112].

In consideration of the literature reviewed above, the next section discusses the methodology employed in this research, before results are disclosed and discussed.

3. Materials and Methods

This paper relied on a quantitative research approach, based on an investigation of a South African case study in Potchefstroom, to analyze residential property values in relation to UGS proximity (relative distance from a UGS) within three areas reflecting characteristics of varying SESs. The following section discusses the choice of Potchefstroom as the case study area and elaborates on the steps followed in analyzing property values in the three areas of differentiated SES.

3.1. Study Area

Potchefstroom (26°42′3″ S 27°05′49″ E), located in South Africa's North West province and within the JB Marks Local Municipality (see Figure 1), was selected as the case study area, as the town has previously been the focus of preliminary research on the proximity principle and UGSs [48–50], offering established methods and data for longitudinal comparison. This paper drew primarily on the research conducted by Cilliers and Cillliers [48], returning to the areas investigated, refining the methodology employed with updated data and expanding on the approach through a broader geographic and socioeconomic focus.

Cilliers and Cilliers [48] employed hedonic price analysis and investigated the proximity principle in relation to a local UGS in five middle-income neighborhoods in Potchefstroom, using as variables municipal valuations of properties surrounding a UGS in Grimbeek Park, van der Hoff Park, Potchefstroom Dam, Heilige Akker and Oewersig (for more detail on neighborhood characteristics and the UGS included in each sample, see Section 3.2). Cilliers and Cilliers [48] identified sample properties in each neighborhood depending on their relative distance to a specific UGS, and classified these into three zones depending on their proximity to the UGS in question. Properties in zone 1 bordered the UGS; those in zone 2 were further away from the UGS, mostly located across the street or one property away from those in zone 1; and those in zone 3 were further away from the UGS than those in zone 2. Figure 2 illustrates an example of a UGS and the properties included in each zone, selected based on their relative distance from the UGS as determined by Cilliers and Cilliers [48].

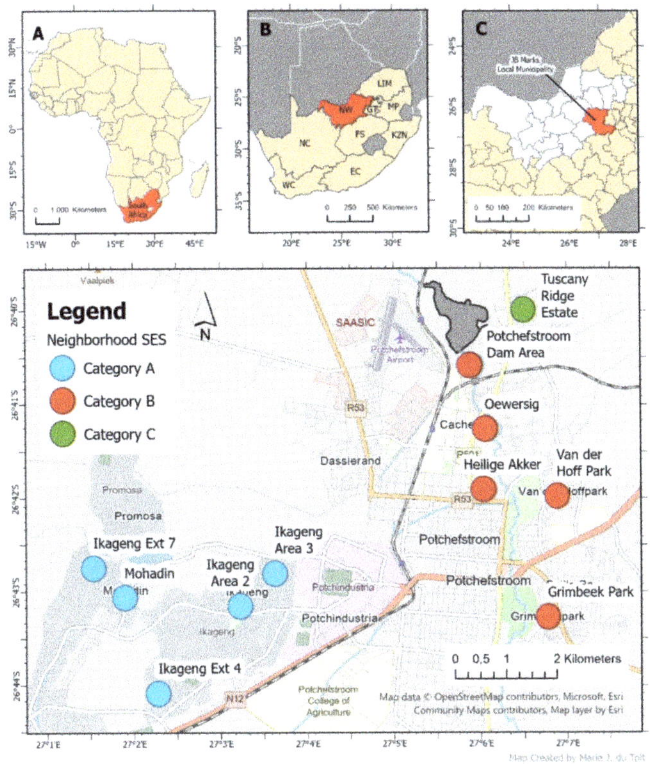

Figure 1. Maps of the study area showing South Africa (**A**); The North-West province (NW) (**B**); and the JB marks Local Municipliaty (**C**) within which Potchefstroom is located, as well as the location of neighborhoods within each SES Category in Potchefstroom. Blue: Category A sample (SES 1); Red: Category B (SES 4) sample; Green: Category C (SES 5) sample. Source: Own construction (2022) by M.J. du Toit.

Figure 2. The Heilige Akker neighborhood as an example of Category B (SES 4), with sample properties in three zones of relative distance from a specific UGS. Source: Own construction (2022) based on Cilliers and Cilliers [48] by M.J. du Toit (2022).

Cilliers and Cilliers [48] employed 2013 municipal valuations to arrive at a mean South African rand price per square meter value (R/m^2) for each property included in the sample. Each zone in each neighborhood was then assigned a mean R/m^2, employed as a variable in statistical analyses to compare zones. In general terms, the findings indicated that properties in zone 1 presented lower average valuations than those further away from a UGS, leading Cilliers and Cilliers [48] to reject the proximity principle in the case of middle-income neighborhoods in Potchefstroom. In 2020, Combrinck et al. [49] returned to the Potchefstroom case study to replicate the study in the same middle-income sample using the latest municipal valuations released in 2019. Again, the results presented by Combrinck et al. [49] rejected the proximity principle. The validation of previous results led to two main questions: firstly, whether the rejection of the proximity principle in the Potchefstroom case study, in comparison with most international findings (see Section 2.3), could be explained by the use of municipal valuations as variables, in contrast with most other studies that have employed market values? Lategan et al. [50] addressed this issue, returning to Cilliers and Cilliers' [48] case study, but employing estimated market values as variables in their analysis. Again, the results generally rejected the proximity principle. The second question related to the impact of a socioeconomic gradient on the confirmation or rejection of the proximity principle, or in other words, whether the proximity principle would also be rejected in low- and high-income samples in Potchefstroom. This question is considered particularly important given the spatial legacy of South Africa's colonial and apartheid history and the inequitable distribution and quality of UGSs discussed previously (see Section 1.1).

Potchefstroom is characterized by an unequal and typically fragmented apartheid-city layout, presenting a steep socioeconomic gradient [57]. The wealthy are located in green suburbs and gated communities, and the previously disadvantaged are still accommodated in peripheral townships that often provide lower-quality housing and access to services and amenities, including UGSs [113]. Lubbe [113] examined the distribution of domestic garden flora across a socioeconomic gradient in Potchefstroom, demarcated according to five SES categories based on employment, household size, access to basic services, number of rooms in dwellings and schooling. These included SES levels 1 to 3 (low-income), SES level 4 (middle income), and SES level 5 (high-income). The results showed that the higher the SES level, the greater the access to higher-quality public and private UGSs, the more species rich the environment and the higher the vegetation and tree cover [57,113], in keeping with international precedents (also see Figure 3).

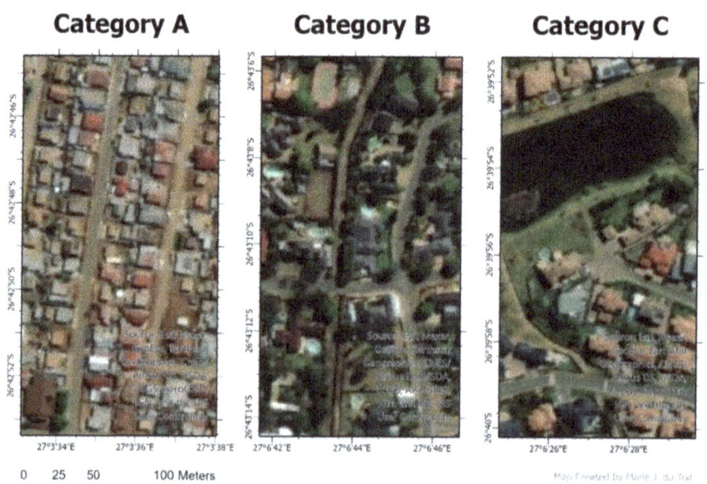

Figure 3. Aerial views of sections of Category A (SES 1), B (SES 4) and C (SES 5) neighborhoods. Source: Own construction (2022) by M.J. du Toit.

3.2. Identification of Samples

This paper replicated the approach followed by Cilliers and Cilliers [48], but like Combrinck et al. [49], employed 2019 municipal valuations to investigate the proximity principle using three zones of proximity to a UGS in three distinct areas of Potchefstroom differentiated by level of SES. This included properties surrounding a single UGS in five lower-income neighborhoods (SES 1), from here on referred to as the Category A sample (Ikageng Extension 7; Mohadin; Ikageng Extension 4; Ikageng Area 2; and Ikageng Area 3); the same five middle-income neighborhoods (SES 4) examined by Cilliers and Cilliers [48], from here on referred to as the Category B sample (Grimbeek Park; van der Hoff Park; Potchefstroom Dam; Heilige Akker; and Oewersig); and a higher-income sample (SES 5), from here on referred to as the Category C sample (Tuscany Ridge Estate) (see Figure 1). As noted in Table 3, the number of properties included in each sample varied as a result of neighborhood layout and size or the availability of municipal valuation data.

Table 3. SES Category neighborhoods and descriptive details.

Sample	Neighborhoods	Zoning and Size for Each UGS	Characteristics/Land Use for Each UGS	Number of Residential Properties Included	Property Size Range (m^2)
Category A	Mohadin	Public open space	Bordering a communal urban park	40	500–800
	Ikageng Extension 7	Public open space	Bordering vacant land connected to the Poortjies Dam.	72	200–500
	Ikageng Extension 4	Public open space	Bordering vacant land with informal squatters.	37	200–500
	Ikageng Area 2	Private open space	Located opposite a sporting field.	36	200–500
	Ikageng Area 3	Private open space	Bordering vacant land.	36	200–500
Category B	Grimbeek Park	Public and private open space	Golf course and areas used for birdwatching and horseback riding.	41	1000–2000
	Van der Hoff Park	Public open space	Wetland with limited tree cover.	43	1000–2000
	Potchefstroom Dam	Public open space	Potchefstroom Dam and Mooirivier with dense vegetation.	27	1000–2000
	Heilige Akker	Educational	UGS with limited vegetation and tree cover.	36	1000–2000
	Oewersig	Public open space	Mooirivier with dense vegetation and tree cover.	41	1000–2000
Category C	Tuscany Ridge	Private open space	UGS with moderate tree and vegetation cover, with well-maintained lawns and water bodies.	72	700–2000

Table 3 provides details regarding each neighborhood included under each SES category, presenting zoning, each UGS's characteristics and uses, the number of properties included in each sample and average property size ranges to inform the discussion.

In further support of Figures 1 and 2 and Table 3, Figure 3 provides an overview of general neighborhood characteristics for each included category area. Note the stark contrasts in density and vegetation cover between Category A and Categories B and C.

Whereas the Category A and Category B areas contained five neighborhood samples each, certain limitations in the context of Potchefstroom, and its high-income gated communities specifically, limited the number of feasible samples within the Category C classification. These included the limited number of high-income gated communities in the town; the lack of large communal green spaces in other options, in favor of large private gardens; the relatively young age of other options in comparison with Tuscany Ridge and its mature public green spaces; challenges related to the layout of other options, complicating the selection of residential properties surrounding communal greens in zones of comparable distance; and the proliferation of townhouses and clusters rather than detached dwellings in other options. Single-case studies are commonly applied and are regarded as particularly useful experimental designs in a variety of situations, for example, when researchers have limited resources, conditions present low incidences or when examining the effects of novel or expensive interventions [114]. However, the inclusion of only one sample in Category C is still acknowledged as a limitation of this research, providing limited representation of high-income demographics and their associated UGSs in Potchefstroom. Results obtained from Category C are thus only interpreted in an exploratory manner.

Following the identification of sample properties and the calculation of mean R/m^2 values, statistical tests were conducted. This paper employed two-way analysis of variance (ANOVA) using Categories A, B and C and the three zones of relative distance to a UGS as factors. For the two-way ANOVA, an effect size of $\approx$0.2 indicates a small, not practically significant difference; an effect size of $\approx$0.5, a medium, practically visible difference; and an effect size of $\approx$0.8, a large, practically significant difference. Post-hoc tests based on Tukey's B were conducted to make pairwise comparisons between groups (category and zone). As a convenience sample was used, p-values were included for the sake of completeness and to add further credence, but were not interpreted. The following section reports and discusses the results obtained from the analysis of municipal valuations in relation to UGS proximity in each SES category.

4. Results and Discussion

The results of the quantitative analysis are provided below, summarizing the data and outcomes of the statistical tests, followed by a discussion that highlights certain findings regarding the proximity principle in the context of each UGS, aided by anecdotal evidence from observations and spatial analyses. Table 4 below provides an overview of each neighborhood included in each category area, the number of properties included, the mean R/m^2 municipal valuations retrieved for each zone and which zone presented the highest, middle and lowest mean R/m^2 in each neighborhood.

Table 4. Summary of individual neighborhoods and three zones of relative distance from a UGS per category area.

Sample	Neighborhoods	Zone	N	Mean R/m^2	Highest, Middle or Lowest Zone (Mean R/m^2)
Category A	Ikageng Extension 7	1	24	526.52	Lowest
		2	24	635.96	Highest
		3	24	588.89	Middle
	Mohadin	1	13	969.21	Highest
		2	15	815.99	Middle
		3	13	765.87	Lowest
	Ikageng Extension 4	1	12	1099.52	Highest
		2	12	824.62	Middle
		3	13	599.57	Lowest
	Ikageng Area 2	1	12	973.41	Highest
		2	12	833.98	Middle
		3	12	758.63	Lowest
	Ikageng Area 3	1	12	898.65	Lowest
		2	12	1083.30	Highest
		3	12	1080.96	Middle

Table 4. *Cont.*

Sample	Neighborhoods	Zone	N	Mean R/m^2	Highest, Middle or Lowest Zone (Mean R/m^2)
Category B	Grimbeek Park	1	14	1260.70	Lowest
		2	14	1611.67	Middle
		3	13	1699.25	Highest
	Van der Hoff Park	1	15	1290.59	Lowest
		2	15	1472.43	Middle
		3	13	1624.30	Highest
	Potchefstroom Dam	1	9	1116.44	Lowest
		2	9	1303.45	Middle
		3	9	1448.64	Highest
	Heilige Akker	1	10	1751.96	Lowest
		2	12	1904.15	Highest
		3	14	1850.28	Middle
	Oewersig	1	14	1668.44	Middle
		2	14	1852.15	Highest
		3	13	1549.20	Lowest
Category C	Tuscany Ridge Estate	1	24	3915.46	Highest
		2	24	3612.63	Middle
		3	24	3058.63	Lowest

Table 5 follows from Table 4, and presents the mean R/m^2 values for each zone per category area, used as the dependent variable in two-way ANOVA testing to determine the relationship between different zones in each category area.

Table 5. Two-way ANOVA based on mean R/m^2 identified for each zone in each category sample using three zones in each category area as factors.

Dependent Variable	R/m^2				Effect Sizes	
Category	**Zone**	**Mean**	**Std. Deviation**	**N**	**1 with**	**2 with**
A	1	834.1791	519.47186	73		
	2	805.4081	411.62278	75	0.06	
	3	729.1787	378.35255	74	0.20	0.19
	Total	789.4590	440.33770	222		
B	1	1418.2978	389.71718	62		
	2	1643.1354	382.48991	64	0.58	
	3	1649.7992	424.23869	62	0.55	0.02
	Total	1571.1845	411.23169	188		
C	1	3915.4593	1018.23754	24		
	2	3612.6349	444.37633	24	0.30	
	3	3058.6274	610.12136	24	0.84	0.91
	Total	3528.9072	805.02298	72		
Total	1	1527.0475	1193.93740	159		
	2	1547.6657	1026.01751	163		
	3	1435.3364	914.53858	160		
	Total	1503.5766	1049.60171	482		

In Table 6, the *p*-values (significance) indicate that the interaction effect between category and zone is statistically significant, seeing that the reported value of <0.000 is below the guideline value of <0.05.

Table 6. Test of between-subject effects.

Dependent Variable Source	R/m^2 Type III Sum of Squares	df	Mean Square	F	Sig.
Corrected Model	421,072,689.835 [a]	8	52,634,086.229	228.765	0.000
Intercept	1,462,376,992.023	1	1,462,376,992.023	6355.965	0.000
Category	409,338,084.457	2	204,669,042.228	889.558	0.000
Zone	4,365,404,467	2	2,182,702,234	9487	0.000
Category * Zone	10,487,761,981	4	2,621,940,495	11.396	0.000
Error	108,827,577,814	473	230,079,446		
Total	1,619,578,126,813	482			
Corrected Total	529,900,267,649	481			

[a] R Squared = 0.795 (Adjusted R Squared = 0.791).

The results of post-hoc testing using Tukey's B are reported below, first by category (Table 7) and then by zone (Table 8).

Table 7. Tukey's B with homogenous subsets: category.

		R/m^2		
Tukey's B [a,b,c]				
Category	N	Subset		
		1	2	3
A	222	789.4590		
B	188		1571.1845	
C	72			3528.9072

Means for groups in homogeneous subsets are displayed. Based on observed means. The error term is Mean Square(Error) = 230,079.446. [a.] Uses Harmonic Mean Sample Size = 126.515. [b.] The group sizes are unequal. The harmonic mean of the group sizes is used. Type I error levels are not guaranteed. [c.] Alpha = 0.05.

Table 8. Tukey's B with homogenous subsets: zone.

	R/m^2	
Tukey B [a,b,c]		
Zone	N	Subset
		1
3	160	1435,3364
1	159	1527,0475
2	163	1547,6657

Means for groups in homogeneous subsets are displayed. Based on observed means. The error term is Mean Square(Error) = 230,079.446. [a.] Uses Harmonic Mean Sample Size = 160.649. [b.] The group sizes are unequal. The harmonic mean of the group sizes is used. Type I error levels are not guaranteed. [c.] Alpha = 0.05.

Whilst the two-way ANOVA (Tables 5 and 6) showed significant differences between the zones in each category, from Tables 7 and 8 it is apparent that differences were found between all categories, whilst zones presented no differences from one to another.

In illustrating the interaction between categories and zones, Chart 1 below provides a graphic representation of the results for ease of reference.

The results captured in Chart 1 present a simplified view of the data, illustrating that in Category A, mean R/m^2 values decreased further away from a UGS; in Category B, mean R/m^2 values increased further away from a UGS; whilst in Category C, mean R/m^2 values decreased as distance from a UGS increased. These observations facilitate deliberation for a verdict on the proximity principle in each category area, as discussed in more detail below.

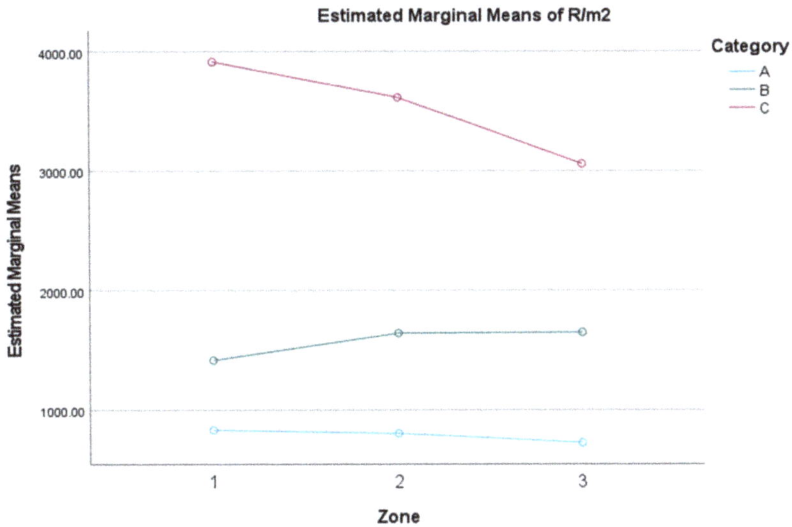

Chart 1. Graphical representation of findings.

4.1. Findings on the Confirmation or Rejection of the Proximity Principle in Category A (Low SES)

As shown in Table 4, the results for Category A indicated that zone 1 (closest to a UGS) presented the highest mean R/m^2 value in three of the five residential areas (Mohadin, Ikageng Extension 4 and Ikageng Area 2) compared to zones 2 and 3 (further away from a UGS). The Mohadin sample presented a mean value of R969.21/m^2 in zone 1, compared to R815.99/m^2 (18,8%) in zone 2 and R765,87/m^2 (12.6%) in zone 3. Likewise, Ikageng Extension 4 presented a mean value of R1099.52/m^2 in zone 1, compared to R824.62/m^2 in zone 2 (33.35%) and R599.57/m^2 in zone 3 (83.52%). Ikageng Area 2 reflected the same trends with a mean value of R973.41/m^2 in zone 1, compared to R833.98/m^2 in zone 2 (16.7%) and R758.63/m^2 in zone 3 (28.3%), confirming the proximity principle. Zone 1 presented the lowest R/m^2 value in two of the five residential areas (Ikageng Extension 7 and Ikageng Area 3), contradicting the premise of the proximity principle. In Ikageng Extension 7, zone 1 presented a mean value of R526.52/m^2, compared to R635.95/m^2 (20.8%) and R588.89/m^2 in zone 3 (11.8%). In Ikageng Area 3, zone 1 delivered a mean value of R898.65/m^2, in comparison with R1083.30/m^2 in zone 2 (20.5%) and R1080.96 in zone 3 (20.3%). The result of the two-way ANOVA test indicated small effect sizes ($d \leq 0.20$) or practically insignificant differences between the mean R/m^2 values for zone 1, zone 2 and zone 3 in Category A, confirming small differences between the mean R/m^2 values presented (Table 5). The results translated to a general confirmation of the proximity principle in the Category A sample (Chart 1).

As shown in Table 4, the proximity principle was confirmed in samples in Mohadin, Ikageng Extension 4 and Ikageng Area 2. The UGS in Mohadin was constituted by a vacant and relatively barren land parcel with a low vegetation cover and no trees, providing limited regulating and supporting ecosystem services, but no provisioning services. The space did present ample room for the realization of cultural ecosystem services, facilitating recreation and social interaction in a quiet environment. These benefits could overshadow other, more negative considerations such as ecosystem disservices, including diminished aesthetic appeal, translating to a willingness to pay for proximity to the UGS. In Ikageng Extension 4, the confirmation of the proximity principle could be attributed to a relatively visually attractive UGS that provided good access to a more park-like setting populated by a number of trees, whilst still providing some open space for recreation and play (cultural ecosystem services). However, a site visit revealed the presence of newly erected informal dwellings, which feasibly produce ecosystem disservices in terms of the loss of aesthetic appeal and the emergence of additional security issues, with the potential to negatively affect more up-to-date property valuations. In comparison with the other neighborhoods included under the Category A sample, the UGS in Ikageng Area 2 seemingly delivered the highest amenity value, specifically in terms of various cultural ecosystem services. The space provided a link to a sporting facility located close by and delivered opportunities for recreation and social gatherings in a quiet and relatively well-maintained, aesthetically pleasing environment, in comparison with the surrounding area. This was reflected in the higher municipal valuations for residential properties located closer to it. The results in Ikageng Extension 7 rejected the proximity principle; however, the highest mean R/m^2 value was identified in zone 2, indicating that closer proximity, but not adjacency, to the UGS was still preferred by the market. On closer inspection, the UGS in question, comprised of vacant land, appeared to provide a quiet environment with adequate open space for recreation and social gatherings, as well as a relative aesthetic appeal contributed by moderate vegetation cover. Vegetation also provided regulating and supporting ecosystem services. However, the UGS's location, linking it to a densely vegetated area with high tree cover surrounding the Poortjies Dam and the potential to harbor criminal activity and vagrants on its periphery (as reported elsewhere by Davoren and Shackleton [97] and Lategan et al. [50]), in conjunction with the potential for disturbances from social gatherings and children at play (ecosystem disservices), could be cited to explain the proximity principle's rejection. Although this area offered regulating services in terms of climate regulation and the removal of pollutants, and supporting services such as high native biodiversity, these benefits are potentially not recognized by all residents.

The results in Ikageng Area 3 merit closer inspection in that, although the proximity principle was rejected, the difference between zone 2 and zone 3 was marginal (0.22%), thus insinuating that increased distance from the UGS did not result in significant increases in mean R/m^2 values. The UGS in Ikageng Area 3 was of a relatively small size, had sparse vegetation cover (therefore presenting low levels of regulating and supporting ecosystems services) and was poorly maintained, thus decreasing its aesthetic appeal (cultural ecosystem service). It was also located adjacent to a very busy road, further limiting its attraction power and use in terms of other cultural ecosystem services, and feasibly accounting for the negative impact on immediately adjacent property values. Interpretations based on ecosystem services, for example, a lack of aesthetic appeal, to account for the rejection of the proximity principle are highly subjective, and studies of such impacts conducted in the Global South have been limited [97]. Studies in sub-Saharan Africa have indicated that aesthetic values in such contexts may be linked to perceptions regarding urban trees, with more focus often placed on negative features such as thorns, poisons and allergic reactions [115].

4.2. Findings on the Confirmation or Rejection of the Proximity Principle in Category B (Medium SES)

For Category B, the mean R/m^2 in four of the five residential areas (Grimbeek Park, Van der Hoff Park, Potchefstroom Dam and Heilige Akker) indicated that zone 1 presented the lowest value in mean R/m^2 compared to zone 2 and zone 3. As illustrated by Table 4, zone 1 in Grimbeek Park presented a value of R1260.70/m^2, compared to a higher value of R1611.67/m^2 in zone 2 (27.8%) and R1699.25/m^2 in zone 3 (34.8%). In Van der Hoff Park, the same trends were presented, with a value of R1290.59/m^2 in zone 1, R1472.43/m^2 in zone 2 and the highest value, R1624.30/m^2, in zone 3. Zone 1 to zone 2 presented a difference of 14%; zone 2 to zone 3, a difference of 10.3%; and zone 1 to zone 3, a difference of 25.9%. In the Potchefstroom Dam sample, zone 1 presented a value of R1116.44/m^2, compared to R1303.45/m^2 in zone 2 (16.8%) and R1448.64/m^2 in zone 3 (29.8%). In Heilige Akker, zone 1 presented the lowest value of R1751.96/m^2, but zone 2 presented the highest value of R1904.15/m^2, followed by the second-highest value of R1850.28/m^2 in zone 3. Thus, zone 1, located closest to the UGS, presented the lowest mean R/m^2 value, 8.7% lower than zone 2 and 5.6% lower than zone 3, in opposition to the proximity principle. In

Oewersig, zone 1 presented a mean value of R1668.44/m^2, 11% lower than the R1852.15/m^2 in zone 2 and 7.7% lower than the R1549.20/m^2 in zone 3. The results of the two-way ANOVA test presented medium effects or practically visible differences (d = 0.58 and 0.55) between zone 1 and the zones further away from the UGS (Table 5), where zone 1's mean R/m^2 value was lower than that of the other two zones (1418.2978 vs. 1643.1354 and 1649.7992). In totality, these results lead to a general rejection of the proximity principle in Category B, as also presented by Cilliers and Cilliers [48] and Combrinck et al. [49] (also see Chart 1).

Despite a rejection of the proximity principle, the majority of the UGSs included under this SES category were of a much higher quality in terms of vegetation and tree cover, amenities and general maintenance compared to those in Category A. The UGS in Grimbeek Park accommodated a golf course and areas for horseback riding and birdwatching, and thus hosted a range of recreational activities and opportunities for social interaction (cultural ecosystem services) [48]. The space could also be perceived as aesthetically pleasing and well maintained and managed, as well as housing high levels of biodiversity, presenting the highest quality UGS in this sample. A relatively low difference in R/m^2 values was also identified between zone 2 and zone 3 here (5.43%), signifying the strong impact of immediate adjacency compared to proximity. In Van der Hoff Park, the sample UGS bordered the Mooi River and included a biodiverse and species rich wetland [116] worthy of conservation [117], presenting a quiet environment that could accommodate recreation and physical activity. Additionally, the urban wetlands provide various regulating ecosystem services such as water regulation, purification and drainage [118,119] and high biodiversity (supporting ecosystem services) [117]. However, the location along the Mooi River, having dense vegetation cover, also provided a potential movement corridor for vagrants and criminals, creating high security risks for residents living close by [48,97]. Other ecosystem disservices observed for this wetland included decreased aesthetic appeal during the winter months when annual plants have died down, unpleasant odors due to polluted water entering through stormwater drains and standing water during the rainy season, providing breeding ground for mosquitos. The UGS in the Heilige Akker neighborhood also linked to the Mooi River and presented high biodiversity, but with a lower tree cover and larger lawn cover [48]. The space provided some opportunities for recreation and social gathering in a relatively quiet and well-maintained environment [48]. Properties in zone 2 of the sample in Heilige Akker presented peak mean R/m^2 values, demonstrating that immediate adjacency may not have been valued, and that relative proximity was preferred. This was also the case in the Oewersig sample. The UGSs in the Oewersig and Potchefstroom Dam neighborhoods were both densely vegetated and tree covered, accommodating high levels of biodiversity (supporting ecosystems services) and fulfilling several regulating ecosystem services [48]. However, as elsewhere, this also generated ecosystem disservices by obscuring surveillance and providing potential hiding places for intruders. Yet, the Oewersig UGS, in particular, provided important cultural ecosystem services, as it included a large lawn fit for recreational purposes and presented significant educational value, owing to a short trail with information boards on bird, reptile and amphibian biodiversity, which was the outcome of a resident initiative to improve the area. The rejection of the proximity principle in Category B (Chart 1), despite the relatively high-quality UGSs included in the sample, which provided several ecosystem services, could be ascribed to some of the ecosystem disservices noted, in conjunction with the generally large and lush private gardens available to home owners in these neighborhoods which fulfil their basic recreational needs [110,112].

4.3. Preliminary Findings on the Confirmation or Rejection of the Proximity Principle in Category C (High SES)

As seen in Table 4, mean R/m^2 values in the Tuscany Ridge Estate indicated that zone 1 presented the highest value in mean R/m^2 compared to zone 2 and zone 3. Zone 1, featuring properties bordering a UGS, presented the highest value of R3915.46/m^2, zone 2 presented the second-highest value of R3612.63/m^2 and zone 3 presented the lowest value of R3058.63/m^2. Zone 1 to zone 2 presented a difference of 8.4%; zone 2 to zone 3, a difference of 18.1%; and zone 1 to zone 3, a difference of 28%. For category C, large effect sizes or practically significant differences (d = 0.84 and 0.91) were reported between zone 1 and the other two zones' mean R/m^2 values (Table 5). This time around, zone 1's mean R/m^2 value was higher than that of the other two zones (3915.4593 vs. 3612.6349 and 3058.6274). The results in the Tuscany Ridge Estate sample thus showed that properties in closer proximity to public green spaces were valued substantially higher than those located further away.

The well-maintained and aesthetically pleasing UGSs in the high-income Tuscany Ridge Estate presented moderate tree and vegetation cover, but accommodated high levels of biodiversity (supporting ecosystem services), including grassland fragments from the vulnerable Rand Highveld Grassland vegetation type [94]. These spaces also presented social benefits, as the UGSs encouraged social interaction and recreation in a safe environment. The UGSs further provided functions as part of sustainable stormwater management systems. The residential properties included in the sample were larger than those included under Category A, with some also being smaller and equal in size to those in Category B. Tuscany Ridge Estate residents could, however, substitute more limited, private UGSs—in some cases due to smaller stand sizes and, in others, to exceptionally large dwelling footprints—with high quality public UGSs, where many of the ecosystem disservices often associated with such spaces in the South African context could be avoided or effectively managed (see Section 1.1).

5. Conclusions

This research fits under the ever-expanding framework of environmental justice, with its strong focus on equality in the distribution and quality of environmental resources, such as UGSs, in relation to socioeconomic differentiations. Within this broad focus, this paper attempted to relate an economic value to UGSs based on international conventions on the use of hedonic price analysis and the proximity principle, not with the intention of monetizing nature or viewing it solely through a capitalist lens [120], but to provide preliminary evidence of the manner in which UGSs may be valued by the market along a socioeconomic gradient. This is intended to inform decision making in support of a more just and greener urban agenda. As presented in Sections 1.1 and 2, improved understanding of such values is critical in the underrepresented contexts of the Global South, where rapid urbanization, its accompanying resource depletion and socioeconomic and environmental inequality demand increased scholarship and understanding. South Africa presents an interesting microcosm in which such issues are particularly pronounced.

Through the case study of Potchefstroom, this paper provided a novel investigation of the market's willingness to pay for proximity to UGS based on the average municipal valuation/m^2 in areas of varied SES, including low (Category A), middle (Category B) and high (Category C) income areas. Broadly speaking, the results verified the research hypothesis, demonstrating that the proximity principle was confirmed in low-income neighborhoods where residential properties were smaller and UGSs were of low to moderate quality, and rejected in middle-income neighborhoods where UGSs were of improved quality, but the residential properties were larger than in low-income neighborhoods. The limited sample in the high-income gated community, with its UGSs of outstanding quality and its number of smaller residential properties compared to Category B neighborhoods, also showed higher property valuations in proximity to UGSs (see Tables 3 and 4 and Chart 1). Given the small size of the sample in Category C, a generalized verdict on the confirmation or rejection of the proximity principle in high-income areas cannot be reached. However, useful observations may still be made. The results and subsequent discussion delivered key conclusions and recommendations.

5.1. Socioeconomic Characteristics and Urban Green Space Functions Impact the Value Attributed to Them

Firstly, the diverse results within each SES level reconfirmed the discussions presented in the literature review on the impact of socioeconomic characteristics, as determinants of context, on the use of and value attributed to UGSs and ecosystem services and disservices [25,26,29,30]. Whilst certain common themes emerged across SES categories, context-appropriate planning responses are justified and demand consideration in UGS planning.

The potential impact of residential property size, and therefore the area available for the cultivation and use of domestic green spaces, as well as the probable effects of crowding and a lack of privacy [58], can be regarded as major influences on the economic value placed on public UGSs. This was evidenced in both the Category A and C samples, representing opposite sides of the SES spectrum, where smaller residential property sizes could be linked to higher property prices when in proximity to UGSs. Thus, this potentially calls for increased areas for public UGSs where smaller residential properties and increased population densities are present. Whilst this may be regarded as a foregone and logical conclusion and recommendation, it opposes the standard practice of prescriptive UGS targets, which may limit adaptable responses in diverse socioeconomic contexts [18]. In the South Africa case, this may apply directly to development standards in low-income and subsidized housing projects where small stand sizes, high population densities and the incursion of informal backyard structures often threaten domestic UGS access [62]. Public UGSs also require additional consideration regarding the improvement of UGS area and quality in both completed and new low-income housing developments. In existing areas, the danger of infill development at the cost of valuable UGSs should be carefully considered, and existing UGSs and vacant land protected and renewed.

Such considerations are not restricted to areas occupied by those of lower SES. In Potchefstroom's Category B sample, the rejection of the proximity principle directed attention to the quality and amenity value of existing public UGSs [48]. Whilst we observed these spaces in middle-income neighborhoods to be of improved quality over those in lower-income areas, without dedicated functions or support for social activities, these spaces may not be used or valued sufficiently. To draw the immediate and broader community, many of whom may have substantial access to private UGSs and transport and can afford to travel to other destinations for leisure and recreation, concerted efforts are required. Public–private partnerships, in which local champions often prove critical, may be needed to capitalize on a neighborhood's resources and compensate for the incapacity and lack of resources or political will of the public sector. Such initiatives should also place a firm focus on ecosystem disservices, especially the social ails potentially related to UGSs, as presented in Section 1.1 and emphasized by Davoren and Shackleton [97] as important issues in the Global South. This may include programs to improve the maintenance of vegetation and clear areas for recreation, and to improve surveillance and lighting. It is, however, critical that public UGSs are retained as accessible and welcoming public spaces that contribute to social cohesion and not division, especially in the South African context.

5.2. Urban Green Spaces Need to Be Valued as Green Infrastructure to Enhance Value

Increased municipal property valuations and their associated property taxes may follow when UGSs are valued by homeowners and prospective buyers, but local authorities need to value UGSs as green infrastructure as well. The lack of dedicated green infrastructural systems in public UGSs in the low- and middle-income samples included in this research attests to a lack of application and a failure to realize the potential multifunctional values UGSs can deliver. It is thus recommended that alternatives combining grey and green infrastructure in the development of new and the retrofitting of existing UGSs and infrastructural networks be more ardently pursued as priorities in planning practice. For example, the UGSs included under Category C constituted important components of the development's stormwater management network. Stormwater management through drainage, retention, filtration, etc., is one of the most commonly derived green infrastructure functions, as part of regulating ecosystem services (see Table 1). Whilst regulating, provisioning and habitat/supporting ecosystem services often bear identifiable and even quantifiable benefits, more obscure advantages related to cultural ecosystem services must also be considered in unison to extract optimal value. Whereas the Category C sample and its abundant, well-maintained public UGSs may present an ideal, far removed from the realities of low-income and even middle-income areas, it provides some perspective on the impact of quality UGSs on property value. In the Category C sample, higher property valuations in relation to UGSs between properties in zone 1 and zone 3, by substantial margins in value (28%), highlight the potential significant impact of access to usable UGSs of high quality.

5.3. More Valuation Studies Are Needed across Socioeconomic Gradients

As noted, the consideration of only one sample in Category C is recognized as a limitation of this study. Other limitations include the lack of quantitative and more structured qualitative evidence on the ecosystem services and disservices delivered, as well as the need for more data on resident perceptions pertaining to the UGSs included in this study, to overcome a reliance on anecdotal observations. It is thus recommended that future research be expanded to other case study locations in South Africa and the Global South where an increased number of samples in high-income areas and gated communities can be included for analysis. It is also recommended that this case study be revisited and supplemented by more quantitative and qualitative analyses of the UGSs included to augment the preliminary findings presented here. Despite its limitations, this research provides a firm foundation from which to investigate the impact of SES on the proximity principle to improve UGS planning in South Africa (and the broader context of the Global South) and advance emerging themes such as environmental justice, context sensitivity and greener urban environments [1].

Author Contributions: Conceptualization, S.S.C., E.J.C. and L.G.L.; methodology, S.S.C. and E.J.C.; formal analysis, Z.S. and L.G.L.; investigation, Z.S.; writing—original draft preparation, L.G.L.; writing—review and editing, L.G.L., Z.S., E.J.C. and S.S.C.; visualization, Z.S.; supervision, E.J.C., S.S.C. and L.G.L.; project administration, L.G.L.; funding acquisition, E.J.C. All authors have read and agreed to the published version of the manuscript.

Funding: This research was funded by National Research Foundation South Africa, grant number 116243.

Data Availability Statement: The data presented in this study are contained in the body of the paper.

Acknowledgments: This research (or parts thereof) was made possible by the financial contributions of the National Research Foundation (NRF), South Africa (Grant No. 116243). Any opinion, findings, conclusions or recommendations expressed in this material are those of the author(s). The NRF does not accept any liability in regard thereto. The authors also thank M.J. du Toit for her assistance in generating maps and figures.

Conflicts of Interest: The authors declare no conflict of interest.

References

1. Du Toit, M.J.; Shackleton, C.M.; Cilliers, S.S.; Davoren, E. Advancing Urban Ecology in the Global South: Emerging Themes and Future Research Directions. In *Urban Ecology in the Global South*; Shackleton, C.M., Cilliers, S.S., Davoren, E., du Toit, M.J., Eds.; Springer: Cham, Switzerland, 2021; pp. 433–461.
2. Van Zyl, B.; Cilliers, E.J.; Lategan, L.G.; Cilliers, S.S. Closing the Gap between Urban Planning and Urban Ecology: A South African Perspective. *Urban Plan.* **2021**, *6*, 122–134. [CrossRef]
3. Cilliers, S.S.; Breed, C.A.; Cilliers, E.J.; Lategan, L.G. Urban Ecological Planning and Design in the Global South. In *Urban Ecological Planning and Design in the Global South*; Shackleton, C.M., Cilliers, S.S., Davoren, E., du Toit, M.J., Eds.; Springer: Cham, Switzerland, 2021; pp. 365–401.
4. Ernstson, H. The social production of ecosystem services: A framework for studying environmental justice and ecological complexity in urbanized landscapes. *Landsc. Urban Plan.* **2013**, *109*, 7–17. [CrossRef]

5. Wolch, J.R.; Byrne, J.; Newell, J.P. Urban green space, public health, and environmental justice: The challenge of making cities 'just green enough'. *Landsc. Urban Plan.* **2014**, *125*, 234–244. [CrossRef]
6. Jennings, V.; Johnson Gaither, C.; Gragg, R.S. Promoting environmental justice through urban green space access: A synopsis. *Environ. Justice* **2012**, *5*, 1–7. [CrossRef]
7. Kabisch, N.; Haase, D. Green justice or just green? Provision of urban green spaces in Berlin, Germany. *Landsc. Urban Plan.* **2014**, *122*, 129–139. [CrossRef]
8. Venter, Z.S.; Shackleton, C.M.; Van Staden, F.; Selomane, O.; Masterson, V.A. Green Apartheid: Urban green infrastructure remains unequally distributed across income and race geographies in South Africa. *Landsc. Urban Plan.* **2020**, *203*, 103889. [CrossRef]
9. Riley, C.B.; Gardiner, M.M. Examining the distributional equity of urban tree canopy cover and ecosystem services across United States cities. *PLoS ONE* **2020**, *15*, e0228499. [CrossRef] [PubMed]
10. McConnachie, M.M.; Shackleton, C.M. Public green space inequality in small towns in South Africa. *Habitat Int.* **2010**, *34*, 244–248. [CrossRef]
11. Hartig, T. Green space, psychological restoration, and health inequality. *Lancet* **2008**, *372*, 1614–1615. [CrossRef]
12. Song, Y.; Chen, B.; Ho, H.C.; Kwan, M.-P.; Liu, D.; Wang, F.; Wang, J.; Cai, J.; Li, X.; Xu, Y. Observed inequality in urban greenspace exposure in China. *Environ. Int.* **2021**, *156*, 106778. [CrossRef]
13. Rigolon, A.; Browning, M.H.; Lee, K.; Shin, S. Access to urban green space in cities of the Global South: A systematic literature review. *Urban Sci.* **2018**, *2*, 67. [CrossRef]
14. Astell-Burt, T.; Feng, X. Does sleep grow on trees? A longitudinal study to investigate potential prevention of insufficient sleep with different types of urban green space. *SSM-Popul. Health* **2020**, *10*, 100497. [CrossRef]
15. Nero, B.F. Urban green space dynamics and socio-environmental inequity: Multi-resolution and spatiotemporal data analysis of Kumasi, Ghana. *Int. J. Remote Sens.* **2017**, *38*, 6993–7020. [CrossRef]
16. Sathyakumar, V.; Ramsankaran, R.; Bardhan, R.J.G.; Sensing, R. Linking remotely sensed Urban Green Space (UGS) distribution patterns and Socio-Economic Status (SES)-A multi-scale probabilistic analysis based in Mumbai, India. *GISci. Remote Sens.* **2019**, *56*, 645–669. [CrossRef]
17. Kimengsi, J.N.; Fogwe, Z.N. Urban green development planning opportunities and challenges in sub-saharan Africa: Lessons from Bamenda city, Cameroon. *Int. J. Glob. Sustain.* **2017**, *1*, 1–17. [CrossRef]
18. Wilkerson, M.L.; Mitchell, M.G.; Shanahan, D.; Wilson, K.A.; Ives, C.D.; Lovelock, C.E.; Rhodes, J.R. The role of socio-economic factors in planning and managing urban ecosystem services. *Ecosyst. Serv.* **2018**, *31*, 102–110. [CrossRef]
19. Heynen, N.; Perkins, H.A.; Roy, P. The political ecology of uneven urban green space: The impact of political economy on race and ethnicity in producing environmental inequality in Milwaukee. *Urban Aff. Rev.* **2006**, *42*, 3–25. [CrossRef]
20. Landry, S.M.; Chakraborty, J. Street trees and equity: Evaluating the spatial distribution of an urban amenity. *Environ. Plan.* **2009**, *41*, 2651–2670. [CrossRef]
21. Nesbitt, L.; Hotte, N.; Barron, S.; Cowan, J.; Sheppard, S.R. The social and economic value of cultural ecosystem services provided by urban forests in North America: A review and suggestions for future research. *Urban For. Urban Green.* **2017**, *25*, 103–111. [CrossRef]
22. Watkins, S.L.; Gerrish, E. The relationship between urban forests and race: A meta-analysis. *J. Environ. Manag.* **2018**, *209*, 152–168. [CrossRef]
23. Apparicio, P.; Séguin, A.-M.; Landry, S.; Gagnon, M. Spatial distribution of vegetation in Montreal: An uneven distribution or environmental inequity? *Landsc. Urban Plan.* **2012**, *107*, 214–224.
24. Pickett, S.T.; Cadenasso, M.L.; Grove, J.M.; Groffman, P.M.; Band, L.E.; Boone, C.G.; Burch, W.R.; Grimmond, C.S.B.; Hom, J.; Jenkins, J.C.J.B. Beyond urban legends: An emerging framework of urban ecology, as illustrated by the Baltimore Ecosystem Study. *BioScience* **2008**, *58*, 139–150. [CrossRef]
25. McDonald, R. Ecosystem service demand and supply along the urban-to-rural gradient. *J. Conserv. Plan.* **2009**, *5*, 1–14.
26. Escobedo, F.J.; Kroeger, T.; Wagner, J.E. Review: Urban forests and pollution mitigation: Analyzing ecosystem services and disservices. *Environ. Pollut.* **2011**, *159*, 2078–2087. [CrossRef] [PubMed]
27. Boone, C.G.; Buckley, G.L.; Grove, J.M.; Sister, C. Parks and people: An environmental justice inquiry in Baltimore, Maryland. *Ann. Assoc. Am. Geogr.* **2009**, *99*, 767–787. [CrossRef]
28. Dai, D. Racial/ethnic and socioeconomic disparities in urban green space accessibility: Where to intervene? *Landsc. Urban Plan.* **2011**, *102*, 234–244. [CrossRef]
29. Madge, C. Public parks and the geography of fear. *Tijdschr. Voor Econ. En Soc. Geogr.* **1997**, *88*, 237–250. [CrossRef]
30. Tinsley, H.E.; Tinsley, D.J.; Croskeys, C.E. Park usage, social milieu, and psychosocial benefits of park use reported by older urban park users from four ethnic groups. *Leis. Sci.* **2002**, *24*, 199–218. [CrossRef]
31. Jones, A.; Hillsdon, M.; Coombes, E. Greenspace access, use, and physical activity: Understanding the effects of area deprivation. *Prev. Med.* **2009**, *49*, 500–505. [CrossRef]
32. Leslie, E.; Cerin, E.; Kremer, P. Perceived neighborhood environment and park use as mediators of the effect of area socio-economic status on walking behaviors. *J. Phys. Act. Health Place* **2010**, *7*, 802–810. [CrossRef]
33. Shackleton, C.M.; Blair, A. Perceptions and use of public green space is influenced by its relative abundance in two small towns in South Africa. *Landsc. Urban Plan.* **2013**, *113*, 104–112. [CrossRef]

34. Guenat, S.; Porras Lopez, G.; Mkwambisi, D.D.; Dallimer, M. Unpacking stakeholder perceptions of the benefits and challenges associated with urban greenspaces in sub-Saharan Africa. *Front. Environ. Sci.* **2021**, *9*, 1–12. [CrossRef]
35. Lyytimäki, J.; Sipilä, M. Hopping on one leg–The challenge of ecosystem disservices for urban green management. *Urban For. Urban Green.* **2009**, *8*, 309–315. [CrossRef]
36. Reis, E.; López-Iborra, G.M.; Pinheiro, R.T. Changes in bird species richness through different levels of urbanization: Implications for biodiversity conservation and garden design in Central Brazil. *Landsc. Urban Plan.* **2012**, *107*, 31–42. [CrossRef]
37. Wolfe, M.K.; Mennis, J. Does vegetation encourage or suppress urban crime? Evidence from Philadelphia, PA. *Landsc. Urban Plan.* **2012**, *108*, 112–122. [CrossRef]
38. Lategan, L.G.; Cilliers, E.J. The value of public green spaces and the effects of South Africa's informal backyard rental sector. *WIT Trans. Ecol. Environ.* **2016**, *191*, 427–438.
39. Myers, G. Urbanisation in the Global South. In *Urban Ecology in the Global South*; Springer: Cham, Switzerland, 2021; pp. 27–49.
40. Asomani-Boateng, R. Urban cultivation in Accra: An examination of the nature, practices, problems, potentials and urban planning implications. *Habitat Int.* **2002**, *26*, 591–607. [CrossRef]
41. Mougeot, L.J. *Growing Better Cities: Urban Agriculture for Sustainable Development*; IDRC: Ottawa, ON, Canada, 2006.
42. Burkhard, B.; Kroll, F.; Nedkov, S.; Müller, F. Mapping ecosystem service supply, demand and budgets. *Ecol. Indic.* **2012**, *21*, 17–29. [CrossRef]
43. Huston, G. Building a Case Study for Transdisciplinary Thinking: The Inclusion of Green Infrastructure as Part of Spatial Planning Education. Master's Thesis, North-West University, Potchefstroom, South Africa, 2018.
44. Van Oijstaeijen, W.; Van Passel, S.; Cools, J. Urban green infrastructure: A review on valuation toolkits from an urban planning perspective. *J. Environ. Manag.* **2020**, *267*, 110603. [CrossRef]
45. Czembrowski, P.; Łaszkiewicz, E.; Kronenberg, J.; Engström, G.; Andersson, E. Valuing individual characteristics and the multifunctionality of urban green spaces: The integration of sociotope mapping and hedonic pricing. *PLoS ONE* **2019**, *14*, e0212277. [CrossRef]
46. Immergluck, D.; Balan, T. Sustainable for whom? Green urban development, environmental gentrification, and the Atlanta Beltline. *Urban Geogr.* **2018**, *39*, 546–562. [CrossRef]
47. Daams, M.N.; Sijtsma, F.J.; Veneri, P. Mixed monetary and non-monetary valuation of attractive urban green space: A case study using Amsterdam house prices. *Ecol. Econ.* **2019**, *166*, 106430. [CrossRef]
48. Cilliers, J.; Cilliers, S. From green to gold: A South African example of valuing urban green spaces in some residential areas in Potchefstroom. *Town Reg. Plan.* **2015**, *2015*, 67.
49. Combrinck, Z.; Cilliers, E.J.; Lategan, L.; Cilliers, S. Revisiting the proximity principle with stakeholder input: Investigating property values and distance to urban green space in potchefstroom. *Land* **2020**, *9*, 235. [CrossRef]
50. Lategan, L.; Cilliers, J.; Huston, Z.; Blaauw, N.; Cilliers, S. Economic Assessment of South African Urban Green Spaces Using the Proximity Principle: Municipal Valuation vs. Market Value. *Urban Plan.* **2021**, *6*, 54–66. [CrossRef]
51. Pauleit, S.; Vasquéz, A.; Maruthaveeran, S.; Liu, L.; Cilliers, S.S. Urban green infrastructure in the Global South. In *Urban Ecology in the Global South*; Springer: Cham, Switzerland, 2021; pp. 107–143.
52. Du Toit, M.J.; Cilliers, S.S.; Dallimer, M.; Goddard, M.; Guenat, S.; Cornelius, S.F. Urban green infrastructure and ecosystem services in sub-Saharan Africa. *Landsc. Urban Plan.* **2018**, *180*, 249–261. [CrossRef]
53. Makakavhule, K.; Landman, K. Towards deliberative democracy through the democratic governance and design of public spaces in the South African capital city, Tshwane. *Urban Des. Int.* **2020**, *25*, 280–292. [CrossRef]
54. Posel, D.; Casale, D.; Grapsa, E. Household variation and inequality: The implications of equivalence scales in South Africa. *Afr. Rev. Econ. Financ.* **2020**, *12*, 102–122.
55. Bigon, L. Garden cities in colonial Africa: A note on historiography. *Plan. Perspect.* **2013**, *28*, 477–485. [CrossRef]
56. Alexander, J.; Smith, D.A.E.; Smith, Y.C.E.; Downs, C.T. Eco-estates: Diversity hotspots or isolated developments? Connectivity of eco-estates in the Indian Ocean Coastal Belt, KwaZulu-Natal, South Africa. *Ecol. Indic.* **2019**, *103*, 425–433. [CrossRef]
57. Cilliers, S.; Cilliers, J.; Lubbe, R.; Siebert, S. Ecosystem services of urban green spaces in African countries—perspectives and challenges. *Urban Ecosyst.* **2013**, *16*, 681–702. [CrossRef]
58. Lategan, L.G. Informality and Sustainability: Reflecting on South Africa's Informal Backyard Rental Sector from a Planning Perspective. Ph.D. Thesis, North-West University, Potchefstroom, South Africa, 2017.
59. Girma, Y.; Terefe, H.; Pauleit, S.; Kindu, M. Urban green spaces supply in rapidly urbanizing countries: The case of Sebeta Town, Ethiopia. *Remote Sens. Appl. Soc. Environ.* **2019**, *13*, 138–149. [CrossRef]
60. Barbosa, O.; Tratalos, J.A.; Armsworth, P.R.; Davies, R.G.; Fuller, R.A.; Johnson, P.; Gaston, K.J. Who benefits from access to green space? A case study from Sheffield, UK. *Landsc. Urban Plan.* **2007**, *83*, 187–195. [CrossRef]
61. Sanesi, G.; Chiarello, F. Residents and urban green spaces: The case of Bari. *Urban For. Urban Green.* **2006**, *4*, 125–134. [CrossRef]
62. Lategan, L.; Cilliers, J. Considering urban green space and informal backyard rentals in South Africa: Disproving the compensation hypothesis. *Town Reg. Plan.* **2016**, *69*, 1–16. [CrossRef]
63. Cilliers, E.J.; Timmermans, W. Approaching value added planning in the green environment. *J. Place Manag. Dev.* **2013**, *6*, 144–154. [CrossRef]

64. De Wit, M.; van Zyl, H.; Crookes, D.; Blignaut, J.; Jayiya, T.; Goiset, V.; Mahumani, B. Including the economic value of well-functioning urban ecosystems in financial decisions: Evidence from a process in Cape Town. *Ecosyst. Serv.* **2012**, *2*, 38–44. [CrossRef]
65. Schäffler, A.; Swilling, M. Valuing green infrastructure in an urban environment under pressure—The Johannesburg case. *Ecol. Econ.* **2013**, *86*, 246–257. [CrossRef]
66. Papageorgiou, M.; Gemenetzi, G. Setting the grounds for the green infrastructure in the metropolitan areas of Athens and Thessaloniki: The role of green space. *Eur. J. Environ. Sci.* **2018**, *8*, 83–92. [CrossRef]
67. Van Zyl, B.; Lategan, L.G.; Cilliers, E.J.; Cilliers, S.S. An Exploratory Case-Study Approach to Understand Multifunctionality in Urban Green Infrastructure Planning in a South African Context. *Front. Sustain.* **2021**, *3*, 725539. [CrossRef]
68. Pauleit, S.; Liu, L.; Ahern, J.; Kazmierczak, A. Multifunctional green infrastructure planning to promote ecological services in the city. In *Handbook of Urban Ecology: Patterns, Processes, and Applications*; Oxford University Press Inc.: New York, NY, USA, 2011.
69. Van den Bosch, M.; Sang, Å.O. Urban natural environments as nature-based solutions for improved public health–A systematic review of reviews. *Environ. Res.* **2017**, *158*, 373–384. [CrossRef]
70. Bendict, M.; McMahon, E. *Green Infrastructure: Linking Landscapes and Communities*; Island Press: Washington, DC, USA, 2006.
71. Davies, C.; MacFarlane, R.; McGloin, C.; Roe, M. Green Infrastructure Planning Guide Project: Final Report. 2006. Available online: http://www.greeninfrastructurenw.co.uk/resources/North_East_Green_Infrastructure_Planning_Guide.pdf (accessed on 20 December 2021).
72. Ghofrani, Z.; Sposito, V.; Faggian, R. A comprehensive review of blue-green infrastructure concepts. *Int. J. Environ. Sustain.* **2017**, *6*, 15–36. [CrossRef]
73. Cilliers, J.; Cilliers, S. *Planning for Green Infrastructure: Options for South African Cities*; South African Cities Network: Johannesburg, South Africa, 2016; p. 56.
74. Grunewald, K.; Bastian, O. Ecosystem Services (ES): More Than Just a Vogue Term? In *Ecosystem Services—Concept, Methods and Case Studies*; Springer: Cham, Switzerland, 2015; pp. 1–11.
75. Bolund, P.; Hunhammar, S. Ecosystem services in urban areas. *Ecol. Econ.* **1999**, *29*, 293–301. [CrossRef]
76. De Groot, R.S.; Alkemade, R.; Braat, L.; Hein, L.; Willemen, L. Challenges in integrating the concept of ecosystem services and values in landscape planning, management and decision making. *Ecol. Complex.* **2010**, *7*, 260–272. [CrossRef]
77. Sukhdev, P.; Wittmer, H.; Schröter-Schlaack, C.; Nesshöver, C.; Bishop, J.; Brink, P.; Gundimeda, H.; Kumar, P.; Simmons, B. *The Economics of Ecosystems and Biodiversity: Mainstreaming the Economics of Nature: A Synthesis of the Approach, Conclusions and Recommendations of TEEB*; Progress Press: Gland, Switzerland, 2010.
78. La Notte, A.; D'Amato, D.; Mäkinen, H.; Paracchini, M.L.; Liquete, C.; Egoh, B.; Geneletti, D.; Crossman, N.D. Ecosystem services classification: A systems ecology perspective of the cascade framework. *Ecol. Indic.* **2017**, *74*, 392–402. [CrossRef]
79. Schlesinger, J.; Drescher, A.; Shackleton, C.J.A.G. Socio-spatial dynamics in the use of wild natural resources: Evidence from six rapidly growing medium-sized cities in Africa. *Appl. Geogr.* **2015**, *56*, 107–115. [CrossRef]
80. Manditsera, F.A.; Lakemond, C.M.; Fogliano, V.; Zvidzai, C.J.; Luning, P.A. Consumption patterns of edible insects in rural and urban areas of Zimbabwe: Taste, nutritional value and availability are key elements for keeping the insect eating habit. *Food Secur.* **2018**, *10*, 561–570. [CrossRef]
81. Garekae, H.; Shackleton, C.M. Urban foraging of wild plants in two medium-sized South African towns: People, perceptions and practices. *Habitat Int. Urban For. Urban Green.* **2020**, *49*, 126581. [CrossRef]
82. Shackleton, C.M. Ecosystem provisioning services in Global South cities. In *Urban Ecology in the Global South*; Springer: Cham, Switzerland, 2021; pp. 203–226.
83. Shackleton, S.; Chinyimba, A.; Hebinck, P.; Shackleton, C.; Kaoma, H.J.L.; Planning, U. Multiple benefits and values of trees in urban landscapes in two towns in northern South Africa. *Landsc. Urban Plan.* **2015**, *136*, 76–86. [CrossRef]
84. Rouget, M.; Davids, R.; Boon, R.; Roberts, D. Identifying ecosystem service hotspots for environmental management in Durban, South Africa. *Bothalia-Afr. Biodivers. Conserv.* **2016**, *46*, 1–18.
85. Sutherland, C.; Sim, V.; Buthelezi, S.; Khumalo, D. Social constructions of environmental services in a rapidly densifying area of eThekwini Municipality. *Bothalia-Afr. Biodivers. Conserv.* **2016**, *46*, 1–18.
86. Ngulani, T.; Shackleton, C. The degree, extent and value of air temperature amelioration by urban green spaces in Bulawayo, Zimbabwe. *S. Afr. Geogr. J.* **2020**, *102*, 344–355. [CrossRef]
87. Simwanda, M.; Ranagalage, M.; Estoque, R.C.; Murayama, Y. Spatial analysis of surface urban heat islands in four rapidly growing African cities. *Remote Sens.* **2019**, *11*, 1645. [CrossRef]
88. Escobedo, F.J. Understanding urban regulating ecosystem services in the Global South. In *Urban Ecology in the Global South*; Springer: Cham, Switzerland, 2021; pp. 227–244.
89. Shackleton, C.; Blair, A.; De Lacy, P.; Kaoma, H.; Mugwagwa, N.; Dalu, M.; Walton, W. How important is green infrastructure in small and medium-sized towns? Lessons from South Africa. *Landsc. Urban Plan.* **2018**, *180*, 273–281. [CrossRef]
90. Munien, S.; Nkambule, S.S.; Buthelezi, H.Z. Conceptualisation and use of green spaces in peri-urban communities: Experiences from Inanda, KwaZulu-Natal, South Africa. *Afr. J. Phys. Health Educ. Recreat.* **2015**, *21*, 155–167.
91. De Lacy, P.; Shackleton, C. Aesthetic and spiritual ecosystem services provided by urban sacred sites. *Sustainability* **2017**, *9*, 1628. [CrossRef]

92. Dobbs, C.; Vasquez, A.; Olave, P.; Olave, M.J. Cultural urban ecosystem services. In *Urban Ecology in the Global South*; Springer: Cham, Switzerland, 2021; pp. 245–264.
93. Bernholt, H.; Kehlenbeck, K.; Gebauer, J.; Buerkert, A. Plant species richness and diversity in urban and peri-urban gardens of Niamey, Niger. *Agrofor. Syst.* **2009**, *77*, 159–179. [CrossRef]
94. Van der Walt, L.; Cilliers, S.; Du Toit, M.; Kellner, K. Conservation of fragmented grasslands as part of the urban green infrastructure: How important are species diversity, functional diversity and landscape functionality? *Urban Ecosyst.* **2015**, *18*, 87–113. [CrossRef]
95. Cilliers, S.; Siebert, S.; Du Toit, M.; Barthel, S.; Mishra, S.; Cornelius, S.; Davoren, E. Garden ecosystem services of Sub-Saharan Africa and the role of health clinic gardens as social-ecological systems. *Landsc. Urban Plan.* **2018**, *180*, 294–307. [CrossRef]
96. TEEB. *The Economics of Ecosystems and Biodiversity: TEEB Manual for Cities—Ecosystem Services in Urban Management*; TEEB: Geneva, Switzerland, 2011.
97. Davoren, E.; Shackleton, C.M. Urban ecosystem disservices in the Global South. In *Urban Ecology in the Global South*; Springer: Cham, Switzerland, 2021; pp. 265–292.
98. Lyytimäki, J. Ecosystem disservices: Embrace the catchword. *Ecosyst. Serv.* **2015**, *12*, 136. [CrossRef]
99. Gómez-Baggethun, E.; Gren, Å.; Barton, D.N.; Langemeyer, J.; McPhearson, T.; O'farrell, P.; Andersson, E.; Hamstead, Z.; Kremer, P. Urban ecosystem services. In *Urbanization, Biodiversity and Ecosystem Services: Challenges and Opportunities*; Springer: Dordrecht, The Netherlands, 2013; pp. 175–251.
100. Von Döhren, P.; Haase, D. Ecosystem disservices research: A review of the state of the art with a focus on cities. *Ecol. Indic.* **2015**, *52*, 490–497. [CrossRef]
101. Shackleton, C.M.; Ruwanza, S.; Sanni, G.S.; Bennett, S.; De Lacy, P.; Modipa, R.; Mtati, N.; Sachikonye, M.; Thondhlana, G. Unpacking Pandora's box: Understanding and categorising ecosystem disservices for environmental management and human wellbeing. *Ecosystems* **2016**, *19*, 587–600. [CrossRef]
102. Panduro, T.E.; Veie, K.L. Classification and valuation of urban green spaces—A hedonic house price valuation. *Landsc. Urban Plan.* **2013**, *120*, 119–128. [CrossRef]
103. Sharmin, F. Towards the Valuation of Open Spaces: A Hedonic Based Investigation for Sustainable Planning in the Dense Urban Context of Dhaka. *Am. J. Environ. Resour. Econ.* **2020**, *5*, 97–103. [CrossRef]
104. Schaubroeck, T. A need for equal consideration of ecosystem disservices and services when valuing nature; countering arguments against disservices. *Ecosyst. Serv.* **2017**, *26*, 95–97. [CrossRef]
105. Farber, S.C.; Costanza, R.; Wilson, M.A. Economic and ecological concepts for valuing ecosystem services. *Ecol. Econ.* **2002**, *41*, 375–392. [CrossRef]
106. Cilliers, E.J. Rethinking Sustainable Development: The Economic Value of Green Spaces. Master's Thesis, North-West University, Potchefstroom, South Africa, 2010.
107. Troy, A.; Grove, J.M. Property values, parks, and crime: A hedonic analysis in Baltimore, MD. *Landsc. Urban Plan.* **2008**, *87*, 233–245. [CrossRef]
108. Mayor, K.; Lyons, S.; Duffy, D.; Tol, R.S. *A Hedonic Analysis of the Value of Parks and Green Spaces in the Dublin Area*; ESRI Working Paper No. 331; The Economic and Social Research Institute (ESRI): Dublin, Ireland, 2009.
109. Nicholls, S. Measuring the Impact of Parks on Property Values: New research shows that green spaces increase the value of nearby housing. *Parks Recreat.* **2004**, *39*, 24–32.
110. Crompton, J.L. The impact of parks on property values: Empirical evidence from the past two decades in the United States. *Manag. Leis.* **2005**, *10*, 203–218. [CrossRef]
111. Crompton, J.L. The impact of parks on property values: A review of the empirical evidence. *J. Leis. Res.* **2001**, *33*, 1–31. [CrossRef]
112. Trojanek, R.; Gluszak, M.; Tanas, J. The effect of urban green spaces on house prices in Warsaw. *Int. J. Strateg. Prop. Manag.* **2018**, *22*, 358–371. [CrossRef]
113. Lubbe, C.S.; Siebert, S.J.; Cilliers, S.S. Political legacy of South Africa affects the plant diversity patterns of urban domestic gardens along a socio-economic gradient. *Sci. Res. Essays* **2010**, *5*, 2900–2910.
114. Lobo, M.A.; Moeyaert, M.; Cunha, A.B.; Babik, I. Single-case design, analysis, and quality assessment for intervention research. *J. Neurol. Phys. Ther. JNPT* **2017**, *41*, 187. [CrossRef]
115. Shackleton, C.; Mograbi, P. Meeting a diversity of needs through a diversity of species: Urban residents' favourite and disliked tree species across eleven towns in South Africa and Zimbabwe. *Urban For. Urban Green.* **2020**, *48*, 126507. [CrossRef]
116. Cilliers, S.; Schoeman, L.; Bredenkamp, G. Wetland plant communities in the Potchefstroom Municipal area, North-west, South Africa. *Bothalia-Afr. Biodivers. Conserv.* **1998**, *28*, 213–229. [CrossRef]
117. Du Toit, M.; Du Preez, C.; Cilliers, S. Plant diversity and conservation value of wetlands along a rural-urban gradient. *Bothalia-Afr. Biodivers. Conserv.* **2021**, *51*, 1–18. [CrossRef]
118. Bateganya, N.L.; Nakalanzi, D.; Babu, M.; Hein, T. Buffering municipal wastewater pollution using urban wetlands in sub-Saharan Africa: A case of Masaka municipality, Uganda. *Environ. Technol.* **2015**, *36*, 2149–2160. [CrossRef]
119. Cimon-Morin, J.; Poulin, M. Setting conservation priorities in cities: Approaches, targets and planning units adapted to wetland biodiversity and ecosystem services. *Landsc. Ecol.* **2018**, *33*, 1975–1995. [CrossRef]
120. Pascual, U.; Balvanera, P.; Díaz, S.; Pataki, G.; Roth, E.; Stenseke, M.; Watson, R.T.; Dessane, E.B.; Islar, M.; Kelemen, E. Valuing nature's contributions to people: The IPBES approach. *Curr. Opin. Environ. Sustain.* **2017**, *26*, 7–16. [CrossRef]

land

MDPI

Article

Spatial Decision-Making for Dense Built Environments: The Logic Scoring of Preference Method for 3D Suitability Analysis

Kendra Munn [1,*], Suzana Dragićević [1] and Rob Feick [2]

1 Spatial Analysis and Modeling Laboratory, Department of Geography, Simon Fraser University, Burnaby, BC V5A 1S6, Canada; suzanad@sfu.ca
2 School of Planning, University of Waterloo, Waterloo, ON N2L 3G1, Canada; rob.feick@uwaterloo.ca
* Correspondence: kendra_munn@sfu.ca

Abstract: As many urban areas undergo increasing densification, there is a growing need for methods that can extend spatial analysis and decision-making for three-dimensional (3D) environments. Traditional multicriteria evaluation (MCE) methods implemented within geographic information systems (GIS) can assist in spatial decision-making but are rarely suited for 3D environments. These methods typically use a simplified decision logic that limits the number of evaluation criteria and variability of output suitability scores. In this study, the logic scoring of preference (LSP) as a generalized MCE method is used for 3D suitability analysis to better represent human reasoning through flexible soft computing stepwise decision logic operators. This research: (1) implements the LSP–MCE method to compare the suitability of high-rise residential units in 3D, and (2) performs criteria weight sensitivity and cost–suitability analyses using datasets for the City of Vancouver, Canada. LSP aggregation structures are developed for unique priorities and requirements of three demographic profiles. The results demonstrate the method's flexibility in representing unique preference sets comprising 2D and 3D criteria, and that cost has a significant effect on residential unit attractiveness in a dense built environment. The proposed 3D LSP–MCE method could be adapted to benefit other stakeholders, such as property tax assessors, urban planners, and developers.

Keywords: 3D spatial multicriteria evaluation (MCE); dense built environment; 3D suitability analysis; logic scoring of preference (LSP); spatial decision-making; three spatial dimensions (3D); geographic information systems (GIS)

Citation: Munn, K.; Dragićević, S.; Feick, R. Spatial Decision-Making for Dense Built Environments: The Logic Scoring of Preference Method for 3D Suitability Analysis. *Land* **2022**, *11*, 443. https://doi.org/10.3390/land11030443

Academic Editors: Elizelle Juanee Cilliers and Sarel Cilliers

Received: 24 February 2022
Accepted: 16 March 2022
Published: 19 March 2022

Publisher's Note: MDPI stays neutral with regard to jurisdictional claims in published maps and institutional affiliations.

1. Introduction

To mitigate the negative environmental and social impacts of increasing urbanization, many cities are seeking to accommodate growing populations while also reducing environmental degradation [1]. One widely implemented practice has been urban densification, which focuses new development within existing urban boundaries rather than converting natural or agricultural lands in urban peripheries. By limiting outward sprawl, populations and economic activities are concentrated in previously serviced land, and urban form shifts to a more compact and higher density model that better supports public transportation and physical activity [2–4]. To achieve higher densities, many municipalities are encouraging high-rise development to maximize the use of limited land [5].

As residential high-rise units become a more common alternative to low-density detached housing, the housing choice problem becomes more complex. Many housing features and environmental qualities, such as viewscapes, vary with respect to not only the neighborhood where a unit is situated within but also the unit's vertical location within a given building. These three-dimensional (3D) attributes can ultimately influence the overall desirability of different units for prospective residents [6]. Multicriteria evaluation (MCE) methods have been used extensively with geographic information systems (GISs) to aid in two-dimensional (2D) spatial decision-making problems and in a wide range of spatial applications [7,8]. However, there is a lack of such methods for accommodating decision problems where alternatives, criteria, or processes are situated in 3D space.

Furthermore, software that can handle spatial visualization and analysis in three spatial dimensions, such as Autodesk Revit [9], ArcGIS Urban [10], and CityEngine [11], generally excel in urban design, visualization, 3D mapping, or selected types of "objective" analyses, such as viewscapes or shadow analysis [12,13]. While these tools can perform certain types of 3D analysis, they are ill-suited to aiding decision-making where stakeholders must consider many factors simultaneously that vary in importance. With increasing 3D data availability, there is a demand to move beyond simple 3D analysis and to develop spatial decision methods, such as MCE, that operate in 3D space. There are no off-the-shelf software packages that combine 3D and MCE currently. Hence, the analytical processes related to suitability analysis must be coded and embedded within existing geospatial software in order to operationalize 3D MCE. Munn and Dragićević [14] recently proposed and applied a 3D MCE approach to residence selection scenarios with 3D decision alternatives using CityEngine software with additional programming routines. However, the core of this 3D GIS–MCE approach is based on weighted linear combination (WLC), a very simple and commonly used MCE that has shortcomings. Typical MCE methods available in common GIS software, such as WLC, analytical hierarchy process (AHP), or ordered weighted average (OWA), are based on the simple additive scoring (SAS) that uses the geometric mean with a linear additive nature [15]. When a larger number of attributes are necessary to resolve a complex spatial decision problem, each individual attribute's importance decreases within the standardized weighting schema [16,17]. This oversimplification insufficiently describes decision-making perspectives typical of human reasoning.

The logic scoring of preference (LSP) is a generalized MCE method that has been developed to overcome these limitations [17]. LSP addresses large and complex decision problems through a variety of logic operators that better depict nuances of decision-making processes and the use of stepwise hierarchies to allow more criteria to be considered. Many of these logic aggregators are not available in any other MCE decision method, thus making the LSP method more appropriate to address complex spatial problems such as 3D suitability analysis in the dense built environment. However, up to this point, LSP–MCE has only been used for 2D spatial decision-making studies. Therefore, the objectives of this research are to: (1) develop an LSP–MCE to operate in 3D as a novel approach applied to/for dense built environments; (2) to implement the 3D LSP–MCE for suitability analysis of high-rise residences using data for downtown Vancouver, Canada; and (3) perform LSP–MCE sensitivity and cost–suitability analysis on decision alternatives in 3D. The novel approach proposed in this study will allow planners, urban developers, real estate, taxation offices, and other end users to interactively and independently address all the 3D components, criteria, and scenarios, and seamlessly follow through all of the analytical processes that are embedded within a 3D MCE analysis.

2. Logic Scoring of Preference Method

The logic scoring of preference (LSP) is a soft-computing-based MCE method characterized by stepwise aggregation and a wide range of flexible decision logic conditions [17]. Stepwise criteria aggregation allows more evaluation criteria to be considered as the significance of each criterion is not diminished by the others through a linear combination function. Furthermore, LSP includes decision logic aggregators representing simultaneity and replaceability, which are intrinsic to most real-world decision-making problems. As a result, LSP–MCE analysis and output suitability maps can better reflect real-world decision logic than traditional MCE methods [18,19].

LSP–MCE analysis consists of five steps [17] and begins with the definition of the decision problem related to the interested stakeholders, experts, or groups. Next, a hierarchical attribute tree is created that contains the attributes that affect the overall suitability of a decision alternative. The structure forms a hierarchy of groups and subgroups of related components that are iteratively decomposed until the simplest components, termed elementary attributes, remain.

The third step involves defining elementary criteria or suitability functions, which transform raw elementary attribute values into standardized unitless elementary preference values. This permits a comparison of performance between attributes with differing units of measurement [20]. Each function describes the relationship between possible attribute values and the associated degree of satisfaction (i.e., elementary preference) established for that attribute [21].

In step 4, the hierarchical LSP aggregation structure is defined to determine how elementary preferences are aggregated to compute the overall suitability of each alternative. The attribute tree serves as the framework for the structure, with each subgroup in the attribute tree system comprising at least one preference aggregation block, depending on whether components in the subgroup are mandatory, desired (nonmandatory), or a combination thereof. Mandatory attributes must be satisfied for the decision alternative to be considered at all, whereas desired attributes are preferred but not necessary. Weights of relative importance are determined by the relevant stakeholders or field experts and are assigned to each component such that all weights in an aggregation block sum up to 1or 100%. In its entirety, the overarching LSP system can be referred to as the LSP criterion function.

A logic aggregator is also applied to each aggregation block. Aggregators are graded preference logic functions that dictate how components are combined to compute the satisfaction degree of each block [16]. They are derived from the fundamental generalized conjunction/disjunction (GCD) function used to model human reasoning across the spectrum of decision logic ranging from full conjunction or ANDness ($\alpha = 1$; $\omega = 0$) to full disjunction or ORness ($\alpha = 0$; $\omega = 1$) [19].

Conjunctive and disjunctive aggregators model logic requirements for simultaneity and replaceability, respectively. The spectrum of partial conjunction/disjunction aggregators can be divided into hard (HPC/HPD) and soft (SPC/SPD) partial conjunction/disjunction aggregators as presented in Figure 1 as per [17]. HPC aggregators model mandatory requirements, where an elementary preference of zero results in an overall suitability score of zero regardless of how completely other attributes are satisfied. Conversely, HPD aggregators represent sufficient requirements, where an elementary preference of 1 results in an overall score of 1 regardless of whether other attributes are satisfied. SPC and SPD model softer versions of simultaneity and replaceability. The SPC aggregators (Figure 1) apply a penalty to preference values when desired attributes are not satisfied, rather than assigning an automatic suitability score of zero. Similarly, SPD aggregators simply apply a reward to preference values when one of the desired attributes is fully satisfied, instead of an overall score of 1. The neutrality aggregator A also exists as the arithmetic mean for cases where a high elementary preference for any attribute can partially offset a low preference of any other [21], making this logical operator parallel to the compensatory nature of MCE, such as WLC. The last step involves calculating the overall suitability scores of each decision alternative and creating the suitability map.

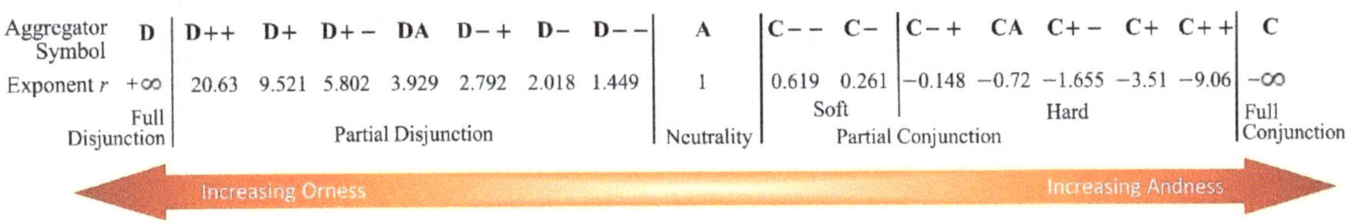

Figure 1. Logic scoring of preference (LSP) logic aggregators with associated symbols and r exponent values.

A hypothetical example of LSP-based site selection for a new college location is shown in Figure 2. Mandatory criteria, such as a sufficient site area and an appropriate slope, are combined using an HPC aggregator. Referencing Figure 1, this can be any aggregator from C−+ to C++, depending on the level of ANDness or simultaneity desired. An SPC aggregator is likewise applied to desired criteria where both are wanted but a location is not automatically rejected if either is unsatisfied (e.g., distance from other schools, availability of nearby parking). Finally, an SPD aggregator is assigned to desired criteria where replaceability is being modeled. The disjunctive aspect allows for train and bus access to partially substitute for the other, while the soft aspect enables rewarding of locations that offer both.

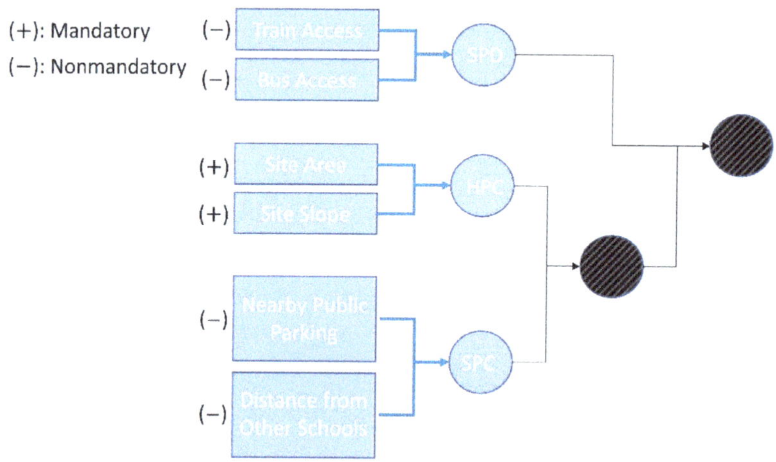

Figure 2. Sample LSP aggregation structure for selecting a location for a new college.

Recent studies have integrated LSP–MCE with GIS to address a number of spatial decision-making problems, including groundwater contamination vulnerability assessment [20], conservation planning [22], and landfill site selection [23]. LSP–MCE analysis has also been used for urban land use suitability [24], residential development [20], and urban densification [25]. While GIS-based LSP–MCE has been used to evaluate the suitability of residences or neighborhoods [17,19,26], these studies have not been applied to high-rise residential units where some attributes vary in magnitude along the vertical gradient. Furthermore, LSP–MCE output suitability maps have been restricted to 2D raster displays of decision alternatives to date. The 3D LSP–MCE approach used in this research is capable of analyzing alternatives spanning 3D space and visualizing their suitability in 3D to aid comprehension of outputs for decision-making.

3. Methods

3.1. Study Area and Datasets

The City of Vancouver, Canada, grew by 10% between 2011 and 2021 and was home to 662,248 residents in 2021 [27]. For that reason, it has become increasingly densified to accommodate population growth while simultaneously meeting municipal and regional sustainability goals [28]. This trend is most pronounced in downtown Vancouver, where a significant proportion of residential and office buildings are high-rises as defined by Statistics Canada [29] with five or more stories. Unlike many North American cities, downtown Vancouver is known as an attractive, livable space and not simply a central work hub [30].

Building footprints from the City of Vancouver [31] were used to generate residential unit data layers for all open market residential high-rise buildings in downtown Vancouver, amounting to 613 buildings comprising 49,130 residential units. Building height, story number, and number of units were available for many buildings [32]. However, as floor plans were not publicly accessible due to privacy and availability issues, hypothetical 3D units were created. Additional geospatial datasets used to derive criteria attribute values were obtained from [31,33–40]. Attribute analyses were conducted in the ArcGIS Desktop suite v.10.7.1 [41].

The residential units are represented in the 3D suitability map as cubelike voxel data structures, here defined as 3D spatial units possessing attribute values for each criterion, located at a position (x, y, z) in a 3D GIS environment. The 3D LSP–MCE analysis was implemented using CityEngine [11] software, and its CGA programming language with scripts developed to voxelize the units (i.e., extrude the 2D data layers to create 3D voxels), run the 3D analysis, and select a color symbology to display the obtained level of suitability for each voxel unit.

3.2. Decision Problem, LSP Attribute Tree, and Elementary Criteria

In this research, the decision problem consists of identifying the most suitable high-rise residential units for prospective residents. LSP attributes and 3D preference scenarios can be developed for various groups and stakeholders, including real-estate tax assessments, urban planners, and developers. For this study, three generalized sociodemographic groups were chosen with characteristics based on existing scientific literature: families with children (Scenario 1), elderly (Scenario 2), and working professionals (Scenario 3). Many 3D attributes, defined here as attributes whose values are affected by the vertical location of the unit, were included as factors that need to be considered in high-rise home selection in addition to the typical 2D factors and amenities. Ultimately, 10 2D and 11 3D elementary attributes influencing residence selection were identified and included in Scenario 1, three of which were omitted from either Scenario 2 or 3. The attribute trees presented for each scenario in Figure 3 are composed of three overarching attribute groups: neighborhood features (11), housing features, and (12) and environmental quality (13). The figure also distinguishes between attributes that are mandatory and nonmandatory and those that are 2D or 3D for each scenario.

The neighborhood features group (11) is composed entirely of 2D attributes. Family areas are specific to Scenario 1 and link the suitability of a neighborhood for raising a family to the number of children available to play and interact with [42]. Similarly, distance to the nearest elementary school is important to families as shorter distances reduce children's travel time and safety risks [43]. Prospective residents may also base their decision on neighborhood crime level, which is negatively correlated with neighborhood satisfaction and mental health [44,45], or population density, which often has positive correlations with neighborhood satisfaction and smaller but stronger social networks [46,47]. Living in close proximity to urban parks is also beneficial for mental health, promotes physical activity, and provides recreational opportunities [48]. The bike, transit, and walk scores for each residential building were included as assessments of multimodal accessibility to desirable neighborhood features based on walking, cycling, or transit travel [49].

The housing features group (12) includes the compass aspect of the unit and the year the building was built, as certain orientations are typically preferred over others [50], and newer homes are generally considered more desirable [51,52]. The remaining attributes are 3D in nature. Units with a larger viewshed are more pleasing and valued higher [53], as are views that encompass desirable features, such as parks or bodies of water [54]. Safety as referred to in this study relates to the floor a unit is on; particularly, residents on higher floors will require more travel time and may encounter more hazards in the event of an emergency evacuation [55]. Privacy is also an important factor impacting home satisfaction [56,57]. Here interunit visibility was calculated as a proxy for privacy.

The environmental quality group (13) entirely comprises 3D attributes. High levels of traffic-related air pollution or tree pollen can exacerbate respiratory conditions and asthmatic attacks [58,59]. Strong winds can cause thermal discomfort or in extreme cases may pose a level of danger [60]. Excessive noise from various sources (e.g., SkyTrain, railways, cars) can negatively affect physical and mental well-being [61], while increased exposure to sunlight improves both mental and physical well-being [62].

	Scenario		
	1	**2**	**3**
1. Residential Unit Suitability			
11. Neighbourhood Features (+)			
111. Family Areas	(−)		
112. Neighbourhood Crime	(+)	(+)	(−)
113. Population Density	(−)	(−)	(−)
114. Proximity to Amenities	(+)	(+)	(+)
1141. Distance to Parks	(+)	(−)	(−)
1142. Distance to School	(+)		
1143. Accessibility	(+)	(+)	(+)
11431. Bike Score	(−)		(−)
11432. Transit Score	(−)	(−)	(−)
11433. Walk Score	(−)	(−)	(−)
12. Housing Features (+)			
121. Aspect	(−)	(−)	(−)
122. Privacy	(−)	(−)	(−)
123. Safety	(+)	(+)	(−)
124. Year Built	(−)	(−)	(+)
125. View	(−)	(−)	(+)
1251. Total View Amount	(−)	(−)	(−)
1252. View Type	(−)	(−)	(−)
13. Environmental Quality (+)			
131. Air Pollution	(+)	(−)	(−)
132. Hours of Sunlight	(−)	(−)	(+)
133. Pollen Level	(−)	(−)	(−)
134. Wind	(−)	(+)	(−)
135. Noise	(+)	(+)	(−)
1351. Railway Noise	(−)	(−)	(−)
1352. SkyTrain Noise	(−)	(−)	(−)
1353. Traffic Noise	(−)	(−)	(−)

SkyTrain = local public rail transit

(+) Mandatory
(−) Nonmandatory

Figure 3. Attribute tree with elementary attributes for the three scenarios with 3D attributes highlighted in bold text.

Most elementary attribute data were not readily available and so were computed or estimated through GIS-based analyses using methods found in the literature (see Table 1). Elementary criteria functions were defined for each attribute using breakpoints based on values found in the literature (Table 1).

Table 1. Elementary attribute criteria in vertex notation format with associated criterion function breakpoints and rationale. Attributes that are 3D in nature are highlighted in bold text.

	Attribute	Analysis Method	Rationale
Neighborhood Features	Family Areas	{(20,0), (80,1)} (percentile of all census tract areas % occupied dwellings with children)	Preferences of 0 and 1 assigned to 20th and 80th percentile values, respectively, as more children means more peers to interact with. Analysis: Percentage of occupied dwellings with families.
	Neighborhood Crime	{(0–0.005,1), (0.005–0.0015,0.8), (0.0015–0.005,0.6), (0.005–0.01,0.4), (0.01–0.02,0.2), (0.02,0)} (crime density)	Crime is negatively correlated with neighborhood satisfaction and mental health [44], so lower crime level is preferred. Analysis: Kernel density of geolocated crime incidents.
	Population Density	{(29,0), (211,1)} (people per ha in each census tract)	Based on the density-satisfaction study of [46], which averaged 29 people/ha for low-density neighborhoods and 211 people/ha for high-density neighborhoods. Analysis: Population density.
	Distance to Parks	{(400,1), (800,0)} (m)	City of Vancouver aimed to have every resident within 5 min' walk (400 m) of a green space [63]; 99% are within 10 min' walk (800 m) [64]. Analysis: Network distance.
	Distance to School	{(400,1), (2000,0)} (m)	Preference of 1 assigned to 5 min' walk, 0 to 25 min' walk (2000 m), putting the 50% mark close to the average distance of 1274 m [65]. Analysis: Network distance.
	Bike Score, Transit Score, Walk Score	{(0–24,0), (25–49,0.25), (50–69,0.5), (70–89,0.75), (90–100,1) (score)	Based on walk score's [49] score class ranges and score values.
Housing Features	Aspect	{(N, 0), (NE/NW, 0.25), (E/W, 0.5), (SE/SW, 0.75), (S, 1)} (aspect)	How property values proportionally increase with different aspects is presented in [66]. Analysis: Determine unit aspect(s) using building and unit footprints and linear directional mean tool.
	Privacy	{(Low, 0), (relatively low, 0.33), (medium, 0.67), (high, 1)} (privacy level)	Four levels of visual exposure are defined in [67]: low (>50 m from nearest view point); relatively low (25–50 m), medium (10–25 m), and high (<10 m). Analysis: Intervisibility.
	Safety	{(1,0.5), (2,0.75), (3,1), (7,1), (18,0)}(floor number)	First few floors [68,69] are more likely to be burglarized, while typical aerial ladder reach limits fire rescue to about the seventh floor [55]. Average time to descend one floor is 16.4 s [70]; 18th floor is equivalent to 5 min' evacuation time.
	Year Built	{(1960, 0), (1983, 0.42), (1984, 0.1), (1998, 0.35), (1999, 0.71), (2015, 1)} (year built)	Newer homes are generally more desirable, but 1984–1998 buildings are from Vancouver's "leaky condo era" and may be prone to leaks [71]. Buildings from 2015 onwards built under the most recent National Building Code. Only 8% of Metro Vancouver high-rises built before 1961 [72].
	Total View Amount	{(10,0), (90,1)} (percentile of view amount (m^2) results)	Larger views are more pleasing and positively correlated with property prices [53,73]. Analysis: Visibility [54].
	View Type	{(0,0), (100,1)} (% desirable view)	Properties with larger views of desirable view features (e.g., water) are priced higher than properties with smaller views [53]. Analysis: Overlay of visibility output and raster of desirable/undesirable view features.

Table 1. *Cont.*

	Attribute	Analysis Method	Rationale
Environmental Quality	**Air Pollution**	{(7,1), (17,0)} (ppb NO_2)	Metro Vancouver's air quality objective for NO_2 is an annual average of 17 ppb. Areas of cleaner air in Metro Vancouver average around 7 ppb [74].
	Hours of Sunlight	{(0,0), (12,1)} (h)	More hours of direct sunlight is considered desirable due to its beneficial effects [62]. Analysis: Points solar radiation [75] for the equinox.
	Pollen Level	{(2,0), (12,1)} (m above ground level)	Although pollen level decreases with height, the most significant decrease generally occurs within 10 m, with concentrations at ground level ~1.5 times higher than those 10 m higher [76].
	Wind	{(low, 1), (moderate, 0.75), (very low, 0.5), (high, 0.25), (very high, 0)} (ventilation potential)	High wind speeds can bring thermal discomfort or even a level of danger [60]. In the mild Vancouver climate, a low to moderate amount of ventilation would be more appreciated. Analysis: Computed relative wind level for three 50 m height increments [77] using most predominant winds in Vancouver.
	Railway Noise, SkyTrain Noise, Traffic Noise	{(45,1), (95,0)} (dBA)	Sleep disruption can occur at a sustained noise level of ≥30 dBA. More severe consequences (e.g., hearing damage) can occur at levels of ≥80 dBA [78]. However, it is assumed that transmission loss at windows is ~15 dBA [79]. Railway analysis: Estimated using similar site noise map digitization and extrapolation [80]. SkyTrain analysis: Estimated using digitization and extrapolation of SkyTrain noise maps [81]. Traffic analysis: Estimated using traffic noise equation in [78] as per [14].

3.3. LSP Aggregation Structures

The LSP aggregation structures created for the three scenarios are depicted in Figure 4. An SPD aggregator is applied to the accessibility aggregation block (node 1143 in Figure 3) to indicate that walk, bike, and transit score criteria are replaceable to an extent, but a higher satisfaction degree is awarded to locations that satisfy all three. Similarly, an SPD aggregator is used for the view aggregation block (node 125) to suggest that while having both a large and desirable view is preferred, either criterion can partially replace the other. The remaining standard aggregators are either SPC or HPC, selected based on whether mandatory or desired components were being aggregated and the level of ANDness desired. For instance, an HPC aggregator is applied to aggregate the mandatory environment block (134–135) in Scenario 2 as they are mandatory requirements (Figure 4B), and C− + in particular is chosen to indicate that only a low to moderate level of simultaneity is needed. Both of the nonmandatory neighborhood (112–113) and nonmandatory housing (121–123) aggregation blocks in Scenario 3 are composed of desired components and thus require SPC aggregators (Figure 4C), but a larger ANDness (C−) is used for nonmandatory housing, denoting the demographic's desire for a slightly higher level of simultaneity in having all of safety, privacy, and a desired aspect.

Criteria weights were selected based on the expected priorities for each demographic. In reality, subject matter experts or decision problem stakeholders should be consulted in determining the appropriate criteria weights, as the resulting suitability scores can be highly sensitive to input weight values. However, the weight estimations used are sufficient for the purposes of this study, as the primary goal is to demonstrate the 3D LSP–MCE approach, and the decision problem itself is subjective in nature. Some weights were identical across the three scenarios. For example, the view type is weighted as 60% of the view block (node 125) value for all prospective resident types, while the total view amount is set at 40% (Figure 4). However, most weights and, more generally, the aggregation blocks vary between scenarios. For instance, aspect is a nonmandatory attribute in all scenarios and varies in weight from 0.2 (Scenario 2) through 0.3 (Scenario 1) to a high of 0.45 for Scenario 3.

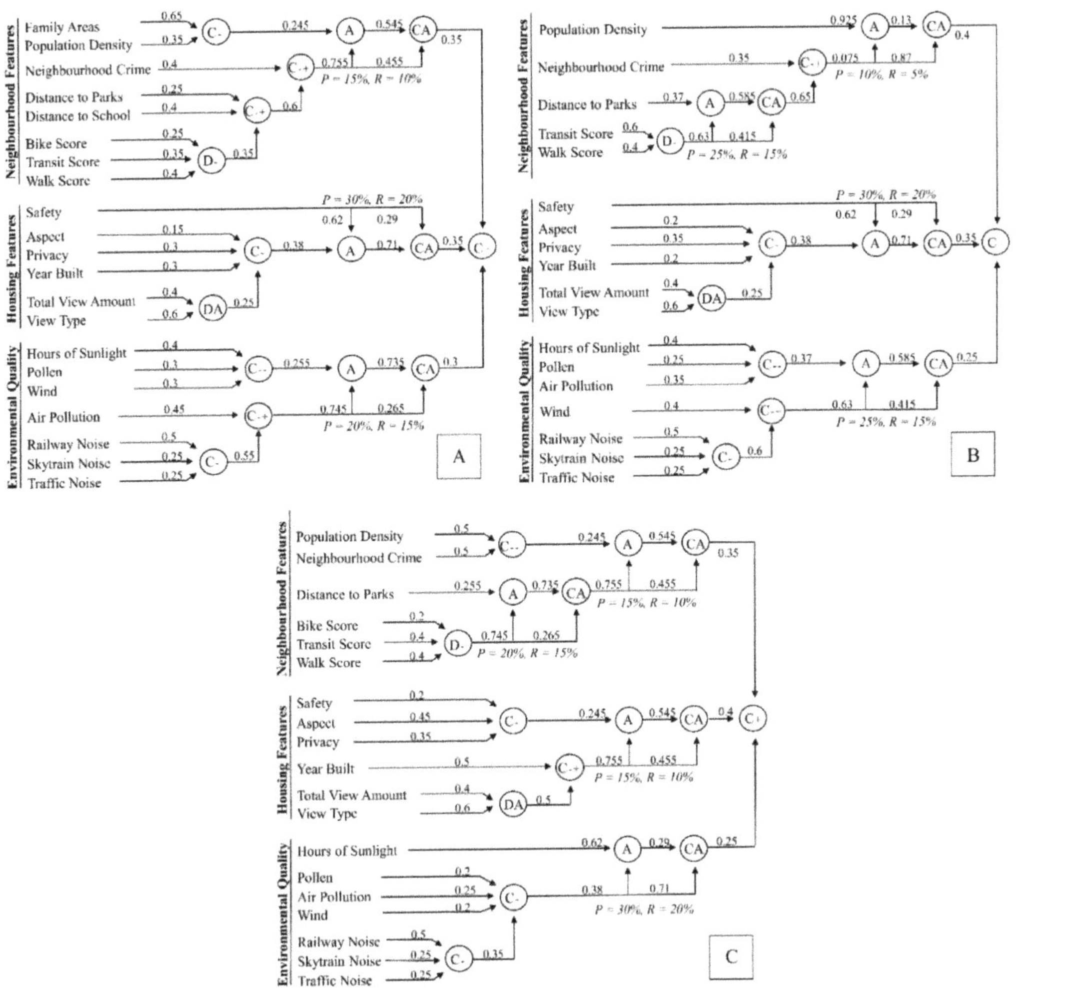

Figure 4. LSP aggregation structures for (**A**) Scenario 1 (family oriented), (**B**) Scenario 2 (elderly oriented), and (**C**) Scenario 3 (working professionals oriented).

Where asymmetric logic occurs in the aggregation structures (i.e., combination of a mandatory and a desired input), a conjunctive partial absorption (CPA) aggregator must be used, in this case an A/CA aggregator combination [17]. The CPA is a compound aggregator that requires the selection of reward (R) and penalty (P) values in order to determine the relationship between the inputs and the weights assigned to them in each aggregation step [21]. R and P are percent values used, respectively, to either increase the elementary preference if the desired input is fully satisfied or decrease the preference if the desired input is not satisfied [19]. Accordingly, the magnitudes of P and R reflect the desirability of the nonmandatory input. For instance, Scenario 2 both penalizes and rewards units much higher on the combined nonmandatory housing features than it does on population density attributes, an indication of the preferences for elderly residents (Figure 4B).

To compute the 3D suitability values, elementary attribute values were converted to elementary preferences that were aggregated in a stepwise manner from the leaves of the tree structure towards the root to calculate the overall suitability score of each voxel unit in the study area buildings. Individual logic aggregators were implemented using the weighted power mean, which can be described as follows [17]:

$$GCD(X_1, \ldots, X_n) = [W_1X_1{}^r + \ldots + W_nX_n{}^r]^{1/r} \tag{1}$$

where $GCD(X_1, \ldots, X_n)$ is the suitability score for an alternative with the input elementary preferences $X_1, \ldots, X_n$ for a set of n attributes; $W_1, \ldots, W_n$ are the weights of relative importance for the preference inputs; and r indicates the degree of simultaneity and replaceability as determined by the aggregator (Figure 1). More broadly, the overall 3D suitability score $S(v_i)$ for a voxel unit i located at position (X_i, Y_i, Z_i) can be expressed as the output of the overarching LSP criterion function, as presented in the following extended formula:

$$S(v_i) = g(a_1(v_i),\ a_2(v_i),\ \ldots,\ a_n(v_i)) \tag{2}$$

where g is the LSP criterion function, and $(a_1(v_i),\ a_2(v_i),\ \ldots,\ a_n(v_i))$ is the array of n elementary attributes associated with the 3D voxel unit i at position (X_i, Y_i, Z_i).

A cost–suitability analysis was also conducted to compute the overall value of each unit alternative, which can be viewed as a function of the suitability and cost of a unit [16,17]. Cost is incorporated separately from suitability in this manner as it is better aligned with human reasoning [26]. The cost–suitability analysis is extended in 3D and implemented to calculate the overall value $V(v_i)$ for a voxel unit i located at a position (X_i, Y_i, Z_i) using the following extended value indicator formula:

$$\begin{gathered} V(v_i) = \left[W_u(U(v_i))^r + W_p(P(v_i))^r\right]^{1/r} \\ 0 \le W_u \le 1,\ 0 \le W_p \le 1,\ W_u + W_p = 1, -\infty \le r \le +\infty \end{gathered} \tag{3}$$

where $U(v_i)$ is the usefulness indicator defining the range of acceptable suitability scores; $P(v_i)$ is the inexpensiveness indicator defining the range of acceptable cost; W_u and W_p are the weights of relative importance assigned to the usefulness and inexpensiveness indicators, respectively; and r indicates the degree of simultaneity and replaceability as determined by the aggregator. Unit conveyance prices were obtained from BC Assessment [82] datasets, with missing values estimated using proximate unit prices.

4. Results and Discussion

4.1. Three-Dimensional Suitability Analysis

The LSP criterion functions and 3D LSP–MCE suitability analyses were coded and run in CityEngine 2019.1 software [11] using the Computer Generated Architecture (CGA) programming language. The suitability scores for the 3D voxel units are presented and classified into seven equal interval classes of suitability: unacceptable (0.0–0.14), very poor (0.14–0.28), poor (0.28–0.43), average (0.43–0.57), good (0.57–0.71), very good (0.71–0.86), and excellent (0.88–1.00). Figure 5 presents the suitability analysis results for each scenario.

The 3D LSP–MCE suitability results vary considerably between the three scenarios. Scenario 1 is the most restrictive scenario with both the most low suitability units and the least high suitability units (Table 2). Some 44% of high-rise residential units fall within the unacceptable category, while no units are classified as excellent. Only 1% exhibit very good suitability with most confined to lower- to midlevel floors of waterfront buildings along the southern downtown boundary (Table 2, Figure 5A). This is due primarily to the scenario's stricter requirements with a greater number of mandatory attributes. For instance, low levels of noise, crime, and air pollution are considered mandatory requirements for families, which eliminates units located in higher crime areas and near high volume streets. Safety is also mandatory, resulting in lower suitability for lower floors and lower suitability or nonsuitability for higher floors.

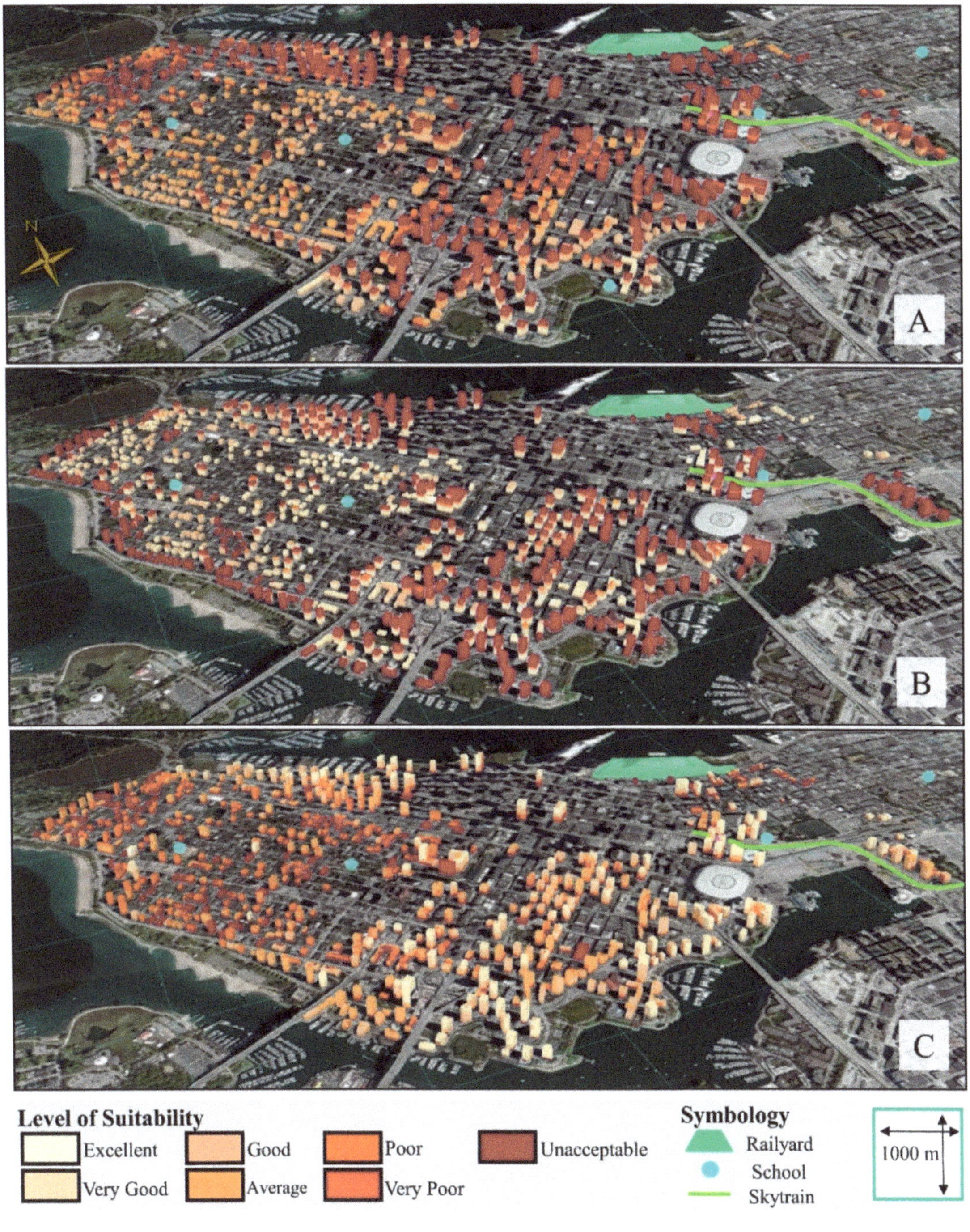

Figure 5. Results of the 3D LSP multicriteria evaluation (MCE) suitability analysis for Scenarios (**A**) 1, (**B**) 2, and (**C**) 3.

Table 2. Number and percentage of units in each suitability class by scenario.

Suitability Class	Scenario 1		Scenario 2		Scenario 3	
	# of Units	% of Units	# of Units	% of Units	# of Units	% of Units
Excellent (0.88–1.00)	0	0.00	999	2.03	9454	19.24%
Very Good (0.71–0.86)	492	1.00	12,563	25.57	6822	13.88%
Good (0.57–0.71)	5187	10.55	8743	17.79	12,180	24.79%
Average (0.43–0.57)	12,718	25.88	4676	9.51	10,526	21.42%
Poor (0.28–0.43)	6678	13.59	1177	2.39	6341	12.90%
Very Poor (0.14–0.28)	2449	4.98	809	1.64	3788	7.71%
Unacceptable (0.00–0.14)	21,606	43.97	20,163	41.04	19	0.03%

Although no voxel units are completely unsatisfied for the mandatory noise attribute, units near busy streets, the railyard, or the SkyTrain receive lower noise preference scores and thus lower suitability scores. Units with good or very good suitability are found primarily along the downtown periphery as they have greater access to higher weighted ($\geq$0.4) desired attributes, such as large desirable water views and sunlight, although there is also a cluster of higher suitability voxel units towards the downtown center due to their proximity to an elementary school. The units exhibiting very good suitability along the southern downtown edge are also influenced by the high family area score of the neighborhood. Moderately suitable units in the northwest downtown region are located in areas with higher crime rates and lower percentages of families and have lower quality views and sunlight due to their central position, shorter stature, and obstructions from neighboring downtown buildings (Figure 5A).

There is considerable overlap in unacceptable units between Scenarios 1 and 2, particularly on higher floors, as both scenarios share mandatory safety, crime, and noise attributes (Figure 5A,B). However, Scenario 2 results depict many additional unacceptable units along the downtown periphery, reflecting the mandatory emphasis on a low-wind environment for the elderly demographic. Overall, Scenario 2 has the greatest number of high suitability units, with 2% and 26% of voxel units displaying excellent and very good suitability, respectively (Table 2). This may be partly attributable to having fewer mandatory attributes and partly to some high-scoring desired attributes, such as distance to parks, transit score, and walk score, being weighted relatively highly. Similar to Scenario 1, many of the highest suitability units are found along the downtown perimeter where views are generally abundant and pollution is less, but a few streets in where the wind is not as strong.

The pattern of suitability is drastically different in Scenario 3 as safety, crime, and noise are not mandatory attributes, and instead, building age, view, and sunlight are prioritized (Figure 5C). As a result, buildings located along high-volume streets or in higher crime areas are not as heavily penalized, and units on higher floors are actually preferred as they are less prone to view and sunlight obstruction. Conversely, most units on lower floors are less suitable due to restricted views and sunlight, except for buildings along the downtown periphery. Unsurprisingly, many of the high suitability units occur in waterfront buildings, which have the extra advantage of having a desirable view type. The influence of building age is also evident, with older buildings predominantly situated in the northwest region of downtown, where the majority of low suitability voxel units are located. Scenario 3 has both the most excellent suitability units and the least unacceptable units (Table 2).

Certain trends are visible across all scenarios. Figure 6 presents a subsection of the study area permitting higher detail. For instance, corner units and units on higher floors tend to exhibit higher levels of suitability as they typically have access to larger views and longer hours of sunlight. Higher levels of near-ground noise, pollen, and air pollutants further contribute to the disparity in suitability between units on higher and lower floors. Privacy also has an influence as units within viewing distance of other units or from street spectators are considered less suitable. Units near parks or water also tend to score higher as they have more desirable views than units with views of surrounding city buildings. South-oriented units (Figure 6C) typically display higher levels of suitability than those oriented north (Figure 6D) as a south aspect is generally preferred and receives more sunlight.

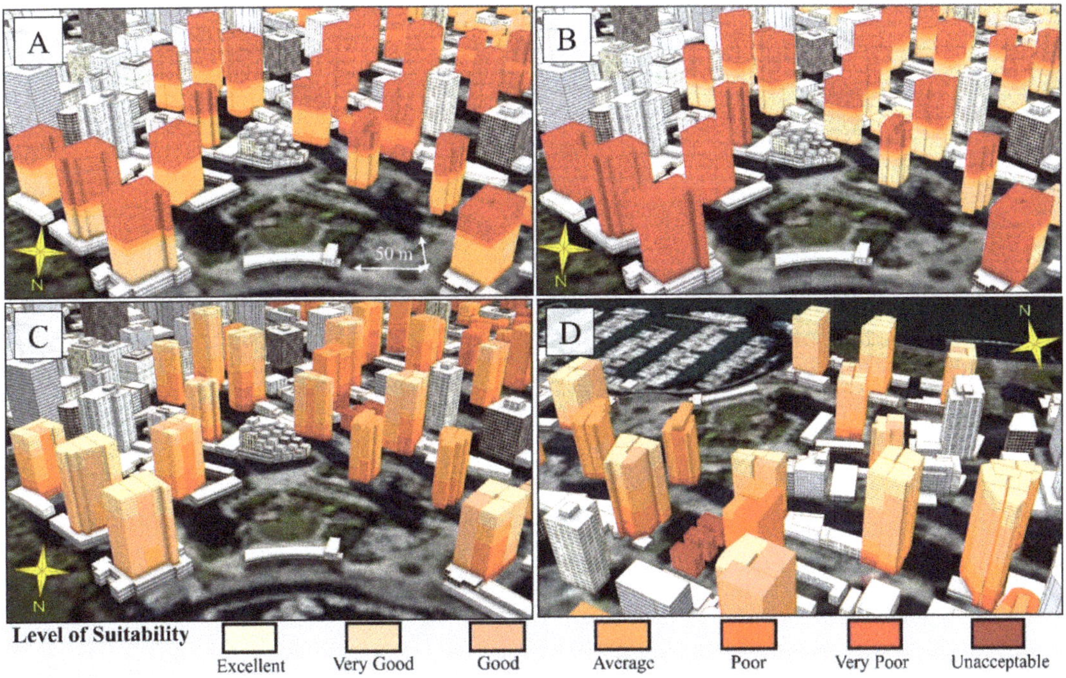

Figure 6. Suitability values obtained from the 3D LSP–MCE analysis for a subset of buildings under Scenarios (**A**) 1, (**B**) 2, and (**C**) 3 facing south and (**D**) Scenario 3 facing north.

4.2. Sensitivity and Cost–Suitability Analysis

In addition to performing the 3D LSP–MCE suitability analysis, the sensitivity of parameters was tested by modifying parameters for Scenario 1 using four separate test scenarios and by comparing the obtained results. The first sensitivity test involved increasing the number of mandatory attributes for Scenario 1, namely, family areas, population density, privacy, and pollen. The second test removed all mandatory attributes from Scenario 1 such that all attributes were desired. The third test weighted all components within each aggregation block in Scenario 1 equally. Finally, the last test implemented a one-at-a-time approach by incrementally increasing the input elementary attribute and elementary preference values for three elementary criteria (traffic noise, family areas, distance to schools) and the ANDness of aggregators for three aggregation subgroups (mandatory environmental, desired environmental, mandatory neighborhood) at three residential unit locations (Figure 7) to derive sensitivity curves and influence ranges.

Figure 7. The three residential units selected for the one-at-a-time sensitivity tests with associated suitability scores for Scenario 1.

To visualize the change in results for the first two tests, 3D difference maps were created by subtracting the original Scenario 1 suitability value for each unit from the corresponding new suitability value. Figure 8 presents the results of the sensitivity analysis with the initial Scenario 1 provided in panel (A) as a reference. Contrasting suitability values for more and no mandatory attributes are found in panels (B) and (C), and the respective difference maps are shown in panels (D) and (E). The results of the third sensitivity analysis test with equal weights provide a very small change in obtained values for 3D suitability ranging from −0.12 to +0.05, which can be explained by the aggregation block components in Scenario 1 not having extreme differences for weights.

The addition of further mandatory attributes generates more unacceptable units compared with the original Scenario 1. This is particularly evident along the west downtown border, where the percentage of family units is relatively low (Figure 8B), although some units' suitability has increased marginally. Overall, the range in suitability increase was minimal with a maximum increase of 0.11, compared with a maximum decrease in suitability of 0.75. The results of the second test with all nonmandatory attributes show a considerably different pattern of many fewer unacceptable units with the majority of units increasing in suitability with some exceptions, most notably in the same section as the first test results (Figure 8C). This can be explained that nonmandatory attributes that score relatively low in this area, such as family neighborhood composition and year built, now have a relatively larger influence on suitability outcome as there are no mandatory attributes. While approximately a quarter of all units have decreased in suitability by as much as −0.5, the overwhelming majority have increased by a maximum of 0.72. These results demonstrate how significantly suitability can change depending on the designation of mandatory or nonmandatory attributes, as mandatory attributes are inherently restrictive. For this reason, the LSP–MCE method should be used with direct consultation with interested groups who can provide direct input in the choice of parameters for decision-making.

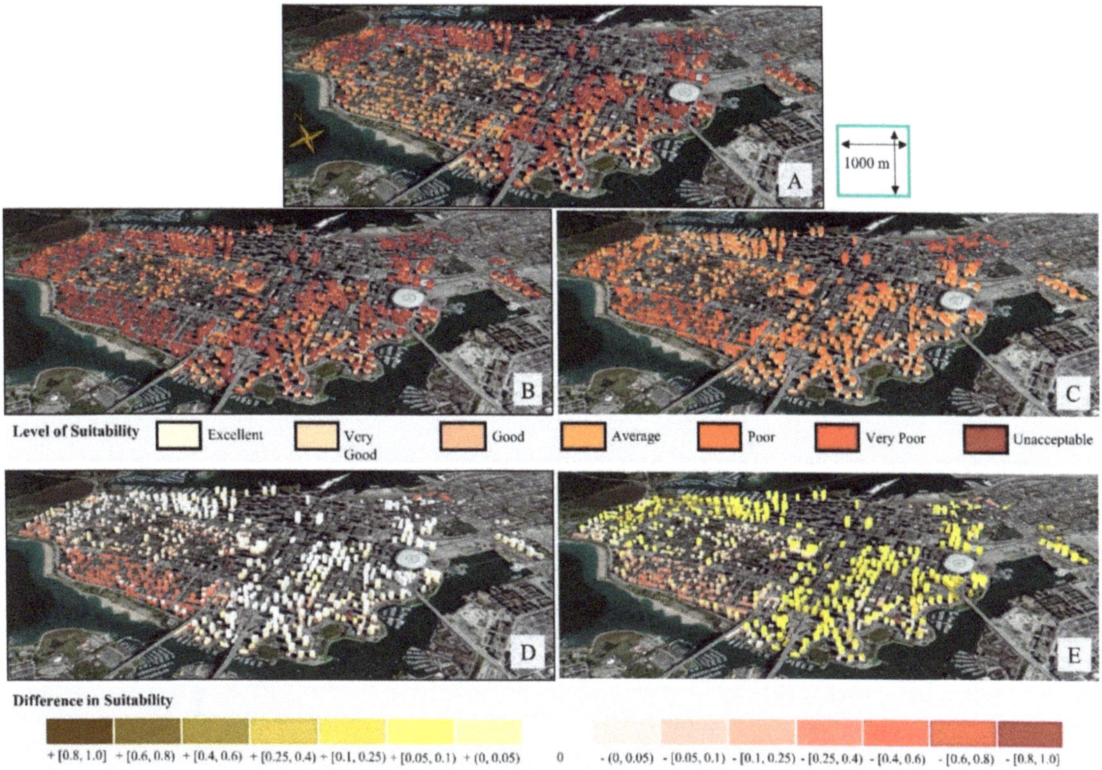

Figure 8. The 3D suitability maps for initial Scenario 1 (**A**) and the sensitivity analysis results for Scenario 1 with (**B**) more mandatory attributes and (**C**) no mandatory attributes and the calculated corresponding 3D difference maps (**D**,**E**), respectively. Change in overall suitability compared with the original Scenario 1 is expressed as the increase or decrease in suitability resulting from parameter variation.

Sensitivity curves from the fourth sensitivity test are presented in Figure 9 along with corresponding influence ranges for each of the three locations. Distance to school has the largest influence range of the three criteria, which can be explained by its designation as a mandatory criterion, whereas the desired criterion family areas has the smallest influence range (Figure 9A). Traffic noise is also a desired criterion, but it serves as an input for a mandatory component, which may explain its intermediate range of influence. The shapes of the curves are also significant with all following increasing concave functions with a progressively slower rate of increase in suitability for higher preference scores, although it is most pronounced for distance to school (Figure 9B). This makes sense as mandatory criteria model hard simultaneity, and therefore, suitability is more strongly limited by other potentially lower-scoring criteria in the same aggregation block. The distance to school curve starts at the origin (0,0) representing its mandatory nature and the automatic rejection of locations that do not satisfy the criterion. Another observation is that Location 1 with the highest overall suitability score also produces the greatest ranges in output suitability, while Location 3 with the lowest suitability produces the smallest ranges.

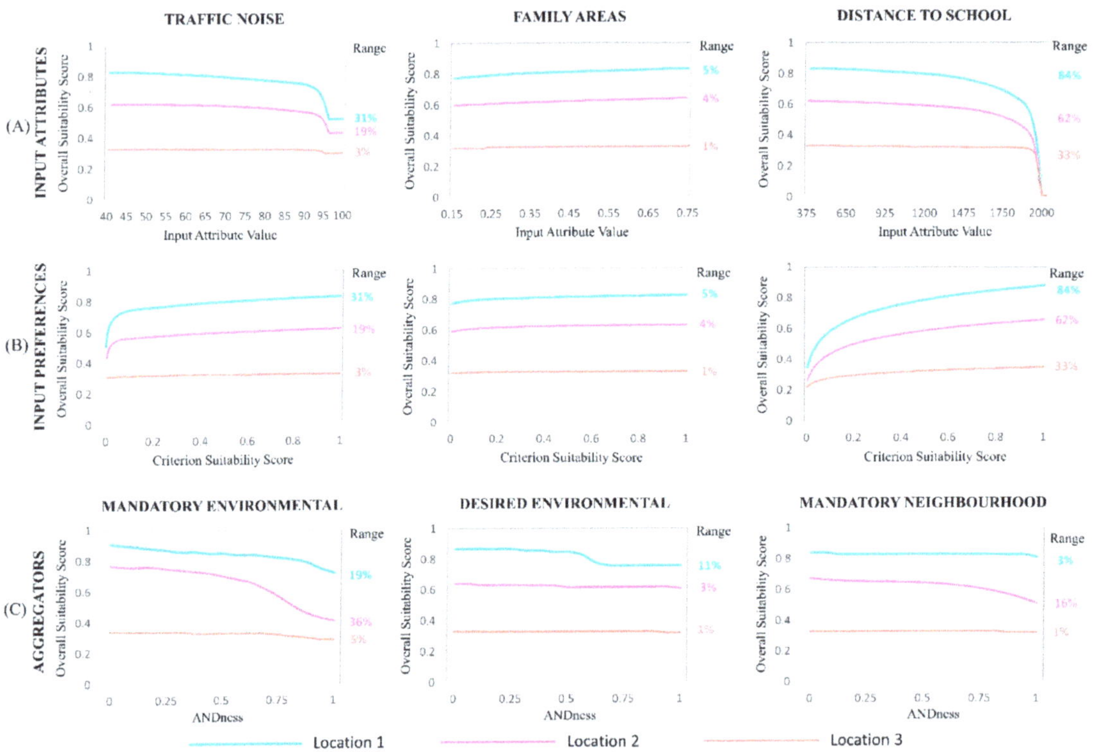

Figure 9. Sensitivity curves obtained for three locations using varying values for (**A**) input elementary attributes and (**B**) input elementary preferences for three elementary criteria and (**C**) aggregators of varying ANDness for three aggregation subgroups.

When the ANDness of aggregators is increased, the overall pattern is one of decreasing suitability as the progression towards simultaneity is more restrictive with fewer opportunities to satisfy user requirements based on other aggregation subgroup component preference scores (Figure 9C). In this case, Location 1 has the largest range in output suitability values under the desired environment subgroup, although Location 2 has the largest range under the two mandatory subgroups. Mandatory environment demonstrates the largest influence ranges of the three subgroups, which may be explained by the smaller number of input components that can ultimately impact the output preference score. The results suggest that extra care should be taken in assigning proper attribute values and calculating preference scores for the most sensitive parameters, namely, mandatory criteria, particularly where elementary preferences are low.

The cost–suitability analysis on the three scenarios was performed using unique usefulness $U(v_i)$ and inexpensiveness $P(v_i)$ indicators and associated importance weights for each scenario to represent the relative importance and range of acceptable cost for each demographic type. The calculated values of the 3D voxel units for each scenario are displayed in Figure 10 and summarized by class distribution in Table 3. The minimum suitability threshold of the usefulness indicator further restricts the overall value scores compared with the suitability scores and results in an increase in the number of unacceptable units across all scenarios (Figure 10, Table 3). Conversely, the number of higher-value units has decreased for Scenarios 2 and 3, but has actually increased significantly for Scenario 1. This can be explained by the scenario's relatively wide range in acceptable cost values for the family demographic's inexpensiveness indicator, which results in relatively high cost scores at voxel unit alternatives that already possessed at least a moderate degree of suitability.

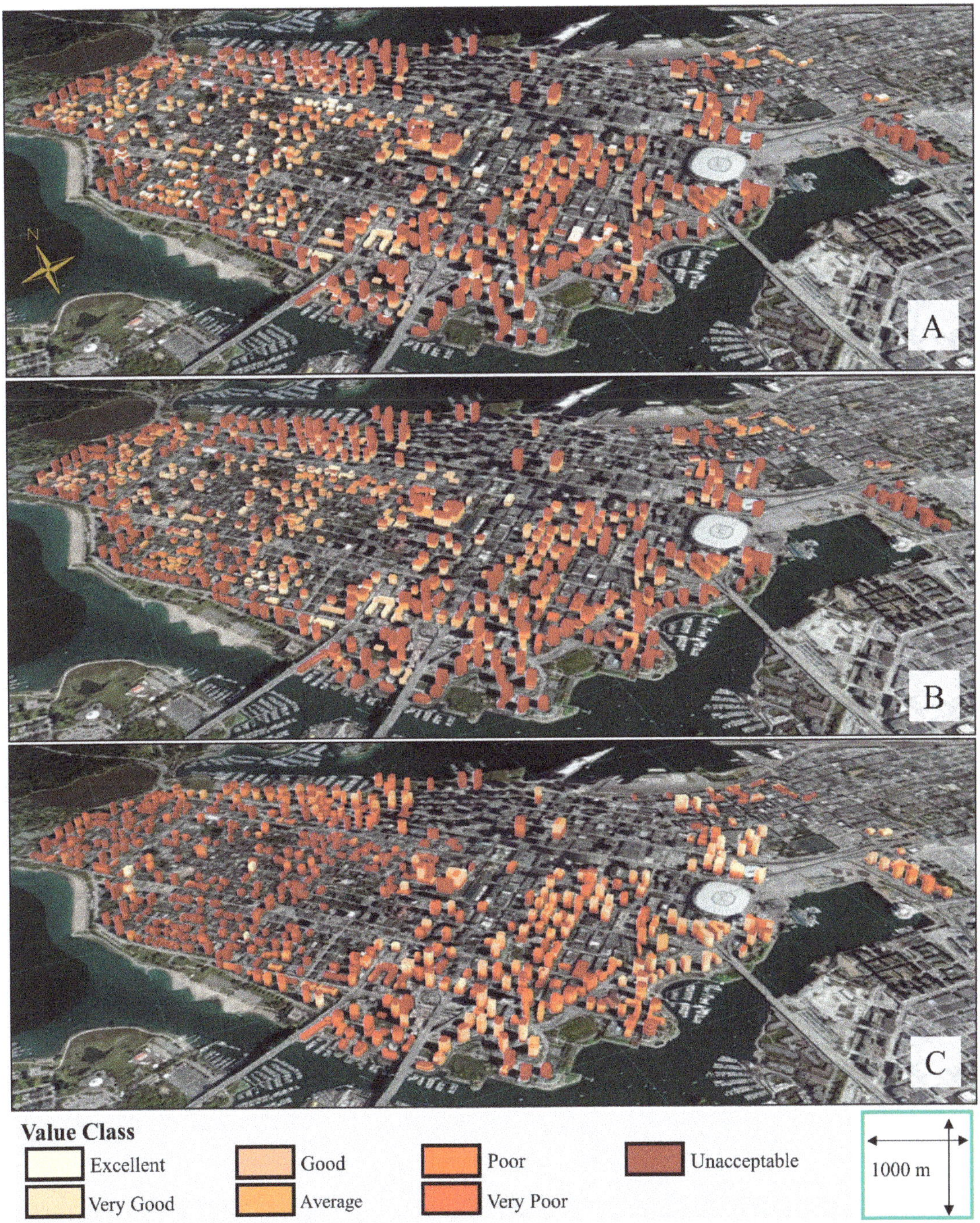

Figure 10. Results of the cost–suitability analysis for (**A**) Scenarios 1, (**B**) 2, and (**C**) 3.

Table 3. Number and percentage of units in each value class by scenario.

Suitability Class	Scenario 1		Scenario 2		Scenario 3	
	# of Units	% of Units	# of Units	% of Units	# of Units	% of Units
Excellent (0.88–1.00)	3878	7.89	82	0.16	8	0.01
Very Good (0.71–0.86)	1770	3.60	2513	5.11	784	1.59
Good (0.57–0.71)	3054	6.21	6812	13.86	3177	6.46
Average (0.43–0.57)	7254	14.76	6939	14.12	3830	7.79
Poor (0.28–0.43)	6010	12.23	4331	8.81	5379	10.94
Very Poor (0.14–0.28)	3151	6.41	1644	3.34	4935	10.04
Unacceptable (0.00–0.14)	24,013	48.87	26,809	54.56	31,017	63.13

In contrast, the number of higher-value voxel units in Scenario 2 has decreased as a lower acceptable price range is assigned to the elderly demographic, which reduces the desirability of higher-priced units. Working professionals are assigned a wider acceptable cost range, but higher-priced units, which typically coincide with highly suitable units in Scenario 3, receive lower inexpensiveness scores. Clearly, incorporating a cost requirement is necessary as both an important factor in residential selection decision problems and a strong influence on the overall unit value abundance and distribution. This analysis can also be interpreted as an indicator of affordability as conveyance price is a proxy for market value and the ability for the three demographic groups to afford a particular unit.

5. Conclusions

This research builds on the work of [14] by implementing a more advanced LSP–MCE method and extending it to operate in a 3D GIS environment. The suitability scenarios used in this study demonstrate the proposed method's ability to effectively analyze the suitability of thousands of 3D unit decision alternatives based on the distinct priorities, requirements, and desires of different prospective resident demographic types. In addition, the sensitivity analysis illustrated the 3D LSP–MCE method's responsiveness to different LSP structures as the results varied greatly across the three scenarios in terms of both the abundance and distribution of units for each suitability class. More broadly, the study provides evidence of the LSP method's ability to accommodate various decision scenarios and flexibility in decision logic for modeling human reasoning across a large number of criteria. Furthermore, the cost–suitability analysis results support the need for incorporating the cost factor into such decision problems, as it is integral to real-world residence selection and can substantially alter alternative value outcomes.

Validation measures were not performed as the residential unit data layers were synthetically created due to privacy and data availability issues, and the LSP criterion function parameters by nature of the decision problem are highly subjective. Thus, the results are not meant necessarily to inform home-buying decisions in downtown Vancouver, but rather to demonstrate the functionality and application of the proposed 3D LSP–MCE method for a dense built environment. However, if residential data, such as precise unit floor plans, were available, the results would be improved through more accurate statistics of units under each suitability class, as well as other potential metrics, such as total floor area or volume. This study can be improved with more detailed residence data, such as information on the number of bedrooms and bathrooms and market pricing. It could also further benefit from more quantitative sensitivity analyses on other parameters, such as surveying prospective residents to refine criteria weights and designation of mandatory/desired attributes, and more accurate attribute analysis methods. For instance, inclusion of 3D vegetation data could change view attribute values considerably where lower units' views are obscured by street trees or privacy shrubbery. Transportation infrastructure could also affect view attributes; however, this is not critical in the current study, where only one SkyTrain platform and rail exists aboveground on the periphery of the study area.

The 3D LSP–MCE method could benefit fields of study that require analysis and modeling of decision alternatives occurring across three dimensions, particularly where large numbers of attributes are involved. The wide range of decision logic aggregators allows for more expressive suitability values in 3D through criterion function structures that better reflect human reasoning in comparison with more traditional MCE methods, such as WLC, that implement a limited decision logic on a smaller number of criteria and produce less variability in output 3D scores. The study could be extended by including more relevant criteria, analyzing a larger study area of a built environment, or by considering different user group requirements. In addition, it could be applied to much denser built environments, such as those in Asian, US, or larger European cities, to examine and compare the effects of different potential evaluation criteria and their aggregations. The proposed method can assist urban planners and developers in assessing various urban development scenarios by analyzing the impacts of new development on neighboring buildings. More broadly, the 3D LSP–MCE approach developed in this research study can support spatial decision-making in real time or in applications that require analysis and representation in 3D, such as management of the housing market, urban property, and tax assessments.

Author Contributions: Conceptualization, K.M., S.D. and R.F.; methodology, K.M., S.D. and R.F.; software, S.D. and K.M.; validation, K.M.; formal analysis, K.M., S.D. and R.F.; investigation, K.M., S.D. and R.F.; resources, S.D.; data curation, K.M.; writing—original draft preparation, K.M.; writing—review and editing, S.D., R.F. and K.M.; visualization, K.M.; supervision, S.D. and R.F.; project administration, S.D.; funding acquisition, S.D. All authors have read and agreed to the published version of the manuscript.

Funding: This research was funded by the Natural Sciences and Engineering Research Council of Canada (NSERC) grant number RGPIN-2017-03939 and the Canada Graduate Scholarships Master's Program (CGSM). The article processing charges (APC) were fully funded by the Simon Fraser University Central Open Access Fund.

Data Availability Statement: This research uses some data derived directly from sources that prohibit redistribution or publication, and so they would not be available for sharing.

Acknowledgments: The authors would like to thank BC Assessment for providing data on residential units in Metro Vancouver. They would also like to thank Simon Fraser University's GIS and Map Librarian Sarah Zhang for her guidance in locating geospatial data. In addition, we are thankful to the four anonymous reviewers for their valuable feedback.

Conflicts of Interest: The authors declare no conflict of interest.

References

1. Shen, L.; Zhou, J.; Skitmore, M.; Xia, B. Application of a hybrid Entropy–McKinsey Matrix method in evaluating sustainable urbanization: A China case study. *Cities* **2015**, *42*, 186–194. [CrossRef]
2. Kaur, M.; Hewage, K.; Sadiq, R. Investigating the impacts of urban densification on buried water infrastructure through DPSIR framework. *J. Clean. Prod.* **2020**, *259*, 120897. [CrossRef]
3. Amer, M.; Mustafa, A.; Teller, J.; Attia, S.; Reiter, S. A methodology to determine the potential of urban densification through roof stacking. *Sustain. Cities Soc.* **2017**, *35*, 677–691. [CrossRef]
4. Boyko, C.T.; Cooper, R. Clarifying and re-conceptualizing density. *Prog. Plan.* **2011**, *76*, 1–61. [CrossRef]
5. Koziatek, O.; Dragićević, S. A local and regional spatial index for measuring three-dimensional urban compactness growth. *Environ. Plan. B Urban Anal. City Sci.* **2019**, *46*, 143–164. [CrossRef]
6. Ying, Y.; Koeva, M.; Kuffer, M.; Asiama, K.O.; Li, X.; Zevenbergen, J. Making the Third Dimension (3D) Explicit in Hedonic Price Modelling: A Case Study of Xi'an, China. *Land* **2021**, *10*, 24. [CrossRef]
7. Malczewski, J.; Jankowski, P. Emerging trends and research frontiers in spatial multicriteria analysis. *Int. J. Geogr. Inf. Sci.* **2020**, *34*, 1257–1282. [CrossRef]
8. Malczewski, J.; Rinner, C. *Multicriteria Decision Analysis in Geographic Information Science*; Springer: Berlin/Heidelberg, Germany, 2015. [CrossRef]

9. Autodesk. *Revit*; Version 2022; Autodesk: San Rafael, CA, USA, 2022; Available online: https://www.autodesk.ca/en/products/revit/overview?term=1-YEAR&tab=subscription (accessed on 10 December 2021).
10. ESRI. ArcGIS Urban. 2022. [Online Software]. Available online: https://www.esri.com/en-us/arcgis/products/arcgis-urban/overview (accessed on 10 December 2021).
11. ESRI. *CityEngine*, Version 2019.1; ESRI: Redlands, CA, USA, 2019. Available online: https://www.esri.com/en-us/arcgis/products/esri-cityengine/overview (accessed on 10 December 2021).
12. Trubka, R.; Glackin, S.; Lade, O.; Pettit, C. A web-based 3D visualisation and assessment system for urban precinct scenario modelling. *ISPRS J. Photogramm. Remote Sens.* **2016**, *117*, 175–186. [CrossRef]
13. Saran, S.; Oberai, K.; Wate, P.; Konde, A.; Dutta, A.; Kumar, K.; Kumar, A.S. Utilities of virtual 3D city models based on CityGML: Various use cases. *J. Indian Soc. Remote Sens.* **2018**, *46*, 957–972. [CrossRef]
14. Munn, K.; Dragićević, S. Spatial multi-criteria evaluation in 3D context: Suitability analysis of urban vertical development. *Cartogr. Geogr. Inf. Sci.* **2021**, *48*, 105–123. [CrossRef]
15. Montgomery, B.; Dragićević, S. Comparison of GIS-based logic scoring of preference and multicriteria evaluation methods: Urban land use suitability. *Geogr. Anal.* **2016**, *48*, 427–447. [CrossRef]
16. Dujmović, J.; de Tré, G.; Dragićević, S. Comparison of multicriteria methods for land-use suitability assessment. In Proceedings of the 2009 IFSA World Congress/EUSFLAT Conference, Lisbon, Portugal, 20–24 July 2009; pp. 1404–1409.
17. Dujmović, J. *Soft Computing Evaluation Logic: The LSP Decision Method and Its Applications*; John Wiley & Sons: Hoboken, NJ, USA, 2018.
18. Dragićević, S.; Dujmović, J.; Minardi, R. Modeling urban land-use suitability with soft computing: The GIS-LSP method. In *GeoComputational Analysis and Modeling of Regional Systems*; Thill, J.-C., Dragicevic, S., Eds.; Springer International Publishing: Cham, Switzerland, 2018; pp. 257–275. [CrossRef]
19. Dujmović, J.; Scheer, D. Logic aggregation of suitability maps. In Proceedings of the 2010 IEEE World Congress on Computational Intelligence, WCCI 2010, Barcelona, Spain, 18–23 July 2010; Volume 1, pp. 1–8. [CrossRef]
20. Rebolledo, B.; Gil, A.; Flotats, X.; Sánchez, J.Á. Assessment of groundwater vulnerability to nitrates from agricultural sources using a GIS-compatible logic multicriteria model. *J. Environ. Manag.* **2016**, *171*, 70–80. [CrossRef]
21. Dujmović, J.; de Tré, G.; van de Weghe, N. LSP suitability maps. *Soft Comput.* **2010**, *14*, 421–434. [CrossRef]
22. Dujmović, J.; Allen, W.L. Soft computing logic decision making in strategic conservation planning for water quality protection. *Ecol. Inform.* **2021**, *61*, 101167. [CrossRef]
23. Pirmanto, D.; Suseno, J.E.; Adi, K. Logic Scoring of Preference Method for Determining Landfill with Geographic Information System. In Proceedings of the 2018 International Seminar on Research of Information Technology and Intelligent Systems (ISRITI), Yogyakarta, Indonesia, 21–22 November 2018; pp. 459–464. [CrossRef]
24. Luan, C.; Liu, R.; Peng, S. Land-use suitability assessment for urban development using a GIS-based soft computing approach: A case study of Ili Valley, China. *Ecol. Indic.* **2021**, *123*, 107333. [CrossRef]
25. Shen, S.; Dragicevic, S.; Dujmović, J. GIS.LSP: A soft computing logic method and tool for geospatial suitability analysis. *Trans. GIS* **2021**, *89*, 101654. [CrossRef]
26. Dujmović, J.; de Tré, G. Multicriteria methods and logic aggregation in suitability maps. *Int. J. Intell. Syst.* **2011**, *26*, 971–1001. [CrossRef]
27. Statistics Canada. Census Data. 2020. Available online: https://www12.statcan.gc.ca/datasets/index-eng.cfm?Temporal=2016 (accessed on 20 July 2021).
28. Metro Vancouver. *Metro Vancouver 2040 Shaping Our Future*; Metro Vancouver: Vancouver, BC, Canada, 2015.
29. Statistics Canada. Classification by Type of Building. 2021. Available online: https://www23.statcan.gc.ca/imdb/p3VD.pl?Function=getVD&TVD=1226827&CVD=1226832&CPV=1.1&CST=01012019&CLV=2&MLV=3 (accessed on 10 December 2020).
30. Macdonald, E. Chapter 5—Urban design for sustainable and livable communities: The case of Vancouver. In *Transportation, Land Use, and Environmental Planning*; Elsevier: Amsterdam, The Netherlands, 2020; pp. 83–104. [CrossRef]
31. City of Vancouver. Open Data Portal. 2021. Available online: https://opendata.vancouver.ca/pages/home/ (accessed on 6 December 2020).
32. BC Assessment. Assessment Search Tool. 2021. Available online: https://www.bcassessment.ca/ (accessed on 6 December 2020).
33. City of Vancouver. VanMap: View, Search, Create, and Print Maps of Vancouver. 2017. Available online: http://vancouver.ca/your-government/vanmap.aspx (accessed on 7 November 2020).
34. DMTI Spatial Inc. Abacus Dataverse Network: Geospatial Data Sets. CanMap RouteLogistics, v.2014.3. 2014. Available online: https://www.lib.sfu.ca/find/other-materials/data-gis/gis/spatial-data (accessed on 1 July 2020).
35. DMTI Spatial Inc. Abacus Dataverse Network: Geospatial Data Sets. CanMap Content Suite, v2019.3. 2019. Available online: https://www.lib.sfu.ca/find/other-materials/data-gis/gis/spatial-data (accessed on 1 July 2020).
36. University of Toronto. CHASS Canadian Census Analyser. 2017. Available online: http://dc1.chass.utoronto.ca/census/index.html (accessed on 11 July 2017).
37. Vancouver Police Department. Vancouver Police Department Crime Data Download. 2021. Available online: https://geodash.vpd.ca/opendata/ (accessed on 7 November 2020).
38. Government of Canada. High Resolution Digital Elevation Model (HRDEM)—CanElevation Series. 2020. Available online: https://open.canada.ca/data/en/dataset/957782bf-847c-4644-a757-e383c0057995 (accessed on 1 July 2020).

39. Metro Vancouver. Metro Vancouver Open Data Catalogue 2016. Generalized Land Use Classification. 2020. Available online: http://www.metrovancouver.org/data (accessed on 7 November 2020).
40. Government of British Columbia. DataBC. 2021. Available online: https://data.gov.bc.ca/ (accessed on 12 December 2020).
41. ESRI. *ArcGIS*, version 10.7.1; ESRI: Redlands, CA, USA, 2019.
42. Mee, K.J. 'Any place to raise children is a good place': Children, housing and neighbourhoods in inner Newcastle, Australia. *Child. Geogr.* **2010**, *8*, 193–211. [CrossRef]
43. Martin, S.; Carlson, S. Barriers to children walking to or from school—United States, 2004. *J. Am. Med. Assoc.* **2005**, *294*, 2160–2162. [CrossRef]
44. Mccrea, R.; Stimson, R.; Western, J. Testing a moderated model of satisfaction with urban living using data for Brisbane-South East Queensland, Australia. *Soc. Indic. Res.* **2005**, *72*, 121–152. [CrossRef]
45. Leslie, E.; Cerin, E. Are perceptions of the local environment related to neighbourhood satisfaction and mental health in adults? *Prev. Med.* **2008**, *47*, 273–278. [CrossRef] [PubMed]
46. Mouratidis, K. Is compact city livable? The impact of compact versus sprawled neighbourhoods on neighbourhood satisfaction. *Urban Stud.* **2018**, *55*, 2408–2430. [CrossRef]
47. Raman, S. Designing a liveable compact city: Physical forms of city and social life in urban neighbourhoods. *Built Environ.* **2010**, *36*, 63–80. [CrossRef]
48. Wood, L.; Hooper, P.; Foster, S.; Bull, F. Public green spaces and positive mental health—Investigating the relationship between access, quantity and types of parks and mental wellbeing. *Health Place* **2017**, *48*, 63–71. [CrossRef] [PubMed]
49. Walk Score. Walk Score: Live Where You Love. 2021. Available online: https://www.walkscore.com/ (accessed on 1 February 2021).
50. Huang, Z.; Chen, R.; Xu, D.; Zhou, W. Spatial and hedonic analysis of housing prices in Shanghai. *Habitat Int.* **2017**, *67*, 69–78. [CrossRef]
51. Zortuk, M. An evaluation of higher education students' apartment preferences: A real world study in a newly urbanized city. *Istanb. Univ. Econom. Stat. e-J.* **2014**, *0*, 1–20.
52. Bina, M.; Warburg, V.; Kockelman, K.M. Location choice vis-à-vis transportation: Apartment dwellers. *Transp. Res. Rec.* **2006**, *1977*, 93–102. [CrossRef]
53. Samarasinghe, O.E.; Sharp, B.M.H. The value of a view: A spatial hedonic analysis. *N. Z. Econ. Pap.* **2008**, *42*, 59–78. [CrossRef]
54. Yasumoto, S.; Jones, A.P.; Nakaya, T.; Yano, K. The use of a virtual city model for assessing equity in access to views. *Comput. Environ. Urban Syst.* **2011**, *35*, 464–473. [CrossRef]
55. Craighead, G. *High-Rise Security and Fire Life Safety*; Butterworth-Heinemann: Oksford, UK, 2009.
56. Ghani, S.A. A changing pattern of house market in Kuala Lumpur. In Proceedings of the National Conference on Affordable Quality Housing, Miri, Sarawak, Malaysia, 24–26 November 2004; pp. 24–26.
57. Hashim, A.H.; Ali, H.M.; Samah, A.A. Urban Malay's user-behaviour and perspective on privacy and spatial organization of housing. *ArchNet-IJAR* **2009**, *3*, 197–208.
58. Xie, X.; Semanjski, I.; Gautama, S.; Tsiligianni, E.; Deligiannis, N.; Rajan, R.T.; Pasveer, F.; Philips, W. A review of urban air pollution monitoring and exposure assessment methods. *ISPRS Int. J. Geo-Inf.* **2017**, *6*, 389. [CrossRef]
59. Kasprzyk, I.; Ćwik, A.; Kluska, K.; Wójcik, T.; Cariñanos, P. Allergenic pollen concentrations in the air of urban parks in relation to their vegetation. *Urban For. Urban Green.* **2019**, *46*, 126486. [CrossRef]
60. Reiter, S. Assessing wind comfort in urban planning. *Environ. Plan. B Plan. Des.* **2010**, *37*, 857–873. [CrossRef]
61. Kim, A.; Sung, J.H.; Bang, J.-H.; Cho, S.W.; Lee, J.; Sim, C.S. Effects of self-reported sensitivity and road-traffic noise levels on the immune system. *PLoS ONE* **2017**, *12*, e0187084. [CrossRef] [PubMed]
62. Swanson, V.; Sharpe, T.; Porteous, C.; Hunter, C.; Shearer, D. Indoor annual sunlight opportunity in domestic dwellings may predict well-being in urban residents in scotland. *Ecopsychology* **2016**, *8*, 121–130. [CrossRef]
63. City of Vancouver. *Greenest City—2020 Action Plan*; City of Vancouver: Vancouver, BC, Canada, 2012; p. 82.
64. Vancouver Board of Parks and Recreation. *Vancouver Park Provision Study*; Vancouver Board of Parks and Recreation: Vancouver, BC, Canada, 2018.
65. Vancouver Board of Education. *Long Range Facilities Plan*; Vancouver Board of Education: Vancouver, BC, Canada, 2015.
66. Lu, J. The value of a south-facing orientation: A hedonic pricing analysis of the Shanghai housing market. *Habitat Int.* **2018**, *81*, 24–32. [CrossRef]
67. Shach-Pinsly, D. Visual openness and visual exposure analysis models used as evaluation tools during the urban design development process. *J. Urban.* **2010**, *3*, 161–184. [CrossRef]
68. Robinson, M.B.; Robinson, C.E. Environmental characteristics associated with residential burglaries of student apartment complexes. *Environ. Behav.* **1997**, *29*, 657–675. [CrossRef]
69. Gordon, B.; Winkler, D.; Barrett, D.; Zumpano, L. The effect of elevation and corner location on oceanfront condominium value. *J. Real Estate Res.* **2013**, *35*, 345–364. [CrossRef]
70. Proulx, G.; Mcqueen, C. *Evacuation Timing in Apartment Buildings*; National Research Council of Canada. Institute for Research in Construction: Ottawa, ON, Canada, 1994.
71. The Condo Group. How to Spot a Leaky Condo. 2011. Available online: https://thecondogroup.com/how-to-spot-a-leaky-condo/ (accessed on 7 November 2020).
72. Metro Vancouver. *Metro Vancouver Housing Data Book*; Metro Vancouver: Vancouver, BC, Canada, 2019.

73. Bourassa, S.C.; Hoesli, M.; Sun, J. What's in a view? *Environ. Plan. A Econ. Space* **2004**, *36*, 1427–1450. [CrossRef]
74. Metro Vancouver. *Caring for the Air 2020*; Metro Vancouver: Vancouver, BC, Canada, 2020.
75. Yasumoto, S.; Jones, A.; Yano, K.; Nakaya, T. Virtual city models for assessing environmental equity of access to sunlight: A case study of Kyoto, Japan. *Int. J. Geogr. Inf. Sci.* **2012**, *26*, 1–13. [CrossRef]
76. Rojo, J.; Oteros, J.; Pérez-Badia, R.; Cervigón, P.; Ferencova, Z.; Gutiérrez-Bustillo, M.A.; Bergmann, K.-C.; Oliver, G.; Thibaudon, M.; Albertini, R.; et al. Near-ground effect of height on pollen exposure. *Environ. Res.* **2019**, *174*, 160–169. [CrossRef] [PubMed]
77. Wong, M.S.; Nichol, J.E.; To, P.H.; Wang, J. A simple method for designation of urban ventilation corridors and its application to urban heat island analysis. *Build. Environ.* **2010**, *45*, 1880–1889. [CrossRef]
78. Wakefield Acoustics Ltd. *City of Vancouver Noise Control Manual*; Wakefield Acoustics Ltd.: Victoria, BC, Canada, 2004.
79. Health Canada. *Guidance for Evaluating Human Health Impacts in Environmental Assessment: Noise*; Health Canada: Ottawa, ON, Canada, 2017; Volume 53.
80. BKL Consultants Ltd. Cargill Rail Expansion Project: Environmental Noise Assessment. North Vancouver. 2015. Available online: https://www.portvancouver.com/wp-content/uploads/2015/03/3057-15a-cargill-rail-expansion-environmental-noise-assessment_final.pdf (accessed on 7 November 2020).
81. SLR Consulting Ltd. SkyTrain Noise Study Vancouver. Vancouver, Canada. 2018. Available online: https://www.translink.ca/-/media/translink/documents/plans-and-projects/skytrain-noise-study/skytrain-noise-report-20181128.pdf (accessed on 7 November 2020).
82. BC Assessment. *Custom Dataset*; BC Assessment: Nanaimo, BC, Canada, 2021.

MDPI
St. Alban-Anlage 66
4052 Basel
Switzerland
www.mdpi.com

Land Editorial Office
E-mail: land@mdpi.com
www.mdpi.com/journal/land

www.ingramcontent.com/pod-product-compliance
Lightning Source LLC
Chambersburg PA
CBHW061614220926
51850CB00005B/506
9783725809158